STUDENT STUDY GUIDE

FOR

BIOLOGY

CAMPBELL • REECE

STUDENT STUDY GUIDE

FOR

BIOLOGY

CAMPBELL • REECE

MARTHA R. TAYLOR
Ithaca, New York

EIGHTH EDITION

PEARSON

Benjamin Cummings

San Francisco   Boston   New York
Capetown   Hong Kong   London   Madrid   Mexico City
Montreal   Munich   Paris   Singapore   Sidney   .   Toronto

Editor-in-Chief: Beth Wilbur
Senior Editorial Manager: Ginnie Simione Jutson
Senior Supplements Project Editor: Susan Berge
Project Editor: Mary Douglas, Rogue Valley Publications
Executive Marketing Manager: Lauren Harp
Managing Editor: Michael Early
Production Supervisor: Jane Brundage
Manufacturing Buyer: Michael Early, Michael Penne
Production Management and Composition: S4Carlisle, Judy Ludowitz
Cover Design: Yvo Riezebos Design
Text and Cover Printer: Courier, Stoughton

Cover Photo Credit: Magnolia Flower—Corbis. Photographer: Chris Fox

ISBN-13: 978-0-321-50156-1
ISBN-10: 0-321-50156-X

Dedication:

To Mary Douglas, my totally excellent project editor, guru, and friend.

Credits
Figures **IQ 5.2, IQ 5.5, IQ 6.4, SYK 6.3, IQ 7.1, SYK 7.2, IQ 8.6, IQ 9.8, IQ 10.1, IQ 10.2, IQ 10.5, IQ 13.3, IQ 16.5, IQ 19.1, IQ 20.3, SYK 25.1, SYK 26.1, SYK 26.2, IQ 28.3, SYK 28.1, IQ 29.5, SYK 29.2, IQ 30.5, SYK 30.3, IQ 34.1, IQ 34.3, IQ 34.8, IQ 35.1, IQ 35.4, IQ 35.5, SYK 35.1, IQ 36.7, IQ 40.2, IQ 40.3, IQ 40.4, IQ 45.1, IQ 47.2, IQ 47.3, IQ 47.4, IQ 47.5, IQ 48.1, IQ 48.3, IQ 48.4, IQ 49.3, IQ 50.3, IQ 50.6, IQ 50.8, IQ 52.5, IQ 52.6:** Adapted from Campbell, N. A., Reece, J. B., *Biology*, 8th ed. (San Francisco, CA: Benjamin Cummings, 2008) © 2008 Pearson Education Inc., publishing as Benjamin Cummings. Photos: [27-2] (Bacillus anthracis), H. S. Pankratz, T. C. Beaman/Biological Photo Service; [27-3] (Anabaena), Susan M. Barns, PhD; [35-2] (top), Ed Reschke; [35-2] (bottom), Carolina Biological Supply/Phototake NYC; [40.02] a–c (bone tissue), Nina Zanetti; [40.02] d–f (loose connective tissue), Nina Zanetti; [40.02] g–i (blood tissue), Gopal Murti/SPL/Photo Researchers, Inc.; [40.03] a (cardiac muscle), Manfred Kage/Peter Arnold, Inc.; [40.03] b (smooth muscle), Gladden Willis, M.D./Visuals Unlimited.

PEARSON

Benjamin Cummings

1 2 3 4 5 6 7 8 9 10 —CRS— 10 09 08 07

www.aw-bc.com

Contents

PREFACE   v

CHAPTER 1
Introduction: Themes in the Study of Life   1

UNIT ONE   THE CHEMISTRY OF LIFE   7

CHAPTER 2
The Chemical Context of Life   7

CHAPTER 3
Water and the Fitness of the Environment   14

CHAPTER 4
Carbon and the Molecular Diversity of Life   20

CHAPTER 5
The Structure and Function of Macromolecules   24

UNIT TWO   THE CELL   35

CHAPTER 6
A Tour of the Cell   35

CHAPTER 7
Membrane Structure and Function   46

CHAPTER 8
An Introduction to Metabolism   53

CHAPTER 9
Cellular Respiration: Harvesting
   Chemical Energy   61

CHAPTER 10
Photosynthesis   69

CHAPTER 11
Cell Communication   78

CHAPTER 12
The Cell Cycle   85

UNIT THREE   GENETICS   91

CHAPTER 13
Meiosis and Sexual Life Cycles   91

CHAPTER 14
Mendel and the Gene Idea   98

CHAPTER 15
The Chromosomal Basis of Inheritance   108

CHAPTER 16
The Molecular Basis of Inheritance   116

CHAPTER 17
From Gene to Protein   125

CHAPTER 18
Regulation of Gene Expression   132

CHAPTER 19
Viruses   143

CHAPTER 20
Biotechnology   149

CHAPTER 21
Genomes and Their Evolution   159

UNIT FOUR   MECHANISMS
OF EVOLUTION   167

CHAPTER 22
Descent with Modification: A Darwinian View of
   Life   167

CHAPTER 23
The Evolution of Populations   174

CHAPTER 24
The Origin of Species   181

CHAPTER 25
The History of Life on Earth   186

UNIT FIVE   THE EVOLUTIONARY HISTORY OF BIOLOGICAL DIVERSITY   193

CHAPTER 26
Phylogeny and the Tree of Life   193

CHAPTER 27
Bacteria and Archaea   200

CHAPTER 28
Protists   207

CHAPTER 29
Plant Diversity I: How Plants Colonized Land   214

CHAPTER 30
Plant Diversity II: The Evolution of Seed Plants   222

CHAPTER 31
Fungi   229

CHAPTER 32
An Introduction to Animal Diversity   235

CHAPTER 33
Invertebrates   240

CHAPTER 34
Vertebrates   249

UNIT SIX   PLANT FORM AND FUNCTION   261

CHAPTER 35
Plant Structure, Growth, and Development   261

CHAPTER 36
Resource Acquisition and Transport in Vascular Plants   270

CHAPTER 37
Soil and Plant Nutrition   278

CHAPTER 38
Angiosperm Reproduction and Biotechnology   284

CHAPTER 39
Plant Responses to Internal and External Signals   291

UNIT SEVEN   ANIMAL FORM AND FUNCTION   301

CHAPTER 40
Basic Principles of Animal Form and Function   301

CHAPTER 41
Animal Nutrition   309

CHAPTER 42
Circulation and Gas Exchange   317

CHAPTER 43
The Immune System   330

CHAPTER 44
Osmoregulation and Excretion   341

CHAPTER 45
Hormones and the Endocrine System   351

CHAPTER 46
Animal Reproduction   359

CHAPTER 47
Animal Development   369

CHAPTER 48
Neurons, Synapses, and Signaling   377

CHAPTER 49
Nervous Systems   384

CHAPTER 50
Sensory and Motor Mechanisms   392

CHAPTER 51
Animal Behavior   403

UNIT EIGHT   ECOLOGY   411

CHAPTER 52
An Introduction to Ecology and the Biosphere   411

CHAPTER 53
Population Ecology   419

CHAPTER 54
Community Ecology   426

CHAPTER 55
Ecosystems   434

CHAPTER 56
Conservation Biology and Restoration Ecology   441

ANSWER SECTION   447

Another name for the *Student Study Guide* for Campbell/Reece, *Biology*, Eighth Edition, could be "A Student Structuring Guide." This guide is designed to help you structure and organize your developing knowledge of biology and create your own personal understanding of the topics covered in the text. Take a few minutes to learn about some strategies of successful students as well as some strategies for using this study guide.

Strategies of Successful Students

What does it take to be a successful student? How can you learn more efficiently and earn better grades? And how can you enjoy your biology course more?

Every successful student tackles learning in his or her own unique way. But there are commonalities that can be shared and results from educational research that can inform you as you develop your own approach. Here is a short list of suggestions that may help.

Make it interesting. Biology is fascinating and so much easier to learn when you love it. Try to bring a sense of wonder to your classes and your studying. Appreciate how much has been discovered and how much we have yet to learn about the living world.

Focus in lecture. What does it mean to focus? Obviously sleeping, text messaging, and daydreaming are out. But so is taking down every word your professor says without thinking. If a lecture outline is provided, take notes on it and/or in a notebook. Many students find that taking notes on only one side of a page allows them to rework, elaborate, or at least make legible what they wrote down in class on the facing page. Doing this shortly after class helps to review the material while it is fresh and develop an organized set of notes that will help in studying for exams.

Make reading active. Research has shown that simply rereading material does not improve test scores, but interacting with the reading assignment does. How do you interact with the printed word? Ask yourself questions and look for answers. Reword every concept heading as a question before you start reading. Make questions out of subheadings. If you are trying to answer a question, you are looking for meaning. Your brain is not just decoding words, it is working toward understanding those words.

Use your resources. What does your course provide—review sessions, office hours, tutors, study groups? Science is a social endeavor; interact with people as well as the material. Use the written resources provided by your course and the Campbell/Reece textbook: Web sites, practice exams, and tutorials. When possible, prepare for exams by practicing with questions from previous exams. Each professor has his/her own approach to writing questions. One way to improve your grade is to study how your professor asks questions.

Learn from your mistakes. Carefully review your exams—certainly the questions you missed, but even those you got right. If possible, go over the exam with an instructor. Learn both *what* you missed and *why* you missed it so you will know *how* to choose the correct answers next exam. Work on improving your test-taking skills. (See the following section on tips for taking objective tests.)

Try something new. If what you are doing isn't working, don't just keep trying to do it harder. Experiment with new strategies. Are you a visual learner? Try drawing pictures and diagrams. If you are an auditory learner, tape your lectures or make study tapes and listen to them as you commute or exercise. Do you need to interact physically with ideas? Build models, move about as you study, and be active in your learning.

Strategies for Using this *Student Study Guide*

This Student Study Guide certainly is not a replacement for your textbook or biology class. However, it should help support and even streamline your learning process. Some students read the assigned text chapters before lecture and then the study guide chapters after class to reinforce and review. Others reverse that order, skimming the study guide chapter before lecture and then carefully dissecting the text after listening to the professor's presentation. This study guide is especially helpful in preparation for tests. With the text open beside it so you can refer to the essential textbook diagrams, read the relevant study guide chapters for a quick review.

Biology is a visual science, and this *Student Study Guide* includes many diagrams taken from *Biology*, Eighth Edition, to help you see the relationships among the parts of structures and processes. Interactive Questions, which are interspersed throughout each chapter's narrative, are designed to help you develop an active learning approach to your study of biology. The numerous multiple choice questions in each chapter will help you test your growing understanding of biological facts and concepts. The Word Roots section helps you learn the large new vocabulary that is a necessary part of a biology course.

The seven divisions of each study guide chapter are as follows:

- The listing of the **Key Concepts** from the text keeps you focused on the major themes of the chapter.
- The **Framework** identifies the overall picture; it provides a conceptual framework into which the chapter information fits.
- The **Chapter Review** is a condensation of each textbook chapter, with a brief summary for each concept heading within the chapter. All text bold terms are also shown in boldface in this section. Interspersed in this summary are **Interactive Questions** that help you stop and synthesize the material just covered.
- The **Word Roots** section presents the derivation of key biological prefixes, suffixes, and word roots. Examples of terms from the chapter are then defined. Breaking a complicated term down to identifiable components will help you to recognize and learn many new biological terms.
- The **Structure Your Knowledge** section directs you to organize and relate the main concepts of the chapter. It helps you to piece together the key ideas into a bigger picture.
- In the **Test Your Knowledge** section, you are provided with objective questions to test your understanding. The multiple choice questions ask you to choose the best answer. Some answers may be partly correct; almost all choices have been written to test your ability to discriminate among alternatives or to point out common misconceptions.
- Suggested answers to the Interactive Questions, the Structure Your Knowledge section, and the Test Your Knowledge questions are provided in the *Answer Section* at the end of the book.

Using Concept Maps. What are the *concept maps* that appear throughout this study guide? A concept map is a diagram that shows how ideas are organized and related. The *structure* of a concept map is a hierarchically organized cluster of concepts, enclosed in boxes and connected with labeled lines, which explicitly show the relationships among the concepts. The *function* of a concept map is to help you structure your understanding of a topic and create meaning. The *value* of a concept map is in the thinking and organizing required to create a map.

Developing a concept map for a group of concepts requires you to evaluate the relative importance of the concepts (Which are most inclusive and important? Which are less important and subordinate to other concepts?), arrange the concepts in meaningful clusters, and draw and label the connections between them.

This book uses concept maps in several ways. A map of a chapter may be presented in the Framework section to show the organization of the key concepts in that chapter. An Interactive Question may provide a skeleton concept map, with some concepts provided and boxes for you to complete. This technique is intended to help you become more familiar with concept maps and to illustrate one possible approach to organizing the concepts of a particular section.

You will also be asked to develop your own concept maps on certain subsets of ideas. In these cases, the Answer Section will present a suggested concept map. A concept map is an individual picture of your understanding at the time you make the map. As your understanding of an area develops, your map will evolve—sometimes becoming more complex and interrelated, sometimes becoming simplified and streamlined. Do not look to the Answer Section for the "right" concept map. After you have organized your own thoughts, look at the answer map to make sure you have included the key concepts (although you may have added more), to check that the connections you have made are reasonable, and perhaps to see another way to organize the information.

Tips for Taking Multiple Choice Tests. Interact with each question. Read the stem of the question carefully, underlining or using a highlighter to identify the key concept. Read each answer slowly. Cross out the ones you know are wrong. Circle the key idea that you think identifies the correct answer. Now read the question and the answer you chose together, making sure your choice really does answer what is asked. If you aren't sure about a question, try rereading the question and each choice individually. Don't do all of your thinking in your head. Write in the margins and blank spaces of your test. Draw yourself diagrams and pictures. Write down what you do know, and it may jog your memory. If you still are not sure, mark the question to come back to later. As you work through related questions, you may find information that helps you figure out that question. And remember, there is no substitute for good preparation, proper rest and nutrition, and a positive attitude.

Biology is a fascinating, broad, and exciting subject. Campbell/Reece, *Biology*, Eighth Edition is filled with terminology and facts organized in a manner that will help you build a conceptual framework of the major themes of modern biology. This *Student Study Guide* is intended to help you learn and recall information and, most importantly, encourage and guide you as you develop your own understanding of and appreciation for biology.

Martha R. Taylor
Ithaca, New York

Chapter 1

Introduction: Themes in the Study of Life

Key Concepts

1.1 Themes help connect the concepts of biology

1.2 *The Core Theme:* Evolution accounts for the unity and diversity of life

1.3 Scientists use two main forms of inquiry in their study of nature

Framework

This chapter outlines the broad scope of biology, describes themes that unify the study of life, and examines the scientific construction of biological knowledge. A course in biology is neither a vocabulary course nor a classification exercise for the diverse forms of life. Biology is a collection of facts and concepts structured within theories and organizing principles. Recognizing the common themes within biology will help you to structure your knowledge of this fascinating and challenging study of life.

Chapter Review

Biology is the scientific study of life, with **evolution**, the process of change that has shaped life from its origin on Earth to today's diversity, as its organizing principle. The properties and processes of life include highly ordered structure, evolutionary adaptation, response to the environment, regulation, energy processing, growth and development, and reproduction.

1.1 Themes help connect the concepts of biology

Evolution, the Overarching Theme of Biology Evolution explains the unity and diversity of life: All the diverse past and present organisms are related through common ancestry.

Theme: New Properties Emerge at Each Level in the Biological Hierarchy The scale of biology extends from the biosphere to molecules.

INTERACTIVE QUESTION 1.1

Write a brief description of each of these levels of biological organization.

a. biosphere

b. ecosystem

c. community

d. population

e. organism

f. organs and organ systems

g. tissues

h. cells

i. organelles

j. molecules

Interactions among components at each level of biological organization lead to the emergence of novel properties at the next level. These **emergent properties** result from the structural arrangement and interaction of parts.

Biology combines the powerful and pragmatic strategy of reductionism, which breaks down complex systems into simpler components, with the study of the highly complex organizational levels of life.

Many researchers are seeking to understand the emergent properties of life by looking at the functional integration of the parts of a system. **Systems biology** models biological systems and predicts their responses as variables change.

INTERACTIVE QUESTION 1.2

Give examples of how systems biology may impact medical practice or environmental policy making.

Theme: Organisms Interact with Their Environments, Exchanging Matter and Energy Both organisms and the environment are affected by the interactions between them. The dynamics of an ecosystem include the cycling of nutrients and the flow of energy from sunlight to producers (photosynthesizers) to consumers. In each energy transformation, some energy is converted to thermal energy, which is dissipated to the surroundings as heat. Thus, energy flows through ecosystems, entering as solar energy and exiting as heat.

Theme: Structure and Function Are Correlated at All Levels of Biological Organization The form of a biological structure fits its function.

Theme: Cells Are an Organism's Basic Units of Structure and Function The cell is the lowest structural level capable of performing all the activities of life. Every cell uses DNA as its genetic information and is enclosed by a membrane. The simpler and smaller **prokaryotic cell,** unique to bacteria and archaea, lacks both a nucleus to enclose its DNA and most cytoplasmic organelles. The **eukaryotic cell,** with a nucleus containing its DNA, and numerous membrane-bound organelles, is typical of all other living organisms.

Theme: The Continuity of Life Is Based on Heritable Information in the Form of DNA The heritable information of a cell is coded in **DNA,** deoxyribonucleic acid, the substance of genes. **Genes** are the units of inheritance, which transmit information from parents to offspring. Genes are located on chromosomes, long DNA molecules that replicate before cell division and provide identical copies to daughter cells.

The biological instructions for the development and functioning of organisms are coded in the arrangement of the four kinds of nucleotides in DNA molecules.

Most genes program the cell's production of proteins, and almost all cellular actions involve one or more proteins. Enzymes are proteins that catalyze a cell's chemical reactions.

All forms of life use essentially the same genetic code of nucleotides. All the genetic instructions an organism inherits are called its **genome.** The human genome (one set of chromosomes) is about 3 billion nucleotides long, and it codes for the production of more than 75,000 proteins, each with a specific function.

INTERACTIVE QUESTION 1.3

Describe the pathway from DNA nucleotides to proteins.

With the sequencing of the human genome and the genomes of many other organisms, research now focuses on the coordination of the proteins coded for by the DNA sequences on a cellular and organismal level. The first step of the systems strategy is to inventory all the known parts of a system. The second step is to explore how each part behaves in relation to others in the working system. Then computers and software combine the data to model a system network.

Three research developments contribute to systems biology: high-throughput technology or mega-data-collection methods such as the automatic DNA-sequencing machines; **bioinformatics,** which provides the computing power, software, and mathematical models to process and integrate data from enormous data sets; and interdisciplinary research teams with diverse specialists from many scientific fields.

Theme: Feedback Mechanisms Regulate Biological Systems Many biological systems self-regulate by a mechanism called feedback. In **negative feedback,** an end product slows down a process, often by inhibiting an enzyme. In **positive feedback,** less common in biological processes, an end product speeds up its own production. Regulatory mechanisms ensure a dynamic balance in living systems.

1.2   The Core Theme: Evolution accounts for the unity and diversity of life

Organizing the Diversity of Life About 1.8 million species, out of an estimated total of 10 to 400 million, have been identified and named. Taxonomy is the branch of biology that names organisms and groups species into ever broader categories from genera to family, order, class, phylum, kingdom, and domain.

The number of kingdoms is an ongoing debate, but all of life is now grouped into three **domains.** The prokaryotes are divided into **domain Archaea** and **domain Bacteria.** The eukaryotes are placed in **domain Eukarya.** Within the Eukarya, the traditional kingdom Protista is being split into several kingdoms to better reflect evolutionary relationships.

Within this diversity, living forms share a universal genetic language of DNA and similarities in cell structure.

INTERACTIVE QUESTION 1.4

What is a commonly used criterion for placing plants, fungi, and animals into separate kingdoms?

Charles Darwin and the Theory of Natural Selection In *The Origin of Species,* published in 1859, Charles Darwin presented his case for "descent with modification," that present forms evolved from a succession of ancestral forms. Darwin synthesized the theory of natural selection as the mechanism of evolution by drawing an inference from several observations: Individuals vary in many heritable traits, the overproduction of offspring sets up a competition, and species are generally matched to their environments. Individuals with traits best suited for an environment leave more offspring than do less fit individuals. This natural selection, or differential reproductive success within a population, results in the gradual accumulation of favorable adaptations to an environment.

The Tree of Life The underlying unity seen in the structures of related species, both living and in the fossil record, reflects the inheritance of that structure from a common ancestor. Their diversity results from natural selection acting over millions of generations in different environments. The tree-like diagrams of evolutionary relationships reflect the branching genealogy extending from ancestral species. Similar species share a common ancestor at a more recent branch point on the tree of life. Distantly related species share a more ancient common ancestor.

INTERACTIVE QUESTION 1.5

Describe in your own words Darwin's theory of natural selection as the mechanism of evolutionary adaptation and the origin of new species.

1.3 Scientists use two main forms of inquiry in their study of nature

Science is a way of knowing that involves **inquiry,** searching for information by asking and endeavoring to answer questions about nature.

Discovery Science Careful and verifiable observation and analysis of data are the basis of **discovery science.** Observations involve our senses and tools that extend our senses; **data,** both quantitative and qualitative, are recorded observations. Using **inductive reasoning,** a generalized conclusion can often be drawn from collections of observations.

Hypothesis-Based Science Observations and inductions often lead to the search for explanations. A **hypothesis** is a tentative answer to a question or explanation of observations. Using "if . . . then" logic, **deductive reasoning** proceeds from the general to the specific, from a general hypothesis to specific predictions of results if the general premise is correct. A hypothesis is usually tested by performing experiments or making observations to see whether predicted results occur.

A hypothesis must be testable and falsifiable—there must be some observation or experiment that could reveal if the hypothesis is actually not true. In hypothesis-based science, the ideal is to frame two or more alternative hypotheses and design experiments to falsify each candidate explanation. Hypotheses cannot be proven; they can simply be not eliminated through falsification. The more attempts to falsify it that fail, however, the more a hypothesis gains credibility.

The scientific method, as outlined by a structured series of steps, is rarely adhered to rigidly in a scientific inquiry. Scientists often backtrack to make more observations, or make progress on answering a question only after other research provides a new context.

A Case Study in Scientific Inquiry: Investigating Mimicry in Snake Populations The scarlet king snake mimics the ringed coloration of the highly poisonous coral snake. D. and K. Pfennig and W. Harcombe tested the hypothesis that predators will attack king snakes more frequently in non–coral snake areas than in areas where predators have adapted to the warning coloration of coral snakes. They placed equal numbers of plain brown and ringed colored artificial king snakes in regions with and without coral snakes. Compared to the brown "control" snakes, the ringed "experimental" snakes were attacked less frequently only in field sites within the range of the poisonous coral snakes.

This experimental design illustrates a **controlled experiment** in which subjects are divided into an experimental group and a control group. Both groups are treated alike except for the one variable that the experiment is trying to test.

INTERACTIVE QUESTION   1.6

a. How did predators "learn" to avoid coral snakes?

b. Why were the results of the mimicry study presented as the percent of attacks on king snakes in each area rather than the total number of attacks?

Limitations of Science   Scientific inquiry is limited by the requirements that hypotheses be testable and falsifiable and that observations and experimental results be repeatable. Science seeks natural causes for natural phenomena.

Theories in Science   A **theory** is broader in scope than a hypothesis and is supported by a large body of evidence. Still, theories can be modified or even rejected through the testing of the specific, falsifiable hypotheses they generate.

Model Building in Science   Scientific **models,** such as diagrams, graphs, computer programs, or mathematical equations, help explain ideas and processes. The test of a good model is that it fits available data, incorporates new observations, and makes accurate predictions of new experiments.

The Culture of Science   Most scientists work in teams and share their results with a broader research community in seminars, publications, and websites. The political and cultural environment influences the ways in which scientists approach their work. But the adherence to the criteria of verifiable observations and hypotheses that are testable and falsifiable sets science apart from other ways of "knowing nature."

Science, Technology, and Society   Science and technology are interdependent as the information generated by science is applied by **technology** in the development of goods and services, and as technological advances are used to extend scientific knowledge. The uses of scientific knowledge and technologies are influenced by and in turn influence politics, economics, and cultural values.

INTERACTIVE QUESTION   1.7

a. Compare hypotheses and theories.

b. Compare science and technology.

Word Roots

bio- = life (*biology:* the scientific study of life; biosphere: all the environments on Earth that are inhabited by life; *bioinformatics:* using information technology to extract useful information from large sets of biological data)

-ell = small (*organelle:* a small *membrane-enclosed* body with a specialized function found in the cytoplasm of eukaryotic cells)

eu- = true (*eukaryotic cell:* a cell that has a true nucleus)

pro- = before; **karyo-** = nucleus (*prokaryotic cell:* a cell that has no nucleus)

Structure Your Knowledge

1. This chapter presents seven unifying themes of biology. Briefly describe each of these in your own words:
   a. evolution
   b. emergent properties
   c. interaction with the environment
   d. structure and function
   e. cells
   f. heritable information
   g. regulation

Test Your Knowledge

MULTIPLE CHOICE: *Choose the one best answer.*

1. The core idea that makes sense of the unity and all the diversity of life is
   a. the scientific method.
   b. inductive reasoning.

c. deductive reasoning.

d. evolution.

e. systems biology.

2. In an experiment similar to the mimicry experiment performed by the Pfennigs, a researcher found that there were more total predator attacks on model king snakes in areas with coral snakes than in areas outside the range of coral snakes. From this the researcher concluded that

   a. the mimicry hypothesis is false.

   b. there were more predators in the areas with coral snakes.

   c. king snakes do not resemble coral snakes enough to protect them from attack.

   d. the data that should be compared to draw a conclusion must include a control—a comparison with the number of attacks on model brown snakes.

   e. more data must be collected before a conclusion can be drawn.

3. Why can a hypothesis never be "proven" to be true?

   a. One can never collect enough data to be 100% sure.

   b. There may always be alternative hypotheses that might account for the results and that were not tested.

   c. Science is limited by our senses.

   d. Experimental error is involved in every research project.

   e. Science "evolves"; hypotheses and even theories are always changing.

4. Which of the following is an example of positive feedback regulation?

   a. The hormones insulin and glucagon regulate blood-sugar levels.

   b. In the birth of a baby, uterine contractions stimulate release of chemicals that stimulate more uterine contractions.

   c. A rise in temperature when you exercise stimulates sweating and increased blood flow to the skin.

   d. When cells have sufficient energy available, the pathways that break down sugars are turned off.

   e. A rise in CO_2 in the atmosphere correlates with increasing global temperature.

5. Which of the following areas is mismatched with its description?

   a. discovery science—data collection and analysis; deductive reasoning

   b. hypothesis-based science—hypothesis generation; predictions; experiments or observations

   c. systems biology—high throughput technology, bioinformatics, interdisciplinary teams

   d. taxonomy—identify and name organisms; place in hierarchical categories

   e. technology—inventing practical uses of scientific knowledge

6. In a pond sample, you find a unicellular organism that has numerous chloroplasts and a whiplike flagella. In which of the following groups do you think it should be classified?

   a. plant

   b. animal

   c. domain Archaea

   d. one of the proposed kingdoms of protists

   e. You cannot tell unless you see if it has a nucleus or not.

7. What is DNA?

   a. the substance of heredity

   b. a double helix made of four types of nucleotides

   c. a code for protein synthesis

   d. a component of chromosomes

   e. all of the above

8. Which of the following represents the correct sequence in life's hierarchical levels, proceeding upward?

   a. organ, tissue, organ system, organism, population

   b. organism, community, population, ecosystem, biosphere

   c. molecule, organelle, cell, tissue, organ, organism

   d. tissue, cell, organ, organism, community

   d. Both b and c are correct sequences.

9. Which of the following themes of biology is most related to the goals and practice of systems biology?

   a. Evolution accounts for the unity and diversity of life.

   b. Cells are an organism's basic units of structure and function.

   c. The continuity of life is based on heritable information in the form of DNA.

   d. Organisms interact with the environments, exchanging matter and energy.

   e. New properties emerge at each level in the biological hierarchy.

1 The Chemistry of Life

Chapter 2

The Chemical Context of Life

Key Concepts

2.1 Matter consists of chemical elements in pure form and in combinations called compounds

2.2 An element's properties depend on the structure of its atoms

2.3 The formation and function of molecules depend on chemical bonding between atoms

2.4 Chemical reactions make and break chemical bonds

Framework

This chapter considers the basic principles of chemistry that explain the behavior of atoms and molecules. You will learn how the subatomic particles—protons, neutrons, and electrons—are organized into atoms and how atoms are connected by covalent or ionic bonds. Weak chemical bonds help to create the shapes and functions of molecules. Emergent properties are associated with each new level of structural organization in the hierarchy from atoms to life.

Chapter Review

2.1 Matter consists of chemical elements in pure form and in combinations called compounds

Elements and Compounds **Matter** is anything that takes up space and has mass. (Although sometimes used interchangeably, mass reflects the amount of matter in an object, whereas weight reflects gravity's pull on that mass.) **Elements** are substances that cannot be chemically broken down to other types of matter. A **compound** is made up of two or more elements combined in a fixed ratio. A compound usually has characteristics quite different from its constituent elements, an example of the emergence of novel properties in higher levels of organization.

Essential Elements of Life Carbon (C), oxygen (O), hydrogen (H), and nitrogen (N) make up 96% of living matter. The seven elements listed in Interactive Question 2.1 make up most of the remaining 4%. Some elements, like iron (Fe) and iodine (I), may be required in very minute quantities and are called **trace elements**.

INTERACTIVE QUESTION   2.1

Fill in the names beside the symbols of the following elements commonly found in living matter.

Ca   Na

P   Cl

K   Mg

S

2.2 An element's properties depend on the structure of its atoms

An **atom** is the smallest unit of an element retaining the properties of that element.

Subatomic Particles   Three stable subatomic particles are important to our understanding of atoms. Uncharged **neutrons** and positively charged **protons** are packed tightly together to form the **atomic nucleus** of an atom. Negatively charged **electrons** form a cloud around the nucleus.

Protons and neutrons have a similar mass of about 1.7×10^{-24} g or close to 1 dalton each. A **dalton** is the measurement unit for atomic mass. Electrons have negligible mass.

Atomic Number and Atomic Mass   Each element has a characteristic **atomic number,** or number of protons in the nucleus of each of its atoms. Unless indicated otherwise, an atom has a neutral electrical charge, and thus the number of protons is equal to the number of electrons. A subscript to the left of the symbol for an element indicates its atomic number; a superscript indicates mass number. The **mass number** is equal to the number of protons and neutrons in the nucleus and approximates the mass of an atom of that element in daltons. The term **atomic mass** refers to the total mass of an atom.

INTERACTIVE QUESTION   2.2

The difference between the mass number and the atomic number of an atom is equal to the number of _____.
An atom of phosphorus, $^{31}_{15}P$, contains _____ protons, _____ electrons, and _____ neutrons. The atomic mass of phosphorus is approximately _____.

Isotopes   Although the number of protons is constant, the number of neutrons can vary among the atoms of an element, creating different **isotopes** that have slightly different masses but the same chemical behavior. Some isotopes are unstable; the nuclei of **radioactive isotopes** spontaneously decay, giving off particles and energy. Radioactive isotopes are important tools in biological research and medicine. Too great an exposure to radiation from decaying isotopes poses a significant hazard.

The Energy Levels of Electrons   **Energy** is defined as the ability to cause change. **Potential energy** is energy stored in matter as a consequence of its position or structure. The potential energy of electrons increases as their distance from the positively charged nucleus increases. Electrons can be located in different **electron shells** surrounding the nucleus.

INTERACTIVE QUESTION   2.3

To move to a shell farther from the nucleus, an electron must _____ energy; energy is _____ when an electron moves to a closer shell.

Electron Distribution and Chemical Properties   The chemical behavior of an atom is a function of the distribution of its electrons—in particular, the number of **valence electrons** in its outermost electron shell, or **valence shell.** A valence shell of eight electrons is complete, resulting in an unreactive or inert atom. Atoms with incomplete valence shells are chemically reactive. Each row in the periodic table of elements is arranged in order of the sequential addition of electrons to electron shells.

INTERACTIVE QUESTION   2.4

Draw the electron shell diagram for these atoms.
a. $_6C$ **c.** $_8O$

b. $_7N$ **d.** $_{12}Mg$

Electron Orbitals   An **orbital** is the three-dimensional space or volume within which an electron is most likely to be found. No more than two electrons can occupy the same orbital. The first electron shell can contain two electrons in a single spherical orbital, called the 1s orbital. The second electron shell can hold a maximum of eight electrons in its four orbitals, which are a 2*s* spherical orbital and three dumbbell-shaped *p* orbitals located along the x, y, and z axes.

INTERACTIVE QUESTION 2.5

Look again at the electron shell diagram you drew for carbon (a) in Interactive Question 2.4. Did you show the outer shell electrons unpaired? Why?

INTERACTIVE QUESTION 2.6

Fill in the blanks in the following concept map to help you review the atomic structure of atoms.

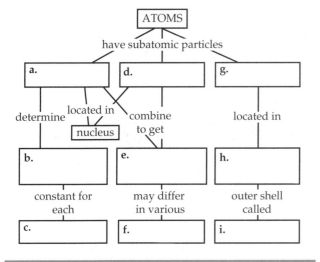

2.3 The formation and function of molecules depend on chemical bonding between atoms

Atoms with incomplete valence shells can either share electrons with or completely transfer electrons to or from other atoms such that each atom is able to complete its valence shell. These interactions usually result in attractions, called **chemical bonds,** that hold the atoms close together.

Covalent Bonds   When two atoms share a pair of valence electrons, a **covalent bond** is formed. A **molecule** consists of two or more atoms held together by covalent bonds. A **structural formula,** such as H—H, indicates both the atoms and bonds in a molecule. The dash indicates a **single bond.** A **molecular formula,** such as O_2, indicates only the kinds and numbers of atoms. In an

oxygen molecule, two pairs of valence electrons are shared between oxygen atoms, forming a double covalent bond, or simply a **double bond** (O=O). The **valence,** or bonding capacity, of an atom usually equals the number of unpaired electrons in its valence shell.

INTERACTIVE QUESTION 2.7

a. What are the valences of the four most common elements of living matter?

b. Draw the Lewis dot structures for the following molecules.

H_2 O_2 N_2 CH_4

Electronegativity is the attraction of a particular type of atom for shared electrons. If the atoms in a molecule have similar electronegativities, the electrons remain equally shared between the two nuclei, and the covalent bond is said to be a **nonpolar covalent bond.** If one element is more electronegative, it pulls the shared electrons closer to itself, creating a **polar covalent bond.** This unequal sharing of electrons results in a slight negative charge ($\delta-$) associated with the more electronegative atom and a slight positive charge ($\delta+$) associated with the atom from which the electrons are pulled.

INTERACTIVE QUESTION 2.8

Explain whether the following molecules contain nonpolar or polar covalent bonds. (Hint: N and O both have high electronegativities.)

a. nitrogen molecule N≡N   **c.** methane

$$H-\overset{\overset{\textstyle H}{|}}{\underset{\underset{\textstyle H}{|}}{C}}-H$$

b. ammonia

$$H\diagup\overset{\overset{\textstyle N}{|}}{\underset{\textstyle H}{}}\diagdown H$$

d. formaldehyde

$$\overset{\textstyle H}{\underset{\textstyle H}{}}\diagup C=O$$

Ionic Bonds   If two atoms are very different in their attraction for the shared electrons, the more electronegative atom may completely transfer an electron from another atom, resulting in the formation of charged atoms called **ions.** The atom that lost the electron is a positively charged **cation.** The negatively charged atom that gained the electron is called an **anion.** An **ionic bond** may hold these ions together because of the attraction of their opposite charges.

Ionic compounds, called **salts,** often exist as three-dimensional crystalline lattice arrangements held together by electrical attractions. The number of ions present in a salt crystal is not fixed, but the atoms are present in specific ratios. Salts have strong ionic bonds when dry, but the crystal dissolves in water.

Ion also refers to entire covalent molecules that are electrically charged.

INTERACTIVE QUESTION 2.9

Calcium ($_{20}$Ca) and chlorine ($_{17}$Cl) can combine to form the salt calcium chloride. Based on the number of electrons in their valence shells and their bonding capacities, what would the molecular formula for this salt be?
a. _____ Which atom becomes the cation? b. _____

Weak Chemical Bonds   Weak bonds, such as ionic bonds in water, form temporary interactions between molecules. Weak bonds within large molecules such as proteins help to create their three-dimensional functional shape.

When a hydrogen atom is covalently bonded with an electronegative atom, and thus has a partial positive charge, it can be attracted to another electronegative atom and form a **hydrogen bond.**

All atoms and molecules are attracted to each other when in close contact by **van der Waals interactions.** Momentary uneven electron distributions produce changing positive and negative regions that create these weak attractions.

INTERACTIVE QUESTION 2.10

Sketch a water molecule, showing oxygen's electron shells and the covalently shared electrons. Indicate the areas with slight negative and positive charges that enable a water molecule to form hydrogen bonds with other polar molecules.

Molecular Shape and Function   A molecule's characteristic size and shape affect how it interacts with other molecules. When atoms form covalent bonds, their *s* and three *p* orbitals hybridize to form four teardrop-shaped orbitals in a tetrahedral arrangement. These hybrid orbitals dictate the specific shapes of different molecules.

INTERACTIVE QUESTION 2.11

Look at your diagram of a water molecule in Interactive Question 2.10. Why should its shape be roughly like a V?

2.4   Chemical reactions make and break chemical bonds

Chemical reactions involve the making or breaking of chemical bonds. Matter is conserved in chemical reactions; the same number and kinds of atoms are present in both **reactants** and **products,** although the rearrangement of electrons and atoms causes the properties of these molecules to be different.

INTERACTIVE QUESTION 2.12

Fill in the missing coefficients for respiration, the conversion of glucose and oxygen to carbon dioxide and water, so that all atoms are conserved in the chemical reaction.

$$C_6H_{12}O_6 + \underline{} O_2 \rightarrow \underline{} CO_2 + \underline{} H_2O$$

Chemical reactions are reversible—the products of the forward reaction can become reactants in the reverse reaction. Increasing the concentrations of reactants can speed up the rate of a reaction. **Chemical equilibrium** is reached when the forward and reverse reactions proceed at the same rate, and the relative concentrations of reactants and products no longer change.

Word Roots

an- = not (*anion:* a negatively charged ion)
co- = together; **-valent** = strength (*covalent bond:* an attraction between atoms that share one or more pairs of outer-shell electrons)
electro- = electricity (*electronegativity:* the tendency for an atom to pull electrons toward itself)

iso- = equal (*isotope:* an element having the same number of protons and electrons but a different number of neutrons)

neutr- =neither (*neutron:* a subatomic particle with a neutral electrical charge)

pro- = before (*proton:* a subatomic particle with a single positive electrical charge)

Structure Your Knowledge

Take the time to write out or discuss your answers to the following questions. Then refer to the suggested answers at the end of the book.

1. Fill in the following chart for the major subatomic particles of an atom.

Particle	Charge	Mass	Location

2. Atoms can have various numbers associated with them.
   a. Define the following and show where each of them is placed relative to the symbol of an element such as C (use the most common isotope, C-12): atomic number, mass number, atomic mass.
   b. Define valence.
   c. Which of these four numbers is most related to the chemical behavior of an atom? Explain.

3. Explain what is meant by saying that the sharing of electrons between atoms falls on a continuum from nonpolar covalent bonds to ionic bonds.

Test Your Knowledge

MULTIPLE CHOICE: *Choose the one best answer.*

1. Each element has its own characteristic atom in which
   a. the atomic mass is constant.
   b. the atomic number is constant.
   c. the mass number is constant.
   d. two of the above are correct.
   e. all of the above are correct.

2. Radioactive isotopes can be used in studies of metabolic pathways because
   a. their half-life allows a researcher to time an experiment.
   b. they are more reactive.
   c. the cell does not recognize the extra protons in the nucleus, so isotopes are readily used in metabolism.
   d. their location or quantity can be experimentally determined because of their radioactivity.
   e. their extra neutrons produce different colors that can be traced through the body.

3. In a reaction in chemical equilibrium,
   a. the forward and reverse reactions are occurring at the same rate.
   b. the reactants and products are in equal concentration.
   c. the forward reaction has gone further than the reverse reaction.
   d. there are equal numbers of atoms on both sides of the equation.
   e. a, b, and d are correct.

4. Oxygen has eight electrons. You would expect the arrangement of these electrons to be
   a. eight in the second electron shell, creating an inert element.
   b. two in the first electron shell and six in the second, creating a valence of six.
   c. two in the 1*s* orbital and two each in the three 2*p* orbitals, creating a valence of two.
   d. two in the 1*s* orbital, one each in the 2*s* and three 2*p* orbitals, and two in the 3*s* orbital, creating a valence of two.
   e. two in the 1*s* orbital, two in both the 2*s* and 2*px* orbitals, and one each in the 2*py* and 2*pz* orbitals, creating a valence of two.

5. A covalent bond between two atoms is likely to be polar if
   a. one of the atoms is much more electronegative than the other.
   b. the two atoms are equally electronegative.
   c. the two atoms are of the same element.
   d. the bond is part of a tetrahedrally shaped molecule.
   e. one atom is an anion.

6. A triple covalent bond would
   a. be very polar.
   b. involve the bonding of three atoms.
   c. involve the bonding of six atoms.
   d. produce a triangularly shaped molecule.
   e. involve the sharing of six electrons.

7. A cation
  a. has gained an electron.
  b. can easily form hydrogen bonds.
  c. is more likely to form in an atom with seven electrons in its valence shell.
  d. has a positive charge.
  e. Both c and d are correct.

8. What types of bonds are identified in the following illustration of a water molecule interacting with an ammonia molecule?

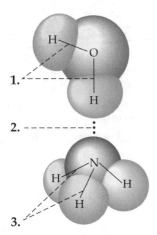

  a. Bonds 1 are polar covalent bonds, bond 2 is an ionic bonds and bonds 3 are nonpolar covalent bond.
  b. Bonds 1 and 3 are polar covalent bonds, and bond 2 is a hydrogen bond.
  c. Bonds 1 and 3 are polar covalent bonds, and bond 2 is an ionic bond.
  d. Bonds 1 and 3 are nonpolar covalent bonds, and bond 2 is a hydrogen bond.
  e. Bonds 1 and 3 are polar covalent bonds, and bond 2 is a nonpolar covalent bond.

9. Which of the following weak bonds may form between any closely aligned molecules?
  a. nonpolar covalent
  b. polar covalent
  c. ionic
  d. hydrogen
  e. van der Waals interactions

10. The ability of morphine to mimic the effects of the body's endorphins is due to
  a. a chemical equilibrium developing between morphine and endorphins.
  b. the one-way conversion of morphine into endorphin.
  c. molecular shape similarities that allow morphine to bind to endorphin receptors.

  d. the similarities between morphine and heroin.
  e. hydrogen bonding and other weak bonds forming between morphine and endorphins.

Use this information to answer questions 11 through 16.

The six elements most common in living organisms are:

$$^{12}_{6}C \quad ^{16}_{8}O \quad ^{1}_{1}H \quad ^{14}_{7}N \quad ^{32}_{16}S \quad ^{31}_{15}P$$

11. How many electrons does phosphorus have in its valence shell?
  a. 3   **c.** 7   **e.** 16
  b. 5   **d.** 15

12. What is the atomic mass of phosphorus?
  a. 15   **c.** 31   **e.** 62
  b. 16   **d.** 46

13. A radioactive isotope of carbon has the mass number 14. How many neutrons does this isotope have?
  a. 2   **c.** 8   **e.** 14
  b. 6   **d.** 12

14. How many covalent bonds is a sulfur atom most likely to form?
  a. 1   **c.** 3   **e.** 5
  b. 2   **d.** 4

15. Based on electron configuration, which of these elements would have chemical behavior most like that of oxygen?
  a. C   **c.** N   **e.** S
  b. H   **d.** P

16. How many of these elements are found next to each other (side by side) on the periodic table?
  a. one group of two
  b. two groups of two
  c. one group of two and one group of three
  d. one group of three
  e. all of them

17. Taking into account the bonding capacities or valences of carbon (C) and oxygen (O), how many hydrogen (H) must be added to complete the structural diagram of this molecule?

$$\overset{O}{\underset{O}{\diagdown}}{}^{\diagup}C-C-C-C=C-C-C$$

  a. 9   **c.** 11   **e.** 13
  b. 10   **d.** 12

18. A sodium ion (Na^+) contains 10 electrons, 11 protons, and 12 neutrons. What is the atomic number of sodium?

    a. 10   c. 12   e. 33

    b. 11   d. 23

19. What type of bond would you expect potassium ($^{39}_{19}K$) to form?

    a. ionic; it would donate one electron and carry a positive charge

    b. ionic; it would donate one electron and carry a negative charge

    c. covalent; it would share one electron and make one covalent bond

    d. covalent; it would share two electrons and form two bonds

    e. none; potassium is an inert element

20. What is the molecular shape of methane (CH_4)?

    a. planar or flat, with the H arranged around the C

    b. pentagonal, or a flat five-sided arrangement

    c. tetrahedral, due to the hybridization of the *s* and three *p* orbits of the C

    d. circular, with the four H attached in a ring around the C

    e. linear, since all the bonds are nonpolar covalent

21. Which of the following is a molecule capable of forming hydrogen bonds?

    a. CH_4

    b. H_2O

    c. NaCl

    d. H_2

    e. a, b, and d can form hydrogen bonds.

22. Chlorine has an atomic number of 17 and a mass number of 35. How many electrons would a chloride ion have?

    a. 16   c. 18   e. 34

    b. 17   d. 33

23. What is the difference between a molecule and a compound?

    a. There is no difference; the terms are interchangeable.

    b. Molecules contain atoms of a single element, whereas compounds contain two or more elements.

    c. A molecule consists of two or more covalently bonded atoms; a compound contains two or more atoms held by ionic bonds.

    d. A compound consists of two or more elements in a fixed ratio; a molecule has two or more covalently bonded atoms of the same or different elements.

    e. Compounds always consist of molecules, but molecules are not always compounds.

24. Which of the following atomic numbers would describe the element that is least reactive?

    a. 1   c. 12   e. 18

    b. 8   d. 16

25. What coefficients must be placed in the blanks to balance this chemical reaction?

    $$C_5H_{12} + \underline{} O_2 \rightarrow \underline{} CO_2 + \underline{} H_2O$$

    a. 5; 5; 5

    b. 6; 5; 6

    c. 6; 6; 6

    d. 8; 4; 6

    e. 8; 5; 6

Water and the Fitness of the Environment

Key Concepts

3.1 The polarity of water molecules results in hydrogen bonding

3.2 Four emergent properties of water contribute to Earth's fitness for life

3.3 Acidic and basic conditions affect living organisms

Framework

Water makes up 70% to 95% of the cell content of living organisms and covers 75% of Earth's surface. Its unique properties make the external environment fit for life and the internal environments of organisms fit for the chemical and physical processes of life.

Hydrogen bonding between polar water molecules creates a cohesive liquid with a high specific heat and high heat of vaporization, both of which help to regulate environmental temperature. Ice floats and protects oceans and lakes from freezing. The polarity of water makes it a versatile solvent. The [H$^+$] in a solution is expressed as pH. Buffers regulate an organism's pH.

Chapter Review

3.1 The polarity of water molecules results in hydrogen bonding

A water molecule consists of two hydrogen atoms each covalently bonded to a more electronegative oxygen atom. This **polar molecule** has a shape like a wide V with a slight positive charge on each hydrogen atom ($\delta+$) and a slight negative charge ($\delta-$) associated with the oxygen. Hydrogen bonds, electrical attractions between the hydrogen atom of one water molecule and the oxygen atom of a nearby water molecule, create a structural organization that leads to the emergent properties of water.

3.2 Four emergent properties of water contribute to Earth's fitness for life

Cohesion   Liquid water is unusually cohesive due to the constant forming and reforming of hydrogen bonds that hold the molecules close together. This **cohesion** creates a more structurally organized liquid and helps water to be pulled upward in plants. The **adhesion** of water molecules to the walls of plant vessels also contributes to water transport. Hydrogen bonding between water molecules produces a high **surface tension** at the interface between water and air.

INTERACTIVE QUESTION   3.1

Draw the four water molecules that can hydrogen-bond to this water molecule. Show the bonds and the slight negative and positive charges that account for the formation of these hydrogen bonds.

Moderation of Temperature   In a body of matter, **heat** is a measure of the total quantity of **kinetic energy,** the energy associated with the movement of atoms and molecules. **Temperature** measures the average kinetic energy of the molecules in a substance.

Temperature is measured using a **Celsius scale.** Water at sea level freezes at 0°C and boils at 100°C. A **calorie (cal)** is the amount of heat energy it takes to raise 1 g of water 1°C. A **kilocalorie (kcal)** is 1,000 calories, the amount of heat required to raise 1 kg of water 1°C. A **joule (J)** equals 0.239 cal; a calorie is 4.184 J.

Specific heat is the amount of heat absorbed or lost when 1 g of a substance changes its temperature by 1°C. Water's specific heat of 1 cal/g/°C is unusually high compared with other common substances. Why does water absorb or release a relatively large quantity of heat as its temperature changes? Heat must be absorbed to break hydrogen bonds before water molecules can move faster and the temperature can rise, and conversely, heat is released when hydrogen bonds form as the temperature of water drops. The high proportion of water in the environment and within organisms keeps temperature fluctuations within limits that permit life.

Vaporization or evaporation occurs when molecules of a liquid with sufficient kinetic energy overcome their attraction to other molecules and escape into the air as gas. The **heat of vaporization** is the quantity of heat that must be absorbed for 1 g of a liquid to be converted to a gas. Water has a high heat of vaporization (580 cal/g at 25°C) because a large amount of heat is needed to break the hydrogen bonds holding water molecules together. This property of water helps moderate the climate on Earth as solar heat is dissipated from tropical seas during evaporation and heat is released when moist tropical air condenses to form rain.

As a liquid vaporizes, the surface left behind loses the kinetic energy of the escaping molecules and cools down. **Evaporative cooling** helps to protect terrestrial organisms from overheating and contributes to the stability of temperatures in lakes and ponds.

Insulation of Water Bodies by Floating Ice   As water cools below 4°C, it expands. By 0°C, each water molecule becomes hydrogen-bonded to four other molecules, creating a crystalline lattice that spaces the molecules apart. Ice is less dense than liquid water; therefore, it floats.

The Solvent of Life   A **solution** is a liquid homogeneous mixture of two or more substances; the dissolving agent is called the **solvent** and the substance that is dissolved is the **solute.** Water is the solvent in an **aqueous solution.** The positive and negative regions of water molecules are attracted to oppositely charged ions or partially charged regions of polar molecules. Thus, solute molecules become surrounded by water molecules (a **hydration shell**) and dissolve into solution.

Ionic and polar substances are **hydrophilic;** they have an affinity for water due to electrical attractions and hydrogen bonding. Large hydrophilic substances may not dissolve but become suspended in an aqueous solution, forming a mixture called a **colloid.** Nonpolar molecules are **hydrophobic;** they will not easily mix with or dissolve in water.

INTERACTIVE QUESTION   3.2

The following concept map is one way to show how the breaking and forming of hydrogen bonds is related to temperature moderation. Fill in the blanks and compare your choice of concepts to those given in the answer section. Or, even better, create your own map to help you understand how water stabilizes temperature.

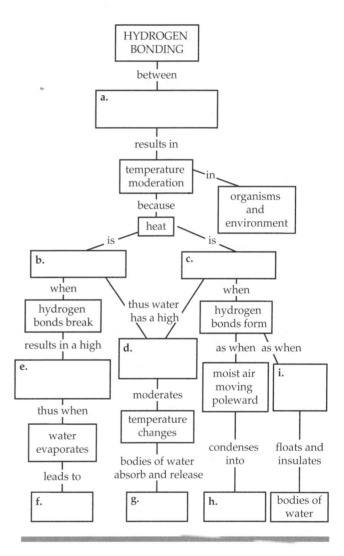

INTERACTIVE QUESTION   3.3

Indicate whether the following are hydrophilic or hydrophobic. Do these substances contain ionic bonds, polar covalent bonds, or nonpolar covalent bonds?

a. olive oil   **c.** salt

b. sugar   **d.** candle wax

Most of the chemical reactions of life take place in water. A **mole (mol)** is the amount of a substance that has a mass in grams numerically equivalent to its **molecular mass** (sum of the mass of all atoms in the molecule or compound) in daltons. A mole of any substance has exactly the same number of molecules—6.02×10^{23}, called Avogadro's number. The **molarity** of a solution (abbreviated M) refers to the number of moles of a solute dissolved in 1 liter of solution.

INTERACTIVE QUESTION 3.4

a. How many grams of lactic acid ($C_3H_6O_3$) are in 1 liter of a 0.5 M solution of lactic acid? (^{12}C, ^1H, ^{16}O)

b. How many molecules of lactic acid are in the above solution?

3.3 Acidic and basic conditions affect living organisms

A water molecule can dissociate into a **hydrogen ion,** H^+ (which binds to another water molecule to form a **hydronium ion,** H_3O^+) and a **hydroxide ion,** OH^-. In pure water at 25°C, the concentrations of H^+ and OH^- ions are the same; both are equal to $10^{-7} M$.

Effects of Changes in pH   When acids or bases dissolve in water, the H^+ and OH^- balance shifts. An **acid** adds H^+ to a solution, whereas a **base** reduces H^+ in a solution by accepting hydrogen ions or by adding hydroxide ions (which then combine with H^+ and thus remove hydrogen ions). A strong acid or strong base may dissociate completely when mixed with water. A weak acid or base reversibly dissociates, either releasing or binding H^+.

In an aqueous solution, the *product* of the $[H^+]$ and $[OH^-]$ is constant at 10^{-14}. Brackets, [ ], indicate molar concentration. If the $[H^+]$ is higher, then the $[OH^-]$ is lower, because the excess hydrogen ions combine with the hydroxide ions in solution and form water. Likewise, an increase in $[OH^-]$ causes an equivalent decrease in $[H^+]$.

The **pH** of a solution is defined as the negative log (base 10) of the $[H^+]$: pH $= -\log [H^+]$. For a neutral aqueous solution, $[H^+]$ is $10^{-7} M$, and the pH equals 7. As the $[H^+]$ increases in an acidic solution, the pH value decreases. The difference between each unit of the pH scale represents a tenfold difference in the concentration of $[H^+]$ and $[OH^-]$.

Most cells have an internal pH close to 7. **Buffers** within the cell maintain a stable pH by accepting excess H^+ ions or donating H^+ ions when H^+ concentration decreases. Weak acid-base pairs that reversibly bind hydrogen ions are typical of most buffering systems.

INTERACTIVE QUESTION 3.5

Complete the following table to review your understanding of pH.

$[H^+]$	$[OH^-]$	pH	Acidic, Basic, or Neutral?
	10^{-11}	3	acidic
10^{-8}			
	10^{-7}		
		1	

INTERACTIVE QUESTION 3.6

The carbonic acid/bicarbonate system is an important biological buffer. Label the molecules and ions in this equation and indicate which is the H^+ donor and which is the acceptor.

$$H_2CO_3 \rightleftharpoons HCO_3^- + H^+$$

In which direction will this reaction proceed

a. when the pH of a solution begins to fall?

b. when the pH rises above normal level?

Threats to Water Quality on Earth   Acid **precipitation,** rain, snow, or fog with a pH lower than normal (pH 5.6), is due to the reaction of water in the atmosphere with the sulfur oxides and nitrogen oxides released by the combustion of fossil fuels. Aquatic life is damaged by acid precipitation, and lowering the pH of the soil solution affects the solubility of minerals needed by plants.

The burning of fossil fuels increases the release of CO_2 to the atmosphere, contributing to the "greenhouse" effect and increasing the absorption of CO_2 by the oceans. The resulting lower pH of seawater decreases the concentration of carbonate (CO_3^{2-}), an important ion needed for coral reef calcification.

INTERACTIVE QUESTION 3.7

a. Use a formula to explain why increasing [CO_2] in water leads to a lower pH.

b. Use a formula to explain how a lower pH would affect the [CO_3^{2-}] in the ocean.

c. Assuming a fairly constant [Ca^{2+}] in the ocean, how would a change in [CO_3^{2-}] affect the calcification rate in a coral reef?

Word Roots

kilo- = a thousand (*kilocalorie:* a thousand calories)
hydro- = water; **-philos** = loving; **-phobos** = fearing (*hydrophilic:* having an affinity for water; *hydrophobic:* having an aversion to water)

Structure Your Knowledge

1. To become proficient in the use of the concepts relating to pH, develop a concept map to organize your understanding of the following terms: pH, [H^+], [OH^-], acidic, basic, neutral, buffer, 1–14, acid-base pair. Remember to label connecting lines and add additional concepts as you need them.

A suggested concept map is given in the answer section, but remember that your concept map should represent your own understanding. The value of this exercise is in organizing these concepts for yourself.

2. Fill in the table at the bottom of the page that summarizes the properties of water that contribute to the fitness of the environment for life.

Test Your Knowledge

MULTIPLE CHOICE: *Choose the one best answer.*

1. Each water molecule is capable of forming
   a. one hydrogen bond.
   b. three hydrogen bonds.
   c. four hydrogen bonds.
   d. two covalent bonds and two hydrogen bonds.
   e. two ionic bonds and two hydrogen bonds.

2. The polarity of water molecules
   a. promotes the formation of hydrogen bonds.
   b. helps water to dissolve nonpolar solutes.
   c. lowers the heat of vaporization and leads to evaporative cooling.
   d. creates a crystalline structure in liquid water.
   e. does all of the above.

3. What accounts for the movement of water up the veins of a tall tree?
   a. cohesion
   b. hydrogen bonding
   c. adhesion
   d. hydrophilic cell walls
   e. all of the above

Property	Explanation of Property	Example of Benefit to Life
a.	Hydrogen bonds hold molecules together and adhere them to hydrophilic surface.	**b.**
High specific heat	**c.**	Temperature changes in environment and organisms are moderated.
d.	Hydrogen bonds must be broken for water to evaporate.	**e.**
f.	Water molecules with high kinetic energy evaporate; remaining molecules are cooler.	**g.**
Less dense as solid	**h.**	**i.**
j.	**k.**	Most chemical reactions in life involve solutes dissolved in water.

4. Climates tend to be moderate near large bodies of water because
   a. a large amount of solar heat is absorbed during the gradual rise in temperature of the water.
   b. water releases heat to the environment as it cools.
   c. the high specific heat of water helps to moderate air temperatures.
   d. a great deal of heat is absorbed and released as hydrogen bonds break or form.
   e. of all of the above.

5. Temperature is a measure of
   a. specific heat.
   b. average kinetic energy of molecules.
   c. total kinetic energy of molecules.
   d. Celsius degrees.
   e. joules.

6. Evaporative cooling is a result of
   a. a low heat of vaporization.
   b. the release of heat during the breaking of hydrogen bonds when water molecules escape.
   c. the absorption of heat as hydrogen bonds break.
   d. the reduction in the average kinetic energy of a liquid after energetic water molecules enter the gaseous state.
   e. both a and b.

7. Ice floats because
   a. air is trapped in the crystalline lattice.
   b. the formation of hydrogen bonds releases heat; warmer objects float.
   c. it has a smaller surface area than liquid water.
   d. it insulates bodies of water so they do not freeze from the bottom up.
   e. hydrogen bonding spaces the molecules farther apart, creating a less dense structure.

8. The molarity of a solution is equal to
   a. Avogadro's number of molecules in 1 liter of solvent.
   b. the number of moles of a solute in 1 liter of solution.
   c. the molecular mass of a solute in 1 liter of solution.
   d. the number of solute molecules in 1 liter of solvent.
   e. 342 g if the solute is sucrose.

9. Some archaea are able to live in lakes with pH values of 11. How does pH 11 compare with the pH 7 typical of your body cells?
   a. It is four times more acidic than pH 7.
   b. It is four times more basic than pH 7.
   c. It is a thousand times more acidic than pH 7.
   d. It is a thousand times more basic than pH 7.
   e. It is ten thousand times more basic than pH 7.

10. A buffer
    a. changes pH by a magnitude of 10.
    b. releases excess OH^-.
    c. releases excess H^+.
    d. is often a weak acid-base pair.
    e. always maintains a neutral pH.

11. Which of the following is least soluble in water?
    a. polar molecules
    b. nonpolar molecules
    c. ionic compounds
    d. hydrophilic molecules
    e. anions

12. Which would be the best method for reducing acid precipitation?
    a. Raise the height of smokestacks so that exhaust enters the upper atmosphere.
    b. Add buffers and bases to bodies of water whose pH has dropped.
    c. Use coal-burning generators rather than nuclear power to produce electricity.
    d. Tighten emission control standards for factories and automobiles.
    e. Reduce the concentration of heavy metals in industrial exhaust.

13. What bonds must be broken for water to vaporize?
    a. polar covalent bonds
    b. nonpolar covalent bonds
    c. hydrogen bonds
    d. ionic bonds
    e. polar covalent and hydrogen bonds

14. How would you make a 0.1 M solution of glucose ($C_6H_{12}O_6$)? The mass numbers for these elements are approximately: C = 12, O = 16, H = 1.
    a. Mix 6 g C, 12 g H, and 6 g O in 1 liter of water.
    b. Mix 72 g C, 12 g H, and 96 g O in 1 liter of water.
    c. Mix 18 g of glucose with enough water to yield 1 liter of solution.

Chapter 3: Water and the Fitness of the Environment

d. Mix 29 g of glucose with enough water to yield 1 liter of solution.
e. Mix 180 g of glucose with enough water to yield 1 liter of solution.

15. How many molecules of glucose would be in the 1 liter 0.1 M solution made in question 14?
    a. 0.1
    b. 6
    c. 60
    d. 6×10^{23}
    e. 6×10^{22}

16. Why is water such an excellent solvent?
    a. As a polar molecule, it can surround and dissolve ionic and polar molecules.
    b. It forms ionic bonds with ions, hydrogen bonds with polar molecules, and hydrophobic interactions with nonpolar molecules.
    c. It forms hydrogen bonds with itself.
    d. It has a high specific heat and a high heat of vaporization.
    e. It is wet and has a great deal of surface tension.

17. Which of the following, when mixed with water, would form a colloid?
    a. a large hydrophobic protein
    b. a large hydrophilic protein
    c. sugar
    d. cotton
    e. NaCl

18. Adding a base to a solution would
    a. raise the pH.
    b. lower the pH.
    c. decrease [H^+].
    d. do both a and c.
    e. do both b and c.

19. A hydration shell is most likely to form around
    a. an ion.
    b. a fat.
    c. a sugar.
    d. both a and c.
    e. both b and c.

20. The following are the pH values for each item: cola–2; orange juice–3; beer–4; coffee–5; human blood–7.4. Which of these liquids has the *highest* molar concentration of OH⁻?
    a. cola
    b. orange juice
    c. beer
    d. coffee
    e. human blood

21. Comparing the [H^+] of orange juice and coffee, the [H^+] of
    a. orange juice is 10 times higher.
    b. orange juice is 100 times higher.
    c. orange juice is 1,000 times higher.
    d. coffee is 2 times higher.
    e. coffee is 100 times higher.

22. The ability of water molecules to form hydrogen bonds accounts for water's
    a. high specific heat.
    b. evaporative cooling.
    c. high heat of vaporization.
    d. cohesiveness and surface tension.
    e. All of the above result from water's hydrogen-bonding capacity.

Chapter 4

Carbon and the Molecular Diversity of Life

Key Concepts

4.1 Organic chemistry is the study of carbon compounds

4.2 Carbon atoms can form diverse molecules by bonding to four other atoms

4.3 Characteristic chemical groups help control how biological molecules function

Framework

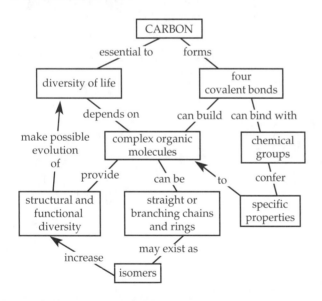

Chapter Review

4.1 Organic chemistry is the study of carbon compounds

Early organic chemists could not synthesize the complex molecules found in living organisms and, therefore, attributed their formation to a life force independent of physical and chemical laws, a belief known as *vitalism. Mechanism*, the philosophy underlying modern **organic chemistry,** holds that physical and chemical explanations are sufficient to account for all natural phenomena.

INTERACTIVE QUESTION 4.1

How did Stanley Miller's classic experiment apply mechanism to the evolution of life?

4.2 Carbon atoms can form diverse molecules by bonding to four other atoms

Formation of Bonds with Carbon How many covalent bonds must carbon (with an atomic number of 6) form to complete its valence shell? Carbon's *tetravalence* is at the center of its ability to form large and complex molecules. When carbon forms four single covalent bonds, its hybrid orbitals create a tetrahedral shape. When two carbons are joined by a double bond, the other carbon bonds are in the same plane, forming a flat molecule.

Molecular Diversity Arising from Carbon Skeleton Variation Carbon skeletons can vary in length, branching, placement of double bonds, and location of atoms of other elements. **Hydrocarbons** consist of only carbon and hydrogen. The nonpolar C—H bonds in hydrocarbon chains account for their hydrophobic properties.

Isomers are compounds with the same molecular formula but different structural arrangements and, thus, different properties. **Structural isomers** differ in the covalent arrangement of atoms and often in the location of double bonds. **Geometric isomers** have the same sequence of covalently bonded atoms but differ in spatial arrangement due to the inflexibility of double bonds. A *cis* isomer has the same atoms attached to double-bonded carbons on the same side of the double bond; a *trans* isomer has these atoms on opposite sides of the double bond. **Enantiomers** are left- and right-handed versions of each other and can differ greatly in their biological activity. An *asymmetric carbon* is one that is covalently bonded to four different kinds of atoms or groups of atoms, whose arrangement can result in mirror images.

INTERACTIVE QUESTION 4.2

Identify the structural isomers, geometric isomers, and enantiomers from the following compounds. Which of the geometric isomers is the *cis* isomer?

4.3 Characteristic chemical groups help control how biological molecules function

The Chemical Groups Most Important in the Processes of Life The properties of organic molecules are largely determined by characteristic groups of atoms attached to the carbon skeleton that behave consistently from one carbon-based molecule to another. The six **functional groups** may participate in chemical reactions. These hydrophilic groups increase the solubility of organic compounds in water. The nonpolar methyl group may serve as a molecular tag on organic molecules.

The **hydroxyl group** consists of an oxygen and hydrogen (—OH). Organic molecules with hydroxyl groups are called **alcohols,** and their names often end in *-ol.*

A **carbonyl group** consists of a carbon double-bonded to an oxygen ($>$CO). If the carbonyl group is at the end of the carbon skeleton, the compound is called an **aldehyde.** Otherwise, the compound is called a **ketone.**

A **carboxyl group** consists of a carbon double-bonded to an oxygen and also attached to a hydroxyl group (—COOH). Compounds with a carboxyl group are called **carboxylic acids** or organic acids because they tend to release H^+, becoming a carboxylate ion (—COO⁻).

An **amino group** consists of a nitrogen atom bonded to two hydrogens (—NH₂). Compounds with an amino group, called **amines,** can act as bases, picking up a hydrogen ion and becoming —NH₃⁺.

The **sulfhydryl group** consists of a sulfur atom bonded to a hydrogen (—SH). **Thiols** are compounds containing sulfhydryl groups.

A **phosphate group** is bonded to the carbon skeleton by an oxygen attached to a phosphorus atom that is bonded to three other oxygen atoms (—OPO₃²⁻). This anion contributes negative charge to **organic phosphates.**

A **methyl group** is a carbon bonded to three hydrogens (—CH₃). **Methylated compounds** may have their function modified due to the addition of the methyl group.

ATP: An Important Source of Energy for Cellular Processes **Adenosine triphosphate,** or **ATP,** consists of the organic molecule adenosine attached to three phosphate groups. When ATP reacts with water, the third phosphate is split off and energy is released.

The Chemical Elements of Life: A Review Carbon, oxygen, hydrogen, nitrogen, and smaller quantities of sulfur and phosphorus, all capable of forming strong covalent bonds, are combined into the complex organic molecules of living matter.

Word Roots

hydro- = water (*hydrocarbon:* an organic molecule consisting only of carbon and hydrogen)

iso- = equal (*isomer:* one of several organic compounds with the same molecular formula but different structures and, therefore, different properties)

enanti- = opposite (*enantiomer:* molecules that are mirror images of each other)

carb- = coal (*carboxyl group:* a functional group present in organic acids, consisting of a carbon atom double-bonded to an oxygen atom and a hydroxyl group)

sulf- = sulfur (*sulfhydryl group:* a functional group that consists of a sulfur atom bonded to an atom of hydrogen)

thio- = sulfur (*thiol:* organic compounds containing sulfhydryl groups)

Structure Your Knowledge

1. Construct a concept map that illustrates your understanding of the characteristics and significance of the three types of isomers. *A suggested map is in the answer section. Comparing and discussing your map with that of a study partner would be most helpful.*

2. Fill in the following table on the important chemical groups of organic compounds.

Chemical Group	Molecular Formula	Names and Characteristics of Organic Compounds Containing Chemical Group
	—OH	
		Aldehyde or ketone; polar group
Carboxyl		
	—NH$_2$	
		Thiols; cross-links stabilize protein structure
Phosphate		
	—CH$_3$	

Test Your Knowledge

MULTIPLE CHOICE: *Choose the one best answer.*

1. The tetravalence of carbon most directly results from
   a. its tetrahedral shape.
   b. its very slight electronegativity.
   c. its four electrons in the valence shell that can form four covalent bonds.
   d. its ability to form single, double, and triple bonds.
   e. its ability to form chains and rings of carbon atoms.

2. Hydrocarbons are not soluble in water because
   a. they are hydrophilic.
   b. the C—H bond is nonpolar.
   c. they do not ionize.
   d. they store energy in the many C—H bonds along the carbon backbone.
   e. they are lighter than water.

3. Which of the following is *not* true of an asymmetric carbon atom?
   a. It is attached to four different atoms or groups.
   b. It results in right- and left-handed versions of a molecule.
   c. It is found in all enantiomers.
   d. Its configuration is in the shape of a tetrahedron.
   e. It is found in all geometric isomers.

4. Which of the following is mismatched with its description?
   a. phosphate—forms bonds that stabilize protein structure
   b. hydroxyl and carbonyl—components of sugars
   c. ATP—source of energy for cellular processes
   d. methyl—addition changes shape and function of molecules
   e. amino and carboxyl—components of amino acids

5. The chemical group that can cause an organic molecule to act as a base is
   a. —COOH.   c. —SH.   e. —CH$_3$.
   b. —OH.   d. —NH$_2$.

6. The chemical group that confers acidic properties to organic molecules is
   a. —COOH.   c. —SH.   e. —CH$_3$.
   b. —OH.   d. —NH$_2$.

7. Which statement is *not* true about structural isomers?
   a. They have different chemical properties.
   b. They have the same molecular formula.
   c. Their atoms and bonds are arranged in different sequences.
   d. They are a result of restricted movement around a carbon double bond.
   e. Their possible numbers increase as carbon skeletons increase in size.

8. The fats stored in your body consist mostly of
   a. methyl groups.
   b. alcohols.
   c. carboxylic acids.
   d. hydrocarbons.
   e. organic phosphates.

9. How many asymmetric carbons are there in the sugar ribose shown below?

   a. 1   c. 3   e. 5
   b. 2   d. 4

MATCHING: *Match the formulas (**a–f**) to the terms at the right. Choices may be used more than once; more than one right choice may be available.*

a.

b.

c.

d.

e.

f.

_____ 1. structural isomers

_____ 2. geometric isomers

_____ 3. can have enantiomers

_____ 4. carboxylic acid

_____ 5. can make cross-link in protein

_____ 6. hydrophilic

_____ 7. hydrocarbon

_____ 8. amino acid

_____ 9. organic phosphate

_____ 10. aldehyde

_____ 11. amine

_____ 12. ketone

Chapter 5

The Structure and Function of Macromolecules

Key Concepts

5.1 **Most macromolecules are polymers, built from monomers**

5.2 **Carbohydrates serve as fuel and building material**

5.3 **Lipids are a diverse group of hydrophobic molecules**

5.4 **Proteins have many structures, resulting in a wide range of functions**

5.5 **Nucleic acids store and transmit hereditary information**

Framework

The central ideas of this chapter are that molecular function relates to molecular structure and that the diversity of molecular structure is the basis for the diversity of life. Combining a small number of monomers or subunits into unique sequences and three-dimensional structures creates a huge variety of macromolecules. The table below briefly summarizes the major characteristics of the four classes of macromolecules.

Chapter Review

Smaller organic molecules are joined together to form carbohydrates, lipids, proteins, and nucleic acids. These molecules, many of which are giant **macromolecules**, represent another level in the hierarchy of biological organization, and their functions derive from their complex and unique architectures.

5.1 Most macromolecules are polymers, built from monomers

Polymers are chainlike molecules formed from the linking together of many similar or identical small molecules, called **monomers**.

Synthesis and Breakdown of Polymers Monomers are joined by a **condensation reaction** (specifically a **dehydration reaction**), in which one monomer provides a hydroxyl group (—OH) and the other contributes a hydrogen (—H) to release a water molecule. **Hydrolysis** is the breaking of bonds between monomers through the addition of water molecules. A hydroxyl group is joined to one monomer while a hydrogen is bonded with the other. **Enzymes** catalyze both dehydration reactions and hydrolysis.

Class	Monomers or Components	Functions
Carbohydrates	Monosaccharides	Energy, raw materials, energy storage, structural compounds
Proteins	Amino acids	Enzymes, transport, movement, receptors, defense, structure
Nucleic acids	Nucleotides	Heredity, code for amino acid sequence
Lipids	Glycerol and fatty acids → fats; phospholipids; steroids (do not form polymers)	Energy storage, membranes, hormones

Diversity of Polymers Macromolecules are constructed from about 40 to 50 common monomers and a few rarer molecules. The seemingly endless variety of polymers arises from the essentially infinite number of possibilities in the sequencing and arrangement of these basic building blocks.

5.2 Carbohydrates serve as fuel and building material

Carbohydrates include sugars and their polymers.

Sugars **Monosaccharides** have the general formula of $(CH_2O)_n$. The number (n) of these units forming a sugar varies from three to seven, with hexoses ($C_6H_{12}O_6$), trioses, and pentoses found most commonly. Sugar molecules may be enantiomers due to the spatial arrangement of parts around asymmetric carbons.

INTERACTIVE QUESTION 5.1

Fill in the blanks to review the structure of monosaccharides.

You can recognize a monosaccharide by its multiple **(a)** _____ groups and its one **(b)** _____ group, whose location determines whether the sugar is an **(c)** _____ or a **(d)** _____. In aqueous solutions, most monosaccharides form **(e)** _____.

Glucose is broken down to yield energy in cellular respiration. Monosaccharides serve also as the raw materials for synthesis of other organic molecules. Two monosaccharides are joined by a **glycosidic linkage** to form a **disaccharide.**

Polysaccharides **Polysaccharides** are storage or structural macromolecules made from a few hundred to a few thousand monosaccharides. **Starch,** a storage molecule in plants, is a polymer made of glucose molecules joined by 1–4 linkages that give starch a helical shape. Animals produce **glycogen,** a highly branched polymer of glucose, as their energy storage form.

Cellulose, the major component of plant cell walls, is the most abundant organic compound on Earth. It differs from starch by the configuration of the ring form of glucose and the resulting geometry of the glycosidic bonds. In a plant cell wall, hydrogen bonds between hydroxyl groups hold parallel cellulose molecules together to form strong microfibrils.

Enzymes that digest the α linkages of starch are unable to hydrolyze the β linkages of cellulose. Only a few organisms (some microbes and fungi) have enzymes that can digest cellulose.

Chitin is a structural polysaccharide formed from glucose monomers with a nitrogen-containing group and found in the exoskeleton of arthropods and the cell walls of many fungi.

INTERACTIVE QUESTION 5.2

Circle the atoms of these two glucose molecules that will be removed by a dehydration reaction. Then draw the resulting maltose molecule with its 1–4 glycosidic linkage (between the number 1 carbon of the first glucose and the number 4 carbon of the second).

GLUCOSE   GLUCOSE

MALTOSE

INTERACTIVE QUESTION 5.3

Fill in the following concept map that summarizes this section on carbohydrates.

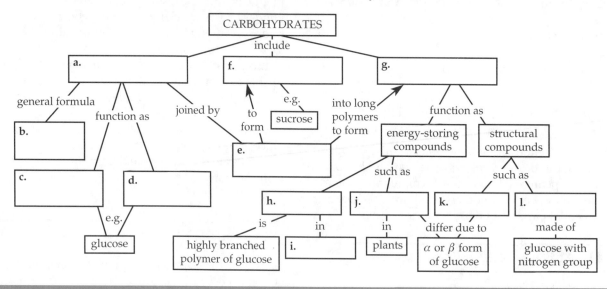

5.3 Lipids are a diverse group of hydrophobic molecules

Fats, phospholipids, and steroids are part of a diverse assemblage of macromolecules that are classed together as lipids based on their hydrophobic behavior. Lipids do not form polymers.

Fats **Fats** are composed of **fatty acids** attached to the three-carbon alcohol, glycerol. A fatty acid consists of a long hydrocarbon chain with a carboxyl group at one end. The nonpolar hydrocarbons make a fat hydrophobic.

A **triacylglycerol**, or fat, consists of three fatty acid molecules, each linked to glycerol by an ester linkage, a bond that forms between a hydroxyl and a carboxyl group. Triglyceride is another name for fats.

Fatty acids with double bonds in their carbon skeletons are called **unsaturated fatty acids.** The *cis* double bonds create a kink in the hydrocarbon chain and prevent fat molecules that contain unsaturated fatty acids from packing closely together and becoming solidified at room temperature. **Saturated fatty acids** have no double bonds in their carbon skeletons. Most animal fats are saturated and solid at room temperature. The fats of plants and fish are generally unsaturated and are called oils. Diets rich in saturated fats and in *trans* fats made in the process of hydrogenating vegetable oils have been linked to cardiovascular disease.

Fats are excellent energy storage molecules, containing twice the energy reserves of carbohydrates such as starch. Adipose tissue, made of fat storage cells, also cushions organs and insulates the body.

Phospholipids **Phospholipids** consist of a glycerol linked to two fatty acids and a negatively charged phosphate group, to which other small molecules are attached. The phosphate head of this molecule is hydrophilic and water soluble, whereas the two fatty acid chains are hydrophobic. The unique structure of phospholipids makes them ideal constituents of cell membranes.

INTERACTIVE QUESTION 5.4

Sketch a section of a phospholipid bilayer of a membrane, and label the hydrophilic head and hydrophobic tails of one of the phospholipids.

Steroids **Steroids** are a class of lipids distinguished by four connected carbon rings with various functional groups attached. **Cholesterol** is an important steroid that is a common component of animal cell membranes and a precursor for other steroids, including many hormones.

INTERACTIVE QUESTION 5.5

Fill in this concept map to help you organize your understanding of lipids.

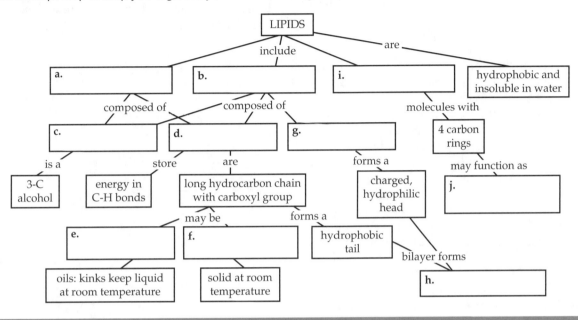

5.4 Proteins have many structures, resulting in a wide range of functions

Proteins are central to almost every function of life. As **catalysts**, enzymatic proteins selectively speed up the chemical reactions of a cell.

Polypeptides   A **polypeptide** is a polymer of amino acids. A **protein** consists of one or more polypeptide chains folded into a specific three-dimensional shape.

Amino acids are composed of an asymmetric carbon (called the α carbon) bonded to a hydrogen, a carboxyl group, an amino group, and a variable side chain called the R group. At the pH in a cell, the amino and carboxyl groups are usually ionized. The R group confers the unique physical and chemical properties of each amino acid. Side chains may be either nonpolar and hydrophobic, or polar or charged (acidic or basic) and thus hydrophilic.

A **peptide bond** links the carboxyl group of one amino acid with the amino group of another.

Protein Structure and Function   A protein has a unique three-dimensional shape, or structure, created by the twisting or folding of one or more polypeptide chains. Protein structure is dependent upon the interactions

INTERACTIVE QUESTION 5.6

a. Draw the amino acids alanine (R group—CH_3) and serine (R group—CH_2OH) and then show how a dehydration reaction will form a peptide bond between them.

b. Which of these amino acids has a polar R group? a nonpolar R group?

c. What does the molecule segment shown below represent? (Note the N—C—C—N—C—C sequence.)

among the amino acids making up the polypeptide chain and usually arises spontaneously as the protein is synthesized in the cell. The unique structure of a protein enables it to recognize and bind to other molecules.

Primary structure is the unique, genetically coded sequence of amino acids within a protein.

Secondary structure involves the coiling or folding of the polypeptide backbone, stabilized by hydrogen bonds between the electronegative oxygen of one peptide bond and the weakly positive hydrogen attached to the nitrogen of another peptide bond. An **α helix** is a coil produced by hydrogen bonding between every fourth amino acid. A **β pleated sheet** is also held by repeated hydrogen bonds along regions of the polypeptide backbone lying parallel to each other.

Interactions between the various side chains (R groups) of the constituent amino acids produce a protein's **tertiary structure. Hydrophobic interactions** between nonpolar side groups clumped in the center of the molecule due to their repulsion by water, van der Waals interactions among those nonpolar side chains, hydrogen bonds between polar side chains, and ionic bonds between negatively and positively charged side chains produce the stable and unique shape of the protein. Strong covalent bonds, called **disulfide bridges,** may occur between the sulfhydryl side groups of cysteine monomers that have been brought close together by the folding of the polypeptide. (See Interactive Question 5.7.)

Quaternary structure occurs in proteins that are composed of more than one polypeptide chain. The individual polypeptide subunits are held together in a precise structural arrangement.

Using automated sequencing, the amino acid sequences of more than 1.2 million proteins have been determined. Researchers have developed methods for following a protein through its intermediate states on the way to its final form and have discovered **chaperonins,** chaperone proteins that assist other proteins during the folding process, perhaps by providing a sheltered environment. Using the techniques of **X-ray crystallography,** nuclear magnetic resonance (NMR) spectroscopy, and bioinformatics, biochemists have established the structure of thousands of proteins.

INTERACTIVE QUESTION   5.7

In the following diagram of a portion of a polypeptide, label the types of interactions that are shown. What level of structure are these interactions producing?

INTERACTIVE QUESTION   5.8

a. Why would a change in pH cause a protein to **denature**?

b. Why would transfer to a nonpolar organic solvent (such as ether) cause denaturation?

c. A denatured protein may re-form its functional shape when returned to its normal environment. What does that indicate about a protein's structure?

INTERACTIVE QUESTION 5.9

Now that you have gained more experience with concept maps, create your own map to help you organize the key concepts you have learned about proteins. Try to include the concepts of structure and function and look for cross-links on your map. You may want to include the functions of proteins. *One version of a protein concept map is included in the answer section, but remember that the real value is in the thinking process you must go through to create your own map.*

5.5 Nucleic acids store and transmit hereditary information

Genes are the units of inheritance that determine the primary structure of proteins. **Nucleic acids** are the polymers that carry and transmit this code.

The Roles of Nucleic Acids DNA, **deoxyribonucleic acid,** is the genetic material that is inherited from one generation to the next and is copied whenever a cell divides so that all cells of an organism contain its DNA. The instructions coded in DNA are transcribed to **RNA, ribonucleic acid,** which directs the synthesis of proteins, the ultimate enactors of the genetic program. In a eukaryotic cell, DNA resides in the nucleus. Messenger RNA carries the instructions for protein synthesis to ribosomes located in the cytoplasm.

INTERACTIVE QUESTION 5.10

Show the flow of genetic information in a cell.

_____   →   _____   →   _____

The Structure of Nucleic Acids **Polynucleotides** are polymers of **nucleotides**—monomers that consist of a pentose (five-carbon sugar) covalently bonded to a phosphate group and a nitrogenous base. A monomer without the phosphate group is called a nucleoside. There are two families of nitrogenous bases. **Pyrimidines,** including cytosine (C), thymine (T), and uracil (U), are characterized by six-membered rings of carbon and nitrogen atoms. **Purines,** adenine (A) and guanine (G), add a five-membered ring to the pyrimidine ring. Thymine is only in DNA; uracil is only in RNA. In DNA, the pentose is **deoxyribose;** in RNA, it is **ribose.**

Nucleotides are linked together into a polynucleotide by phosphodiester linkages, which join the sugar of one nucleotide with the phosphate of the next. The polymer has two distinct ends: a 5′ end with a phosphate attached to a 5′ carbon of a sugar and a 3′ end with a hydroxyl group on a 3′ carbon of a sugar. The nitrogenous bases extend from this backbone of repeating sugar-phosphate units. The unique sequence of bases in a gene codes for the specific amino acid sequence of a protein.

INTERACTIVE QUESTION 5.11

a. Label the three parts of this nucleotide. Indicate with an arrow where the phosphate group of the next nucleotide would attach to build a polynucleotide. Number the carbons of the pentose.

b. Is this a purine or a pyrimidine? How do you know?

c. Is this a DNA or RNA nucleotide? How do you know?

The DNA Double Helix DNA molecules consist of two chains of polynucleotides spiraling in a **double helix.** The two chains run in opposite 5′ to 3′ directions, an arrangement called **antiparallel.** In 1953, Watson and Crick first proposed this double-helix arrangement, which consists of two sugar-phosphate backbones on the outside of the helix with their nitrogenous bases pairing and hydrogen-bonding together in the inside. Adenine pairs only with thymine, guanine always pairs with cytosine. Thus, the sequences of nitrogenous bases on the two strands of DNA are *complementary.* Because of this specific base-pairing property, DNA can replicate itself and precisely copy the genes of inheritance.

DNA and Proteins as Tape Measures of Evolution Genes form the hereditary link between generations. Closely related members of the same species share many common DNA sequences and proteins. More closely related species have a larger proportion of their DNA and proteins in common. This "molecular genealogy" provides evidence of evolutionary relationships.

INTERACTIVE QUESTION 5.12

Take the time to create a concept map that summarizes what you have just reviewed about nucleic acids. Compare your map with that of a study partner or explain it to a friend. One version of a map on nucleic acids is included in the answer section. Refer to Figures 5.27 and 5.28 in your textbook to help you visualize polynucleotides and the double helix of DNA.

The Theme of Emergent Properties in the Chemistry of Life: A Review   At each stage in the hierarchy of levels from atoms through macromolecules, we have seen that novel properties arise with increasing structural organization.

Structure Your Knowledge

1. Describe the four structural levels in the functional shape of a protein.

2. Identify the type of monomer or group shown by the formulas a–g. Then match the chemical formulae with their description. Answers may be used more than once.

   1. molecules that would combine to form a fat

   2. molecule that would be attached to other monomers by a peptide bond

   3. molecules or groups that would combine to form a nucleotide

   4. molecules that are carbohydrates

   5. molecule that is a purine

   6. monomer of a protein

   7. groups that would be joined by phosphodiester bonds

   8. most nonpolar (hydrophobic) molecule

Word Roots

con- = together (*condensation reaction*: a reaction in which two molecules become covalently bonded to each other through the loss of a small molecule, usually water)

di- = two (*disaccharide*: two monosaccharides joined together)

glyco- = sweet (*glycogen*: a polysaccharide sugar used to store energy in animals)

hydro- = water; **-lyse** = break (*hydrolysis*: breaking chemical bonds by adding water)

macro- = large (*macromolecule*: a large molecule)

meros- = part (*polymer*: a chain made from smaller organic molecules)

mono- = single; **-sacchar** = sugar (*monosaccharide*: simplest type of sugar)

poly- = many (*polysaccharide*: many monosaccharides joined together)

tri- = three (*triacylglycerol*: three fatty acids linked to one glycerol molecule)

b. _____

c. _____   d. _____

e. _____

f. _____

a. _____

g. _____

Test Your Knowledge

MATCHING: *Match the molecule with its class of molecule.*

_____ 1. glycogen   **A.** carbohydrate

_____ 2. cholesterol   **B.** lipid

_____ 3. RNA   **C.** protein

_____ 4. collagen   **D.** nucleic acid

_____ 5. hemoglobin

_____ 6. a gene

_____ 7. triacylglycerol

_____ 8. enzyme

_____ 9. cellulose

_____ 10. chitin

MULTIPLE CHOICE: *Choose the one best answer.*

1. Polymerization is a process that
   a. creates bonds between amino acids in the formation of a peptide chain.
   b. often involves the removal of a water molecule.
   c. links the sugar of one nucleotide with the phosphate of the next.
   d. requires a condensation or dehydration reaction.
   e. may involve all of the above.

2. Which of the following is *not* true of a pentose?
   a. It can be found in nucleic acids.
   b. It can occur in a ring structure.
   c. It has the formula $C_5H_{12}O_5$.
   d. It has one carbonyl and four hydroxyl groups.
   e. It may be an aldose or a ketose.

3. Disaccharides can differ from each other in all of the following ways except
   a. in the number of their monosaccharides.
   b. as enantiomers.
   c. in the monomers involved.
   d. in the location of their glycosidic linkage.
   e. in their structural formulas.

4. Which of the following is *not* true of cellulose?
   a. It is the most abundant organic compound on Earth.
   b. It differs from starch because of the configuration of glucose and the geometry of the glycosidic linkage.

   c. It may be hydrogen-bonded to neighboring cellulose molecules to form microfibrils.
   d. Few organisms have enzymes that hydrolyze its glycosidic linkages.
   e. Its monomers are glucose with nitrogen-containing appendages.

5. Plants store most of their energy for later use as
   a. unsaturated fats.   c. starch.   e. cellulose.
   b. saturated fats.   d. sucrose.

6. What happens when a protein denatures?
   a. Its primary structure is disrupted.
   b. Its secondary and tertiary structures are disrupted.
   c. It becomes irreversibly insoluble and precipitates.
   d. It hydrolyzes into component amino acids.
   e. Its hydrogen bonds, ionic bonds, hydrophobic interactions, disulfide bridges, and peptide bonds are disrupted.

7. The α helix of proteins is
   a. part of the tertiary structure and is stabilized by disulfide bridges.
   b. a double helix.
   c. stabilized by hydrogen bonds and commonly found in fibrous proteins.
   d. found in some regions of globular proteins and stabilized by hydrophobic interactions.
   e. a complementary sequence to messenger RNA.

8. A fatty acid that has the formula $C_{16}H_{32}O_2$ is
   a. saturated.   d. hydrophilic.
   b. unsaturated.   e. part of a steroid molecule.
   c. branched.

9. Three molecules of the fatty acid in question 8 are joined to a molecule of glycerol ($C_3H_8O_3$). The resulting molecule has the formula
   a. $C_{48}H_{96}O_6$.   c. $C_{51}H_{102}O_8$.   e. $C_{51}H_{104}O_9$.
   b. $C_{48}H_{98}O_9$.   d. $C_{51}H_{98}O_6$.

10. β pleated sheets are characterized by
    a. disulfide bridges between cysteine amino acids.
    b. parallel regions of the polypeptide chain held together by hydrophobic interactions.
    c. folds stabilized by hydrogen bonds between segments of the polypeptide backbone.
    d. membrane sheets composed of phospholipids.
    e. hydrogen bonds between adjacent cellulose molecules.

11. Cows can derive nutrients from cellulose because
    a. they can produce the enzymes that break the β linkages between glucose molecules.
    b. they chew and rechew their cud so that cellulose fibers are finally broken down.
    c. its rumin contains prokaryotes that can hydrolyze the bonds of cellulose.
    d. their intestinal tract contains termites, which harbor microbes that hydrolyze cellulose.
    e. they can convert cellulose to starch and then hydrolyze starch to glucose.

12. Which of these molecules would provide the most energy (kcal/g) when eaten?
    a. glucose   c. glycogen   e. protein
    b. starch   d. fat

13. What *determines* the sequence of the amino acids in a particular protein?
    a. its primary structure
    b. the sequence of nucleotides in RNA, which was determined by the sequence of nucleotides in the gene for that protein
    c. the sequence of nucleotides in DNA, which was determined by the sequence of nucleotides in RNA
    d. the sequence of RNA nucleotides making up the ribosome
    e. the three-dimensional shape of the protein

14. Sucrose is made from joining a glucose and a fructose molecule in a dehydration reaction. What is the molecular formula for this disaccharide?
    a. $C_6H_{12}O_6$ c. $C_{12}H_{22}O_{11}$ e. $C_{12}H_{24}O_{13}$
    b. $C_{10}H_{20}O_{10}$ d. $C_{12}H_{24}O_{12}$

15. How are the nucleotide monomers connected to form a polynucleotide?
    a. hydrogen bonds between complementary nitrogenous base pairs
    b. ionic attractions between phosphate groups
    c. disulfide bridges between cysteine amino acids
    d. covalent bonds between the sugar of one nucleotide and the phosphate of the next
    e. ester linkages between the carboxyl group of one nucleotide and the hydroxyl group on the ribose of the next

16. Which of the following would be the most hydrophobic molecule?
    a. cholesterol   d. chitin
    b. nucleotide   e. glucose
    c. amino acid

17. What is the best description of this molecule?

    a. chitin   d. nucleotide
    b. amino acid   e. protein
    c. polypeptide (tripeptide)

18. Which number(s) in the molecule in question 17 refer(s) to a peptide bond?
    a. 1   c. 3   e. both 2 and 4
    b. 2   d. 4

19. If the nucleotide sequence of one strand of a DNA helix is 5′GCCTAA3′, what would be the 3′–5′ sequence on the complementary strand?
    a. GCCTAA   c. CGGATT   e. TAAGCC
    b. CGGAUU   d. ATTCGG

20. Monkeys and humans share many of the same DNA sequences and have similar proteins, indicating that
    a. the two groups belong to the same species.
    b. the two groups share a relatively recent common ancestor.
    c. humans evolved from monkeys.
    d. monkeys evolved from humans.
    e. the two groups first appeared on Earth at about the same time.

21. Which of the following would be the major component of the cell membrane of a fungus?
    a. cellulose   d. phospholipids
    b. chitin   e. unsaturated fatty acids
    c. cholesterol

22. Hydrophobic as well as hydrophilic interactions would be important for which of the following types of molecules?
    a. proteins   d. polynucleotides
    b. unsaturated fats   e. all of the above
    c. glycogen and cellulose

23. What are *trans* fats?
    a. hydrogenated vegetable oils that have been identified with health risks
    b. fats made from cholesterol that are components of plaques in the walls of blood vessels
    c. fats that are derived from animal sources and are associated with cardiovascular disease
    d. fats that contain *trans* double bonds and may contribute to atherosclerosis
    e. polyunsaturated fats produced by removing H from fatty acids and forming *cis* double bonds

24. Which of the following is *not* one of the many functions performed by proteins?
    a. signals and receptors
    b. enzymatic catalyst for metabolic reactions
    c. protection against disease
    d. contractile components of muscle
    e. transport of proteins by chaperonins

Chapter 6

A Tour of the Cell

▮ Key Concepts

6.1 To study cells, biologists use microscopes and the tools of biochemistry

6.2 Eukaryotic cells have internal membranes that compartmentalize their functions

6.3 The eukaryotic cell's genetic instructions are housed in the nucleus and carried out by the ribosomes

6.4 The endomembrane system regulates protein traffic and performs metabolic functions in the cell

6.5 Mitochondria and chloroplasts change energy from one form to another

6.6 The cytoskeleton is a network of fibers that organizes structures and activities in the cell

6.7 Extracellular components and connections between cells help coordinate cellular activities

▮ Framework

The cell is the fundamental unit of life. The complexities in the processes of life are reflected in the complexities of the structure of the cell. It is easy to become overwhelmed by the number of new vocabulary terms for this array of cell organelles and membranes. The following diagram provides an organizational framework for the wealth of detail found in "a tour of the cell."

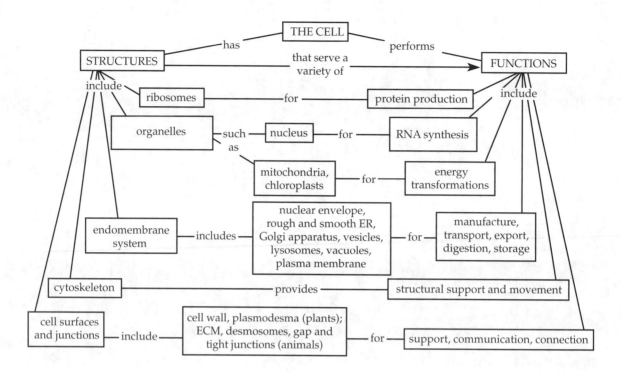

Chapter Review

The cell is the basic structural and functional unit of all living organisms. In the hierarchy of biological organization, the capacity for life emerges from the structural order of the cell. All cells are related through common descent, but evolution has shaped diverse adaptations.

6.1   To study cells, biologists use microscopes and the tools of biochemistry

Microscopy   The glass lenses of **light microscopes (LMs)** refract (bend) the visible light passing through a specimen such that the projected image is magnified. *Magnification* is the ratio of the size of this projected image to the real size of the object. *Resolution* is a measure of how clear an image is, and is determined by the minimum distance two points must be separated to be distinguished. The resolution of the light microscope is limited by the shortest wavelength of visible light, so that details finer than 0.2 μm (micrometers = 10^{-3} mm) cannot be resolved. Staining of specimens and using techniques such as fluorescence, phase-contrast, and confocal microscopy improve visibility by increasing contrast between structures that are large enough to be resolved.

Most subcellular structures, including membrane-enclosed **organelles,** cannot be resolved by the light microscope. Cells were discovered by Robert Hooke in 1665, but their ultrastructure was largely unknown until the development of the **electron microscope (EM)** in the 1950s. The electron microscope focuses a beam of electrons through the specimen. The short wavelength of the electron beam allows a resolution of about 2 nm (nanometers = 10^{-3} μm), a hundred times greater than that of the light microscope.

In a **scanning electron microscope (SEM),** an electron beam scans the surface of a specimen usually coated with a thin gold film, exciting electrons from the specimen, which are detected and translated into an image on a video screen. This image appears three-dimensional.

In a **transmission electron microscope (TEM),** a beam of electrons is passed through a thin section of a specimen stained with atoms of heavy metals. Electromagnets, acting as lenses, then focus the image onto a screen or film.

Modern cell biology integrates *cytology* with *biochemistry* to understand relationships between cellular structure and function.

INTERACTIVE QUESTION   6.1

a. Define cytology.

b. What do cell biologists use a TEM to study?

c. What does an SEM show best?

d. What advantages does light microscopy have over both TEM and SEM?

Cell Fractionation **Cell fractionation** is a technique that separates organelles and other structures of a cell so that they can be identified and their functions studied. Cells are broken apart and the homogenate is separated into component fractions by centrifugation at increasing speeds.

6.2 Eukaryotic cells have internal membranes that compartmentalize their functions

Comparing Prokaryotic and Eukaryotic Cells All cells are bounded by a plasma membrane, which encloses a semifluid medium called **cytosol.** All cells contain chromosomes and ribosomes.

Only members of the domains Bacteria and Archaea have **prokaryotic cells,** which are cells with no nucleus or membrane-enclosed organelles. The DNA of prokaryotic cells is concentrated in a region called the **nucleoid. Eukaryotic cells** have a true nucleus enclosed in a nuclear envelope and numerous organelles suspended in cytosol. **Cytoplasm** refers to the entire region between the nucleus and the plasma membrane, and also to the interior of a prokaryotic cell.

Most bacterial cells range from 1 to 10 µm in diameter, whereas eukaryotic cells are 10 times larger, ranging from 10 to 100 µm.

The small size of cells is dictated by geometry and the requirements of metabolism. Area is proportional to the square of linear dimension, while volume is proportional to its cube. The **plasma membrane** surrounding every cell must provide sufficient surface area for exchange of oxygen, nutrients, and wastes relative to the volume of the cell.

INTERACTIVE QUESTION 6.2

a. Describe the molecular structure of the plasma membrane.

b. If a eukaryotic cell has a diameter that is 10 times that of a bacterial cell, proportionally how much more surface area would the eukaryotic cell have?

c. Proportionally how much more volume would it have?

A Panoramic View of the Eukaryotic Cell Membranes compartmentalize the eukaryotic cell, providing local environments for specific metabolic functions and participating in metabolism through membrane-bound enzymes. Like the plasma membrane, internal membranes are composed of a bilayer of phospholipid molecules associated with diverse proteins.

6.3 The eukaryotic cell's genetic instructions are housed in the nucleus and carried out by the ribosomes

The Nucleus: Information Central The **nucleus** is surrounded by the **nuclear envelope,** a double membrane perforated by pores that regulate the movement of materials between the nucleus and the cytoplasm. The inner membrane is lined by the **nuclear lamina,** a layer of protein filaments that helps to maintain the shape of the nucleus.

Most of the cell's DNA is located in the nucleus, where it is organized into units called **chromosomes,** which are made up of a complex of DNA and proteins called **chromatin.** Each eukaryotic species has a characteristic chromosomal number. Individual chromosomes are visible only when condensed in a dividing cell.

The **nucleolus,** a dense structure visible in the nondividing nucleus, synthesizes ribosomal RNA and combines it with protein to assemble ribosomal subunits, which then pass through nuclear pores to the cytoplasm.

Ribosomes: Protein Factories **Ribosomes** are composed of protein and ribosomal RNA. Most of the proteins produced on *free ribosomes* are used within the cytosol. *Bound ribosomes,* attached to the endoplasmic reticulum or nuclear envelope, usually make proteins that will be included within membranes, packaged into organelles, or exported from the cell.

INTERACTIVE QUESTION 6.3

How does the nucleus control protein synthesis in the cytoplasm?

6.4 The endomembrane system regulates protein traffic and performs metabolic functions in the cell

The **endomembrane system** of a cell consists of the nuclear envelope, endoplasmic reticulum, Golgi apparatus, lysosomes, vacuoles, and the plasma membrane. These membranes are all related either through direct contact or by the transfer of membrane segments by membrane-bound sacs called **vesicles.**

The Endoplasmic Reticulum: Biosynthetic Factory The **endoplasmic reticulum (ER)** is continuous with the nuclear envelope and encloses a network of interconnected tubules or compartments called cisternae. Ribosomes are attached to the cytoplasmic surface of **rough ER; smooth ER** lacks ribosomes.

Smooth ER serves diverse functions in different cells: Its enzymes are involved in phospholipid and steroid (including sex hormone) synthesis, carbohydrate metabolism, and detoxification of drugs and poisons. Barbiturates, alcohol, and other drugs increase a liver cell's production of smooth ER, thus leading to an increased tolerance (and thus reduced effectiveness) for these and other drugs. Smooth ER also functions in storage and release of calcium ions during muscle contraction.

Proteins intended for secretion are manufactured by membrane-bound ribosomes and then threaded into the lumen of the rough ER. Many are covalently bonded to small carbohydrates to form **glycoproteins.** Secretory proteins are transported from the rough ER in membrane-bound **transport vesicles.**

Rough ER manufactures membranes for the cell. Enzymes built into the membrane assemble phospholipids, and membrane proteins formed by bound ribosomes are inserted into the ER membrane. Transport vesicles transfer ER membrane to other parts of the endomembrane system.

The Golgi Apparatus: Shipping and Receiving Center The **Golgi apparatus** consists of a stack of flattened sacs. Vesicles that bud from the ER join to the *cis* face of a Golgi stack, adding to it their contents and membrane. According to the *cisternal maturation model,* Golgi products are processed and tagged as the cisternae themselves progress from the *cis* to the *trans* face. Glycoproteins often have their attached carbohydrates modified. The Golgi of plant cells manufacture some polysaccharides, such as pectins. Golgi products are sorted into vesicles, which pinch off from the *trans* face of the Golgi apparatus. These vesicles may have surface molecules that help direct them to the plasma membrane or to other organelles.

Lysosomes: Digestive Compartments **Lysosomes** are membrane-enclosed sacs of hydrolytic enzymes used by animal cells to digest macromolecules. Lysosomes provide an acidic pH for these enzymes.

In some protists, lysosomes fuse with food vacuoles to digest material ingested by **phagocytosis.** Macrophages, a type of white blood cell, use lysosomes to destroy ingested bacteria. Lysosomes also recycle a cell's own macromolecules by engulfing damaged organelles or small bits of cytosol, a process known as *autophagy.*

Vacuoles: Diverse Maintenance Compartments **Food vacuoles** are formed as a result of phagocytosis. **Contractile vacuoles** pump excess water out of freshwater protists. A large **central vacuole** is found in mature plant cells, surrounded by a vacuolar membrane (tonoplast) and enclosing a solution called cell sap. This vacuole stores organic compounds and inorganic ions for the cell. Dangerous metabolic by-products and poisonous or unpalatable compounds, which may protect the plant from predators, may also be contained in the vacuole. Plant cells increase in size with a minimal addition of new cytoplasm as their vacuoles absorb water and expand.

The Endomembrane System: A Review As membranes move from the ER to the Golgi and then to other organelles, their compositions, functions, and contents are modified.

6.5 Mitochondria and chloroplasts change energy from one form to another

Cellular respiration, the metabolic processing of fuels to produce ATP, occurs within the **mitochondria** of eukaryotic cells. Photosynthesis occurs in the **chloroplasts**

INTERACTIVE QUESTION 6.4

Name the components of the endomembrane system shown in this diagram and list the functions of each of these membranes.

a.

b.

c.

d.

e.

f.

g.

of plants and algae, which produce organic compounds from carbon dioxide and water by absorbing solar energy. The membrane proteins of mitochondria and chloroplasts are made by ribosomes either free in the cytosol or contained within these organelles. They also contain a small amount of DNA that directs the synthesis of some of their proteins.

Mitochondria: Chemical Energy Conversion Two membranes, each a phospholipid bilayer with unique embedded proteins, enclose a mitochondrion. A narrow intermembrane space exists between the smooth outer membrane and the convoluted inner membrane. The folds of the inner membrane, called **cristae,** create a large surface area and enclose the **mitochondrial matrix.** Many respiratory enzymes, mitochondrial DNA, and ribosomes are housed in this matrix. Other respiratory enzymes and proteins are built into the inner membrane.

Chloroplasts: Capture of Light Energy **Plastids** are plant organelles that include amyloplasts, which store starch; chromoplasts, which contain pigments; and chloroplasts, which contain the green pigment chlorophyll and function in photosynthesis.

Chloroplasts are bounded by two membranes separated by a thin intermembrane space. Inside the inner membrane is a fluid called the **stroma** surrounding a membranous system of flattened sacs called **thylakoids,** inside of which is the thylakoid space. Photosynthetic enzymes are embedded in the thylakoids, which may be stacked together to form structures called **grana.** Chloroplast DNA, ribosomes, and many enzymes are contained in the stroma.

INTERACTIVE QUESTION 6.5

Sketch a mitochondrion and a chloroplast and label their membranes and compartments.

Peroxisomes: Oxidation **Peroxisomes** are oxidative organelles filled with enzymes that function in a variety of metabolic pathways, such as breaking down fatty acids for energy or detoxifying alcohol and other poisons. An enzyme that converts hydrogen peroxide (H_2O_2), a toxic by-product of these pathways, to water is also packaged into peroxisomes. *Glyoxyomes,* found in plant seeds, contain enzymes that convert fatty acids to sugars for growing seedlings.

INTERACTIVE QUESTION 6.6

Why are peroxisomes not considered part of the endomembrane system?

6.6 The cytoskeleton is a network of fibers that organizes structures and activities in the cell

Roles of the Cytoskeleton: Support, Motility, and Regulation The **cytoskeleton** is a network of protein fibers that give mechanical support, function in cell motility (of both internal structures and the cell as a whole), and transmit mechanical signals from the cell's surface to its interior. The cytoskeleton interacts with special proteins called **motor proteins** to produce cellular movements.

Components of the Cytoskeleton The three main types of fibers involved in the cytoskeleton are **microtubules, microfilaments,** and **intermediate filaments.**

All eukaryotic cells have microtubules, which are hollow rods constructed of columns of globular proteins called tubulins. Microtubules change length through the addition or subtraction of tubulin dimers. In addition to providing the supporting framework of the cell, microtubules serve as tracks along which organelles move with the aid of motor molecules.

In many animal cells, microtubules grow out from a region near the nucleus called a **centrosome.** A pair of **centrioles,** each composed of nine sets of triplet microtubules arranged in a ring, is associated with the centrosome and replicates before cell division. Yeast and plant cells lack centrosomes and must have some other microtubule-organizing center.

Cilia and **flagella** are locomotor extensions of some eukaryotic cells. Cilia are numerous and short; flagella occur one or two to a cell and are longer. Many protists use cilia or flagella to move through aqueous media. Cilia or flagella attached to stationary cells of a tissue move fluid past the cell. Cilia may act to transmit environmental signals to the cell's nucleus.

Both cilia and flagella are composed of two single microtubules surrounded by a ring of nine doublets of microtubules (a nearly universal "9 + 2" arrangement), all of which are enclosed in an extension of the plasma membrane. A **basal body,** structurally similar to a centriole, anchors the tubules in the cell. ATP drives the sliding of the microtubule doublets past each other as arms, composed of motor proteins called **dyneins,** alternately attach to adjacent doublets, pull down, release, and reattach. In conjunction with anchoring cross-linking proteins and radial spokes, this action causes the bending of the flagellum or cilium.

Microfilaments, probably present in all eukaryotic cells, are solid rods consisting of a twisted double chain of molecules of the globular protein **actin.** Also called actin filaments, microfilaments function in support, forming a network just inside the plasma membrane and the core of small cytoplasmic extensions called microvilli.

In muscle cells, thousands of actin filaments interdigitate with thicker filaments made of the protein **myosin.** The sliding of actin and myosin filaments past each other causes the contraction of muscles.

Actin and myosin also interact in localized contractions such as cleavage furrows in animal cell division and amoeboid movements. Actin subunits reversibly assemble into microfilaments and then networks, driving the conversion of cytoplasm from sol to gel during the extension and retraction of **pseudopodia.** Actin filaments interacting with myosin may propel cytoplasm forward into pseudopodia. **Cytoplasmic streaming** in plant cells appears to involve both actin–myosin interactions and sol–gel conversions.

Intermediate filaments are intermediate in size between microtubules and Microfilaments and are more diverse in their composition. Intermediate fibers appear to be important in maintaining cell shape. The nucleus is securely held in a web of intermediate filaments, and the nuclear lamina lining the inside of the nuclear envelope is composed of intermediate filaments.

INTERACTIVE QUESTION   6.7

Fill in the following table to organize what you have learned about the components of the cytoskeleton. You may wish to refer to the textbook for additional details.

Cytoskeleton	Structure and Monomers	Functions
Microtubules	a.	b.
Microfilaments (actin filaments)	c.	d.
Intermediate filaments	e.	f.

6.7   Extracellular components and connections between cells help coordinate cellular activities

Cell Walls of Plants   Plant **cell walls** are composed of microfibrils of cellulose embedded in a matrix of polysaccharides and protein.

The **primary cell wall** secreted by a young plant cell is relatively thin and flexible. Microtubules appear to guide the path of cellulose synthase, determining the pattern of cellulose fibril deposition and thus, the direction of cell expansion. Adjacent cells are glued together by the **middle lamella,** a thin layer of polysaccharides (called pectins). When they stop growing, some cells secrete a thicker and stronger **secondary cell wall** between the plasma membrane and the primary cell wall.

INTERACTIVE QUESTION   6.8

Sketch two adjacent plant cells, and show the location of the primary and secondary cell walls and the middle lamella.

The Extracellular Matrix (ECM) of Animal Cells
Animal cells secrete an **extracellular matrix (ECM)** composed primarily of glycoproteins. **Collagen** forms strong fibers that are embedded in a network of proteoglycan complexes. **Proteoglycans** consist of a small core protein with many attached carbohydrate chains. Cells may be attached to the ECM by **fibronectins** and other glycoproteins that bind to **integrins,** receptor proteins that span the plasma membrane and bind, via other proteins, to microfilaments of the cytoskeleton. Thus, information about changes inside and outside the cell can be exchanged through a mechanical signaling pathway involving fibronectins, integrins, and the microfilaments of the cytoskeleton. Signals from the ECM appear to influence the activity of genes in the nucleus.

Intercellular Junctions **Plasmodesmata** are channels in plant cell walls through which the plasma membranes of bordering cells connect, thus linking most cells of a plant into a living continuum. Water, small solutes, and even some proteins and RNA molecules can move through these channels.

There are three main types of intercellular junctions between animal cells. At *tight junctions,* proteins hold adjacent cell membranes tightly together, creating an impermeable seal across a layer of epithelial cells. *Desmosomes* (also called *anchoring junctions*) are reinforced by intermediate filaments and rivet cells into strong sheets. *Gap junctions* (also called *communicating junctions*) are cytoplasmic connections that allow for the exchange of ions and small molecules between cells through protein-lined pores.

INTERACTIVE QUESTION 6.9

Return to your sketch of plant cells in Interactive Question 6.8 and draw in a plasmodesma.

The Cell: A Living Unit Greater Than the Sum of Its Parts The compartmentalization and the many specialized organelles typical of cells exemplify the principle that structure correlates with function. The intricate functioning of a living cell emerges from the complex interactions of its multiple parts.

Word Roots

centro- = the center; **-soma** = a body (*centrosome:* structure present in the cytoplasm of all animal cells, important during cell division)

chloro- = green (*chloroplast:* the site of photosynthesis in plants and algae)

cili- = hair (*cilium:* a short hair-like cellular appendage with a microtubule core)

cyto- = cell (*cytosol:* a semifluid medium in a cell in which are located organelles)

-ell = small (*organelle:* a small membrane-enclosed structure with a specialized function found in the cytoplasm of eukaryotic cells)

endo- = inner (*endomembrane system:* the system of membranes within a cell that includes the nuclear envelope, endoplasmic reticulum, Golgi apparatus, lysosomes, vacuoles, and the plasma membrane)

eu- = true (*eukaryotic cell:* a cell that has a true nucleus)

extra- = outside (*extracellular matrix:* the substance in which animal tissue cells are embedded)

flagell- = whip (*flagellum:* a long whip-like cellular appendage that moves cells)

glyco- = sweet (*glycoprotein:* a protein covalently bonded to a carbohydrate)

lamin- = sheet/layer (*nuclear lamina:* a netlike array of protein filaments that maintains the shape of the nucleus)

lyso- = loosen (*lysosome:* a membrane-enclosed sac of hydrolytic enzymes that a cell uses to digest macromolecules)

micro- = small; **-tubul** = a little pipe (*microtubule:* a hollow rod of tubulin protein in the cytoplasm of almost all eukaryotic cells)

nucle- = nucleus; **-oid** = like (*nucleoid:* the region where the genetic material is concentrated in prokaryotic cells)

phago- = to eat; **-kytos** = vessel (*phagocytosis:* a form of cell eating in which a cell engulfs a smaller organism or food particle)

plasm- = molded; **-desma** = a band or bond (*plasmodesmata:* an open channel in a plant cell wall)

pro- = before; **karyo-** = nucleus (*prokaryotic cell:* a cell that has no nucleus)

pseudo- = false; **-pod** = foot (*pseudopodium:* a cellular extension of amoeboid cells used in moving and feeding)

thylaco- = sac or pouch (*thylakoid:* a series of flattened sacs within chloroplasts)

tono- = stretched; **-plast** = molded (*tonoplast:* the membrane that encloses a large central vacuole in a mature plant cell)

trans- = across; **-port** = a harbor (*transport vesicle:* a membranous compartment used to enclose and transport materials from one part of a cell to another)

ultra- = beyond (*ultracentrifuge:* a machine that spins test tubes at the fastest speeds to separate liquids and particles of different densities)

vacu- = empty (*vacuole:* sac that buds from the ER, Golgi, or plasma membrane)

Structure Your Knowledge

1. The following table lists the general functions performed by an animal cell. List the cellular structures associated with each of these functions.

Functions	Associated Organelles and Structures
Cell division	a.
Information storage and transferal	b.
Energy conversions	c.
Manufacture of membranes and products	d.
Lipid synthesis, drug detoxification	e.
Digestion, recycling	f.
Conversion of H_2O_2 to water	g.
Structural integrity	h.
Movement	i.
Exchange with environment	j.
Cell to cell connections	k.

2. This table lists structures that are found in plant cells but not animal cells. Fill in the functions of these structures.

Plant Cell Structures	Functions
Cell wall	a.
Central vacuole	b.
Chloroplast	c.
Amyloplast	d.
Plasmodesmata	e.

3. Label the indicated structures in this diagram of an animal cell.

4. Create a diagram or flowchart in the space below to trace the development of a secretory product (such as a digestive enzyme) from the DNA code to its export from the cell.

Test Your Knowledge

MULTIPLE CHOICE: *Choose the one best answer.*

1. Which of the following is/are not found in a prokaryotic cell?
   a. ribosomes
   b. plasma membrane
   c. mitochondria
   d. a and c
   e. a, b, and c

2. Resolution of a microscope is
   a. the distance between two separate points.
   b. the sharpness or clarity of an image.
   c. the degree of magnification of an image.
   d. the depth of focus on a specimen's surface.
   e. the wavelength of light.

3. Which of the following is *not* a similarity among the nucleus, chloroplasts, and mitochondria?
   a. They contain DNA.
   b. They are bounded by two phospholipid bilayer membranes.
   c. They can divide to reproduce themselves.
   d. They are derived from the endoplasmic reticulum system.
   e. Their membranes are associated with specific proteins.

4. The pores in the nuclear envelope provide for the movement of
   a. proteins into the nucleus.
   b. ribosomal subunits out of the nucleus.
   c. mRNA out of the nucleus.
   d. signal molecules into the nucleus.
   e. all of the above.

5. The ultrastructure of a chloroplast could be seen with the best resolution using
   a. transmission electron microscopy.
   b. scanning electron microscopy.
   c. phase-contrast light microscopy.
   d. cell fractionation.
   e. fluorescence microscopy.

6. Which of the following is *incorrectly* paired with its function?
   a. peroxisome—contains enzymes that break down H_2O_2
   b. nucleolus—produces ribosomal RNA, assembles ribosome subunits
   c. Golgi apparatus—processes, tags, and ships cellular products
   d. lysosome—food sac formed by phagocytosis
   e. ECM (extracellular matrix)—supports and anchors cells, communicates information with inside of cell

7. The cells of an ant and an elephant are, on average, the same size; an elephant just has more cells. What is the main advantage of small cell size?
   a. Small cells are easier to organize into tissues and organs.
   b. A small cell has a larger plasma membrane surface area than does a large cell, facilitating the exchange of sufficient materials with its environment.
   c. A small cell has a smaller cytoplasmic volume relative to its surface area, which helps to ensure the exchange of sufficient materials across its plasma membrane.
   d. Small cells require less oxygen than do large cells.
   e. The cytoskeleton of a large cell would have to be so large that cells would be too heavy.

8. A growing plant cell elongates primarily by
   a. increasing the number of vacuoles.
   b. synthesizing more cytoplasm.
   c. taking up water into its central vacuole.
   d. synthesizing more cellulose.
   e. producing a secondary cell wall.

9. The innermost portion of the cell wall of a plant cell specialized for support is the
   a. primary cell wall.
   b. secondary cell wall.
   c. middle lamella.
   d. plasma membrane.
   e. plasmodesmata.

10. Contractile elements of muscle cells are
    a. intermediate filaments.
    b. centrioles.
    c. microtubules.
    d. actin filaments (microfilaments).
    e. fibronectins.

11. Microtubules are components of all of the following *except*
    a. centrioles.
    b. the spindle apparatus for separating chromosomes in cell division.
    c. tracks along which organelles can move using motor molecules.
    d. flagella and cilia.
    e. the cleavage furrow that pinches apart cells in animal cell division.

12. Of the following, which is probably the most common route for membrane flow in the endomembrane system?
    a. rough ER → Golgi → lysosomes → nuclear membrane → plasma membrane
    b. rough ER → transport vesicles → Golgi → vesicles → plasma membrane
    c. nuclear envelope → rough ER → Golgi → smooth ER → lysosomes
    d. rough ER → vesicles → Golgi → smooth ER → plasma membrane
    e. smooth ER → vesicles → Golgi → vesicles → peroxisomes

13. Proteins to be used within the cytosol are generally synthesized
    a. by ribosomes bound to rough ER.
    b. by free ribosomes.
    c. by the nucleolus.
    d. within the Golgi apparatus.
    e. by mitochondria and chloroplasts.

14. Plasmodesmata in plant cells are similar in function to
    a. desmosomes.
    b. tight junctions.
    c. gap junctions.
    d. the extracellular matrix.
    e. integrins.

15. In an animal cell fractionation procedure, the first pellet formed would most likely contain
    a. the extracellular matrix.
    b. ribosomes.
    c. mitochondria.
    d. lysosomes.
    e. nuclei.

Use the cells described below to answer, for questions 16–20, the following question:

In which cell would you expect to find . . .
    a. muscle cell in the thigh muscle of a long-distance runner
    b. pancreatic cell that manufactures digestive enzymes
    c. macrophage (white blood cell) that engulfs bacteria
    d. epithelial cell lining digestive tract
    e. ovarian cell that produces estrogen (a steroid hormone)

16. the most tight junctions?

17. the most lysosomes?

18. the most smooth endoplasmic reticulum?

19. the most bound ribosomes?

20. the most mitochondria?

Chapter 7

Membrane Structure and Function

Key Concepts

7.1 **Cellular membranes are fluid mosaics of lipids and proteins**

7.2 **Membrane structure results in selective permeability**

7.3 **Passive transport is diffusion of a substance across a membrane with no energy investment**

7.4 **Active transport uses energy to move solutes against their gradients**

7.5 **Bulk transport across the plasma membrane occurs by exocytosis and endocytosis**

Framework

This chapter presents the fluid mosaic model of membrane structure, relating the molecular structure of biological membranes to their function of regulating the passage of substances into and out of the cell.

Chapter Review

The plasma membrane is the boundary of life; this **selectively permeable** membrane allows some materials to cross it more easily than others and enables the cell to maintain a unique internal environment.

7.1 Cellular membranes are fluid mosaics of lipids and proteins

According to the **fluid mosaic model,** the structure of biological membranes consists of various proteins that are attached to or embedded in a bilayer of **amphipathic** phospholipids.

Membrane Models: Scientific Inquiry Early chemical analysis of membranes revealed a lipid and protein composition. In 1925 two Dutch scientists suggested that cell membranes must be a phospholipid bilayer, with the hydrophobic hydrocarbon tails in the center and the hydrophilic heads facing the aqueous solution on both sides of the membrane. In 1935 H. Davson and J. Danielli proposed a model in which a bilayer of phospholipids is covered with a coat of hydrophilic proteins.

This sandwich model was consistent with the first views of membranes seen with electron microscopy, and by the 1960s, the Davson–Danielli model was widely accepted for all cellular membranes.

In 1972 S. J. Singer and G. Nicolson proposed the fluid mosaic model in which amphipathic membrane proteins are embedded in the phospholipid bilayer with their hydrophilic regions extending out into the aqueous environment. The phospholipid bilayer is envisioned as fluid, with a mosaic of protein molecules floating in it.

The fluid mosaic model is supported by evidence from freeze-fracture electron microscopy. This technique involves freezing a specimen, fracturing it with a cold knife, and examining the exposed interior of the membrane.

This membrane model is currently the most useful model for organizing existing knowledge and extending research on membrane structure. Some of this research now indicates that membranes may be more packed with proteins than originally thought.

INTERACTIVE QUESTION 7.1

Label the components in this diagram of the fluid mosaic model of membrane structure. Indicate the regions that are hydrophobic and those that are hydrophilic.

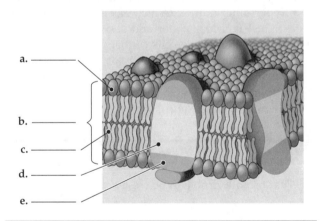

a.
b.
c.
d.
e.

The Fluidity of Membranes Membranes are held together primarily by weak hydrophobic interactions that allow the lipids and some of the proteins to drift laterally. Some membrane proteins seem to be held rigid by attachments to the cytoskeleton; others appear to be directed in their movements by cytoskeletal fibers.

Phospholipids with unsaturated hydrocarbon tails maintain membrane fluidity at lower temperatures. The steroid cholesterol, common in plasma membranes of animals, restricts movement of phospholipids and thus reduces fluidity at warmer temperatures. Cholesterol

also prevents the close packing of lipids and thus enhances fluidity at lower temperatures.

INTERACTIVE QUESTION 7.2

Cite some experimental evidence that shows that membrane proteins drift.

Membrane Proteins and Their Functions Each membrane has its own unique complement of membrane proteins, which determine most of the specific functions of that membrane. **Integral proteins** often extend through the membrane (transmembrane), with two hydrophilic ends. The hydrophobic midsection usually consists of one or more α helical stretches of hydrophobic amino acids. **Peripheral proteins** are attached to the surface of the membrane, often to integral proteins. Attachments of membrane proteins to the cytoskeleton on the cytoplasmic side and fibers of the extracellular matrix on the exterior provide a supportive framework for the plasma membrane.

INTERACTIVE QUESTION 7.3

List the six major kinds of functions that membrane proteins may perform.

The Role of Membrane Carbohydrates in Cell–Cell Recognition The ability of a cell to distinguish other cells is based on recognition of membrane carbohydrates. The **glycolipids** and **glycoproteins** attached to the outside of plasma membranes vary from species to species, from individual to individual, and even among cell types.

Synthesis and Sidedness of Membranes Membranes have distinct inner and outer faces, related to the composition of the lipid layers, the directional orientation of their proteins, and the attachment of carbohydrates to the extracellular surface. Carbohydrates are attached to membrane proteins as they are synthesized in the ER and are modified in the Golgi. Carbohydrates are attached to lipids in the Golgi. When transport vesicles fuse with the plasma membrane, these interior glycoproteins and glycolipids become located on the extracellular face of the membrane.

7.2   Membrane structure results in selective permeability

The plasma membrane permits a regular exchange of nutrients, waste products, oxygen, and inorganic ions. Biological membranes are selectively permeable; the ease and rate at which small molecules pass through them differ.

The Permeability of the Lipid Bilayer   Hydrophobic, nonpolar molecules, such as hydrocarbons, CO_2, and O_2, can dissolve in and cross a membrane.

INTERACTIVE QUESTION   7.4

What types of molecules have difficulty crossing the plasma membrane? Why?

Transport Proteins   Ions and polar molecules may move across the plasma membrane with the aid of **transport proteins.** Hydrophilic passageways through a membrane are provided for specific molecules by *channel proteins,* such as **aquaporins,** which facilitate water passage. *Carrier proteins* may physically bind and transport a specific molecule.

7.3   Passive transport is diffusion of a substance across a membrane with no energy investment

Diffusion is the movement of a substance down its **concentration gradient** due to random thermal motion. The diffusion of one solute is unaffected by the concentration gradients of other solutes. The cell does not expend energy when substances diffuse across membranes down their concentration gradient; therefore, the process is called **passive transport.**

Effects of Osmosis on Water Balance   Osmosis is the diffusion of water across a selectively permeable membrane. Water diffuses down its own concentration gradient, which is affected by the solute concentration. Binding of water molecules to solute particles lowers the proportion of unbound water that is free to cross the membrane.

INTERACTIVE QUESTION   7.5

A solution of 1 *M* glucose is separated by a selectively permeable membrane from a solution of 0.2 *M* fructose and 0.7 *M* sucrose. The membrane is not permeable to the sugar molecules. Indicate which side initially has more free water molecules and which has fewer. Show the direction of osmosis.

Tonicity, the tendency of a cell to gain or lose water in a given solution, is affected by the relative concentrations of solutes that cannot cross the membrane in the solution and in the cell. An animal cell will neither gain nor lose water in an **isotonic** environment. An animal cell placed in a **hypertonic** solution (which has more nonpenetrating solutes) will lose water and shrivel. If placed in a **hypotonic** solution, the cell will gain water, swell, and possibly lyse (burst). Cells without rigid walls must either live in an isotonic environment, such as salt water or isotonic body fluids, or have adaptations for **osmoregulation,** the control of water balance.

INTERACTIVE QUESTION   7.6

a. What osmotic problems do freshwater protists face?

b. What adaptations may help them osmoregulate?

The cell walls of plants, fungi, prokaryotes, and some protists play a role in water balance within hypotonic environments. Water moving into the cell causes the cell to swell against its cell wall. **Turgid** cells provide mechanical support for nonwoody plants. Plant cells in an isotonic surrounding are **flaccid.** In a hypertonic medium, a plant cell undergoes **plasmolysis,** the pulling away of the plasma membrane from the cell wall as water leaves and the cell shrivels.

INTERACTIVE QUESTION 7.7

a. The ideal osmotic environment for animal cells is _____.

b. The ideal osmotic environment for plant cells is _____.

Facilitated Diffusion: Passive Transport Aided by Proteins **Facilitated diffusion** involves the diffusion of polar molecules and ions across a membrane with the aid of transport proteins, either channel proteins or carrier proteins. Aquaporins greatly speed up diffusion of water. Many **ion channels** are **gated channels,** which open or close in response to electrical or chemical stimuli.

The binding of the solute to a carrier protein may cause a change in shape that serves to translocate the binding site and attached solute across the membrane.

INTERACTIVE QUESTION 7.8

Why is facilitated diffusion considered passive transport?

7.4 Active transport uses energy to move solutes against their gradients

The Need for Energy in Active Transport **Active transport,** requiring the expenditure of energy to transport a solute against its concentration gradient, is essential for a cell to maintain internal concentrations of small molecules that differ from environmental concentrations. The terminal phosphate group of ATP may be transferred to a carrier protein, inducing it to change its shape and translocate the bound solute across the membrane. The **sodium-potassium pump** works this way to exchange N^+ and K^+ across animal cell membranes, creating a higher concentration of potassium ions and a lower concentration of sodium ions within the cell.

Membrane Potential Maintenance by Ion Pumps Cells have a **membrane potential,** a voltage across the plasma membrane due to the unequal distribution of ions. This electrical potential energy results from the separation of opposite charges: The cytoplasm of a cell is negatively charged relative to the extracellular fluid. The membrane potential favors the diffusion of cations into the cell and anions out of the cell. Both the membrane potential and the concentration gradient affect the diffusion of an ion; thus, an ion diffuses down its **electrochemical gradient.**

Electrogenic pumps are membrane proteins that generate voltage across a membrane by the active transport of ions. A **proton pump** that transports H^+ out of the cell generates voltage across membranes in plants, fungi, and bacteria.

INTERACTIVE QUESTION 7.9

The sodium-potassium pump, the major electrogenic pump in animal cells, exchanges sodium ions for potassium ions, both of which are cations. How does this exchange generate a membrane potential?

Cotransport: Coupled Transport by a Membrane Protein **Cotransport** is a mechanism through which the active transport of a solute is indirectly driven by an ATP-powered pump that transports another substance against its gradient. As that transported substance then diffuses back down its concentration gradient through a cotransporter, the solute is carried against its concentration gradient across the membrane.

7.5 Bulk transport across the plasma membrane occurs by exocytosis and endocytosis

Bulk transport, like active transport, requires energy to transport larger biological molecules, packaged in vesicles, across the membrane.

Exocytosis In **exocytosis,** the cell secretes large molecules by the fusion of vesicles with the plasma membrane.

Endocytosis In **endocytosis,** a region of the plasma membrane sinks inward and pinches off to form a vesicle containing material that had been outside the cell. **Phagocytosis** is a form of endocytosis in which pseudopodia wrap around a food particle, creating a vacuole that then fuses with a lysosome containing hydrolytic enzymes. In **pinocytosis,** droplets of extracellular fluid are taken into the cell in small vesicles. **Receptor-mediated endocytosis** allows a cell to acquire specific substances from extracellular fluid. **Ligands,** molecules that bind specifically to receptor sites, attach to proteins usually clustered in coated pits on the cell surface and are carried into the cell when a vesicle forms.

INTERACTIVE QUESTION   7.10

a. How is cholesterol, which is used for the synthesis of other steroids and membranes, transported into human cells?

b. Explain why cholesterol accumulates in the blood of individuals with the disease familial hypercholesteremia.

Word Roots

amphi- = dual (*amphipathic molecule:* a molecule that has both a hydrophobic and a hydrophilic region)

aqua- = water; **-pori** = a small opening (*aquaporin:* a transport protein in the plasma membrane of a plant or animal cell that specifically facilitates the diffusion of water across the membrane)

co- = together; **trans-** = across (*cotransport:* the coupling of the "downhill" diffusion of one substance to the "uphill" transport of another against its own concentration gradient)

electro- = electricity; **-genic** = producing (*electrogenic pump:* an ion transport protein generating voltage across a membrane)

endo- = inner; **cyto-** = cell (*endocytosis:* the movement of materials into a cell; cell eating)

exo- = outer (*exocytosis:* the movement of materials out of a cell)

hyper- = exceeding; **-tonus** = tension (*hypertonic:* a solution with a higher concentration of solutes)

hypo- = lower (*hypotonic:* a solution with a lower concentration of solutes)

iso- = same (*isotonic:* solutions with equal concentrations of solutes)

phago- = eat (*phagocytosis:* cell eating)

pino- = drink (*pinocytosis:* cell drinking)

plasm- = molded; **-lyso** = loosen (*plasmolysis:* a phenomenon in walled cells in which the cytoplasm shrivels and the plasma membrane pulls away from the cell wall when the cell loses water to a hypertonic environment)

Structure Your Knowledge

1. Create a concept map to illustrate your understanding of osmosis. This exercise will help you practice using the words *hypotonic, isotonic,* and *hypertonic,* and it will help you focus on the effect of these osmotic environments on plant and animal cells. Explain your map to a friend.

2. The following diagram illustrates passive and active transport across a plasma membrane. Use it to answer questions a–d.

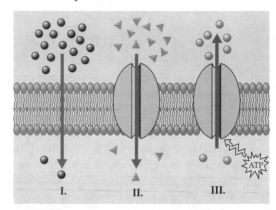

a. Which section represents facilitated diffusion? How can you tell? Does the cell expend energy in this transport? Why or why not? What types of solute molecules may be moved by this type of transport?

b. Which section shows active transport? List two ways that you can tell.

c. Which section shows diffusion? What types of solute molecules may be moved by this type of transport?

d. Which of these sections are considered passive transport?

Test Your Knowledge

MULTIPLE CHOICE: *Choose the one best answer.*

1. Glycoproteins and glycolipids are important for
   a. facilitated diffusion.
   b. active transport.
   c. cell–cell recognition.
   d. cotransport.
   e. signal-transduction pathways.

2. A single layer of phospholipid molecules coats the water in a beaker. Which part of the molecules will face the air?
   a. the phosphate groups
   b. the hydrocarbon tails
   c. both head and tail because the molecules are amphipathic and will lie sideways
   d. the glycolipid regions
   e. The phospholipids would dissolve in the water and not form a membrane coat.

3. Which of the following is *not* true about osmosis?
   a. It is a passive process in cells without walls, but an active one in cells with walls.
   b. Water moves from a hypotonic to a hypertonic solution.

c. Solute molecules bind to water and decrease the water available to move.

d. It can occur more rapidly through channel proteins known as aquaporins.

e. There is no net osmosis between isotonic solutions.

4. Support for the fluid mosaic model of membrane structure comes from

   a. the freeze-fracture technique of electron microscopy.

   b. the movement of proteins in hybrid cells.

   c. the amphipathic nature of many membrane proteins.

   d. both a and c.

   e. all of the above.

5. Facilitated diffusion of ions across a cellular membrane requires _____ and the ions move _____.

   a. energy and transport proteins . . . against their electrochemical gradient

   b. energy and transport proteins . . . against their concentration gradient

   c. cotransport proteins . . . against their electrochemical gradient

   d. transport proteins . . . down their electrochemical gradient

   e. transport proteins . . . down their concentration gradient

6. Which of the following is the most probable description of an integral, transmembrane protein?

   a. amphipathic with a hydrophilic head and a hydrophobic tail region

   b. a globular protein with hydrophobic amino acids in the interior and hydrophilic amino acids arranged around the outside

   c. a fibrous protein coated with hydrophobic fatty acids

   d. a glycolipid attached to the portion of the protein facing the exterior of the cell and cytoskeletal elements attached to the portion facing inside the cell

   e. a middle region composed of α helical stretches of hydrophobic amino acids, with hydrophilic regions at both ends of the protein

7. The fluidity of membranes in a plant in cold weather may be maintained by increasing the

   a. number of phospholipids with saturated hydrocarbon tails.

   b. action of an H^+ pump.

   c. concentration of cholesterol in the membrane.

   d. proportion of peripheral proteins.

   e. number of phospholipids with unsaturated hydrocarbon tails.

8. An animal cell placed in a hypotonic environment will

   a. plasmolyze.

   b. shrivel.

   c. become turgid.

   d. become flaccid.

   e. burst (lyse).

9. Which of the following is *not* true of carrier molecules involved in facilitated diffusion?

   a. They increase the speed of transport across a membrane.

   b. They can concentrate solute molecules on one side of the membrane.

   c. They may have specific binding sites for the molecules they transport.

   d. They may undergo a change in shape upon binding of solute.

   e. They do not require an energy investment from the cell to operate.

10. The membrane potential of a cell favors

    a. the movement of cations into the cell.

    b. the movement of anions into the cell.

    c. the action of an electrogenic pump.

    d. the movement of sodium out of the cell.

    e. both b and d.

11. Cotransport may involve

    a. active transport of two solutes through a transport protein.

    b. passive transport of two solutes through a transport protein.

    c. ion diffusion against the electrochemical gradient created by an electrogenic pump.

    d. a pump such as the sodium-potassium pump that moves ions in two different directions.

    e. transport of one solute against its concentration gradient in tandem with another that is diffusing down its concentration gradient.

12. Exocytosis may involve all of the following *except*

    a. ligands and coated pits.

    b. the fusion of a vesicle with the plasma membrane.

    c. a mechanism to export some carbohydrates during the formation of plant cell walls.

    d. a mechanism to rejuvenate the plasma membrane.

    e. a means of exporting large molecules.

13. The proton pump in plant cells is the functional equivalent of an animal cell's
    a. cotransport mechanism.
    b. sodium-potassium pump.
    c. contractile vacuole for osmoregulation.
    d. receptor-mediated endocytosis of cholesterol.
    e. ATP pump.

14. Pinocytosis involves
    a. the fusion of a newly formed food vacuole with a lysosome.
    b. receptor-mediated endocytosis that involves binding of a ligand.
    c. the pinching in of the plasma membrane around small droplets of external fluid.
    d. the secretion of cell fluid.
    e. the accumulation of specific molecules in a cell.

15. Watering a houseplant with too concentrated a solution of fertilizer can result in wilting because
    a. the uptake of ions into plant cells makes the cells hypertonic.
    b. the soil solution becomes hypertonic, causing the cells to lose water.
    c. the plant will grow faster than it can transport water and maintain proper water balance.
    d. diffusion down the electrochemical gradient will cause a disruption of membrane potential and accompanying loss of water.
    e. the plant will suffer fertilizer burn due to a caustic soil solution.

16. A cell is manufacturing receptor proteins for cholesterol. How would those proteins be oriented in the following membranes before they reach the plasma membrane?
    a. facing inside the ER lumen but outside the transport vesicle membrane
    b. facing inside the ER lumen and inside the transport vesicle
    c. attached outside the ER and outside the transport vesicle
    d. attached outside the ER but facing inside the transport vesicle
    e. embedded in the hydrophobic center of both the ER and transport vesicle membranes

Use the U-tube setup to answer questions 17 through 19.

The solutions in the two arms of this U-tube are separated by a membrane that is permeable to water and glucose but not to sucrose. Side A is filled with a solution of 2.0 *M* sucrose and 1.0 *M* glucose. Side B is filled with 1.0 *M* sucrose and 2.0 *M* glucose.

2.0 *M* sucrose
1.0 *M* glucose

1.0 *M* sucrose
2.0 *M* glucose

selectively permeable membrane

17. *Initially,* the solution in side A, with respect to that in side B,
    a. has a lower solute concentration.
    b. has a higher solute concentration.
    c. has an equal solute concentration.
    d. is lower in the tube.
    e. is higher in the tube.

18. During the period *before* equilibrium is reached, which molecule(s) will show net movement through the membrane?
    a. water
    b. glucose
    c. sucrose
    d. water and sucrose
    e. water and glucose

19. *After* the system reaches equilibrium, what changes are observed?
    a. The water level is higher in side A than in side B.
    b. The water level is higher in side B than in side A.
    c. The molarity of glucose is higher in side A than in side B.
    d. The molarity of sucrose has increased in side A.
    e. Both a and c have occurred.

20. You observe plant cells under a microscope as they are placed in an unknown solution. First the cells plasmolyze; after a minute, the plasmolysis reverses and the cells appear normal. What would you conclude about the unknown solution?
    a. It is hypertonic to the plant cells, and its solute cannot cross the plant cell membranes.
    b. It is hypotonic to the plant cells, and its solute cannot cross the plant cell membranes.
    c. It is isotonic to the plant cells, but its solute can cross the plant cell membranes.
    d. It is hypertonic to the plant cells, but its solute can cross the plant cell membranes.
    e. It is hypotonic to the plant cells, but its solute can cross the plant cell membranes.

Chapter 8

An Introduction to Metabolism

Key Concepts

8.1 An organism's metabolism transforms matter and energy, subject to the laws of thermodynamics

8.2 The free-energy change of a reaction tells us whether or not the reaction occurs spontaneously

8.3 ATP powers cellular work by coupling exergonic reactions to endergonic reactions

8.4 Enzymes speed up metabolic reactions by lowering energy barriers

8.5 Regulation of enzyme activity helps control metabolism

Framework

This chapter considers metabolism, the totality of the chemical reactions that take place in living organisms. Two key topics are emphasized: the energy transformations that underlie all chemical reactions and the role of enzymes in the "cold chemistry" of the cell. Enzymes are macromolecules that act as catalysts, lowering the activation energy of a reaction and thus greatly speeding metabolic processes. Intricate control and feedback mechanisms produce the metabolic integration necessary for life.

The reactions in a cell either release or consume energy. The following illustration summarizes some of the components of the energy changes within a cell.

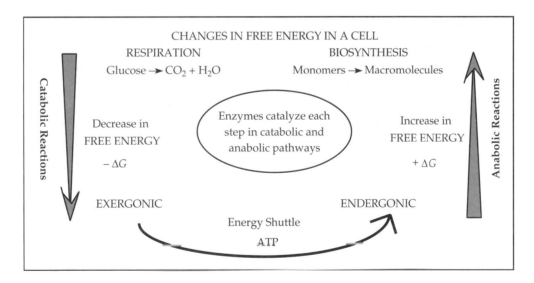

Chapter Review

8.1 An organism's metabolism transforms matter and energy, subject to the laws of thermodynamics

Organization of the Chemistry of Life into Metabolic Pathways Metabolism includes all of the chemical reactions in an organism. These reactions are ordered into **metabolic pathways,** a sequence of steps, each controlled by an enzyme, which convert a specific molecule to a product.

Catabolic pathways release the energy stored in complex molecules through the breakdown of these molecules into simpler compounds. **Anabolic pathways,**

sometimes called biosynthetic pathways, require energy to combine simpler molecules into more complicated ones. The energy released from catabolic pathways drives the anabolic pathways in a cell. The study of how organisms transform energy, called **bioenergetics,** is central to understanding metabolism.

Forms of Energy   **Energy** has been defined as the capacity to cause change. Some forms of energy can do work, such as moving matter against an opposing force. **Kinetic energy** is the energy of motion, of matter that is moving. This matter does its work by transferring its motion to other matter. The kinetic energy of randomly moving molecules is **heat,** or **thermal energy. Potential energy** is the capacity of matter to cause change as a consequence of its location or arrangement. **Chemical energy** is a form of potential energy that is available for release in chemical reactions.

Energy can be converted from one form to another. Plants convert light energy to the chemical energy in sugar, and cells release this potential energy to drive cellular processes.

The Laws of Energy Transformation   **Thermodynamics** is the study of energy transformations. In an *open system,* energy and matter may be exchanged between the *system* and the *surroundings.*

The **first law of thermodynamics** states that energy can be neither created nor destroyed. According to this *principle of conservation of energy,* energy can be transferred between matter and transformed from one kind to another, but the total energy of the universe is constant.

The **second law of thermodynamics** states that every energy transformation or transfer results in an increasing disorder within the universe. **Entropy** is used as a measure of disorder or randomness. In every energy transfer or transformation, some of the energy is converted to heat, the least ordered form of energy. For a process to occur spontaneously, without the input of external energy, it must result in an increase in entropy. A nonspontaneous process will occur only if energy is added to the system.

A cell may become more ordered, but it does so with an attendant increase in the entropy of its surroundings. An organism takes in and uses highly ordered organic molecules as a source of the energy needed to create and maintain its own organized structure, but it returns heat and the simple molecules of carbon dioxide and water to the environment.

INTERACTIVE QUESTION   8.1

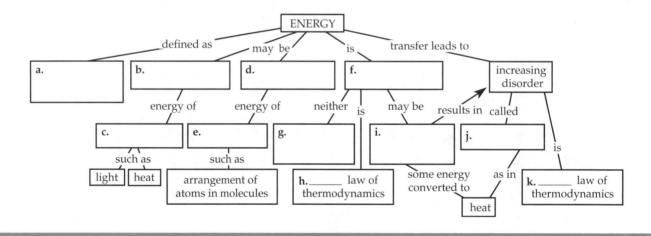

8.2   The free-energy change of a reaction tells us whether or not the reaction occurs spontaneously

Free-Energy Change, **DG**   The portion of a system's energy available to perform work when the system's temperature and pressure are uniform is defined as **free energy.** The change in free energy during a reaction is represented by $\Delta G = \Delta H - T\Delta S$. Enthalpy (*H*) is the total energy of a system; entropy (*S*) is a measure of its disorder, and *T* is absolute temperature (°Kelvin, or

°C + 273). The free energy of the system must decrease (−*G*) for a reaction to be spontaneous: The system must lose energy (*H* must decrease), become more disordered (*TS* must increase), or both.

Free Energy, Stability, and Equilibrium   When Δ*G* is negative, the final state has less free energy than the initial state; thus, the final state is less likely to change and is more stable. A system rich in free energy has a tendency to change spontaneously to a more stable state, a state of *equilibrium.* At equilibrium in a chemical reaction, the forward and backward reactions are proceeding at the same rate and the relative concentrations of products and reactants stays the same. Moving toward equilibrium is spontaneous; the ΔG of the reaction is negative. Once at equilibrium, a system is at a minimum of free energy and will not spontaneously change.

INTERACTIVE QUESTION 8.2

Complete the following table to show how the free energy of a system or reaction relates to its stability, tendency for spontaneous change, equilibrium, and capacity to do work.

	System with High Free Energy	System with Low Free Energy
Stability		
Spontaneous		
Equilibrium		
Work capacity		

Free Energy and Metabolism An **exergonic reaction** (−ΔG) proceeds with a net release of free energy and is spontaneous. The magnitude of ΔG indicates the maximum amount of work the reaction can do. **Endergonic reactions** (+ΔG) are nonspontaneous; they must absorb free energy from the surroundings. The energy released by an exergonic reaction (−ΔG) is equal to the energy required by the reverse reaction (+ΔG).

Metabolic disequilibrium is essential to life. Metabolic reactions are reversible and could reach equilibrium if the cell did not maintain a steady supply of reactants and siphon off the products (as reactants for new processes or as waste products to be expelled).

INTERACTIVE QUESTION 8.3

Develop a concept map on free energy and ΔG. The value in this exercise is for you to wrestle with and organize these concepts for yourself. Do not turn to the suggested concept map until you have worked on your own understanding. Remember that the concept map in the answer section is only one way of structuring these ideas—have confidence in your own organization.

8.3 ATP powers cellular work by coupling exergonic reactions to endergonic reactions

Central to a cell's bioenergetics is **energy coupling**, using exergonic processes to power endergonic ones. A cell usually uses ATP as the immediate source of energy for its mechanical, transport, and chemical work.

The Structure and Hydrolysis of ATP ATP (adenosine triphosphate) consists of the nitrogenous base adenine bonded to the sugar ribose, which is connected to a chain of three phosphate groups. ATP can be hydrolyzed to ADP (adenosine diphosphate) and an inorganic phosphate molecule (P_i), releasing 7.3 kcal (30.5 kJ) of energy per mole of ATP when measured under standard conditions. The ΔG of the reaction in the cell is estimated to be closer to 13 kcal/mol.

INTERACTIVE QUESTION 8.4

Label the three components (a through c) of the ATP molecule shown below.

a. _____

c. _____   b. _____

d. Indicate which bond is likely to break. By what chemical mechanism is the bond broken?

e. Explain why this reaction releases so much energy.

How ATP Performs Work In a cell, the free energy released from the hydrolysis of ATP is used to transfer the phosphate group to another molecule, producing a **phosphorylated** molecule that is more reactive (less stable). The phosphorylation of other molecules by ATP forms the basis for almost all cellular work.

The Regeneration of ATP A cell regenerates ATP at a phenomenal rate. The formation of ATP from ADP and P_i is endergonic, with a ΔG of +7.3 kcal/mol (standard conditions). Cellular respiration (the catabolic processing of glucose and other organic molecules) provides the energy for the regeneration of ATP. Plants can also produce ATP using light energy.

8.4 Enzymes speed up metabolic reactions by lowering energy barriers

Enzymes are biological **catalysts**—agents that speed the rate of a reaction but are unchanged by the reaction.

The Activation Energy Barrier Chemical reactions involve both the breaking and forming of chemical bonds. Energy must be absorbed to make bonds unstable enough that they will break, and energy is released when bonds form and molecules return to stable, lower energy states. **Activation energy,** or the **free energy of activation** (E_A), is the energy that must be absorbed by reactants to reach the unstable *transition state*, in which bonds are likely to break, and from which the reaction can proceed.

How Enzymes Lower the E_A Barrier The E_A barrier is essential to life because it prevents the energy-rich macromolecules of the cell from decomposing spontaneously. For metabolism to proceed in a cell, however, E_A must be reached. Enzymes are able to lower E_A for specific reactions so that metabolism can proceed at cellular temperatures. Enzymes do not change the ΔG for a reaction.

Substrate Specificity of Enzymes Protein enzymes are macromolecules with characteristic three-dimensional shapes. The specificity of an enzyme for a particular **substrate** is determined by this shape. The substrate attaches at the enzyme's **active site,** a pocket or groove found on the surface of the enzyme that has a shape complementary to the substrate. The substrate is temporarily bound to its enzyme, forming an **enzyme-substrate complex.** Interactions between substrate and active site cause the enzyme to change shape slightly, creating what is called an **induced fit** that enhances the ability of the enzyme to catalyze the chemical reaction.

Catalysis in the Enzyme's Active Site The substrate is often held in the active site by hydrogen or ionic bonds. The side chains (R groups) of some of the surrounding amino acids in the active site facilitate the conversion of substrate to product. The product then leaves the active site, and the catalytic cycle repeats, often at astonishing speed.

Whether an enzyme catalyzes the forward or backward reaction is influenced by the relative concentrations of reactants and products and the ΔG of the reactions. Enzymes catalyze reactions moving toward equilibrium.

Enzymes catalyze reactions involving the joining of two reactants by binding the substrates closely together and making sure they are properly oriented. An induced fit can stretch critical bonds in the substrate molecule and make them easier to break. An active site may provide a microenvironment, such as a lower pH, that is necessary for a particular reaction. Enzymes may also actually participate in a reaction by forming brief covalent bonds with the substrate.

The rate of a reaction will increase with increasing substrate concentration up to the point at which all enzyme molecules are *saturated* with substrate molecules and working at full speed. Only adding more enzyme molecules will increase the rate of the reaction at that point.

INTERACTIVE QUESTION   8.5

In this graph of an exergonic reaction with and without an enzyme catalyst, label parts a through e.

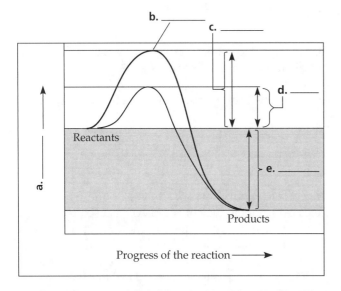

INTERACTIVE QUESTION 8.6

Outline a catalytic cycle using the following diagrammatic enzyme. Sketch two appropriate substrate molecules and two products, identify the enzyme-substrate complex, and describe the key steps of the cycle.

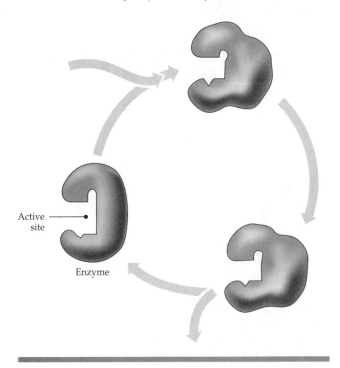

Active site

Enzyme

Effects of Local Conditions on Enzyme Activity   The velocity of an enzyme-catalyzed reaction may increase with rising temperature up to the point at which increased thermal agitation begins to disrupt the bonds and interactions that stabilize protein shape. Each enzyme has *optimal* conditions that include a temperature and pH optimum at which it is most active.

Cofactors are small molecules that bind either permanently or reversibly with enzymes and are necessary for enzyme function. They may be inorganic, such as various metal ions, or organic molecules called **coenzymes.** Most vitamins are coenzymes or precursors of coenzymes.

Enzyme inhibitors selectively disrupt the action of enzymes, either reversibly by binding with the enzyme with weak bonds or irreversibly by attaching with covalent bonds. **Competitive inhibitors** compete with the substrate for the active site of the enzyme. Increasing the concentration of substrate molecules may overcome this type of inhibition. **Noncompetitive inhibitors** bind to a part of the enzyme separate from the active site and change the shape of the enzyme, thus impeding enzyme action.

INTERACTIVE QUESTION 8.7

Return to your diagram in Interactive Question 8.6. Draw a competitive and a noncompetitive inhibitor, and indicate where each would bind to the enzyme molecule.

8.5   Regulation of enzyme activity helps control metabolism

Allosteric Regulation of Enzymes   In **allosteric regulation,** molecules may inhibit or activate enzyme activity when they bind to a site separate from the active site. Enzymes made of two or more polypeptides, each with its own active site, may have binding sites (sometimes called allosteric sites) located where subunits join. The entire unit may oscillate between two forms. The binding of an *activator* stabilizes the catalytically active shape, whereas an *inhibitor* reinforces the inactive form of the enzyme. Allosteric enzymes may be critical regulators of metabolic pathways.

Through a phenomenon called **cooperativity,** the induced-fit binding of a substrate molecule to one subunit can change the shape such that the active sites of all subunits are more active.

Metabolic pathways are commonly regulated by **feedback inhibition,** in which the product of a pathway acts as an allosteric inhibitor of an enzyme early in the pathway.

INTERACTIVE QUESTION 8.8

Both ATP and ADP serve as regulators of enzyme activity. In catabolic pathways, which of these molecules would you predict would act as an inhibitor?

Which molecule would you expect to act as an activator of anabolic pathways?

Specific Localization of Enzymes within the Cell   Enzymes for several steps of a metabolic pathway may be associated in a multienzyme complex, facilitating the sequence of reactions. Specialized cellular compartments may contain high concentrations of the enzymes and substrates needed for a particular pathway. Enzymes are often incorporated into the membranes of cellular compartments. The complex internal structures of the cell facilitate metabolic order.

Word Roots

allo- = different (*allosteric site:* a specific receptor site on some part of an enzyme molecule remote from the active site)

ana- = up (*anabolic pathway:* a metabolic pathway that consumes energy to build complex molecules from simpler ones)

bio- = life (*bioenergetics:* the study of how organisms manage their energy resources)

cata- = down (*catabolic pathway:* a metabolic pathway that releases energy by breaking down complex molecules into simpler ones)

endo- = within (*endergonic reaction:* a reaction that absorbs free energy from its surroundings)

ex- = out (*exergonic reaction:* a reaction that proceeds with a net release of free energy)

kinet- = movement (*kinetic energy:* the energy of motion)

therm- = heat (*thermodynamics:* the study of the energy transformations that occur in a collection of matter)

Structure Your Knowledge

This chapter introduced many new and complex concepts. See if you can step back from the details and answer the following general questions.

1. Relate the concept of free energy to metabolism.

2. What role do enzymes play in metabolism?

Test Your Knowledge

MULTIPLE CHOICE: *Choose the one best answer.*

1. Catabolic and anabolic pathways are often coupled in a cell because
   a. the intermediates of a catabolic pathway are used in the anabolic pathway.
   b. both pathways use the same enzymes.
   c. the free energy released from one pathway is used to drive the other.
   d. the activation energy of the catabolic pathway can be used in the anabolic pathway.
   e. their enzymes are controlled by the same activators and inhibitors.

2. When glucose and O_2 are converted to CO_2 and H_2O, changes in total energy, entropy, and free energy are as follows:
   a. $-\Delta H, -\Delta S, -\Delta G$
   b. $-\Delta H, +\Delta S, -\Delta G$
   c. $-\Delta H, +\Delta S, +\Delta G$
   d. $+\Delta H, +\Delta S, +\Delta G$
   e. $+\Delta H, -\Delta S, +\Delta G$

3. When amino acids join to form a protein, the following energy and entropy changes apply:
   a. $+\Delta H, -\Delta S, +\Delta G$
   b. $+\Delta H, +\Delta S, -\Delta G$
   c. $+\Delta H, +\Delta S, +\Delta G$
   d. $-\Delta H, +\Delta S, +\Delta G$
   e. $-\Delta H, -\Delta S, +\Delta G$

4. A negative ΔG means that
   a. the quantity G of energy is available to do work.
   b. the reaction is spontaneous.
   c. the reactants have more free energy than the products.
   d. the reaction is exergonic.
   e. all of the above are true.

5. According to the first law of thermodynamics,
   a. for every action there is an equal and opposite reaction.
   b. every energy transfer results in an increase in disorder or entropy.
   c. the total amount of energy in the universe is conserved or constant.
   d. energy can be transferred or transformed, but disorder always increases.
   e. potential energy is converted to kinetic energy, and kinetic energy is converted to heat.

6. What is meant by an induced fit?
   a. The binding of the substrate is an energy-requiring process.
   b. A competitive inhibitor can outcompete the substrate for the active site.
   c. The binding of the substrate changes the shape of the active site, which can stress or bend substrate bonds.
   d. The active site creates a microenvironment ideal for the reaction.
   e. Substrates are held in the active site by hydrogen and ionic bonds.

7. One way in which a cell maintains metabolic disequilibrium is to
   a. siphon products of a reaction off to the next step in a metabolic pathway.
   b. provide a constant supply of enzymes for critcal reactions.
   c. use feedback inhibition to turn off pathways.
   d. use allosteric enzymes that can bind to activators or inhibitors.
   e. use the energy from anabolic pathways to drive catabolic pathways.

8. In an experiment, changing the pH from 7 to 6 resulted in an increase in product formation. From this we could conclude that
   a. the enzyme became saturated at pH 6.
   b. the enzyme's optimal pH is 6.
   c. this enzyme works best in a neutral pH.
   d. the temperature must have increased when the pH was changed to 6.
   e. the enzyme was in a more active shape at pH 6.

9. When substance A was added to an enzyme reaction, product formation decreased. The addition of more substrate did not increase product formation. From this we conclude that substance A could be
   a. product molecules.
   b. a cofactor.
   c. an allosteric enzyme.
   d. a competitive inhibitor.
   e. a noncompetitive inhibitor.

10. The formation of ATP from ADP and inorganic phosphate
    a. is an exergonic process.
    b. transfers the phosphate to another intermediate that becomes more reactive.
    c. produces an unstable energy compound that can drive cellular work.
    d. has a ΔG of -7.3 kcal/mol under standard conditions.
    e. involves the hydrolysis of a phosphate bond.

11. At equilibrium,
    a. no enzymes are functioning.
    b. free energy is at a minimum.
    c. the forward and backward reactions have stopped.
    d. the products and reactants have equal values of H.
    e. a reaction has a $+\Delta G$.

12. In cooperativity,
    a. a cellular organelle contains all the enzymes needed for a metabolic pathway.
    b. a product of a pathway serves as a competitive inhibitor of an early enzyme in the pathway.
    c. a molecule bound to the active site of one subunit of an enzyme affects the active site of other subunits.
    d. the allosteric site is filled with an activator molecule.
    e. the product of one reaction serves as the substrate for the next in intricately ordered metabolic pathways.

13. When a cell breaks down glucose, only about 40% of the energy is captured in ATP molecules. The remaining 60% of the energy is
    a. used to increase the order necessary for life to exist.
    b. lost as heat, in accordance with the second law of thermodynamics.
    c. used to increase the entropy of the system by converting kinetic energy into potential energy.
    d. stored in starch or glycogen for use later by the cell.
    e. released when the ATP molecules are hydrolyzed.

14. An endergonic reaction could be described as one that will
    a. proceed spontaneously with the addition of activation energy.
    b. produce products with more free energy than the reactants.
    c. not be able to be catalyzed by enzymes.
    d. release energy.
    e. produce ATP for energy coupling.

15. What is most directly responsible for the specificity of a protein enzyme?
    a. its primary structure
    b. its secondary and tertiary structure
    c. the shape and characteristics of its allosteric site
    d. its cofactors
    e. the R groups of the amino acids in its active site

16. Which of the following parameters does an enzyme raise?
    a. ΔG
    b. ΔH
    c. equilibrium of a reaction
    d. speed of a reaction
    e. free energy of activation

17. Zinc, an essential trace element, may be found bound to the active site of some enzymes. What would be the most likely function of such zinc ions?

    a. coenzyme derived from a vitamin
    b. a cofactor necessary for catalysis
    c. a substrate of the enzyme
    d. a competitive inhibitor of the enzyme
    e. an allosteric activator of the enzyme

Use this diagram to answer questions 18–20.

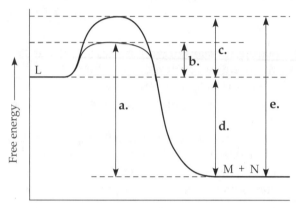

18. Which line in the diagram indicates the ΔG of the enzyme-catalyzed reaction of $L \rightarrow M + N$?

19. Which line in the diagram indicates the activation energy of the noncatalyzed reaction?

20. Which of the following terms would best describe this reaction?

    a. nonspontaneous
    b. $-\Delta G$
    c. endergonic
    d. coupled reaction
    e. anabolic reaction

21. A reaction that is spontaneous

    a. has a $+\Delta G$.
    b. occurs very rapidly.
    c. does not require enzyme catalysis in a cell.
    d. will decrease the entropy of a system.
    e. is exergonic.

22. In the metabolic pathway, $A \rightarrow B \rightarrow C \rightarrow D \rightarrow E$, what effect would molecule E likely have on the enzyme that catalyzes $A \rightarrow B$?

    a. allosteric inhibitor
    b. allosteric activator
    c. competitive inhibitor
    d. feedback activator
    e. coenzyme

FILL IN THE BLANKS

_____ 1. the totality of an organism's chemical processes

_____ 2. pathways that require energy to combine molecules together

_____ 3. the energy of motion

_____ 4. enzymes that change between two shapes, depending on whether an activator or inhibitor is bound to them

_____ 5. term for the measure of disorder or randomness

_____ 6. the energy that must be absorbed by molecules to reach the transition state

_____ 7. inhibitors that decrease an enzyme's activity by binding to the active site

_____ 8. organic molecules that bind to enzymes and are necessary for their functioning

_____ 9. regulatory device in which the product of a pathway binds to an enzyme early in the pathway

_____ 10. more reactive molecules created by the transfer of a phosphate group from ATP

Chapter 9

Cellular Respiration: Harvesting Chemical Energy

Key Concepts

9.1 Catabolic pathways yield energy by oxidizing organic fuels

9.2 Glycolysis harvests chemical energy by oxidizing glucose to pyruvate

9.3 The citric acid cycle completes the energy-yielding oxidation of organic molecules

9.4 During oxidative phosphorylation, chemiosmosis couples electron transport to ATP synthesis

9.5 Fermentation and anaerobic respiration enable cells to produce ATP without the use of oxygen

9.6 Glycolysis and the citric acid cycle connect to many other metabolic pathways

Framework

The catabolic pathways of glycolysis and cellular respiration release the chemical energy in glucose and other fuels and store it in ATP. Glycolysis, occurring in the cytosol, produces ATP, pyruvate, and NADH; the latter two may then enter the mitochondria for respiration. A mitochondrion consists of a matrix in which the enzymes of the citric acid cycle are localized, a highly folded inner membrane in which enzymes and the molecules of the electron transport chain are embedded, and an intermembrane space between the two membranes to temporarily house H^+ that has been pumped across the inner membrane during the redox reactions of the electron transport chain. ATP is produced by oxidative phosphorylation, which involves a chemiosmotic mechanism in which a proton-motive force drives protons through ATP synthases located in the membrane.

Chapter Review

9.1 Catabolic pathways yield energy by oxidizing organic fuels

Catabolic Pathways and Production of ATP **Fermentation,** which occurs without oxygen, is the partial degradation of sugars to release energy. **Aerobic respiration** uses oxygen in the breakdown of glucose (or other energy-rich organic compounds) to yield carbon dioxide and water and release energy as ATP and heat. The *anaerobic respiration* of some prokaryotes does not use oxygen as a reactant but is a similar process. **Cellular respiration** is usually used to refer to the aerobic process. This exergonic process has a free energy change of –686 kcal/mol of glucose.

INTERACTIVE QUESTION   9.1

Fill in this summary equation for cellular respiration.

$$\text{\_\_\_\_\_} + 6\,O_2 \rightarrow \text{\_\_\_\_\_} + 6\,H_2O + \text{\_\_\_\_\_}$$

Redox Reactions: Oxidation and Reduction Oxidation-reduction or **redox reactions** involve the partial or complete transfer of one or more electrons from one reactant to another. **Oxidation** is the loss of electrons from one substance; **reduction** is the addition of electrons to another substance. The substance that loses electrons becomes oxidized and acts as a **reducing agent** (electron donor) to the substance that gains electrons. By gaining electrons, a substance acts as an **oxidizing agent** (electron acceptor) and becomes reduced.

INTERACTIVE QUESTION   9.2

Fill in the appropriate terms in this equation.

X is the reducing agent; it **c.** _____ electrons.
Y is the **d.** _____ ; it **e.** _____ electrons.

Oxygen strongly attracts electrons and is one of the most powerful oxidizing agents. As electrons shift toward a more electronegative atom, they give up potential energy. Thus, chemical energy is released in a redox reaction that relocates electrons closer to oxygen.

Organic molecules with an abundance of hydrogen are rich in "hilltop" electrons that release their potential energy when they "fall" closer to oxygen.

INTERACTIVE QUESTION   9.3

a. In the conversion of glucose and oxygen to carbon dioxide and water, which molecule becomes reduced?

b. Which molecule becomes oxidized?

c. What happens to the energy that is released in this redox reaction?

At certain steps in the oxidation of glucose, two hydrogen atoms are removed by enzymes called dehydrogenases, and the two electrons and one proton are passed to a coenzyme, **NAD$^+$** (nicotinamide adenine dinucleotide), reducing it to NADH.

Energy from respiration is slowly released in a series of small steps as electrons are passed from NADH down an **electron transport chain,** a group of carrier molecules located in the inner mitochondrial membrane (or plasma membrane of aerobic prokaryotes), to a stable location close to a highly electronegative oxygen atom, forming water.

INTERACTIVE QUESTION   9.4

a. NAD$^+$ is called an _____ .

b. Its reduced form is _____ .

The Stages of Cellular Respiration: A Preview **Glycolysis,** occurring in the cytosol, breaks glucose into two molecules of pyruvate. The **citric acid cycle,** located in the mitochondrial matrix or in the cytoplasm of prokaryotes, oxidizes a derivative of pyruvate into carbon dioxide. In some of the steps of glycolysis and the citric acid cycle, dehydrogenase enzymes transfer electrons to NAD$^+$. NADH passes electrons to the electron transport chain, from which they eventually combine with hydrogen ions and oxygen to form water. The energy released through this chain of redox reactions is used to synthesize ATP by **oxidative phosphorylation,** a process that includes electron transport and chemiosmosis.

Up to 38 molecules of ATP may be generated for each molecule of glucose oxidized to carbon dioxide. About 10% of this ATP is produced by **substrate-level phosphorylation,** in which an enzyme transfers a phosphate group from a substrate molecule to ADP.

INTERACTIVE QUESTION   9.5

Fill in the three stages of respiration (a–c). Indicate whether ATP is produced by substrate-level or oxidative phosphorylation (d–f). Label the arrows indicating electrons carried by NADH.

9.2 Glycolysis harvests chemical energy by oxidizing glucose to pyruvate

Glycolysis, a 10-step process occurring in the cytosol, has an energy-investment phase and an energy-payoff phase. Two molecules of ATP are consumed as glucose is split into two three-carbon sugars (glyceraldehyde-3-phosphate). The conversion of these molecules to pyruvate produces two NADH and four ATP by substrate-level phosphorylation. For each molecule of glucose, glycolysis yields a net gain of two ATP and two NADH.

Enzymes catalyze each step in glycolysis. Kinases transfer phosphate groups; a dehydrogenase oxidizes glyceraldehyde-3-phosphate and reduces NAD^+; and other enzymes cleave the six-carbon sugar and re-arrange atoms in substrate molecules.

INTERACTIVE QUESTION 9.6

Fill in the blanks in this summary diagram of glycolysis.

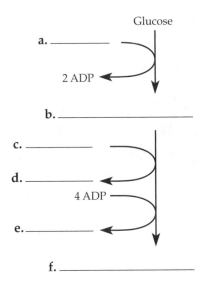

9.3 The citric acid cycle completes the energy-yielding oxidation of organic molecules

Pyruvate is actively transported into the mitochondrion in eukaryotes. Before the citric acid cycle begins, a series of steps occurs within a multienzyme complex. A carboxyl group is removed from the three-carbon pyruvate and released as CO_2; the remaining two-carbon group is oxidized to form acetate with the accompanying reduction of NAD^+ to NADH; and coenzyme A is attached to the acetate by an unstable bond, forming **acetyl CoA.**

In the citric acid cycle (also called the Krebs cycle), the acetyl group of acetyl CoA is added to oxaloacetate to form citrate, which is progressively decomposed back to oxaloacetate. For each turn of the citric acid cycle, two carbons enter in reduced form from acetyl CoA; two carbons exit completely oxidized as CO_2; three NADH and one $FADH_2$ are formed; and one ATP (GTP in most animal cells) is made by substrate-level phosphorylation. There are two turns of the citric acid cycle for each glucose molecule oxidized.

INTERACTIVE QUESTION 9.7

Fill in the blanks in this diagram of the citric acid cycle. Balls represent carbon atoms.

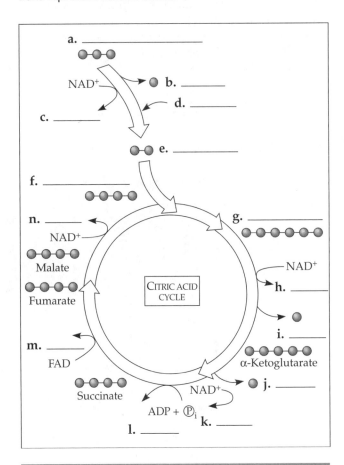

9.4 During oxidative phosphorylation, chemiosmosis couples electron transport to ATP synthesis

The Pathway of Electron Transport Thousands of electron transport chains are embedded in the cristae (infoldings) of the inner mitochondrial membrane (or the plasma membrane of prokaryotes). Most components of the electron transport chain are proteins, with tightly bound, nonprotein *prosthetic groups*. The electron carriers shift between reduced and oxidized states as they accept and donate electrons. The components are organized into four complexes. The transfer of electrons goes from NADH to a flavoprotein to an iron-sulfur protein (Complex I) to a mobile hydrophobic molecule, ubiquinone (Q or CoQ). Next electrons pass down a series of molecules called **cytochromes,** proteins with an iron-containing heme group. The last cytochrome, cyt a_3, passes electrons to oxygen, which picks up a pair of H^+ and forms water.

FADH$_2$ adds its electrons to the chain at a lower energy level (at Complex II); thus, one-third less energy is provided for ATP synthesis by FADH$_2$ as compared to NADH.

Chemiosmosis: The Energy-Coupling Mechanism **ATP synthase,** a protein complex embedded in the inner membrane, uses the energy of a proton (H^+) gradient to make ATP, an example of the process called **chemiosmosis.**

The flow of H^+ ions down their gradient through the rotor and stator part of the ATP synthase complex causes the rotor and attached rod to rotate, activating catalytic sites on the knob portion where ADP and inorganic phosphate join to make ATP.

The electron transport chain creates the proton gradient. When some members of the chain pass electrons, they also accept and release protons, which are pumped into the intermembrane space at three points. The resulting proton gradient stores potential energy, referred to as the **proton-motive force.**

In mitochondria, exergonic redox reactions produce the H^+ gradient that drives the production of ATP. Chloroplasts use light energy to create the proton-motive force used to make ATP by chemiosmosis. Prokaryotes use H^+ gradients generated across their plasma membranes to transport molecules, make ATP, and rotate flagella.

INTERACTIVE QUESTION   9.8

Label this diagram of oxidative phosphorylation in a mitochondrial membrane.

An Accounting of ATP Production by Cellular Respiration   About 36 or 38 ATPs may be produced per glucose molecule oxidized. This number is only an estimate for three reasons: The 3 ATPs/NADH and 2 ATPs/FADH$_2$ are rounded off numbers; the electrons from the NADH produced by glycolysis may be passed across the inner mitochondrial membrane to NAD$^+$ or FAD, depending on the type of shuttle used in the cell; and the proton-motive force generated by the electron transport chain is also used to power other work in the mitochondrion.

The efficiency of respiration in its energy conversions is approximately 40%. The rest of the energy is released as heat.

INTERACTIVE QUESTION 9.9

Fill in the tally for maximum ATP yield from the oxidation of one molecule of glucose to six molecules of carbon dioxide.

Process	# ATP
Initial phosphorylation of glucose:	**a.**
Substrate-level phosphorylation: in glycolysis	**b.** _____
in **c.**	2
Oxidative phosphorylation:*	**d.** _____
Maximum Total	**e.**

*approximately 3 ATP for each of the **f.** _____ NADH from pyruvate → acetyl CoA and the **g.** _____ NADH from citric acid cycle; **h.** _____ ATP for each of the **i.** _____ FADH$_2$ from citric acid cycle; 2 or 3 ATP for the **j.** _____ NADH from glycolysis, depending on whether the shuttle passes electrons across the membrane to NAD$^+$ or FAD.

9.5 Fermentation and anaerobic respiration enable cells to produce ATP without the use of oxygen

Organisms that generate ATP through anaerobic respiration have an electron transport chain that does not use oxygen as the final electron acceptor. Some marine bacteria use the sulfate ion to accept electrons, generating H$_2$S instead of H$_2$O. Fermentation generates ATP without using oxygen or an electron transport chain. In fermentation, oxidation of glucose in glycolysis produces a net of 2 ATP by substrate-level phosphorylation, and NADH is recycled to NAD$^+$ by the transfer of electrons to pyruvate or derivatives of pyruvate.

Types of Fermentation In **alcohol fermentation,** pyruvate is converted into acetaldehyde, and CO$_2$ is released. Acetaldehyde is then reduced by NADH to form ethanol (ethyl alcohol), and NAD$^+$ is regenerated. In **lactic acid fermentation,** pyruvate is reduced directly by NADH to form lactate and recycle NAD$^+$. Muscle cells make ATP by lactic acid fermentation when energy demand is high and oxygen supply is low.

Fermentation and Cellular Respiration Compared Both fermentation and respiration use glycolysis with NAD$^+$ as the oxidizing agent to convert glucose and other organic fuels to pyruvate. To oxidize NADH back to NAD$^+$, fermentation uses pyruvate or acetaldehyde as the final electron acceptor, whereas respiration uses oxygen, via the electron transport chain.

Facultative anaerobes, such as yeasts and many bacteria, can make ATP by fermentation or respiration, depending upon the availability of oxygen. **Obligate anaerobes** make ATP by fermentation or anaerobic respiration.

The Evolutionary Significance of Glycolysis Glycolysis is common to fermentation and respiration. This most widespread of all metabolic processes probably evolved in ancient prokaryotes before oxygen was available. The cytosolic location of glycolysis is also evidence of its antiquity.

INTERACTIVE QUESTION 9.10

How much more ATP can be generated by respiration than by fermentation? Explain why.

9.6 Glycolysis and the citric acid cycle connect to many other metabolic pathways

The Versatility of Catabolism Fats, proteins, and carbohydrates can all be used by cellular respiration to make ATP. Proteins are digested into amino acids, which are then deaminated (amino group removed) and can enter into respiration at several sites. The digestion of fats yields glycerol, which is converted to an intermediate of glycolysis, and fatty acids, which are broken down by **beta oxidation** to two-carbon fragments that enter the citric acid cycle as acetyl CoA.

Biosynthesis (Anabolic Pathways) The organic molecules of food also provide carbon skeletons for biosynthesis. Some monomers, such as amino acids, can be directly incorporated into the cell's macromolecules. Intermediates of glycolysis and the citric acid cycle serve as precursors for anabolic pathways. The molecules of carbohydrates, fats, and proteins can all be interconverted to provide for a cell's needs.

Regulation of Cellular Respiration via Feedback Mechanisms Through feedback inhibition, the end product of a pathway inhibits an enzyme early in the pathway, thus preventing a cell from producing an excess of a particular substance. The supply of ATP in the cell regulates respiration. The allosteric enzyme that catalyzes the third step of glycolysis, phosphofructokinase, is inhibited by ATP and activated by AMP (derived from ADP). Phosphofructokinase is also inhibited by citrate released from the mitochondria into the cytosol, thus synchronizing the rates of glycolysis and the citric acid cycle. Other enzymes located at key intersections help to maintain metabolic balance.

Word Roots

aero- = air (*aerobic*: chemical reaction using oxygen)
an- = not (*anaerobic*: chemical reaction not using oxygen)
chemi- = chemical (*chemiosmosis*: the production of ATP using the energy of hydrogen ion gradients across membranes to phosphorylate ADP)
glyco- = sweet; **-lysis** = split (*glycolysis*: the splitting of glucose into pyruvate)

Structure Your Knowledge

1. This chapter describes how the catabolic pathways of glycolysis and respiration release chemical energy and store it in ATP. One of the best ways to learn the three main components of cellular respiration is to explain them to someone. Find two study partners and have each person be responsible for learning and explaining the important concepts and steps of either glycolysis, the citric acid cycle, or oxidative phosphorylation (electron transport and chemiosmosis). You may want to involve a fourth person to teach fermentation, or learn that together as a group. Use diagrams and sketches to help you explain your process to your partners.

2. Fill in the following table to summarize the major inputs and outputs of glycolysis, the citric acid cycle, oxidative phosphorylation, and fermentation. Base inputs and outputs on one glucose molecule.

3. Create a concept map to organize your understanding of oxidative phosphorylation.

Process	Main Function	Inputs	Outputs
Glycolysis			
Pyruvate to acetyl CoA			
Citric acid cycle			
Oxidative phosphorylation (electron transport and chemiosmosis)			
Fermentation			

Test Your Knowledge

MULTIPLE CHOICE: *Choose the one best answer.*

1. When electrons move closer to a more electronegative atom,
   a. energy is released.
   b. energy is consumed.
   c. a proton gradient is established.
   d. water is produced.
   e. ATP is synthesized.

2. In the reaction $C_6H_{12}O_6 + 6 O_2 \rightarrow 6 CO_2 + 6 H_2O$,
   a. oxygen becomes reduced.
   b. glucose becomes reduced.
   c. oxygen becomes oxidized.
   d. water is a reducing agent.
   e. oxygen is a reducing agent.

3. Some prokaryotes use anaerobic respiration, a process that
   a. does not involve an electron transport chain.
   b. produces ATP solely by substrate-level phosphorylation.
   c. uses a substance other than oxygen as the final electron acceptor.
   d. does not rely on chemiosmosis for the production of ATP.
   e. both a and b are correct.

4. Which of the following is *not* true of oxidative phosphorylation?
   a. It produces approximately three ATP for every NADH that is oxidized.
   b. It involves the redox reactions of the electron transport chain.
   c. It involves an ATP synthase located in the inner mitochondrial membrane.
   d. It uses oxygen as the initial electron donor.
   e. It is an example of chemiosmosis.

5. Substrate-level phosphorylation
   a. involves the shifting of a phosphate group from ATP to a substrate.
   b. can use NADH or $FADH_2$.
   c. takes place only in the cytosol.
   d. accounts for 10% of the ATP formed by fermentation.
   e. is the energy source for facultative anaerobes under anaerobic conditions.

6. The *major* reason that glycolysis is not as energy-productive as respiration is that
   a. NAD^+ is regenerated by alcohol or lactate production, without the high-energy electrons passing through the electron transport chain.
   b. it is the pathway common to fermentation and respiration.
   c. it does not take place in a specialized membrane-bound organelle.
   d. pyruvate still contains much of the energy from glucose.
   e. substrate-level phosphorylation is not as energy efficient as oxidative phosphorylation.

7. Which of the following conversions represents a reduction reaction?
   a. pyruvate $\rightarrow$ acetyl CoA + CO_2
   b. $C_6H_{12}O_6 \rightarrow 6\ CO_2$
   c. NADH + $H^+ \rightarrow NAD^+ +$ 2 H
   d. glucose $\rightarrow$ pyruvate
   e. acetaldehyde (C_2H_4O) $\rightarrow$ ethanol (C_2H_6O)

8. What is the role of oxygen in cellular respiration?
   a. It is reduced in glycolysis as glucose is oxidized.
   b. It provides electrons to the electron transport chain.
   c. It provides the activation energy needed for oxidation to occur.
   d. It is the final electron acceptor for the electron transport chain.
   e. It combines with the carbon removed during the citric acid cycle to form CO_2.

9. When pyruvate is converted to acetyl CoA,
   a. CO_2 and ATP are released.
   b. a multienzyme complex removes a carboxyl group, transfers electrons to NAD^+, and attaches a coenzyme.
   c. one turn of the citric acid cycle is completed.
   d. NAD^+ is regenerated so that glycolysis can continue to produce ATP by substrate-level phosphorylation.
   e. phosphofructokinase is activated and glycolysis continues.

10. How many molecules of CO_2 are generated for each molecule of acetyl CoA introduced into the citric acid cycle?
    a. 1   c. 3   e. 6
    b. 2   d. 4

11. In the chemiosmotic mechanism,
    a. ATP production is linked to the proton gradient established by the electron transport chain.
    b. the difference in pH between the intermembrane space and the cytosol drives the formation of ATP.
    c. the flow of H^+ through ATP synthases from the matrix to the intermembrane space drives the phosphorylation of ADP.
    d. the energy released by the reduction and subsequent oxidation of components of the electron transport chain is transferred as a phosphate to ADP.
    e. the production of water in the matrix by the reduction of oxygen leads to a net flow of water out of a mitochondrion.

12. Which of the following reactions is *incorrectly* paired with its location?
    a. ATP synthesis—inner membrane of the mitochondrion, matrix, and cytosol
    b. fermentation—cell cytosol
    c. glycolysis—cell cytosol
    d. substrate-level phosphorylation—cytosol and matrix
    e. citric acid cycle—cristae of mitochondrion

13. When glucose is oxidized to CO_2 and water, approximately 40% of its energy is transferred to
    a. heat.
    b. ATP.
    c. acetyl CoA.
    d. water.
    e. the citric acid cycle.

14. From an energetic viewpoint, what do muscle cells in oxygen deprivation gain from the reduction of pyruvate?
    a. ATP
    b. ATP and recycled NAD^+
    c. CO_2 and NAD^+
    d. ATP, alcohol, and NAD^+
    e. ATP and CO_2

15. Glucose, made from six radioactively labeled carbon atoms, is fed to yeast cells in the absence of oxygen. How many molecules of radioactive alcohol (C_2H_5OH) are formed from each molecule of glucose?
    a. 0   c. 2   e. 6
    b. 1   d. 3

16. Which of the following produces the most ATP per gram?
    a. glucose, because it is the starting place for glycolysis
    b. glycogen or starch, because they are polymers of glucose
    c. fats, because they are highly reduced compounds
    d. proteins, because of the energy stored in their tertiary structure
    e. amino acids, because they can be fed directly into the citric acid cycle

17. Fats and proteins can be used as fuel in the cell because they
    a. can be converted to glucose by enzymes.
    b. can be converted to intermediates of glycolysis or the citric acid cycle.
    c. can pass through the mitochondrial membrane to enter the citric acid cycle.
    d. contain phosphate groups.
    e. contain more energy than glucose.

18. Which is *not* true of the enzyme phosphofructokinase? It is
    a. an allosteric enzyme.
    b. inhibited by citrate.
    c. the pacemaker of glycolysis and respiration.
    d. inhibited by ADP.
    e. an early enzyme in the glycolytic pathway.

19. Substrate-level phosphorylation accounts for approximately what percentage of ATP formation when glucose is oxidized to CO_2 and water?
    a. 0%   c. 10%   e. 20%
    b. 4%   d. 15%

20. Cyanide is a poison that blocks the passage of electrons along the electron transport chain. Which of the following is a metabolic effect of this poison?
    a. The pH of the intermembrane space becomes much lower than normal.
    b. Electrons are passed directly to oxygen, causing cells to explode.
    c. Alcohol would build up in the cells.
    d. NADH supplies would be exhausted, and ATP synthesis would cease.
    e. No proton gradient would be produced, and ATP synthesis would cease.

21. Which enzyme would use NAD^+ as a coenzyme?
    a. phosphofructokinase
    b. phosphoglucoisomerase
    c. triose phosphate dehydrogenase
    d. hexokinase
    e. phosphoglyceromutase

22. The metabolic function of fermentation is to
    a. oxidize NADH to NAD^+ so that glycolysis can continue in the absence of oxygen.
    b. reduce NADH so that more ATP can be produced by the electron transport chain.
    c. produce lactate during aerobic exercise.
    d. oxidize pyruvate, thus releasing more energy.
    e. make beer.

23. Which compound will produce the most ATP when oxidized?
    a. acetyl CoA
    b. glucose
    c. pyruvate
    d. fructose bisphosphate
    e. glyceraldehyde-3-phosphate

24. Why is glycolysis considered one of the first metabolic pathways to have evolved?
    a. It relies on fermentation, which is characteristic of the archaea and bacteria.
    b. It is found only in prokaryotes, whereas eukaryotes use their mitochondria to produce ATP.
    c. It produces much less ATP than does the electron transport chain and chemiosmosis.
    d. It relies totally on enzymes that are produced by free ribosomes, and bacteria have only free ribosomes and no bound ribosomes.
    e. It is nearly universal, is located in the cytosol, and does not involve O_2.

Chapter 10

Photosynthesis

Key Concepts

10.1 Photosynthesis converts light energy to the chemical energy of food

10.2 The light reactions convert solar energy to the chemical energy of ATP and NADPH

10.3 The Calvin cycle uses ATP and NADPH to convert CO_2 to sugar

10.4 Alternative mechanisms of carbon fixation have evolved in hot, arid climates

Framework

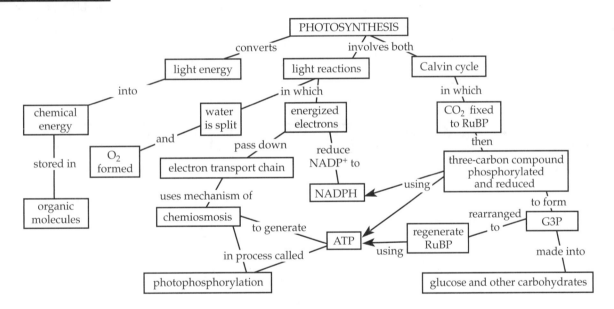

Chapter Review

In **photosynthesis,** the light energy of the sun is converted into chemical energy stored in organic molecules. **Autotrophs** "feed themselves" in the sense that they make their own organic molecules from inorganic raw materials. Plants, algae, some other protists, and some prokaryotes are photoautotrophs. Autotrophs are the *producers* of the biosphere.

Heterotrophs are *consumers.* They may eat plants or animals or decompose organic litter, but almost all are ultimately dependent on photoautotrophs for food and oxygen.

10.1 Photosynthesis converts light energy to the chemical energy of food

The first photosynthetic organisms may have been bacteria with clusters of photosynthetic enzymes and other molecules embedded in infoldings of the plasma membrane.

Chloroplasts: The Sites of Photosynthesis in Plants
Chloroplasts, found mainly in the **mesophyll** tissue of the leaf, contain **chlorophyll,** the green pigment that absorbs the light energy that drives photosynthesis. CO_2 enters and O_2 exits the leaf through **stomata.** Veins carry water from the roots to leaves and distribute sugar to nonphotosynthetic tissue.

A chloroplast consists of a double membrane surrounding a dense fluid called the **stroma** and an elaborate membrane system called **thylakoids,** enclosing the *thylakoid space.* Thylakoid sacs may be stacked to form *grana.* Chlorophyll is embedded in the thylakoid membrane.

INTERACTIVE QUESTION   10.1

Label the indicated parts in this diagram of a chloroplast.

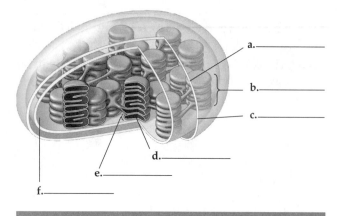

a.————

b.————

c.————

d.————

e.————

f.————

Tracking Atoms through Photosynthesis: Scientific Inquiry   If only the net consumption of water is considered and glucose is the product (even though the direct product is a three-carbon sugar), the equation for photosynthesis is the reverse of respiration:

$$6\ CO_2 + 6\ H_2O + \text{Light energy} \rightarrow C_6H_{12}O_6 + 6\ O_2$$

The simplest equation is $CO_2 + H_2O \rightarrow [CH_2O] + O_2$.

Using evidence from bacteria that utilize hydrogen sulfide (H_2S) for photosynthesis, C. B. van Niel hypothesized that all photosynthetic organisms need a hydrogen source and that plants split water as a source of electrons from hydrogen, releasing oxygen. Scientists confirmed this hypothesis by using a heavy isotope of oxygen (^{18}O). Labeled O_2 was produced in photosynthesis only when water, rather than carbon dioxide, contained the labeled oxygen.

Photosynthesis is a redox process like respiration but differs in the direction of electron flow. The electrons increase their potential energy when they travel from water to reduce CO_2 into sugar, and light provides the energy for this endergonic process.

The Two Stages of Photosynthesis: A Preview   Solar energy is converted into chemical energy in the **light reactions.** Light energy absorbed by chlorophyll drives the transfer of electrons and hydrogen ions from water

---

INTERACTIVE QUESTION   10.2

Fill in the blanks in this overview of photosynthesis in a chloroplast. Indicate the locations of the processes c and h.

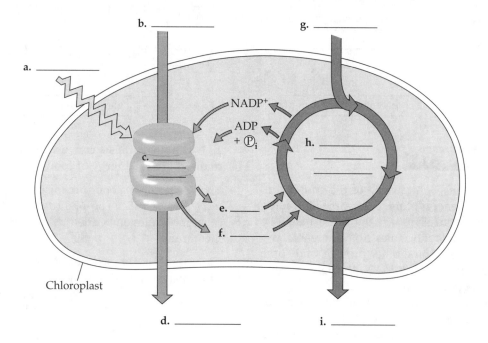

b. ————

g. ————

a. ————

NADP$^+$

ADP + P_i

c. ————

h. ————

e. ————

f. ————

Chloroplast

d. ————

i. ————

to the electron acceptor **NADP⁺**, which is reduced to NADPH and temporarily stores electrons. Oxygen is released when water is split. ATP is formed during the light reactions, using chemiosmosis in a process called **photophosphorylation.**

In the **Calvin cycle,** carbon dioxide is incorporated into existing organic compounds by **carbon fixation,** and these compounds are then reduced to form carbohydrate. NADPH and ATP from the light reactions supply the reducing power and chemical energy needed for the Calvin cycle.

10.2   The light reactions convert solar energy to the chemical energy of ATP and NADPH

The Nature of Sunlight   Electromagnetic energy, also called electromagnetic radiation, travels as rhythmic wave disturbances of electrical and magnetic fields. The distance between the crests of the electromagnetic waves, their **wavelength,** ranges across the **electromagnetic spectrum,** from short gamma waves to long radio waves. The small band of radiation from about 380 to 750 nm is called **visible light.**

Light also behaves as if it consists of discrete particles called **photons,** which have a fixed quantity of energy. The amount of energy in a photon is inversely related to its wavelength.

Photosynthetic Pigments: The Light Receptors   A **spectrophotometer** measures the amounts of light of different wavelengths absorbed by a pigment. The **absorption spectrum** of **chlorophyll *a*,** the pigment that participates directly in the light reactions, shows that it absorbs violet-blue and red light best. Accessory pigments such as **chlorophyll *b*** and some **carotenoids** absorb light of different wavelengths and broaden the spectrum of colors useful in photosynthesis. Some carotenoids function in *photoprotection* by absorbing excessive light energy that might damage chlorophyll or interact with oxygen to form reactive molecules. (These pigments act as antioxidants.)

Excitation of Chlorophyll by Light   When a pigment molecule absorbs energy from a photon, one of the molecule's electrons is elevated to an orbital where it has more potential energy. Only photons whose energy is equal to the difference between the ground state and the excited state for that molecule are absorbed.

The excited state is unstable. Energy is released as heat as the electron drops back to its ground-state orbital. Isolated chlorophyll molecules also emit photons of light, called fluorescence, as their electrons return to ground state.

An **action spectrum** shows the relative rates of photosynthesis under different wavelengths of light. On the following graph, label the line that represents the absorption spectrum for chlorophyll *a* and the line for the action spectrum for photosynthesis. Why are these lines different?

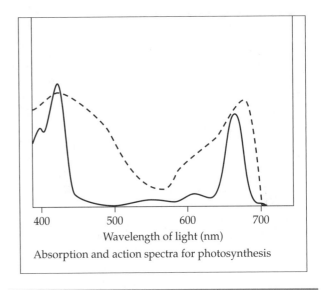

Absorption and action spectra for photosynthesis

A Photosystem: A Reaction Center Complex Associated with Light-Harvesting Complexes   **Photosystems,** located in the thylakoid membrane, contain a number of **light-harvesting complexes** and a **reaction center complex,** which is a protein complex with two special chlorophyll *a* molecules and a **primary electron acceptor.** When a pigment molecule in a light-harvesting complex absorbs a photon, the energy is passed from pigment to pigment until it reaches the reaction center. In a redox reaction, an excited electron of a reaction-center chlorophyll *a* is trapped by the primary electron acceptor before it can return to the ground state.

There are two types of photosystems in the thylakoid membrane. The chlorophyll *a* molecule at the reaction center of **photosystem II (PSII)** is called P680, after the wavelength of light (680 nm) it absorbs best. At the reaction center of **photosystem I (PSI)** is a chlorophyll *a* molecule called P700.

INTERACTIVE QUESTION   10.4

Describe the components of a photosystem.

Linear Electron Flow   Through a sequence called **linear electron flow,** electrons pass through the two photosystems from water to NADP⁺. A pigment molecule absorbs a photon of light and passes energy through a light-harvesting complex to the P680 pair of chlorophyll *a* molecules in the PS II reaction center complex. An excited electron of P680 is trapped by the primary electron acceptor. P680⁺ is a strong oxidizing agent, and its electron hole is filled when an enzyme removes electrons from water, splitting it into two electrons, two H⁺ and an oxygen atom that immediately combines with another oxygen to form O_2.

The primary electron acceptor passes the photoexcited electron to an electron transport chain made up of plastoquinone (Pq), a cytochrome complex, and plastocyanin (Pc). The energy released as electrons "fall" through the electron transport chain pumps protons to build a proton gradient used for the synthesis of ATP.

At the bottom of the electron transport chain, the electron passes to P700⁺ in photosystem I to replace the photoexcited electron captured by its primary electron acceptor. This primary electron acceptor passes the electron down a second electron transport chain through ferredoxin (Fd), from which the enzyme NADP⁺ reductase transfers electrons to reduce NADP⁺ to NADPH.

INTERACTIVE QUESTION   10.5

Fill in the steps of electron flow in the following diagram. Circle the important products that will be used to provide chemical energy and reducing power to the Calvin cycle.

Cyclic Electron Flow   In **cyclic electron flow,** electrons excited from P700 in PSI are passed from Fd to the cytochrome complex and back to P700. Several groups of photosynthetic bacteria have only photosystem I, and photosynthesis may have evolved with cyclic electron flow. Cyclic electron flow may be *photoprotective* in eukaryotic photosynthesizers.

INTERACTIVE QUESTION   10.6

a. On the diagram in Interactive Question 10.5, sketch the path that electrons from P700 take during cyclic electron flow.

b. Why is neither oxygen nor NADPH generated by cyclic electron flow?

c. How, then, is ATP produced by cyclic electron flow?

A Comparison of Chemiosmosis in Chloroplasts and Mitochondria   Chemiosmosis in mitochondria and in chloroplasts is very similar. Electron transport chains built into a membrane pump protons across the membrane as electrons are passed down the chain in a series of redox reactions. The key difference is that in respiration, organic molecules provide the electrons and chemical energy is transferred to ATP, whereas in chloroplasts water provides the electrons as light energy is transformed to the chemical energy of ATP.

In chloroplasts, the electron transport chain pumps protons from the stroma into the thylakoid space. As H^+ diffuses back through ATP synthase, ATP is formed on the stroma side, where it is available for the Calvin cycle.

INTERACTIVE QUESTION   10.7

a. In the light, the proton gradient across the thylakoid membrane is as great as 3 pH units. On which side is the pH lowest?

b. What three factors contribute to the formation of this large difference in H^+ concentration between the thylakoid space and the stroma?

10.3   The Calvin cycle uses ATP and NADPH to convert CO_2 to sugar

The Calvin cycle turns three times to fix three molecules of CO_2 and produce one molecule of the three-carbon sugar **glyceraldehyde-3-phosphate (G3P).** The cycle can be divided into three phases:

1. *Carbon fixation:* CO_2 is added to a five-carbon sugar, ribulose bisphosphate (RuBP), in a reaction catalyzed by the enzyme RuBP carboxylase **(rubisco).** The unstable six-carbon intermediate that is formed splits into two molecules of 3-phosphoglycerate.

2. *Reduction:* Each molecule of 3-phosphoglycerate is then phosphorylated by ATP to form 1,3-bisphosphoglycerate. Two electrons from NADPH reduce this compound to create G3P. The cycle must turn three times to create a net gain of one molecule of G3P.

3. *Regeneration of CO_2 acceptor (RuBP):* The rearrangement of five molecules of G3P into the three molecules of RuBP requires three more ATP.

Nine molecules of ATP and six of NADPH are required to synthesize one G3P.

INTERACTIVE QUESTION   10.8

Label the three phases (a through c) and key molecules (d through o) in this diagram of the Calvin cycle.

10.4   Alternative mechanisms of carbon fixation have evolved in hot, arid climates

Photorespiration: An Evolutionary Relic?   In most plants, CO_2 enters the Calvin cycle and the first product of carbon fixation is 3-phosphoglycerate. When these **C₃ plants** close their stomata on hot, dry days to limit water loss, CO_2 concentration in the leaf air spaces falls, slowing the Calvin cycle. As more O_2 than CO_2 accumulates, rubisco adds O_2 in place of CO_2 to RuBP. The product splits, and a two-carbon compound leaves the chloroplast and is broken down to release CO_2. This seemingly wasteful process is called **photorespiration.**

INTERACTIVE QUESTION   10.9

What are two possible explanations for photorespiration, a process that can result in the loss of as much as 50% of the carbon fixed in the Calvin cycle?

C₄ Plants   In **C₄ plants,** CO_2 is first added to a three-carbon compound, PEP, with the aid of an enzyme

(PEP carboxylase) that has a high affinity for CO_2. The resulting four-carbon compound formed in the mesophyll cells of the leaf is transported to **bundle-sheath cells** tightly packed around the veins of the leaf. The compound is broken down to release CO_2, which rubisco then fixes into the Calvin cycle.

INTERACTIVE QUESTION 10.10

a. Where does the Calvin cycle take place in C_4 plants?

b. How can C_4 plants successfully perform the Calvin cycle in hot, dry conditions when C_3 plants would be undergoing photorespiration?

c. C_4 photosynthesis requires more ATP than does C_3 photosynthesis. Why?

CAM Plants Many succulent plants close their stomata during the day to prevent water loss, but open them at night to take up CO_2 and incorporate it into a variety of organic acids. **CAM plants** break these compounds down to release CO_2 during daylight so that the Calvin cycle can proceed. Unlike the C_4 pathway, the **crassulacean acid metabolism (CAM)** pathway does not structurally separate carbon fixation from the Calvin cycle; instead, the two processes are separated in time.

The Importance of Photosynthesis: A Review About 50% of the organic material produced by photosynthesis is used as fuel for cellular respiration in the mitochondria of plant cells; the rest is used as carbon skeletons for synthesis of organic molecules (proteins, lipids, and a great deal of cellulose), stored as starch, or lost through photorespiration. About 160 billion metric tons of carbohydrate per year are produced by photosynthesis.

Word Roots

auto- = self; **-troph** = food (*autotroph*: an organism that obtains organic food molecules without eating other organisms)

chloro- = green; **-phyll** = leaf (*chlorophyll*: photosynthetic pigment in chloroplasts)

electro- = electricity; **magnet-** = magnetic (*electromagnetic spectrum*: the entire spectrum of radiation)

hetero- = other (*heterotroph*: an organism that obtains organic food molecules by eating other organisms or their by-products)

meso- = middle (*mesophyll*: the green tissue in the middle, inside of a leaf)

photo- = light (*photosystem*: cluster of pigment molecules)

Structure Your Knowledge

1. You have already filled in the blanks in several diagrams of photosynthesis. To really understand this process, however, you should create your own representation. In a diagrammatic form, outline the key events of photosynthesis. Trace the flow of electrons through photosystems II and I, the production of ATP and NADPH by the light reactions and their transfer into the Calvin cycle, and the major steps in the production of G3P. Note where these reactions occur in the chloroplast. You can compare your creation to the sketch in the answer section. Then talk a friend through the steps in your diagram. Perhaps your study group can combine several representations into a clear, concise summary of this chapter.

2. Create a concept map to confirm your understanding of the chemiosmotic synthesis of ATP in photophosphorylation.

Test Your Knowledge

MULTIPLE CHOICE: *Choose the one best answer.*

1. Which of the following is mismatched with its location?
   a. light reactions—grana
   b. electron transport chain—thylakoid membrane
   c. Calvin cycle—stroma
   d. ATP synthase—double membrane surrounding chloroplast
   e. splitting of water—thylakoid space

2. Photosynthesis is a redox process in which
   a. CO_2 is reduced and water is oxidized.
   b. $NADP^+$ is reduced and RuBP is oxidized.
   c. CO_2, $NADP^+$, and water are reduced.
   d. O_2 acts as an oxidizing agent and water acts as a reducing agent.
   e. G3P is reduced and the electron transport chain is oxidized.

3. Which of the following statements is *false*?
   a. When isolated chlorophyll molecules absorb photons, their electrons fall back to ground state, giving off heat and light.
   b. Accessory pigments, cyclic electron flow, and photorespiration may all contribute to photoprotection, protecting plants from the detrimental effects of intense light.
   c. In the cyclic electron flow of purple sulfur bacteria, the electron transport chain would pump H^+ across the plasma membrane from inside to outside the cell.
   d. In both photosynthetic prokaryotes and eukaryotes, ATP synthases catalyze the production of ATP within the cytoplasm of the cell.
   e. In sulfur bacteria, H_2S provides the hydrogen (and thus electron) source for photosynthesis.

4. A spectrophotometer can be used to measure
   a. the absorption spectrum of a substance.
   b. the action spectrum of a substance.
   c. the amount of energy in a photon.
   d. the wavelength of visible light.
   e. the efficiency of photosynthesis.

5. Accessory pigments within chloroplasts are responsible for
   a. driving the splitting of water molecules.
   b. absorbing photons of different wavelengths of light and passing that energy to P680 or P700.
   c. providing electrons to the reaction-center chlorophyll after photoexcited electrons pass to $NADP^+$.
   d. pumping H^+ across the thylakoid membrane to create a proton-motive force.
   e. anchoring chlorophyll *a* within the reaction center.

6. The following diagram is an absorption spectrum for an unknown pigment molecule. What color would this pigment appear to you?

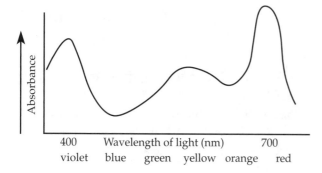

   a. violet   c. green   e. red
   b. blue   d. yellow

7. Linear electron flow along with chemiosmosis in the chloroplast results in the production of
   a. ATP only.
   b. ATP and NADPH.
   c. ATP and G3P.
   d. ATP and O_2.
   e. ATP, NADPH, and O_2.

8. The chlorophyll known as $P680^+$ has its electron "holes" filled by electrons from
   a. photosystem I.
   b. photosystem II.
   c. water.
   d. NADPH.
   e. accessory pigments.

9. CAM plants avoid photorespiration by
   a. fixing CO_2 into organic acids during the night; these acids then release CO_2 during the day.
   b. performing the Calvin cycle at night.
   c. fixing CO_2 into four-carbon compounds in the mesophyll, which release CO_2 in the bundle-sheath cells.
   d. using PEP carboxylate to fix CO_2 to ribulose bisphosphate (RuBP).
   e. keeping their stomata closed during the day.

10. What are the final electron acceptors for the electron transport chains in the light reactions of photosynthesis and in cellular respiration?
    a. O_2 in both
    b. CO_2 in both
    c. H_2O in the light reactions and O_2 in respiration
    d. P700 and NAD^+ in the light reactions and NAD^+ or FAD in respiration
    e. $NADP^+$ in the light reactions and O_2 in respiration

11. Chloroplasts can make carbohydrate in the dark if provided with
    a. ATP, NADPH, and CO_2.
    b. an artificially induced proton gradient.
    c. organic acids or four-carbon compounds.
    d. a source of hydrogen.
    e. photons and CO_2.

12. In the chemiosmotic synthesis of ATP in a chloroplast, H^+ diffuses through the ATP synthase
    a. from the stroma into the thylakoid space.
    b. from the thylakoid space into the stroma.
    c. from the intermembrane space into the matrix.
    d. from the cytoplasm into the intermembrane space.
    e. from the matrix into the stroma.

13. In C_4 plants, the Calvin cycle
    a. takes place at night.
    b. only occurs when the stomata are closed.
    c. takes place in the mesophyll cells.
    d. takes place in the bundle-sheath cells.
    e. uses PEP carboxylase instead of rubisco because of its greater affinity for CO_2.

14. How many "turns" of the Calvin cycle would it take to produce one molecule of glucose?
    a. 1   c. 3   e. 12
    b. 2   d. 6

15. In green plants, most of the ATP for synthesis of proteins, cytoplasmic streaming, and other cellular activities comes directly from
    a. photosystem I.
    b. photosystem II.
    c. the Calvin cycle.
    d. oxidative phosphorylation.
    e. photophosphorylation.

16. Six molecules of G3P formed from the fixation of $3 CO_2$ in the Calvin cycle are used to produce
    a. three molecules of glucose.
    b. three molecules of RuBP and one G3P.
    c. one molecule of glucose and four molecules of 3-phosphoglycerate.
    d. one G3P and three four-carbon intermediates.
    e. none of the above, since three molecules of G3P result from three turns of the Calvin cycle.

17. A difference between electron transport in photosynthesis and respiration is that in photosynthesis
    a. NADPH rather than NADH passes electrons to the electron transport chain.
    b. ATP synthase releases ATP into the stroma rather than into the cytosol.
    c. light provides the energy to push electrons to the top of the electron chain, rather than energy from the oxidation of food molecules.
    d. an H^+ concentration gradient rather than a proton-motive force drives the phosphorylation of ATP.
    e. both a and c are correct.

18. NADPH and ATP from the light reactions are both needed
    a. in the carbon fixation stage to provide energy and reducing power to rubisco.
    b. to regenerate three RuBP from five G3P (glyceraldehyde-3-phosphate).
    c. to combine two molecules of G3P to produce glucose.

    d. to reduce 3-phosphoglycerate to G3P.
    e. to reduce the H^+ concentration in the stroma and contribute to the proton-motive force.

19. What portion of an illuminated plant cell would you expect to have the lowest pH?
    a. nucleus
    b. cytosol
    c. chloroplast
    d. stroma of chloroplast
    e. thylakoid space

20. How does cyclic electron flow differ from linear electron flow?
    a. No NADPH is produced by cyclic electron flow.
    b. No O_2 is produced by cyclic electron flow.
    c. The cytochrome complex in the electron transport chain is not involved in cyclic electron flow.
    d. Both a and b are correct.
    e. a, b, and c are correct.

21. What does rubisco do?
    a. reduces CO_2 to G3P
    b. regenerates RuBP with the aid of ATP
    c. combines electrons and H^+ to reduce $NADP^+$ to NADPH
    d. adds CO_2 to RuBP in the carbon fixation stage
    e. transfers electrons from NADPH to 1,3-bisphosphoglycerate to produce G3P

Use the following for questions 22 through 27.

Indicate if the following events occur during
    a. respiration
    b. photosynthesis
    c. both respiration and photosynthesis
    d. neither respiration nor photosynthesis

22. Chemiosmotic synthesis of ATP
23. Reduction of oxygen
24. Reduction of CO_2
25. Reduction of NAD^+
26. Oxidation of $NADP^+$
27. Oxidative phosphorylation

Chapter 11

Cell Communication

Key Concepts

11.1 External signals are converted into responses within the cell

11.2 Reception: A signal molecule binds to a receptor protein, causing it to change shape

11.3 Transduction: Cascades of molecular interactions relay signals from receptors to target molecules in the cell

11.4 Response: Cell signaling leads to regulation of cytoplasmic activities or transcription

11.5 Apoptosis (programmed cell death) integrates multiple cell signaling pathways

Framework

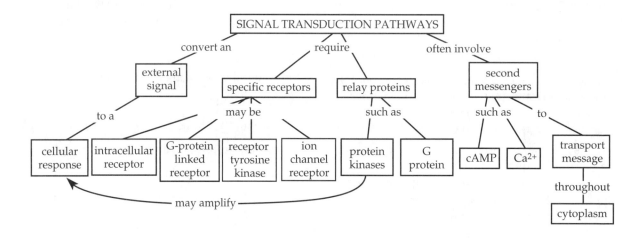

Chapter Review

Cell-to-cell communication is critical to the development and functioning of multicellular organisms and also to communication between unicellular organisms. Similar mechanisms of cellular interaction provide evidence for the evolutionary relatedness of all life.

1.1 External signals are converted into responses within the cell

Evolution of Cell Signaling The two mating types of yeast secrete chemical factors that bind to receptors on the other cell type, initiating fusion (mating) of the cells. The series of steps involved in the conversion of a signal on a cell's surface to a cellular response is called a **signal transduction pathway.** Similarities among these pathways in bacteria, yeast, plants, and animals suggest an early evolution of cell-signaling mechanisms.

The concentration of signaling molecules allows some bacteria to sense their local density, called *quorum sensing*. Aggregations of bacterial cells called *biofilms* may form in response to signaling within a population.

Local and Long-Distance Signaling   Chemical signals may be communicated between cells through direct cytoplasmic connections (gap junctions or plasmodesmata) or through contact of surface molecules (cell–cell recognition in animal cells).

In *paracrine signaling* in animals, a signaling cell releases messenger molecules into the interstitial fluid, and these **local regulators** influence nearby cells. In another type of local signaling called *synaptic signaling*, a nerve cell releases neurotransmitter molecules into the narrow synapse separating it from its target cell.

Hormones are chemical signals that travel to more distant cells. In hormonal or endocrine signaling in animals, the circulatory system transports hormones throughout the body to reach target cells with appropriate receptors.

Transmission of electrical and chemical signals through the nervous system is also a type of long-distance signaling.

INTERACTIVE QUESTION   11.1

How do plant hormones travel between secreting cells and target cells?

The Three Stages of Cell Signaling: A Preview   E. W. Sutherland's work studying epinephrine's effect on the hydrolysis of glycogen in liver cells established that cell signaling involves three stages: **reception** of a chemical signal by binding to a receptor protein either inside the cell or on its surface; **transduction** of the signal, often by a signal transduction pathway—a sequence of changes in relay molecules; and the final **response** of the cell.

11.2   Reception: A signal molecule binds to a receptor protein, causing it to change shape

A signaling molecule acts as a **ligand**, which specifically binds to a receptor protein and usually induces a change in the receptor protein's shape.

Receptors in the Plasma Membrane   There are three major types of membrane receptors that bind with water-soluble signaling molecules and transmit information into the cell.

The various receptors that work with the aid of a G protein, called **G-protein-linked receptors,** are structurally similar, with seven helices spanning the plasma membrane. Binding of the appropriate extracellular signaling molecule to a G-protein-linked receptor activates the receptor, which binds to and activates a specific **G protein** located on the cytoplasmic side of the membrane. This activation occurs when a GTP nucleotide replaces the GDP bound to the G protein. The G protein then activates a membrane-bound enzyme, after which it hydrolyzes its GTP and becomes inactive again. The activated enzyme triggers the next step in the pathway to the cell's response.

G-protein-linked receptor systems are used by many hormones and neurotransmitters and are involved in embryological development and sensory reception. Many bacteria produce toxins that interfere with G-protein systems; up to 60% of all medicines influence G-protein pathways.

INTERACTIVE QUESTION   11.2

Explain why G-protein-regulated pathways shut down rapidly in the absence of a signal molecule.

Receptor tyrosine kinases are receptor proteins with enzymatic activity that can trigger several pathways at once. Part of the receptor protein is tyrosine kinase, an enzyme that transfers phosphate groups from ATP to the amino acid tyrosine on a protein. Many receptor tyrosine receptors exist as single transmembrane polypeptides with a ligand binding site, a transmembrane α helix, and a cytoplasmic tail with a series of tyrosine amino acids. Ligand binding causes two receptors to form a dimer, which causes the tyrosine kinases to phosphorylate the tyrosines on each other's cytoplasmic tails. Different relay proteins now bind to specific phosphorylated tyrosines and become activated, triggering many different transduction pathways in response to one type of signal.

INTERACTIVE QUESTION   11.3

Label the parts in this diagram of an activated receptor tyrosine kinase dimer.

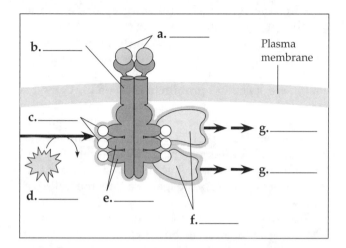

The binding of a chemical signal to a **ligand-gated ion channel** opens or closes a "gate," thus allowing or blocking the flow of specific ions through the receptor channel. The resulting change in ion concentration inside the cell triggers a cellular response. Neurotransmitters often bind to ligand-gated ion channels in the transmission of nervous signals. Some voltage-gated ion channels respond to electrical signals.

Intracellular Receptors   Hydrophobic chemical messengers may cross a cell's plasma membrane and bind to receptors in the cytoplasm or nucleus of target cells. Steroid hormones activate receptors in target cells that function as *transcription factors* to regulate gene expression.

11.3   Transduction: Cascades of molecular interactions relay signals from receptors to target molecules in the cell

Multistep pathways allow a small number of extracellular signaling molecules to be amplified to produce a large cellular response and also provide opportunity for regulation and coordination.

Signal Transduction Pathways   The relay molecules in a signal transduction pathway are usually proteins, which interact as they pass the message from the extracellular signaling molecule to the protein that produces the cellular response.

Protein Phosphorylation and Dephosphorylation   **Protein kinases** are enzymes that transfer phosphate groups from ATP to proteins, often to the amino acid serine or threonine. Relay molecules in signal transduction pathways are often protein kinases, which are sequentially phosphorylated, producing a shape change that activates each enzyme. Hundreds of different kinds of protein kinases regulate the activity of a cell's proteins.

Protein phosphatases are enzymes that remove phosphate groups from proteins. They effectively shut down signaling pathways when the initial signal is no longer present.

INTERACTIVE QUESTION   11.4

a. What does a protein kinase do?

b. What does a protein phosphatase do?

c. What is a phosphorylation cascade?

Small Molecules and Ions as Second Messengers   Small, water-soluble molecules or ions often function as **second messengers,** which rapidly relay the signal from the membrane-receptor-bound "first messenger" into a cell's interior.

Binding of an extracellular signal to a G-protein-linked receptor activates a G protein that may activate **adenylyl cyclase,** a membrane protein that converts ATP to cyclic adenosine monophosphate (**cyclic AMP** or **cAMP**). The cAMP often activates *protein kinase A,* which phosphorylates other proteins. A cytoplasmic enzyme, phosphodiesterase, converts cAMP to inactive AMP, thus removing the second messenger.

Some signal molecules may activate an inhibitory G protein that inhibits adenylyl cyclase.

INTERACTIVE QUESTION   11.5

Label the components in this diagram of the steps in a signal transduction pathway that uses cAMP as a second messenger.

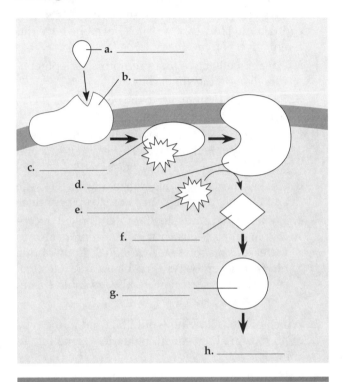

a. _____

b. _____

c. _____

d. _____

e. _____

f. _____

g. _____

h. _____

Calcium ions are widely used as a second messenger in many G-protein-linked and receptor tyrosine kinase pathways. Cytosolic concentration of Ca^{2+} is kept low by active transport of calcium out of the cell and into the endoplasmic reticulum. The release of Ca^{2+} from the ER involves another second messenger, **inositol trisphosphate (IP_3).** In response to the reception of a signal, the enzyme phospholipase C cleaves a membrane phospholipid into two second messengers, IP_3 and **diacylglycerol (DAG).** IP_3 binds to and opens IP_3-gated calcium channels in the ER. The calcium ions then activate cellular responses.

INTERACTIVE QUESTION   11.6

Fill in the blanks to review a G-protein-linked pathway that uses Ca^{2+} as a second messenger.

A **a.** _____ binds to a G-protein-linked receptor. An activated **b.** _____ activates the enzyme phospholipase C, which cleaves a **c.** _____ into DAG and **d.** _____ , which binds to and opens a ligand-gated channel, releasing **e.** _____ from the **f.** _____ .

11.4   Response: Cell signaling leads to regulation of cytoplasmic activities or transcription

Nuclear and Cytoplasmic Responses   Signal transduction pathways may lead to the activation of transcription factors, which regulate the expression of specific genes. Signaling pathways may also activate existing cytoplasmic enzymes, open or close protein channels in membranes, or rearrange the cytoskeleton.

Fine-Tuning of the Response   A signal transduction pathway can amplify a signal in an enzyme cascade, as each successive enzyme in the pathway can process multiple molecules that then activate the next step.

As a result of their particular set of receptor proteins, relay proteins, and response proteins, different cells can respond to different signals or can exhibit different responses to the same molecular signal. Pathways may branch to produce multiple responses, or two pathways may interact ("cross-talk") to mediate a single response.

Scaffolding proteins are large relay proteins to which other relay proteins attach, increasing the efficiency of signal transduction in a pathway.

INTERACTIVE QUESTION   11.7

How do the following mechanisms or molecules discontinue a cell's response to a signal and maintain a cell's ability to respond to fresh signals?

a. reversible binding of signaling molecules

b. GTPase activity of G protein

c. phosphodiesterase

d. protein phosphatases

11.5   Apoptosis (programmed cell death) integrates multiple cell signaling pathways

In the process of controlled cell death called **apoptosis**, cellular components are chopped up and packaged into vesicles that are released as "blebs" and then engulfed by scavenger cells.

Apoptosis in the Soil Worm Caenorhabditis elegans   Apoptosis occurs numerous times during normal development in *C. elegans*. The *ced-9* gene produces a protein that inhibits the activity of Ced-3 and Ced-4, protein products of genes *ced-3* and *ced-4*. When a cell receives a death signal on its membrane receptor,

Ced-9 protein becomes inactivated, and the apoptosis pathway activates proteases and nucleases that hydrolyze the cell's proteins and DNA. Ced-3 is the main *caspase* (protease) of apoptosis in the nematode.

Apoptotic Pathways and the Signals That Trigger Them   One type of apoptotic pathway in mammals involves proteins such as cytochrome c that are released through pores formed in mitochondrial membranes. In other cases, a death-signaling ligand bound to a cell-surface receptor activates caspases that carry out apoptosis. Other signals can come from the nucleus when its DNA has suffered irreparable damage or from the endoplasmic reticulum in response to protein misfolding. Programmed apoptosis is part of normal development. Faulty cell suicide programs have been implicated in some neurological diseases and cancer.

INTERACTIVE QUESTION   11.8

a. What can one conclude from the fact that the mitochondrial apoptosis proteins in mammals are homologous to the Ced-3, Ced-4, and Ced-9 proteins of nematodes?

b. Give some examples of programmed cell death in humans.

Word Roots

liga- = bound or tied (*ligand:* a small molecule that specifically binds to a larger one)

trans- = across (*signal transduction pathway:* the process by which a signal on a cell's surface is converted into a specific cellular response inside the cell)

-yl = substance or matter (*adenylyl cyclase:* an enzyme built into the plasma membrane that converts ATP to cAMP)

Structure Your Knowledge

1. Why is cell signaling such an important component of a cell's life?

2. Briefly describe the three stages of cell signaling.

3. Some signaling pathways alter a protein's activity; others may result in the production of new proteins. Explain the mechanisms for these two different responses.

4. How does an enzyme cascade produce an amplified response to a signal molecule?

Test Your Knowledge

MULTIPLE CHOICE: *Choose the one best answer.*

1. When epinephrine binds to cardiac (heart) muscle cells, it speeds their contraction. When it binds to muscle cells of the small intestine, it inhibits their contraction. How can the same hormone have different effects on muscle cells?
   a. Cardiac cells have more receptors for epinephrine than do intestinal cells.
   b. Epinephrine circulates to the heart first and is in higher concentration around cardiac cells.
   c. The two types of muscle cells have different signal transduction pathways for epinephrine and thus have different cellular responses.
   d. Cardiac muscle is stronger than intestinal muscle and thus has a stronger response to epinephrine.
   e. Epinephrine binds to G-protein-linked receptors in cardiac cells, and these receptors always increase a response to the signal. Epinephrine binds to receptor tyrosine kinases in intestinal cells, and these receptors always inhibit a response to the signal.

2. Which of the following would be used in the type of local signaling called paracrine signaling in animals?
   a. the neurotransmitter acetylcholine
   b. the hormone epinephrine
   c. the neurotransmitter norepinephrine
   d. a local regulator such as a growth factor
   e. both a and c

3. A signaling molecule that binds to a plasma-membrane protein functions as a
   a. ligand.
   b. second messenger.
   c. protein phosphatase.
   d. protein kinase.
   e. receptor protein.

4. What is a G protein?
   a. a specific type of membrane-receptor protein
   b. a protein on the cytoplasmic side of a membrane that becomes activated by a receptor protein
   c. a membrane-bound enzyme that converts ATP to cAMP
   d. a tyrosine kinase relay protein
   e. a guanine nucleotide that converts between GDP and GTP to activate and inactivate relay proteins

5. How do receptor tyrosine kinases transduce a signal?

   a. They transport the signaling molecule into the cell, where it binds to and activates a transcription factor. The transcription factor then alters gene expression.

   b. Signaling molecule binding causes a shape change that activates membrane-bound tyrosine kinase relay proteins that phosphorylate serine and threonine amino acids.

   c. Their activated tyrosine kinases convert ATP to cAMP; cAMP acts as a second messenger to activate other protein kinases.

   d. When activated, they cleave a membrane phospholipid into two second-messenger molecules. One of the molecules opens Ca^{2+} ion channels on the endoplasmic reticulum.

   e. They form a dimer; they phosphorylate each other's tyrosines; specific proteins bind to and are activated by these phosphorylated tyrosines.

6. Which of the following can activate a protein by transferring a phosphate group to it?

   a. G protein

   b. phosphodiesterase

   c. protein phosphatase

   d. protein kinase

   e. both a and c

7. Many signal transduction pathways use second messengers to

   a. transport a signaling molecule through the lipid bilayer portion of the plasma membrane.

   b. relay a signal from the outside to the inside of the cell.

   c. relay the message from the inside of the membrane throughout the cytoplasm.

   d. amplify the message by phosphorylating cascades of proteins.

   e. dampen the message once the signaling molecule has left the receptor.

8. What is a function of the second messenger IP_3?

   a. bind to and activate protein kinase A

   b. activate transcription factors

   c. activate other membrane-bound relay molecules

   d. convert ATP to cAMP

   e. bind to and open ligand-gated calcium channels on the ER

9. Many human diseases, including bacterial infections, and also many medicines used to treat these diseases produce their effects by influencing which of the following?

   a. cAMP concentrations in the cell

   b. Ca^{2+} concentrations in the cell

   c. G-protein-linked receptor pathways

   d. gene expression

   e. receptor tyrosine kinases

10. Signal amplification is most often achieved by

   a. an enzyme cascade involving multiple protein kinases.

   b. the binding of multiple signaling molecules.

   c. branching pathways that produce multiple cellular responses.

   d. activating transcription factors that affect gene expression.

   e. the action of adenylyl cyclase in converting ATP to ADP.

11. From studying the effects of epinephrine on liver cells, Sutherland concluded that

   a. there is a one-to-ten correlation between the number of epinephrine molecules bound to receptors and the number of glucose molecules released from glycogen.

   b. epinephrine enters liver cells and binds to receptors that function as transcription factors to turn on the gene for glycogen phosphorylase.

   c. there is a "second messenger" that transmits the signal of epinephrine binding on the plasma membrane to the enzymes involved in glycogen breakdown inside the cell.

   d. the signal transduction pathway through which epinephrine signals glycogen breakdown involves receptor tyrosine kinases and Ca^{2+} that activate glycogen phosphorylase.

   e. epinephrine functions as a ligand to open ion channels in the plasma membrane that allow Ca^{2+} to enter and initiate a response.

12. Which of the following is a similarity between G protein-linked receptors and receptor tyrosine kinases?

   a. signaling molecule-binding sites specific for steroid hormones

   b. formation of a dimer following binding of a signaling molecule

   c. activation that results from binding of GTP

   d. phosphorylation of specific amino acids in direct response to ligand binding

   e. α-helix regions of the receptor that span the plasma membrane

13. Which of the following is *incorrectly* matched with its description?

    a. scaffolding protein—large relay protein that may bind with several other relay proteins to increase the efficiency of a signaling pathway

    b. protein phosphatase—enzyme that transfers a phosphate group from ATP to a protein, causing a shape change that usually activates that protein

    c. adenylyl cyclase—enzyme attached to plasma membrane that converts ATP to cAMP in response to an extracellular signal

    d. phospholipase C—enzyme that may be activated by a G protein or receptor tyrosine kinase and cleaves a plasma-membrane phospholipid into the second messengers IP_3 and DAG

    e. G protein—relay protein attached to the inside of plasma membrane that, when activated by an activated G-protein-linked receptor, binds GTP and then usually activates another membrane-attached protein

14. The gene *ced-9* codes for a protein that inactivates the proteins of suicide genes found in *C. elegans*. For development to proceed normally, *ced-9*

    a. should be activated in all cells, but its protein product must remain inactive.

    b. should be activated in all cells, but its product will be inactivated when cells programmed to die receive the proper signal.

    c. should be inactivated in all cells except when a cell receives a signal to die.

    d. should be inactivated only in those cells that must die for proper development to occur.

    e. should code for a transcription factor that attaches to the enhancers of other suicide genes.

Chapter 12

The Cell Cycle

Key Concepts

12.1 Cell division results in genetically identical daughter cells

12.2 The mitotic phase alternates with interphase in the cell cycle

12.3 The eukaryotic cell cycle is regulated by a molecular control system

Framework

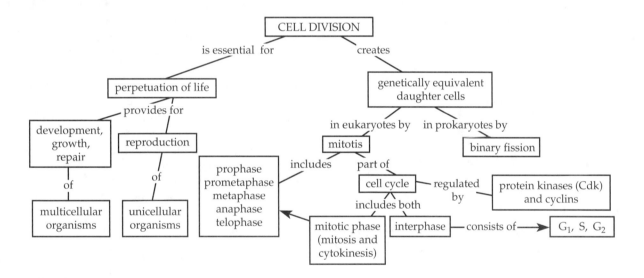

Chapter Review

Cell division creates duplicate offspring in unicellular organisms and provides for growth, development, and repair in multicellular organisms. The **cell cycle** extends from the creation of a new cell by the division of its parent cell to its own division into two cells.

12.1 Cell division results in genetically identical daughter cells

The process of recreating a structure as intricate as a cell necessitates the exact duplication and equal division of the DNA containing the cell's genetic program.

Cellular Organization of the Genetic Material A cell's complete complement of DNA is called its **genome.** Each diploid eukaryotic species has a characteristic number of **chromosomes** in each **somatic cell;** reproductive cells, or **gametes** (egg and sperm), have half that number of chromosomes.

Each chromosome is a very long DNA molecule with associated proteins that help to structure the chromosome and control the activity of the genes. This DNA–protein complex is called **chromatin.**

Distribution of Chromosomes During Eukaryotic Cell Division   Prior to cell division, a cell copies its DNA and each chromosome densely coils and shortens. Duplicated chromosomes consist of two identical **sister chromatids,** initially attached by proteins called *cohesins,* which hold them together in *sister chromatid cohesion.* A duplicated chromosome has a **centromere** where the chromatids are most closely attached. The two sister chromatids separate during **mitosis** (the division of the nucleus), and then the cytoplasm divides during **cytokinesis,** producing two separate, genetically equivalent daughter cells.

A type of cell division called **meiosis** produces daughter cells that have half the number of chromosomes (one set) of the parent cell. With the fertilization of egg and sperm, which were formed (in animals) by meiosis, the chromosome number is restored as the somatic cells of the new offspring again have two sets of chromosomes.

INTERACTIVE QUESTION   12.1

a. How many chromosomes do you have in your somatic cells?

b. How many chromosomes do you have in your gametes?

c. How many chromatids are in one of your body cells that has duplicated its chromosomes prior to mitosis?

12.2   The mitotic phase alternates with interphase in the cell cycle

Phases of the Cell Cycle   The cell cycle consists of the **mitotic (M) phase,** which includes mitosis and cytokinesis, and **interphase,** during which the cell grows and duplicates its chromosomes. Interphase, usually lasting 90% of the cell cycle, includes the **G_1 phase,** the **S phase,** and the **G_2 phase.** Mitosis is conventionally described in five subphases: **prophase, prometaphase, metaphase, anaphase,** and **telophase.**

INTERACTIVE QUESTION   12.2

a. How are the three subphases of interphase alike?

b. What key event happens during the S phase?

The Mitotic Spindle: A Closer Look   The **mitotic spindle** consists of fibers made of microtubules and associated proteins. The assembly of the spindle begins in the **centrosome,** or *microtubule-organizing center.* A pair of centrioles is centered in each centrosome of an animal cell, but centrioles are not required for normal spindle operation. The single centrosome replicates during interphase. As spindle microtubules grow out from them, the two centrosomes move to opposite poles of the cell. Radial arrays of shorter microtubules, called **asters,** extend from the centrosomes.

During prophase, the nucleoli disappear and the chromatin fibers coil and fold into visible chromosomes, consisting of sister chromatids joined by sister chromatid cohesion. During prometaphase, some of the spindle microtubules attach to each chromatid's **kinetochore,** a structure of protein associated with DNA located at the centromere region. Alternate tugging on the chromosome by opposite kinetochore microtubules moves the chromosome to the midline of the cell. At metaphase, the chromosomes are aligned at the **metaphase plate,** across the midline of the spindle. Nonkinetochore microtubules (or "polar" microtubules) extend out from each centrosome and overlap at the midline. Aster microtubules contact the plasma membrane.

The cohesins joining sister chromatids are cleaved in anaphase, and the now separate chromosomes move toward the poles. Motor proteins appear to "walk" a chromosome along the kinetochore microtubules as these shorten by depolymerizing at their kinetochore end. Motor proteins at the spindle poles also appear to "reel in" the chromosomes, with the microtubules depolymerizing at the poles. The extension of the spindle poles away from each other as an animal cell elongates is probably due to the overlapping nonkinetochore microtubules walking past each other, also using motor proteins.

During telophase, equivalent sets of chromosomes gather at the two poles of the cell. Nuclear envelopes form, nucleoli reappear, cytokinesis begins, and the spindle disassembles.

Cytokinesis: A Closer Look   **Cleavage** is the process that separates the two daughter cells in animals. A **cleavage furrow** forms, as a ring of actin microfilaments interacting with myosin proteins begins to contract on the cytoplasmic side of the membrane. The cleavage furrow deepens until the dividing cell is pinched in two.

In plant cells, a **cell plate** forms from the fusion of membrane vesicles derived from the Golgi apparatus. The membrane of the enlarging cell plate joins with the plasma membrane, separating the two daughter cells. A new cell wall develops between the cells from the contents of the cell plate.

INTERACTIVE QUESTION   12.3

These diagrams will depict interphase and the five subphases of mitosis in an animal cell, after you draw in the missing chromosomes. For simplicity, assume that this cell has four chromosomes. Sketch the chromosomes as they would appear in each subphase. Identify the stages and label the indicated structures.

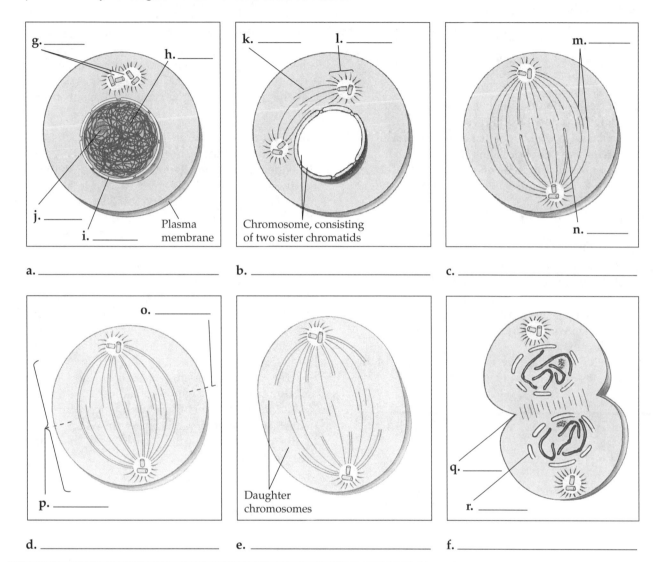

g. _____

h. _____

j. _____

i. _____

Plasma membrane

a. _____

k. _____   l. _____

Chromosome, consisting of two sister chromatids

b. _____

m. _____

n. _____

c. _____

o. _____

p. _____

d. _____

Daughter chromosomes

e. _____

q. _____

r. _____

f. _____

Binary Fission   Single-celled eukaryotes reproduce asexually by a process known as **binary fission,** which includes mitosis. The binary fission of prokaryotes does not include mitosis. The *bacterial chromosome,* a single circular DNA molecule, begins to replicate at the **origin of replication** and one of these duplicated origins moves to the opposite pole of the cell. Replication is completed as the cell doubles in size, and the plasma membrane grows inward to divide the two identical daughter cells. The mechanism of chromosome movement is not fully understood.

The Evolution of Mitosis   Evidence for the evolution of mitosis from prokaryotic cell division includes the relatedness of several proteins involved in both types of division and the possible intermediate stages seen in some unicellular algae, in which chromosomal division takes place within an intact nuclear envelope.

12.3   The eukaryotic cell cycle is regulated by a molecular control system

Normal growth, development, and maintenance depend on proper control of the timing and rate of cell division.

Evidence for Cytoplasmic Signals   Experiments that fuse two cells at different phases of the cell cycle indicate that cytoplasmic chemical signals drive the cell cycle.

The Cell Cycle Control System   A **cell cycle control system,** consisting of a set of molecules that function cyclically, coordinates the events of the cell cycle.

Important internal and external signals are monitored to determine whether the cell cycle will proceed past the three main **checkpoints** in the G_1, G_2, and M phases. If a mammalian cell does not receive a go-ahead signal at the G_1 checkpoint, called the "restriction point," the cell will usually exit the cell cycle to a nondividing state called the **G_0 phase.**

Protein kinases are enzymes that activate or inactivate other proteins by phosphorylating them. The changing concentrations of **cyclins,** regulatory proteins that attach to these kinases, affect the activity of **cyclin-dependent kinases,** or **Cdks.**

A cyclin–Cdk complex called **MPF,** for maturation or M-phase-promoting factor, triggers passage past the G_2 checkpoint into M phase. In addition to phosphorylating proteins and other kinases that initiate mitotic events, MPF activates a protein breakdown process that destroys its cyclin and thus MPF activity during anaphase. The Cdk portion of the complex remains to associate with new cyclin synthesized during S and G_2 phases of the next cycle.

Other Cdk proteins and cyclins appear to control the movement of a cell past the G_1 checkpoint.

INTERACTIVE QUESTION   12.4

a. What is MPF?

b. Describe the relative concentrations of MPF and its constituent molecules throughout the cell cycle:

MPF

Cdk

cyclin

An internal signal is required to move past the M phase checkpoint into anaphase. The cohesins holding sister chromatids together are not cleaved until all chromosomes are attached at their kinetochores to spindle microtubules.

Growing cells in cell culture has allowed researchers to identify chemical and physical factors that affect cell division. Certain nutrients and regulatory proteins

called **growth factors** have been found to be essential for cells to divide in culture. Mammalian fibroblast cells have receptors on their plasma membranes for *platelet-derived growth factor (PDGF),* which is released from blood platelets at the site of an injury. Binding of PDGF initiates a signal-transduction pathway that enables cells to pass the G_1 checkpoint.

Density-dependent inhibition of cell division appears to involve the binding of cell-surface proteins of adjacent cells, which sends a growth-inhibiting signal to both cells. Most animal cells also show **anchorage dependence** and must be attached to a substratum in order to divide.

Loss of Cell Cycle Controls in Cancer Cells   When grown in cell culture, cancer cells do not exhibit density-dependent inhibition, do not depend on growth factors to divide, and may continue to divide indefinitely instead of stopping after the typical 20 to 50 divisions of normal mammalian cells.

When a normal cell is **transformed** or converted to a cancer cell, the body's immune system may destroy it. If it proliferates, a mass of abnormal cells develops within a tissue. **Benign tumors** remain at their original site and can be removed by surgery. **Malignant tumors** cause cancer as they invade and disrupt the functions of one or more organs. Malignant tumor cells may have abnormal metabolism and unusual numbers of chromosomes. They lose their attachments to other cells and may **metastasize,** entering the blood and lymph systems and spreading to other sites. Radiation and chemicals are used to treat tumors that metastasize. Much remains to be learned about control of the cell division processes of both normal and cancerous cells.

Word Roots

ana- = up, throughout, again (*anaphase:* the mitotic stage in which the chromatids of each chromosome have separated and the daughter chromosomes are moving to the poles of the cell)

bi- = two (*binary fission:* a type of cell division in which a cell divides in half)

centro- = the center; **-mere** = a part (*centromere:* the narrow "waist" of a condensed chromosome)

chroma- = colored (*chromatin:* DNA and the various associated proteins that form eukaryotic chromosomes)

cyclo- = a circle (*cyclin:* a regulatory protein whose concentration fluctuates cyclically)

cyto- = cell; **-kinet** = move (*cytokinesis:* division of the cytoplasm)

gamet- = a wife or husband (*gamete:* a haploid egg or sperm cell)

gen- = produce (*genome:* a cell's endowment of DNA)

inter- = between (*interphase:* time when a cell metabolizes and performs its various functions)

mal- = bad or evil (*malignant tumor:* a cancerous tumor that is invasive enough to impair functions of one or more organs)

meio- = less (*meiosis:* a variation of cell division that yields daughter cells with half as many chromosomes as the parent cell)

meta- = between (*metaphase:* the mitotic stage in which the chromosomes are aligned in the middle of the cell, at the metaphase plate)

mito- = a thread (*mitosis:* the division of the nucleus)

pro- = before (*prophase:* the first mitotic stage in which the chromatin is condensing)

soma- = body (*centrosome:* a nonmembranous organelle that functions throughout the cell cycle to organize the cell's microtubules)

telos- = an end (*telophase:* the final stage of mitosis in which daughter nuclei are forming and cytokinesis has typically begun)

trans- = across; **-form** shape (*transformation:* the process that converts a normal cell into a cancer cell)

Structure Your Knowledge

1. Describe the life of one chromosome as it proceeds through an entire cell cycle, starting with G_1 of interphase and ending with telophase of mitosis.

2. Draw a sketch of one half of a mitotic spindle. Identify and list the functions of the components.

3. In this photomicrograph of cells in an onion root tip, identify the cell cycle phases for the indicated cells.

a. _____

b. _____

c. _____

d. _____

Test Your knowledge

FILL IN THE BLANK: *Identify the appropriate phase of the cell cycle.*

_____ 1. most cells that will no longer divide are in this phase

_____ 2. sister chromatids separate and chromosomes move apart

_____ 3. mitotic spindle begins to form

_____ 4. cell plate forms or cleavage furrow pinches cells apart

_____ 5. chromosomes duplicate

_____ 6. chromosomes line up at equatorial plane

_____ 7. nuclear membranes form around separated chromosomes

_____ 8. chromosomes become visible

_____ 9. kinetochore–microtubule interactions move chromosomes to midline

_____ 10. restriction point occurs in this phase

MULTIPLE CHOICE: *Choose the one best answer.*

1. One of the major differences in the binary fission of prokaryotic cells compared to mitosis of eukaryotic cells is that

   a. cytokinesis does not occur in prokaryotic cells.

   b. eukaryotic cell division involves microtubules composed of tubulin and actin microfilaments, whereas no such molecules are involved in prokaryotic binary fission.

   c. the duplicated chromosomes are attached to the plasma membrane in prokaryotic cells and are separated from each other as the membrane splits apart.

   d. the duplicated single chromosome does not separate along a mitotic spindle in prokaryotic cells.

   e. the chromosome number is reduced by half in eukaryotic cells but not prokaryotic cells.

2. A plant cell has 12 chromosomes at the end of mitosis. How many *chromosomes* would it have in the G_2 phase of its next cell cycle?

   a. 6

   b. 9

   c. 12

   d. 24

   e. It depends on whether it is undergoing mitosis or meiosis.

3. How many *chromatids* would this plant cell have in the G_2 phase of its cell cycle?

   a. 6   **d.** 24

   b. 9   **e.** 48

   c. 12

4. The longest part of the cell cycle is

   a. prophase.   **d.** mitosis.

   b. G_1 phase.   **e.** interphase.

   c. G_2 phase.

5. In animal cells, cytokinesis involves

   a. the separation of sister chromatids.

   b. the contraction of a ring of actin microfilaments.

   c. depolymerization of kinetochore microtubules.

   d. a protein kinase that phosphorylates other enzymes.

   e. sliding of nonkinetochore microtubules past each other.

6. Humans have 46 chromosomes. That number of chromosomes will be found in

   a. cells in anaphase.

   b. the egg and sperm cells.

   c. the somatic cells.

   d. all the cells of the body.

   e. only cells in G_1 of interphase.

7. Sister chromatids

   a. have one-half the amount of genetic material as does the original chromosome.

   b. start to move along kinetochore microtubules toward opposite poles during telophase.

   c. each have their own kinetochore.

   d. are formed during S phase but do not join by sister chromatid cohesion until prophase.

   e. slide past each other as nonkinetochore microtubules elongate.

8. Which of the following would *not* be exhibited by cancer cells?

   a. changing levels of MPF concentration

   b. passage through the restriction point

   c. density-dependent inhibition

   d. metastasis

   e. G_1 phase of the cell cycle

9. Which of the following is *not* true of a cell plate?

   a. It forms at the site of the metaphase plate.

   b. It results from the fusion of microtubules.

   c. It fuses with the plasma membrane.

   d. A cell wall is laid down between its membranes.

   e. It forms during telophase in plant cells.

10. A cell that passes the restriction point in G_1 will most likely

   a. undergo chromosome duplication.

   b. have just completed cytokinesis.

   c. continue to divide only if it is a cancer cell.

   d. show a drop in MPF concentration.

   e. move into the G_0 phase.

11. The rhythmic changes in cyclin concentration in a cell cycle are due to

   a. its increased production once the restriction point is passed.

   b. the cascade of increased production once its enzyme is phosphorylated by MPF.

   c. its degradation, which is initiated by active MPF.

   d. the correlation of its production with the production of Cdk.

   e. the binding of the growth factor PDGF.

12. A cell in which of the following phases would have the *least* amount of DNA?

   a. G_0 **d.** metaphase

   b. G_2 **e.** anaphase

   c. prophase

13. What initiates the separation of sister chromatids in anaphase?

   a. the drop in MPF concentration

   b. a rapid rise in Cdk concentration

   c. movement past the G_2 checkpoint

   d. a signal pathway initiated by the binding of a growth factor

   e. activation of a regulatory protein following the attachment of all kinetochores to microtubules

14. Knowledge of the cell cycle control system will be most beneficial to the area of

   a. human reproduction.

   b. plant genetics.

   c. prokaryotic growth and development.

   d. cancer prevention and treatment.

   e. prevention and treatment of cardiovascular disease.

UNIT 3 Genetics

Chapter 13

Meiosis and Sexual Life Cycles

Key Concepts

13.1 Offspring acquire genes from parents by inheriting chromosomes

13.2 Fertilization and meiosis alternate in sexual life cycles

13.3 Meiosis reduces the number of chromosome sets from diploid to haploid

13.4 Genetic variation produced in sexual cycles contributes to evolution

Framework

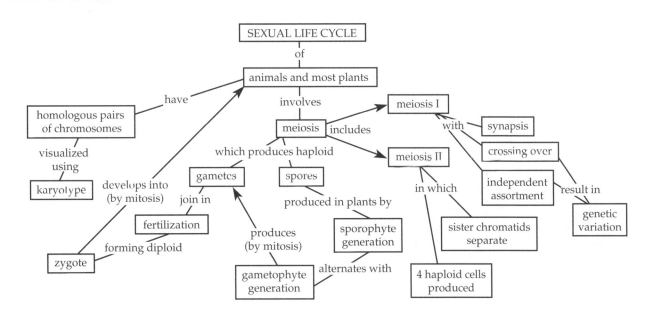

Chapter Review

Genetics is the scientific study of the transmission of traits from parents to offspring **(heredity)** and the **variation** between and within generations.

13.1   Offspring acquire genes from parents by inheriting chromosomes

Inheritance of Genes   The inheritance of traits from parents to offspring involves the transmission of discrete units of information coded in segments of DNA known as **genes.** Most genes contain instructions for synthesizing enzymes and other proteins that then guide the development of inherited traits.

Precise copies of an organism's genes are packaged into **gametes** (sperm and eggs). Upon fertilization, genes from both parents are passed on to offspring. The DNA of a eukaryotic cell is packaged along with various proteins into a species-specific number of chromosomes. The entire complement of DNA is called the genome. A gene's **locus** is its location on a chromosome.

Comparison of Asexual and Sexual Reproduction   In **asexual reproduction,** a single parent passes copies of all its genes to its offspring. A **clone** is a group of genetically identical offspring. In **sexual reproduction,** an individual receives a unique combination of genes inherited from two parents.

13.2   Fertilization and meiosis alternate in sexual life cycles

An organism's **life cycle** is the sequence of stages from conception to production of its own offspring.

Sets of Chromosomes in Human Cells   In **somatic cells,** there are two chromosomes of each type, known as **homologous chromosomes** or homologs. A gene controlling a particular character is found at the same locus on each chromosome of a homologous pair.

A **karyotype** is an ordered display of an individual's condensed chromosomes. Isolated somatic cells are stimulated to undergo mitosis, arrested in metaphase, and stained. A computer uses a digital photograph to arrange chromosomes into homologous pairs by size and shape.

Sex chromosomes determine the sex of a person: Females have two homologous X chromosomes; males have nonhomologous X and Y chromosomes. Chromosomes other than the sex chromosomes are called **autosomes.**

Somatic cells contain a set of chromosomes from each parent. These are **diploid cells,** each with a diploid number of chromosomes, abbreviated $2n$.

Gametes, egg and sperm, are **haploid cells** and contain a single set of chromosomes. The haploid number (n) of chromosomes for humans is 23.

INTERACTIVE QUESTION   13.1

a. If $2n = 14$, how many chromosomes will be present in the somatic cells of an animal? _____
How many chromosomes will be found in gametes? _____

b. If $n = 14$, how many chromosomes will be found in diploid somatic cells? _____
How many sets of homologous chromosomes will be found in gametes? _____

c. If $2n = 28$, how many chromatids will be found in a cell in which DNA synthesis has occurred prior to cell division? _____ What is the difference between sister and nonsister chromatids?

Behavior of Chromosome Sets in the Human Life Cycle   **Fertilization,** or fusion of sperm and ovum (egg), produces a **zygote** containing both paternal and maternal sets of chromosomes. The diploid zygote then divides by mitosis to produce the somatic cells of the body, all of which contain the diploid number ($2n$) of chromosomes.

Meiosis is a special type of cell division that halves the chromosome number and provides a haploid set of chromosomes to each gamete. Gametes are produced by meiosis from specialized *germ cells* in the gonads. An alternation between diploid and haploid numbers of chromosomes, involving the processes of fertilization and meiosis, is characteristic of sexually reproducing organisms.

The Variety of Sexual Life Cycles   In most animals, meiosis occurs in the formation of gametes, which are the only haploid cells in the life cycle.

Plants and some species of algae have a type of life cycle called **alternation of generations** that includes both diploid and haploid multicellular stages. The multicellular diploid **sporophyte** produces haploid **spores** by meiosis. These spores undergo mitosis and develop into a multicellular haploid organism, the **gametophyte,** which produces gametes by mitosis. Gametes fuse to form a diploid zygote, which develops into the next sporophyte generation.

In most fungi and some protists, the only diploid stage is the zygote. Meiosis occurs after the gametes fuse, producing haploid cells that divide by mitosis to create a unicellular or multicellular haploid organism. Gametes are produced by mitosis in these organisms.

INTERACTIVE QUESTION 13.2

Complete these three diagrams of sexual life cycles by filling in the type of cell division, the type of cells, and the ploidy (*n* or 2*n*).

Animals

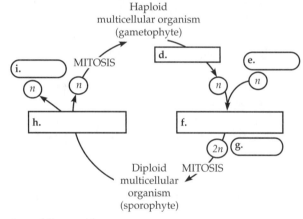

Plants and Some Algae

Most Fungi and Some Protists

13.3 Meiosis reduces the number of chromosome sets from diploid to haploid

In meiosis, chromosome replication is followed by two consecutive cell divisions: **meiosis I** and **meiosis II**, producing four haploid daughter cells, each with one set of chromosomes.

The Stages of Meiosis In **interphase,** each chromosome replicates, producing two genetically identical sister chromatids that remain attached at the centromere and along their length (called *sister chromatid cohesion*). During prophase I, homologous chromosomes synapse. Crossing over may occur between nonsister chromatids, forming chiasmata.

In metaphase I, homologous pairs line up on the metaphase plate with their kinetochores attached to spindle fibers from opposite poles. Each pair separates in anaphase I, with one homolog moving toward each pole. In telophase I, a haploid set of chromosomes, each composed of two sister chromatids, reaches each pole. Cytokinesis usually occurs during telophase I. There is no replication of genetic material prior to the second division of meiosis.

Meiosis II looks like a regular mitotic division, in which chromosomes line up individually on the metaphase plate, and sister chromatids separate and move apart in anaphase II. These sister chromatids, however, are not genetically identical due to crossing over in prophase I. At the end of telophase II, there are four haploid daughter cells.

A Comparison of Mitosis and Meiosis Mitosis produces daughter cells that are genetically identical to the parent cell. Meiosis produces haploid cells that differ genetically from their parent cell and from each other.

The three unique events that produce this result occur during meiosis I: In prophase I, when homologous chromosomes are held together along their lengths by the *synaptonemal complex* (**synapsis**), genetic material is rearranged by **crossing over** between nonsister chromatids. Crossovers are visible later in this stage as X-shaped regions called **chiasmata** where sister chromatid cohesion holds the original sister chromatids together. In metaphase I, chromosomes line up in pairs, not as individuals, on the metaphase plate. During anaphase I, the homologous pairs separate and one homolog (with sister chromatids still attached at the centromere) goes to each pole.

Sister chromatids are held together by protein complexes called *cohesins*. Enzymes cleave the cohesins along the arms of sister chromatids so that homologs can separate in anaphase I. A protein named shugoshin protects the cohesins at the centromere from cleavage until anaphase II, when the sister chromatids separate.

Meiosis I is called a *reductional division* because it reduces the chromosome sets from two (diploid) to one (haploid). The sister chromatids of each homolog do not separate until meiosis II, sometimes called the *equational division*.

INTERACTIVE QUESTION   13.3

The following diagrams represent some of the stages of meiosis (not in the right order). Label these stages.

a._____   b. _____   c. _____   d. _____   e. _____   f._____

Now place these stages in the proper sequence.

_____ _____ _____ _____ _____ _____

13.4   Genetic variation produced in sexual life cycles contributes to evolution

Origins of Genetic Variation Among Offspring   Mutations that result in different alleles are the original source of genetic variation. The three mechanisms that generate genetic variation in sexual reproduction are independent assortment of chromosomes, crossing over, and random fertilization.

Each homologous pair lines up independently at the metaphase plate—the orientation of the maternal and paternal chromosomes is random. The number of possible combinations of maternal and paternal chromosomes in gametes is 2^n, where n is the haploid number.

In prophase I, homologous segments of nonsister chromatids are exchanged by crossing over, forming **recombinant chromosomes** with new genetic combinations of maternal and paternal genes on the same chromosome. These no-longer-equivalent sister chromatids assort independently during meiosis II.

The random nature of fertilization adds to the genetic variability established in meiosis.

INTERACTIVE QUESTION   13.4

a. How many different assortments of maternal and paternal chromosomes are possible in a human gamete?

b. In a human zygote, how many diploid combinations of parental chromosomes are possible?

c. Why does the number in part b still underestimate the possible genetic variations in a zygote?

Evolutionary Significance of Genetic Variation within Populations   In Darwin's theory of evolution by natural selection, genetic variations present in a population result in adaptation as the individuals with the variations best suited to an environment produce the most offspring. The process of sexual reproduction and mutation are the sources of this variation.

Word Roots

a- = not or without (*asexual:* type of reproduction not involving fertilization)

-apsis = juncture (*synapsis:* the pairing of replicated homologous chromosomes during prophase I of meiosis)

auto- = self (*autosome:* the chromosomes that do not determine gender)

chiasm- = marked crosswise (*chiasma:* the X-shaped microscopically visible region representing homologous chromosomes that have exchanged genetic material through crossing over during meiosis)

di- = two (*diploid:* cells that contain two homologous sets of chromosomes)

fertil- = fruitful (*fertilization:* process of fusion of a haploid sperm and a haploid egg cell)

haplo- = single (*haploid:* cells that contain only one chromosome of each homologous pair)

homo- = like (*homologous:* like chromosomes that form a pair)

karyo- = nucleus (*karyotype:* a display of the chromosomes of a cell)

meio- = less (*meiosis:* a variation of cell division that yields daughter cells with half as many chromosomes as the parent cell)

soma- = body (*somatic:* body cells with 46 chromosomes in humans)

sporo- = a seed; **-phyte** = a plant (*sporophyte:* the multicellular diploid form in organisms undergoing alternation of generations that results from a union of gametes and that meiotically produces haploid spores that grow into the gametophyte generation)

syn- = together; **gam-** = marriage (*syngamy:* the process of cellular union during fertilization)

Structure Your Knowledge

1. Label the following diagram to review the terms that describe replicated chromosomes in a diploid cell.

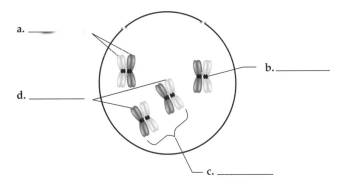

2. Describe the key events of these stages of meiosis.

a. Interphase	
b. Prophase I	
c. Metaphase I	
d. Anaphase I	
e. Metaphase II	
f. Anaphase II	

3. Create a concept map to help you organize your understanding of the similarities and differences between mitosis and meiosis. *Compare your map with those of some classmates to see different ways of organizing the material.*

Test Your Knowledge

MULTIPLE CHOICE: *Choose the one best answer.*

1. What is a karyotype?
   a. a genotype of an individual
   b. a pictorial display of an individual's chromosomes
   c. a blood type determination of an individual
   d. a unique combination of chromosomes found in a gamete
   e. a species-specific diploid number of chromosomes

2. A synaptonemal complex would be found during
   a. prophase of mitosis.
   b. fertilization or syngamy of gametes.
   c. metaphase II of meiosis.
   d. prophase of meiosis I.
   e. anaphase I of meiosis.

3. What are autosomes?
   a. sex chromosomes
   b. chromosomes that occur singly
   c. chromosomal abnormalities that result in genetic defects
   d. chromosomes found in mitochondria and chloroplasts
   e. none of the above

4. Which of the following statements is *not* true?
   a. The restoration of the diploid chromosome number after halving in meiosis is due to fertilization.
   b. Sister chromatid cohesion at areas where crossing over has occurred holds homologous chromosomes together until anaphase I.
   c. Recombinant chromosomes are produced when both maternal and paternal chromosomes independently assort into the same gamete.
   d. In mitosis, separation of sister chromatids in anaphase results in two identical cells.
   e. Separation of sister chromatids in anaphase II does not usually result in two identical cells.

5. During the first meiotic division (meiosis I),
   a. homologous chromosomes separate.
   b. the chromosome number is reduced in half.
   c. crossing over between nonsister chromatids occurs.
   d. paternal and maternal chromosomes assort randomly.
   e. all of the above occur.

6. A cell with a diploid number of 6 could produce gametes with how many different combinations of maternal and paternal chromosomes?
   a. 6   c. 12   e. 128
   b. 8   d. 64

7. The DNA content of a diploid cell is measured in the G_1 phase. After meiosis I, the DNA content of one of the two cells produced would be
   a. equal to that of the G_1 cell.
   b. twice that of the G_1 cell.
   c. one-half that of the G_1 cell.
   d. one-fourth that of the G_1 cell.
   e. impossible to estimate due to independent assortment of homologous chromosomes.

8. In most fungi and some protists,
   a. the zygote is the only haploid stage.
   b. gametes are formed by meiosis.
   c. the multicellular organism is haploid.
   d. the gametophyte generation produces gametes by mitosis.
   e. reproduction is exclusively asexual.

9. In the alternation of generations found in plants,
   a. the sporophyte generation produces spores by mitosis.
   b. the gametophyte generation produces gametes by mitosis.
   c. the zygote will develop into a sporophyte generation by meiosis.
   d. spores develop into the haploid sporophyte generation.
   e. the gametophyte generation produces spores by meiosis.

10. Which of the following is least likely to be a source of genetic variation in sexually reproducing organisms?
    a. crossing over
    b. replication of DNA during S phase before meiosis I
    c. independent assortment of chromosomes
    d. random fertilization of gametes
    e. mutation

11. Meiosis II is similar to mitosis because
    a. sister chromatids separate.
    b. homologous chromosomes separate.
    c. DNA replication precedes the division.
    d. they both take the same amount of time.
    e. haploid cells are almost always produced.

12. Pairs of homologous chromosomes
    a. have identical DNA sequences in their genes.
    b. have genes for the same characters at the same loci.
    c. are found in gametes.
    d. separate in meiosis II.
    e. have all of the above characteristics.

13. Asexual reproduction of a diploid organism would
    a. be impossible.
    b. involve meiosis.
    c. produce identical offspring.
    d. show variation among sibling offspring.
    e. involve spores produced by meiosis.

14. In a sexually reproducing species with a diploid number of 8, how many different diploid combinations of chromosomes would be possible in the *offspring*?
    a. 8
    b. 16
    c. 64
    d. 256
    e. 512

15. The calculation of offspring in question 14 includes only variation resulting from
    a. crossing over.
    b. random fertilization.
    c. independent assortment of chromosomes.
    d. a, b, and c.
    e. only b and c.

16. How many *chromatids* are present in metaphase II in a cell undergoing meiosis from an organism in which $2n = 24$?
    a. 12
    b. 24
    c. 36
    d. 48
    e. 96

17. Which of the following would *not* be considered a haploid cell?
    a. daughter cell after meiosis II
    b. gamete
    c. daughter cell after mitosis in gametophyte generation of a plant
    d. cell in prophase I
    e. cell in prophase II

18. Which of the following is *not* true of homologous chromosomes?
    a. They behave independently in mitosis.
    b. They synapse during the S phase of meiosis.
    c. They travel together to the metaphase plate in meiosis I.
    d. They are held together during synapsis by a synaptonemal complex.
    e. Crossing over between nonsister chromatids of homologous chromosomes is indicated by the presence of chiasmata.

19. Which of the following describes why or how recombinant chromosomes add to genetic variability?
    a. They are formed as a result of random fertilization when two sets of chromosomes combine in a zygote.
    b. They are the result of mutations that change alleles.
    c. They randomly orient during metaphase II and the nonequivalent sister chromatids separate in anaphase II.
    d. Genetic material from two parents is combined on the same chromosome.
    e. Both c and d are true.

20. A cell in G_2 before meiosis compared with one of the four cells produced by that meiotic division has
    a. twice as much DNA and twice as many chromosomes.
    b. four times as much DNA and twice as many chromosomes.
    c. four times as much DNA and four times as many chromosomes.
    d. half as much DNA but the same number of chromosomes.
    e. half as much DNA and half as many chromosomes.

Chapter 14

Mendel and the Gene Idea

Key Concepts

14.1 Mendel used the scientific approach to identify two laws of inheritance

14.2 The laws of probability govern Mendelian inheritance

14.3 Inheritance patterns are often more complex than predicted by simple Mendelian genetics

14.4 Many human traits follow Mendelian patterns of inheritance

Framework

Through his work with garden peas in the 1860s, Mendel developed the fundamental principles of inheritance and the laws of segregation and independent assortment. This chapter describes the basic monohybrid and dihybrid crosses that Mendel performed to establish that inheritance involves particulate genetic factors (genes) that segregate independently in the formation of gametes and recombine to form offspring. The laws of probability can be applied to predict the outcome of genetic crosses.

The phenotypic expression of genotype may be affected by such factors as incomplete dominance, codominance, multiple alleles, pleiotropy, epistasis, and polygenic inheritance, as well as the environment. Genetic screening and counseling for recessively inherited disorders use new technologies and Mendelian principles to analyze human pedigrees.

Chapter Review

14.1 Mendel used the scientific approach to identify two laws of inheritance

Mendel's Experimental, Quantitative Approach Gregor Mendel worked with garden peas, a good choice of study organism because they are available in many varieties, their fertilization is easily controlled, and the characteristics of their many offspring can be quantified.

Mendel studied seven **characters,** or heritable features, that occurred in alternative forms called **traits.** He used **true-breeding** varieties of pea plants, which means that self-fertilizing parents always produce offspring with the parental form of the character. To follow the transmission of these well-defined traits, Mendel performed **hybridizations** in which he cross-pollinated contrasting true-breeding varieties, and then allowed the next generation to self-pollinate. The true-breeding parental plants are the **P generation** (parental); the offspring of the first cross are the **F$_1$ generation** (first filial); and the next generation, from the self-cross of the F$_1$, is known as the **F$_2$ generation.**

The Law of Segregation Mendel found that the F$_1$ offspring did not show a blending of the parental traits. Instead, only one of the parental traits of the character was found in the hybrid offspring. In the F$_2$ generation, however, the missing parental trait reappeared in the ratio of 3:1—three offspring with the *dominant* trait shown by the F$_1$ to one offspring with the reappearing *recessive* trait.

Mendel's explanation for this phenomenon contains four parts: (1) alternate forms of genes, now called **alleles,** account for variations in characters;

(2) an organism has two alleles for each inherited trait, one received from each parent; (3) when two different alleles occur together, one of them, called the **dominant allele,** determines the organism's appearance while the other, the **recessive allele,** has no observable effect on appearance; and (4) allele pairs separate (segregate) during the formation of gametes (Mendel's **law of segregation**), so an egg or sperm carries only one allele for each inherited character. This explanation is consistent with the behavior of chromosomes during meiosis.

A **Punnett square** can be used to predict the results of simple genetic crosses. Dominant alleles are often symbolized by a capital letter, recessive alleles by a small letter.

Fill in this diagram of a cross of round- and wrinkled-seeded pea plants. The round allele (R) is dominant and the wrinkled allele (r) is recessive.

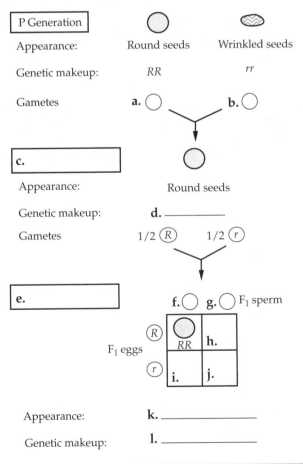

An organism that has a pair of identical alleles is said to be **homozygous** for that gene. If the organism has two different alleles, it is said to be **heterozygous.** Homozygotes are true-breeding; heterozygotes are not, since they produce gametes with one or the other allele that can combine to produce offspring that are dominant homozygotes, heterozygotes, and recessive homozygotes. **Phenotype** is an organism's observable traits; **genotype** is its genetic makeup.

In a **testcross,** an organism expressing the dominant phenotype is crossed with a recessive homozygote to determine the genotype of this phenotypically dominant organism.

A tall pea plant is crossed with a recessive dwarf pea plant. What will the phenotypic and genotypic ratio of offspring be

a. if the tall plant was TT?

b. if the tall plant was Tt?

The Law of Independent Assortment   In a *monohybrid cross,* the inheritance of a single character is followed through the crossing of **monohybrids,** F_1 offspring that are heterozygous for one character. Mendel used *dihybrid crosses* between F_1 **dihybrids,** which are heterozygous for two characters, to determine whether the two characters were transmitted independently from the parent plants.

If the two pairs of alleles segregate independently, then gametes from an F_1 hybrid generation ($AaBb$) should contain four combinations of alleles in equal quantities (AB, Ab, aB, ab). The random fertilization of these four classes of gametes should result in 16 (4×4) gamete combinations that produce four phenotypic categories in a ratio of 9:3:3:1 (nine offspring showing both dominant traits, three showing one dominant trait and one recessive, three showing the opposite dominant and recessive traits, and one showing both recessive traits). Mendel obtained these ratios in the F_2 progeny of dihybrid crosses, providing evidence for his **law of independent assortment.** This principle states that each pair of alleles segregates independently from other pairs in the formation of gametes. This law applies to genes located on different chromosomes.

INTERACTIVE QUESTION 14.3

A true-breeding tall, purple-flowered pea plant (*TTPP*) is crossed with a true-breeding dwarf, white-flowered plant (*ttpp*).

a. What is the phenotype of the F_1 generation?

b. What is the genotype of the F_1 generation?

c. What four types of gametes are formed by F_1 plants?

_____   _____   _____   _____

d. Fill in the following Punnett square to show the offspring of the F_2 generation. Shade each phenotype a different color so you can see the ratio of offspring.

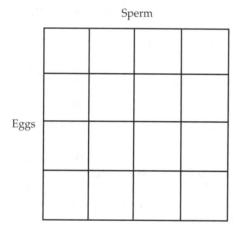

Sperm

Eggs

e. List the phenotypes and ratios found in the F_2 generation.

f. What is the ratio of tall to dwarf plants? _____

Of purple- to white-flowered plants? _____

(Note that the alleles for each individual character segregate as in a monohybrid cross.)

14.2   The laws of probability govern Mendelian inheritance

General rules of probability apply to the laws of segregation and independent assortment. The probability scale goes from 0 to 1; the probabilities of all possible outcomes must add up to 1. The probability of an event occurring is the number of times that event could occur over all the possible events; for example, the probability of drawing an ace of any suit from a deck of cards is $4/52$ The outcome of independent events is not affected by previous or simultaneous trials.

The Multiplication and Addition Rules Applied to Monohybrid Crosses The *multiplication rule* states that the probability that a certain combination of independent events will occur together is equal to the product of the separate probabilities of the independent events. The probability of a particular genotype being formed by fertilization is equal to the product of the probabilities of forming each type of gamete needed to produce that genotype.

If a genotype can be formed in more than one way, then the *addition rule* states that its probability is equal to the sum of the separate probabilities of the different, mutually exclusive ways the event can occur. For example, a heterozygote offspring can occur if the egg contains the dominant allele and the sperm the recessive ($1/2 \times 1/2 = 1/4$ probability) or vice versa ($1/4$) Therefore, a heterozygote offspring would be the predicted result from a monohybrid cross half of the time ($1/4 + 1/4 = 1/2$).

INTERACTIVE QUESTION   14.4

Apply the multiplication rule to a dihybrid cross. How would you determine the probability of getting an F_2 offspring that is homozygous recessive for both traits?

Solving Complex Genetics Problems with the Rules of Probability Fairly complex genetics problems can be solved by applying the multiplication and addition rules. The probability of a particular genotype arising from a cross can be determined by considering each gene involved as a separate monohybrid cross and then multiplying the probabilities of all the independent events involved in the final genotype. When more than one outcome is involved, the addition rule is also used.

The larger the sample size, the more closely the results will conform to statistical predictions.

INTERACTIVE QUESTION 14.5

a. In the following cross, what is the probability of obtaining offspring that show all three dominant traits?

$$AaBbcc \times AabbCC$$

probability of offspring that are $A\_B\_C\_$ = _____

(_ indicates that the second allele can be either dominant or recessive without affecting the phenotype determined by the first dominant allele.)

b. What is the probability that the offspring of this cross will show at least two dominant traits? _____

c. What is the probability of offspring that show only one dominant trait? _____

14.3 Inheritance patterns are often more complex than predicted by simple Mendelian genetics

Extending Mendelian Genetics for a Single Gene In the case of an allele showing **complete dominance,** the phenotype of the heterozygote is indistinguishable from that of the dominant homozygote. Intermediate phenotypes are characteristic of alleles showing **incomplete dominance.** The F_1 hybrids have a phenotype intermediate between that of the parents. The F_2 show a 1:2:1 phenotypic and genotypic ratio. Alleles that exhibit **codominance** will both affect the phenotype in separate, distinguishable ways, as in the case of the MN blood group.

Tay-Sachs disease is a lethal disorder in which brain cells are unable to metabolize certain lipids that then accumulate and damage the brain. In a heterozygote, the Tay-Sachs allele is recessive at the organismal level; at the biochemical level, the enzyme activity level is intermediate between that of both homozygotes; and at the molecular level, the alleles are codominant in that each produces its enzyme product, either normal or dysfunctional.

Whether or not an allele is dominant or recessive has no relation to how common it is in a population.

Most genes exist in more than two allelic forms. The gene that determines human blood groups has three alleles. The alleles I^A and I^B are codominant with each other; each codes for an enzyme that attaches a carbohydrate to the surface of red blood cells. The allele i results in no attached carbohydrate.

INTERACTIVE QUESTION 14.6

List the possible genotypes for the following blood groups.

a. A _____

b. B _____

c. AB _____

d. O _____

Pleiotropy is the characteristic of a single gene having multiple phenotypic effects in an individual. Certain hereditary diseases with complex sets of symptoms are caused by pleiotropic alleles.

Extending Mendelian Genetics for Two or More Genes In **epistasis,** a gene at one locus may affect the expression of another gene. F_2 ratios that differ from the typical 9:3:3:1 often indicate epistasis.

INTERACTIVE QUESTION 14.7

A dominant allele *M* is necessary for the production of the black pigment melanin; mm individuals are white. A dominant allele *B* results in the deposition of a lot of pigment in an animal's hair, producing a black color. The genotype *bb* produces brown hair. Two black animals heterozygous for both genes are bred. Fill in the following table for the offspring of this cross.

Phenotype	Genotype	Ratio
Black		
	M_bb	

Quantitative characters, such as height or skin color, vary along a continuum in a population. Such phenotypic gradations are usually due to **polygenic inheritance,** in which two or more genes have an additive effect on one character. Each dominant allele contributes one "unit" to the phenotype. A polygenic character may result in a normal distribution (forming a bell-shaped curve) of the character within a population.

INTERACTIVE QUESTION   14.8

The height of spike weed is a result of polygenic inheritance involving three genes, each of which can contribute an additional 5 cm to the base height of the plant, which is 10 cm. The tallest plant (*AABBCC*) can reach a height of 40 cm.

a. If a tall plant (*AABBCC*) is crossed with a base-height plant (*aabbcc*), what is the height of the F_1 plants?

b. How many phenotypic classes will there be in the F_2?

Nature and Nurture: the Environmental Impact on Phenotype   The phenotype of an individual is the result of complex interactions between its genotype and the environment. Genotypes have a phenotypic range called a **norm of reaction** within which the environment influences phenotypic expression. Polygenic characters are often **multifactorial,** meaning that a combination of genetic and environmental factors affect phenotype.

Integrating a Mendelian View of Heredity and Variation   The phenotypic expression of most genes is influenced by other genes and by the environment. The theory of particulate inheritance and Mendel's principles of segregation and independent assortment, however, still form the basis for modern genetics.

14.4   Many human traits follow Mendelian patterns of inheritance

Pedigree Analysis   A family **pedigree** is a family tree with the history of a particular trait shown across the generations. By convention, circles represent females, squares are used for males, and solid symbols indicate individuals that express the trait in question. Parents are joined by a horizontal line, and offspring are listed below parents from left to right in order of birth. The genotype of individuals in the pedigree can often be deduced by following the patterns of inheritance.

Recessively Inherited Disorders   Only homozygous recessive individuals express the phenotype for the thousands of genetic disorders that are inherited as simple recessive traits. **Carriers** of the disorder are heterozygotes, who are phenotypically normal but may transmit the recessive allele to their offspring.

INTERACTIVE QUESTION   14.9

Consider this pedigree for the trait albinism (lack of skin pigmentation) in three generations of a family. (Solid symbols represent individuals who are albinos.) From your knowledge of Mendelian inheritance, answer the questions that follow.

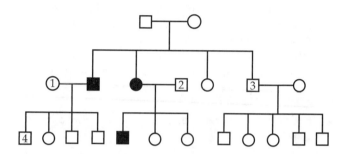

a. Is this trait caused by a dominant or recessive allele? How can you tell?

b. Determine the genotypes of the parents in the first generation. (Let *A* and *a* represent the alleles.) Genotype of father _____; of mother _____.

c. Determine the probable genotypes of the mates of the albino offspring in the second generation and the grandson 4 in the third generation. Genotypes: mate 1 _____; mate 2 _____; grandson 4 _____.

d. Can you determine the genotype of son 3 in the second generation? Why or why not?

The likelihood of two mating individuals carrying the same rare deleterious allele increases when the individuals have common ancestors. Consanguineous matings, between siblings or close relatives, are indicated on pedigrees by double lines.

Genetic diseases are unevenly distributed among groups of humans. Tay-Sachs disease is more common among Ashkenazic Jews.

Cystic fibrosis is the most common lethal genetic disease in the United States; it is found more frequently in people of European descent than in other groups. This recessive allele results in defective chloride channels in certain cell membranes. Accumulating extracellular chloride leads to the buildup of thickened mucus in various organs and a predisposition to bacterial infections.

Sickle-cell disease is the most common inherited disease among African Americans. Due to a single amino acid substitution in the hemoglobin protein, red blood cells deform into a sickle shape when blood-oxygen concentration is low, triggering blood clumping and other pleiotropic effects. Heterozygous individuals are said to have sickle-cell trait but are usually healthy.

INTERACTIVE QUESTION 14.10

a. What is the probability that a mating between two carriers will produce an offspring with a recessively inherited disorder? _____

b. What is the probability that a phenotypically normal child produced by a mating of two heterozygotes will be a carrier? _____

Dominantly Inherited Disorders A few human disorders are due to dominant genes. In achondroplasia, dwarfism is due to a single copy of a mutant allele. Dominant lethal alleles are more rare than recessive lethals because the harmful allele cannot be masked in the heterozygote.

A late-acting lethal dominant allele can be passed on if the symptoms do not develop until after reproductive age. Molecular geneticists have developed a method to detect the lethal gene for **Huntington's disease,** a degenerative disease of the nervous system that does not develop until later in life.

Multifactorial Disorders Many diseases have genetic (usually polygenic) and environmental components. These multifactorial disorders include heart disease, diabetes, cancer, and others.

Genetic Testing and Counseling The probability of a child having a genetic defect may be determined by considering the family history of the disease.

INTERACTIVE QUESTION 14.11

If two prospective parents both have siblings who have a recessive genetic disorder, what is the chance that they would have a child who inherits the disorder?

Tests to identify carriers have been developed for a number of heritable disorders.

Amniocentesis is a procedure that extracts a small amount of amniotic fluid from the sac surrounding the fetus. The fluid is analyzed biochemically, and fetal cells are cultured for several weeks and then tested for certain genetic disorders and karyotyped to check for chromosomal defects.

Chorionic villus sampling (CVS) is a technique in which a small amount of fetal tissue is suctioned from the placenta. These rapidly growing cells can be karyotyped immediately, and this procedure can be performed earlier in pregnancy. A new technique isolates fetal cells from maternal blood, which can then be cultured and tested.

Ultrasound is a simple, noninvasive procedure that can reveal major abnormalities. *Fetoscopy*, the insertion of a needle-thin viewing scope and light into the uterus, allows the fetus to be checked for anatomical problems.

Some genetic disorders can be detected at birth. Routine screening tests for phenylketonuria (PKU) in newborns. If detected, dietary adjustments can lead to normal development.

Word Roots

-centesis = a puncture (*amniocentesis*: a technique for determining genetic abnormalities in a fetus by the presence of certain chemicals or defective fetal cells in the amniotic fluid, obtained by aspiration from a needle inserted into the uterus)

co- = together (*codominance*: phenotype in which both dominant alleles are expressed in the heterozygote)

di- = two (*dihybrid cross*: a breeding experiment in which offspring of a cross of parental varieties differing in two traits are mated)

epi- = beside; **-stasis** = standing (*epistasis*: a phenomenon in which one gene alters the expression of another gene that is independently inherited)

geno- = offspring (*genotype*: the genetic makeup of an organism)

hetero- = different (*heterozygous*: having two different alleles for a trait)

homo- = alike (*homozygous*: having two identical alleles for a trait)

mono- = one (*monohybrid cross*: a breeding experiment that crosses offspring of a cross of parental varieties differing in a single character)

pedi- = a child (*pedigree*: a family tree describing the occurrence of heritable characters in parents and offspring across as many generations as possible)

pheno- = appear (*phenotype*: the physical and physiological traits of an organism)

pleio- = more (*pleiotropy:* when a single gene impacts more than one characteristic)

poly- = many; **gene-** = produce (*polygenic:* an additive effect of two or more gene loci on a single phenotypic character)

Structure Your Knowledge

1. Relate Mendel's two laws of inheritance to the behavior of chromosomes in meiosis that you studied in Chapter 13.

2. How many different types of gametes can be formed by the following genotypes? What generalized formula describes this number?

   Aa =   *AaBb* =

   AaBbCc =   *AABbCc* =

3. How can you tell whether alleles are completely dominant/recessive, incompletely dominant, or codominant?

Genetics Problems

One of the best ways to learn genetics is to work problems. You can't memorize genetic knowledge; you have to practice using it. Work through the problems presented below, methodically setting down the information you are given and what you are to determine. Write down the symbols used for the alleles and genotypes, and the phenotypes resulting from those genotypes. Avoid using Punnett squares; while useful for learning basic concepts, they are much too laborious and mistake-prone. Instead, break complex crosses into their monohybrid components and rely on the rules of multiplication and addition. Always look at the answers you get to see if they make logical sense. And remember that study groups are great for going over your problems and helping each other "see the light." Answers and explanations are provided at the end of the book.

1. Summer squash are either white or yellow. To get white squash, at least one of the parental plants must be white. The allele for which color is dominant?

2. For the following crosses, determine the probability of obtaining the indicated genotype in an offspring.

Cross	Offspring	Probability
AAbb × *AaBb*	*AAbb*	a.
AaBB × *AaBb*	*aaBB*	b.
AABbcc × *aabbCC*	*AaBbCc*	c.
AaBbCc × *AaBbcc*	*aabbcc*	d.

3. True-breeding tall red-flowered plants are crossed with dwarf white-flowered plants. The resulting F_1 generation consists of all tall pink-flowered plants. Assuming that height and flower color are each determined by a single gene locus on different chromosomes, predict the results of an F_1 cross of dihybrid plants. Choose appropriate symbols for the alleles of the height and flower color genes. List the phenotypes and predicted ratios for the F_2 generation.

4. Blood typing has been used as evidence in paternity cases, when the blood type of the mother and child may indicate that a man alleged to be the father could not possibly have fathered the child. For the following mother and child combinations, indicate which blood groups of potential fathers would be exonerated.

Blood Group of Mother	Blood Group of Child	Man Exonerated If He Belongs to Blood Group(s)
AB	A	a.
O	B	b.
A	AB	c.
O	O	d.
B	A	e.

5. In rabbits, the homozygous *CC* is normal, *Cc* results in rabbits with deformed legs, and *cc* is lethal. For a gene for coat color, the genotype *BB* produces black, *Bb* brown, and *bb* a white coat. Give the phenotypic ratio of offspring from a cross of a deformed-leg, brown rabbit with a deformed-leg, white rabbit.

6. Polydactyly (extra fingers and toes) is due to a dominant gene. A father is polydactyl, the mother has the normal phenotype, and they have had one normal child. What is the genotype of the father? Of the mother? What is the probability that a second child will have the normal number of digits?

7. In dogs, black (*B*) is dominant to chestnut (*b*), and solid color (*S*) is dominant to spotted (*s*). What are the genotypes of the parents in a mating that produced ⅜ black solid, ⅜ black spotted, ⅛ chestnut solid, and ⅛ chestnut spotted puppies? (*Hint:* First determine what genotypes the offspring must have before you deal with the fractions.)

8. When hairless hamsters are mated with normal-haired hamsters, about one-half the offspring are

hairless and one-half are normal. When hairless hamsters are crossed with each other, the ratio of normal-haired to hairless is 1:2. How do you account for the results of the first cross? How would you explain the unusual ratio obtained in the second cross?

9. Two pigs whose tails are exactly 25 cm in length are bred over 10 years and they produce 96 piglets with the following tail lengths: 6 piglets at 15 cm, 25 at 20 cm, 37 at 25 cm, 23 at 30 cm, and 5 at 35 cm.

   a. How many pairs of genes are regulating the tail length character? *Hint:* Count the number of phenotypic classes, or determine the sum of the ratios of the classes. In a monohybrid cross, the F_2 ratios add up to 4 (3:1 or 1:2:1). In a dihybrid cross, the F_2 ratios add up to 16 (9:3:3:1 or some variation if the genes are epistatic or quantitative).

   b. What offspring phenotypes would you expect from a mating between a 15-cm and a 30-cm tail-length pig?

10. Fur color in rabbits is determined by a single gene locus, for which there are four alleles. Four phenotypes are possible: black, Chinchilla (gray color caused by white hairs with black tips), Himalayan (white with black patches on extremities), and white. The black allele (C) is dominant over all other alleles, the Chinchilla allele (C^{ch}) is dominant over Himalayan (C^h), and the white allele (c) is recessive to all others.

    a. A black rabbit is crossed with a Himalayan, and the F_1 consists of a ratio of 2 black to 2 Chinchilla. Can you determine the genotypes of the parents?

    b. A second cross was done between a black rabbit and a Chinchilla. The F_1 contained a ratio of 2 black to 1 Chinchilla to 1 Himalayan. Can you determine the genotypes of the parents of this cross?

11. In Labrador retriever dogs, the dominant gene *B* determines black coat color and *bb* produces brown. A separate gene *E*, however, shows dominant epistasis over the *B* and *b* alleles, resulting in a golden coat color. The recessive *e* allows expression of *B* and *b*. A breeder wants to know the genotypes of her three dogs, so she breeds them and makes note of the offspring of several litters. Determine the genotypes of the three dogs.

    a. golden female (Dog 1) × golden male (Dog 2) offspring: 7 golden, 1 black, 1 brown

    b. black female (Dog 3) × golden male (Dog 2) offspring: 8 golden, 5 black, 2 brown

12. The ability to taste phenylthiocarbamide (PTC) is controlled in humans by a single dominant allele (*T*). A woman nontaster married a man taster, and they had three children, two boy tasters and a girl

nontaster. All the grandparents were tasters. Create a pedigree for this family for this trait. (Solid symbols should signify nontasters (*tt*).) Where possible, indicate whether tasters are *TT* or *Tt*.

13. Two true-breeding varieties of garden peas are crossed. One parent had red, axial flowers, and the other had white, terminal flowers. All F_1 individuals had red, terminal flowers. If 100 F_2 offspring were counted, how many of them would you expect to have red, axial flowers?

14. You cross true-breeding red-flowered plants with true-breeding white-flowered plants, and the F_1 are all red-flowered plants. The F_2, however, occur in a ratio of 9 red:6 pale purple:1 white. How many genes are involved in the inheritance of this color character? Explain why the F_1 are all red and how the 9:6:1 ratio of phenotypes in the F_2 occurred.

Test Your Knowledge

MATCHING: *Match the definition with the correct term.*

_____ 1. codominance

_____ 2. homozygote

_____ 3. heterozygote

_____ 4. phenotype

_____ 5. polygenic (quantitative)

_____ 6. pleiotropy

_____ 7. epistasis

_____ 8. testcross

_____ 9. dihybrid cross

_____ 10. incomplete dominance

A. true-breeding variety

B. cross between two monohybrids

C. cross between hybrids that are heterozygous for two genes

D. cross with recessive homozygote to determine genotype of unknown

E. the physical characteristics of an individual

F. genotype with two different alleles

G. gene with multiple alleles

H. one gene influences the expression of another gene

I. both alleles are fully expressed in heterozygote

J. single gene with multiple phenotypic effects

K. heterozygote intermediate between phenotypes of homozygotes

L. two or more genes with additive effect on phenotype

MULTIPLE CHOICE: *Choose the one best answer.*

1. According to Mendel's law of segregation,
   a. there is a 50% probability that a gamete will get a dominant allele.
   b. gene pairs segregate independently of other genes in gamete formation.
   c. allele pairs separate in gamete formation.
   d. the laws of probability determine gamete formation.
   e. there is a 3:1 ratio in the F_2 generation.

2. After obtaining two heads from two tosses of a coin, the probability of tossing the coin and obtaining a head is
   a. $1/2$ c. $1/8$ e. $3/4$
   b. $1/4$ d. $3/8$

3. The probability of tossing three coins simultaneously and obtaining three heads is
   a. $1/2$ c. $1/8$ e. $3/4$
   b. $1/4$ d. $3/8$

4. The probability of tossing three coins simultaneously and obtaining two heads and one tail is
   a. $1/2$ c. $1/8$ e. $3/4$
   b. $1/4$ d. $3/8$

5. The F_2 generation
   a. has a phenotypic ratio of 3:1.
   b. is the result of the self-fertilization or crossing of F_1 individuals.
   c. can be used to determine the genotype of individuals with the dominant phenotype.
   d. has a phenotypic ratio that equals its genotypic ratio.
   e. has 16 different genotypic possibilities.

6. The base height of the dingdong plant is 10 cm. Four genes contribute to the height of the plant, and each dominant allele contributes 3 cm to height. If you cross a 10-cm plant (quadruply homozygous recessive) with a 34-cm plant, how many phenotypic classes will there be in the F_2?
   a. 4   c. 8   e. 64
   b. 5   d. 9

7. A 1:1 phenotypic ratio in a testcross indicates that
   a. the alleles are dominant.
   b. one parent must have been homozygous dominant.
   c. the dominant phenotype parent was a heterozygote.
   d. the alleles segregated independently.
   e. the alleles are codominant.

8. Carriers of a genetic disorder
   a. are indicated by solid symbols on a family pedigree.
   b. are involved in consanguineous matings.
   c. will produce children with the disease.
   d. are heterozygotes for the gene that can cause the disorder.
   e. have a homozygous recessive genotype.

9. If both parents are carriers of a lethal recessive gene, the probability that their child will inherit and express the disorder is
   a. $1/8$ d. $1/2 \times 1/2 \times 1/4$, or $1/16$
   b. $1/4$ e. $2/3 \times 2/3 \times 1/4$, or $1/9$
   c. $1/2$

10. Which phase of meiosis is most directly related to the law of independent assortment?
    a. prophase I   d. metaphase II
    b. prophase II   e. anaphase II
    c. metaphase I

11. You think that two alleles for coat color in mice show incomplete dominance. What is the best and simplest cross to perform in order to support your hypothesis?
    a. a testcross of a homozygous recessive mouse with a mouse of unknown genotype
    b. a cross of F_1 mice to look for a 1:2:1 ratio in the offspring
    c. a reciprocal cross in which the sex of the mice of each coat color is reversed
    d. a cross of two true-breeding mice of different colors to look for an intermediate phenotype in the F_1
    e. a cross of F_1 mice to look for a 9:7 ratio in the offspring

12. Which of the following human diseases is inherited as a simple recessive trait?
    a. Tay-Sachs disease
    b. cancer
    c. diabetes
    d. Alzheimer's disease
    e. cardiovascular disease

13. A multifactorial disorder
    a. can usually be traced to consanguineous matings.
    b. is caused by recessively inherited lethal genes.
    c. has both genetic and environmental causes.
    d. has a collection of symptoms traceable to an epistatic gene.
    e. is usually associated with quantitative traits.

14. A mother with type B blood has two children, one with type A blood and one with type O blood. Her husband has type O blood. Which of the following could you conclude from this information?

a. The husband could not have fathered either child.

b. The husband could have fathered both children.

c. The husband must be the father of the child with type O blood and could be the father of the type A child.

d. The husband could be the father of the child with type O blood, but not the type A child.

e. Neither the mother nor the husband could be the biological parent of the type A child.

15. In guinea pigs, the brown coat color allele (B) is dominant over red (b), and the solid color allele (S) is dominant over spotted (s). The F_1 offspring of a cross between true-breeding brown, solid-colored guinea pigs and red, spotted pigs are crossed. What proportion of their offspring (F_2) would be expected to be red and solid colored?

a. $1/9$ c. $3/16$ e. $3/4$

b. $1/16$ d. $9/16$

16. A dominant allele P causes the production of purple pigment; pp individuals are white. A dominant allele C is also required for color production; cc individuals are white. What proportion of offspring will be purple from a $ppCc \times PpCc$ cross?

a. $1/8$ c. $1/2$ e. None

b. $3/8$ d. $3/4$

Chapter 15

The Chromosomal Basis of Inheritance

Key Concepts

15.1 Mendelian inheritance has its physical basis in the behavior of chromosomes

15.2 Sex-linked genes exhibit unique patterns of inheritance

15.3 Linked genes tend to be inherited together because they are located near each other on the same chromosome

15.4 Alterations of chromosome number or structure cause some genetic disorders

15.5 Some inheritance patterns are exceptions to the standard chromosome theory

Framework

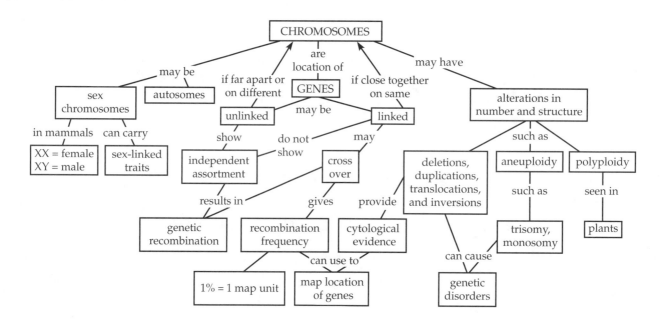

Chapter Review

15.1 Mendelian inheritance has its physical basis in the behavior of chromosomes

Mendel's laws, combined with cytological evidence of the process of meiosis, led to the **chromosome theory of inheritance:** Genes occupy specific positions (loci) on chromosomes, and chromosomes undergo segregation and independent assortment during the process of meiosis in gamete formation.

Morgan's Experimental Evidence: Scientific Inquiry T. H. Morgan worked with fruit flies, *Drosophila melanogaster,* which are prolific and rapid breeders. Fruit flies have only four pairs of chromosomes; the sex chromosomes occur as XX in females and XY in male flies.

The normal phenotype found most commonly in nature for a character is called the **wild type,** whereas alternative traits, assumed to have arisen as mutations, are called *mutant phenotypes.*

Morgan discovered a mutant white-eyed male fly that he mated with a wild-type red-eyed female. The F_1 were all red-eyed. In the F_2, however, all female flies were red-eyed, whereas half of the males were red-eyed and half were white-eyed. Morgan deduced that the gene for eye color was located on the X chromosome. Males have only one X, so their phenotype is determined by the eye-color allele they inherit from their mother. This association of a specific gene with a chromosome provided evidence for the chromosome theory of inheritance.

INTERACTIVE QUESTION 15.1

Complete this summary of Morgan's crosses with the mutant white-eyed fly by filling in the Punnett square and showing the phenotypes and genotypes of the F_2 generation. (X^w stands for the mutant recessive white allele; X^{w+} for the wild-type red allele.) The Y indicates that this eye color gene is not present on the Y chromosome.

P Generation			
Phenotype	red-eyed (wild-type) female	×	white-eyed male
Genotype	$X^{w+}X^{w+}$		X^wY

F₁ Generation			
Phenotype	red-eyed female	×	red-eyed male
Genotype	$X^{w+}X^w$		$X^{w+}Y$

F₂ Generation

Sperm

Ova

Phenotypes

_____ _____ _____ _____

Genotypes

_____ _____ _____ _____

15.2 Sex-linked genes exhibit unique patterns of inheritance

The Chromosomal Basis of Sex Sex is a phenotypic character usually determined by sex chromosomes. In humans and other mammals, females, who are XX, produce ova that each contain an X chromosome. Males, who are XY, produce two kinds of sperm, each with either an X or a Y chromosome. Whether the gonads of an embryo develop into testes or ovaries depends on the presence or absence of the gene *SRY,* found on the Y chromosome, whose protein product regulates many other genes. A few other genes on the Y chromosome are necessary for normal functioning of the testes.

Other sex-determination systems include X-0 (in grasshoppers and some other insects), Z-W (in birds and some fishes and insects), and haplo-diploid (in bees and ants).

Inheritance of Sex-Linked Genes Sex chromosomes may carry genes, called **sex-linked genes,** for traits that are not related to sex. In humans, *sex-linked* typically refers to a gene on the X chromosome. Males inherit sex-linked alleles from their mothers; daughters inherit sex-linked alleles from both parents. Recessive sex-linked traits are seen more often in males, since they are *hemizygous* for sex-linked genes.

Duchenne muscular dystrophy is a sex-linked disorder resulting from the lack of a key muscle protein. **Hemophilia** is a sex-linked trait characterized by excessive bleeding due to the absence of one or more blood-clotting proteins.

X Inactivation in Female Mammals Only one of the X chromosomes is fully active in most mammalian female somatic cells. The other X chromosome is contracted into a **Barr body** located inside the nuclear membrane. M. Lyon demonstrated that the selection of which X chromosome is inactivated is a random event occurring independently in embryonic cells. As a result of X-inactivation, both males and females have an equal dosage of X chromosome genes. A gene called *XIST* is active on the X chromosome that forms the Barr body. Its RNA product may trigger DNA methylation and X-inactivation.

INTERACTIVE QUESTION   15.2

Two normal color-sighted individuals produce the following children and grandchildren. Fill in the probable genotype of the indicated individuals in this pedigree. *Squares are males, circles are females, and solid symbols represent color blindness. Use the superscript N for the normal allele and n for the recessive allele for color blindness.*

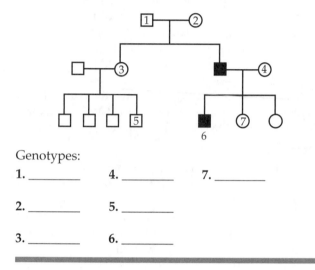

Genotypes:

1. _____   4. _____   7. _____

2. _____   5. _____

3. _____   6. _____

15.3   Linked genes tend to be inherited together because they are located near each other on the same chromosome

Genes located on the same chromosome that tend to be inherited together are called **linked genes.**

How Linkage Affects Inheritance   Morgan performed a testcross of F_1 dihybrid wild-type flies with flies that were homozygous recessive for black bodies and vestigial wings and found that the offspring were not in the predicted 1:1:1:1 phenotypic classes. Rather, most of the offspring were the same phenotypes as the P generation parents—either wild type (gray, normal wings) or double mutant (black, vestigial). Morgan deduced that these traits were inherited together because their genes were located on the same chromosome.

Genetic recombination results in offspring with combinations of traits that differ from those of the parents.

Genetic Recombination and Linkage   In a cross between a dihybrid heterozygote and a recessive homozygote, one-half of the offspring will be **parental types** and have phenotypes like one or the other of the parents, and one-half of the offspring, called **recombinant types (recombinants),** will have combinations of the two traits that are unlike the parents. This 50% frequency of recombination is observed when two genes are located on different chromosomes, and it results from the random alignment of homologous

chromosomes at metaphase I and the resulting independent assortment of alleles.

INTERACTIVE QUESTION   15.3

In a testcross between a heterozygote tall, purple-flowered pea plant and a dwarf, white-flowered plant,

a. what are the phenotypes of offspring that are parental types?

b. what are the phenotypes of offspring that are recombinants?

Linked genes do not assort independently, and one would not expect to see recombination of parental traits in the offspring. Recombination of linked genes does occur, however, due to **crossing over,** the reciprocal trade between nonsister (a maternal and a paternal) chromatids of synapsed homologous chromosomes during prophase of meiosis I. The percentage of recombinant offspring is called the *recombination frequency.*

INTERACTIVE QUESTION   15.4

With unlinked genes, an equal number of parental and recombinant offspring are produced. With linked genes, (more/fewer) parentals than recombinants are produced. (Circle and then explain your answer.)

Mapping the Distance Between Genes Using Recombination Data: Scientific Inquiry   A **genetic map** is an ordered list of genes on a chromosome. A. H. Sturtevant suggested that recombination frequencies reflect the relative distance between genes; genes located farther apart have a greater probability that a crossover event will occur between them. Sturtevant used recombination data to create a **linkage map,** defining one **map unit** (or centimorgan) as equal to a 1% recombination frequency.

The sequence of genes on a chromosome can be determined by finding the recombination frequency between different pairs of genes. Linkage cannot be determined if genes are so far apart that crossovers between them are almost certain. They would then have the 50% recombination frequency typical of unlinked genes. Such genes are *physically linked* but *genetically unlinked.* Distant genes on the same chromosome may be mapped by adding the recombination frequencies determined between them and intermediate genes.

Sturtevant and his colleagues found that the genes for the various known mutations of *Drosophila* clustered into four groups of linked genes, providing additional evidence that genes are located on chromosomes, in this case, on the four *Drosophila* chromosomes.

The frequency of crossing over may vary along the length of a chromosome, and a linkage map provides the sequence but not the exact location of genes on chromosomes. **Cytogenetic maps** locate gene loci in reference to visible chromosomal features.

INTERACTIVE QUESTION 15.5

Recombination frequency is given below for several gene pairs. Create a linkage map for these genes, showing the map unit distance between loci.

j, k 12%   *j, m* 9%   *k, l* 6%   *l, m* 15%

15.4 Alterations of chromosome number or structure cause some genetic disorders

Abnormal Chromosome Number **Nondisjunction** occurs when a pair of homologous chromosomes does not separate properly in meiosis I or sister chromatids do not separate in meiosis II. As a result, a gamete receives either two or no copies of that chromosome. A zygote formed with one of these aberrant gametes has a chromosomal alteration known as **aneuploidy,** a nontypical number of a particular chromosome. The zygote will be either **trisomic** for that chromosome (chromosome number is $2n + 1$) or **monosomic** ($2n - 1$). Aneuploid organisms usually have a set of symptoms caused by the abnormal dosage of genes. A mitotic nondisjunction early in embryonic development is also likely to be harmful.

Polyploidy is a chromosomal alteration in which an organism has more than two complete chromosomal sets, as in *triploidy* ($3n$) or *tetraploidy* ($4n$). Polyploidy is common in the plant kingdom and has played an important role in the evolution of plants.

INTERACTIVE QUESTION 15.6

a. What is the difference between an organism with a trisomy and a triploid organism?

b. Which of these is likely to show the most deleterious effects of its chromosomal anomaly?

Alterations of Chromosome Structure Chromosome breakage can result in chromosome fragments that are lost, called **deletion;** that join to a sister chromatid (or a nonsister chromatid), called **duplication;** that rejoin the original chromosome in the reverse orientation, called **inversion;** or that join a nonhomologous chromosome, called **translocation.** A *nonreciprocal* crossover can result in a deletion and duplication in nonsister chromatids, caused by nonequal exchange between chromatids.

INTERACTIVE QUESTION 15.7

Two nonhomologous chromosomes have gene orders, respectively, of *A-B-C-D-E-F-G-H-I-J* and *M-N-O-P-Q-R-S-T.* What types of chromosome alterations would have occurred if daughter cells were found to have a gene sequence on the first chromosome of *A-B-C-O-P-Q-G-J-I-H?*

Human Disorders Due to Chromosomal Alterations The frequency of aneuploid zygotes may be fairly high in humans, but development is usually so disrupted that the embryos spontaneously abort. Some genetic disorders, expressed as *syndromes* of characteristic traits, are the result of aneuploidy.

Down syndrome, caused by trisomy of chromosome 21, results in characteristic facial features, short stature, heart defects, and mental retardation. The incidence of Down syndrome increases for older mothers.

XXY males exhibit *Klinefelter syndrome,* a condition in which the individual has abnormally small testes, is sterile, and is usually of normal intelligence. Males with an extra Y chromosome do not exhibit any well-defined syndrome.

Trisomy X results in females who are healthy and distinguishable only by karyotype. Monosomy X individuals (X0) exhibit *Turner syndrome* and are phenotypically female, sterile individuals with short stature and usually normal intelligence.

Structural alterations of chromosomes, such as deletions or translocations, may be associated with specific human disorders, such as *cri du chat* syndrome and *chronic myelogenous leukemia.*

Why do most sex chromosome aneuploidies have less deleterious effects than do autosomal aneuploidies?

15.5   Some inheritance patterns are exceptions to the standard chromosome theory

Genomic Imprinting   A few dozen traits in mammals depend on which parent supplied the alleles for the trait. **Genomic imprinting,** which occurs during gamete formation, determines whether an allele will be expressed or not in the offspring. When the next generation makes gametes, old maternal and paternal imprints are removed, and alleles are again imprinted according to the sex of the parent. Most of the mammalian genes subject to imprinting identified so far are involved in embryonic development. The addition of methyl groups may inactivate the imprinted gene, assuring that the developing embryo has only one active copy.

Inheritance of Organelle Genes   Exceptions to Mendelian inheritance are found in the case of *extranuclear* (or *cytoplasmic*) *genes* located on small circles of DNA in mitochondria and plant plastids, which are transmitted to offspring in the cytoplasm of the ovum. Some rare human disorders are caused by mitochondrial mutations, and maternally inherited mitochondrial defects may contribute to diabetes, heart disease, and Alzheimer's disease.

Word Roots

aneu- = without (*aneuploidy:* a chromosomal aberration in which certain chromosomes are present in extra copies or are deficient in number)

cyto- = cell (*cytogenetic maps:* charts of chromosomes that locate genes with respect to chromosomal features)

hemo- = blood (*hemophilia:* a human genetic disease caused by a sex-linked recessive allele, characterized by excessive bleeding following injury)

mono- = one (*monosomic:* a chromosomal condition in which a particular cell has only one copy of a chromosome, instead of the normal two; the cell is said to be monosomic for that chromosome)

non- = not; **dis-** = separate (*nondisjunction:* an accident of meiosis or mitosis in which both members of a pair of homologous chromosomes or both sister chromatids fail to move apart properly)

poly- = many (*polyploidy:* a chromosomal alteration in which the organism possesses more than two complete chromosome sets)

re- = again; **com-** = together; **bin-** = two at a time (*recombinant:* an offspring whose phenotype differs from that of the parents)

trans- = across (*translocation:* attachment of a chromosomal fragment to a nonhomologous chromosome)

tri- = three; **soma-** = body (*trisomic:* a chromosomal condition in which a particular cell has an extra copy of one chromosome, instead of the normal two; the cell is said to be trisomic for that chromosome)

Structure Your Knowledge

1. Mendel's law of independent assortment applies to genes that are on different chromosomes. However, a few of the genes Mendel studied were actually located on the same chromosome. Explain why genes located more than 50 map units apart behave as though they are not linked. How can one determine whether these genes are linked and what the relative distance is between them?

2. You have found a new mutant phenotype in fruit flies that you suspect is recessive and sex-linked. What is the single, best cross you could make to confirm your predictions?

3. Various human disorders or syndromes are related to chromosomal abnormalities. What explanation can you give for the adverse phenotypic effects associated with these chromosomal alterations?

Genetics Problems

Again, one of the best ways to learn genetics is to do problems.

1. The following pedigree traces the inheritance of a genetic trait.

a. What type of inheritance does this trait show?

b. Chose an appropriate allele labeling system and give the predicted genotype for the following individuals:

1. ———   5. ———

2. ———   6. ———

3. ———   7. ———

4. ———

c. What is the probability that a child of individual #6 and a phenotypically normal male will have this trait?

2. Use the following recombination frequencies to determine the order of these genes on the chromosome.

a, c 10%   a, d 30%   b, c 24%   b, d 16%

3. In guinea pigs, black (*B*) is dominant to brown (*b*), and solid color (*S*) is dominant to spotted (*s*). A heterozygous black, solid-colored pig is mated with a brown, spotted pig. The total offspring from several litters are black solid = 16, black spotted = 5, brown solid = 5, and brown spotted = 14. Are these genes linked or unlinked? If they are linked, how many map units are they apart?

4. A woman is a carrier for a sex-linked lethal allele that causes an embryo with the allele to spontaneously abort. She has nine children. How many of these children do you expect to be boys?

5. A dominant sex-linked allele *B* produces white bars on black chickens, as seen in the Barred Plymouth Rock breed. A clutch of chicks has equal numbers of black and barred chicks. (Remember that sex is determined by the Z-W system in birds: ZZ are males, ZW are females.)

a. If only the females are found to be black, what were the genotypes of the parents?

b. If males and females are evenly represented in the black and barred chicks, what were the genotypes of the parents?

Test Your Knowledge

MULTIPLE CHOICE: *Choose the one best answer.*

1. Which of the following is/are part of the chromosomal theory of inheritance?

a. Genes are located on chromosomes.

b. Homologous chromosomes and their associated genes undergo segregation during meiosis.

c. Chromosomes and their associated genes undergo independent assortment in gamete formation.

d. Mendel's laws of inheritance relate to the behavior of chromosomes in meiosis.

e. All of the above are included.

2. A wild type is

a. the phenotype found most commonly in nature.

b. the dominant allele.

c. designated by a small letter if it is recessive or a capital letter if it is dominant.

d. a trait found on the X chromosome.

e. your basic party animal.

3. Sex-linked traits

a. are carried on an autosome but expressed only in males.

b. are coded for by genes located on a sex chromosome.

c. are found in only one or the other sex, depending on the sex-determination system of the species.

d. are always inherited from the mother in mammals and fruit flies.

e. depend on whether the gene was inherited from the mother or the father.

4. Linkage and cytogenetic maps for the same chromosome

a. are both based on mutant phenotypes and recombination data.

b. may have different orders of genes.

c. have both the same order of genes and intergenic distances.

d. have the same order of genes but different intergenic distances.

e. are created using chromosomal abnormalities.

5. The genetic event that results in Turner syndrome (X0) is probably

a. nondisjunction.

b. deletion.

c. parental imprinting.

d. monoploidy.

e. independent assortment.

6. A 1:1:1:1 ratio of offspring from a dihybrid testcross indicates that

a. the genes are linked.

b. the dominant organism was homozygous.

c. crossing over has occurred.

d. the genes are 25 map units apart.

e. the genes are not linked or are more than 50 map units apart.

7. Genes *A* and *B* are linked and 12 map units apart. A heterozygous individual, whose parents were *AAbb* and *aaBB,* would be expected to produce gametes in the following frequencies:

    a. 44% *AB*   6% *Ab*   6% *aB*   44% *ab*
    b. 6% *AB*   44% *Ab*   44% *aB*   6% *ab*
    c. 12% *AB*   38% *Ab*   38% *aB*   12% *ab*
    d. 6% *AB*   6% *Ab*   44% *aB*   44% *ab*
    e. 38% *AB*   12% *Ab*   12% *aB*   38% *ab*

8. A female tortoiseshell cat is heterozygous for the gene that determines black or orange coat color, which is located on the X chromosome. A male tortoiseshell cat

    a. could not occur.
    b. is hemizygous at this locus.
    c. must have resulted from a nondisjunction and has a Barr body in each of his cells.
    d. must have two alleles for coat color in each of his cells, one from his father and one from his mother.
    e. would be hermaphroditic.

9. What do we mean when we say that a few of the genes for Mendel's pea characters were physically linked but genetically unlinked?

    a. The genes are on the same chromosome, but they are more than 50 map units apart.
    b. The genes assort independently even though the chromosomes they are on travel to the metaphase plate together.
    c. Their alleles segregate in anaphase I, and each gamete receives a single allele for all of these genes.
    d. Dihybrid crosses with these genes produce more than 50% recombinant offspring even though they are on the same chromosome.
    e. Mendel could not determine that the genes were on the same chromosome because he did not perform dihybrid crosses with these particular gene pairs.

10. Genomic imprinting

    a. explains cases in which the phenotypic effect of an allele depends on the gender of the parent from whom that allele is inherited.
    b. may involve the silencing of an allele by methylation so that offspring inherit only one active copy of a gene.
    c. occurs more often in females because of the larger maternal contribution of cytoplasm.
    d. is more likely to occur in offspring of older mothers.
    e. involves both a and b.

11. A cross of a wild-type red-eyed female *Drosophila* with a violet-eyed male produces all red-eyed offspring. If the gene is sex-linked, what should the reciprocal cross (violet-eyed female × red-eyed male) produce? (Assume that the red allele is dominant to the violet allele.)

    a. all violet-eyed flies
    b. 3 red-eyed flies to 1 violet-eyed fly
    c. a 1:1 ratio of red and violet eyes in both males and females
    d. red-eyed females and violet-eyed males
    e. all red-eyed flies

12. Which of the following chromosomal alterations does not alter genic balance but may alter phenotype because of differences in gene expression?

    a. deletion
    b. inversion
    c. duplication
    d. nonidentical duplication
    e. nondisjunction

13. Two true-breeding *Drosophila* are crossed: a normal-winged, red-eyed female and a miniature-winged, vermilion-eyed male. The F_1 all have normal wings and red eyes. F_1 offspring are crossed with miniature-winged, vermilion-eyed flies. The following offspring of that cross were "scored":

    233 normal wing, red eye

    247 miniature wing, vermilion eye

    7 normal wing, vermilion eye

    13 miniature wing, red eye

    From these results, you could conclude that the alleles for miniature wings and vermilion eyes are

    a. both X-linked and dominant.
    b. located on autosomes and dominant.
    c. recessive, and these genes are located 4 map units apart.
    d. recessive, and these genes are located 20 map units apart.
    e. recessive, and the deviation from the expected 9:3:3:1 ratio is due to epistasis.

14. Tests of social behavior given to Turner syndrome volunteers (who are X0) found a correlation between scores on "behavioral inhibition" tasks and the source of the lone X chromosome. These test results appear to be an example of

a. parental imprinting of a gene on the X chromosome.

b. Y-linked inheritance.

c. meiotic nondisjunction.

d. chromosomal reciprocal translocation.

e. nonreciprocal crossover.

15. Suppose that alleles for a sex-linked character for wing shape in flies show incomplete dominance. The X^+ allele codes for pointed wings, the X^r for round wings, and X^+X^r individuals have oval wings. In a cross between an oval-winged female and a round-winged male, the following offspring were observed: oval-winged females, round-winged females, pointed-winged males, and round-winged males. A rare pointed-winged female was noted. Cytological study revealed that she had 2 X chromosomes. What could account for this unusual offspring?

a. a crossover between the two X chromosomes

b. a crossover between the X and Y chromosomes

c. a nondisjunction in meiosis II between two X^+ chromatids

d. a nondisjunction between the X and Y chromosomes, producing some sperm with no sex chromosome

e. Both c and d together could produce an X^+X^+ female when an XX ovum was fertilized by a sperm in which there was no sex chromosome.

16. Some girls who fail to undergo puberty are found to have Swyer syndrome, a condition in which they are externally female but have an XY genotype. Which of the following may explain the origin of this syndrome?

a. A mutation in the *XIST* gene, which codes for RNA molecules that coat the X chromosome and initiate X-inactivation, must have occurred.

b. A nondisjunction in the ovum from the mother resulted in both sex chromosomes coming from the father. Genomic imprinting of the father's X chromosome then caused the development of a female phenotype.

c. These individuals are actually XXY; the second X is not seen because it is condensed into a Barr body. They have small testes and are sterile, but otherwise appear female.

d. A mutation or deletion of the *SRY* gene on the Y chromosome prevented development of testes and production of male sex hormones required for a male phenotype.

e. A translocation of part of an X chromosome to the Y chromosome resulted in a double dose of female-determining genes.

17. A mutation in a mitochondrial gene has been linked to a rare muscle-wasting disease in humans. This disease is

a. found more often in males than females.

b. found more often in females than males.

c. inherited in a simple Mendelian fashion.

d. caused by a translocation of a nuclear gene to the mitochondria.

e. inherited from the mother and equally as likely to occur in male as female offspring.

18. In which of the following would you expect to find a Barr body?

a. an ovum

b. a sperm

c. a liver cell of a man

d. a liver cell of a woman

e. a mitochondrion

19. A cross between a wild-type mouse and a dwarf mouse homozygous for a recessive mutation in the *Igf2* gene produces heterozygous offspring that are normal if the dwarf parent was the mother, but dwarf if the dwarf parent was the father. Which of the following explains these results?

a. sex-linked inheritance

b. monosomy

c. genomic imprinting

d. inheritance of mitochondrial genes

e. a mutant *XIST* allele

20. Which of the following statements is *not* true about genetic recombination?

a. Recombination of linked genes occurs by crossing over.

b. Recombination of unlinked genes occurs by independent assortment of chromosomes.

c. Genetic recombination results in offspring with combinations of traits that differ from the phenotypes of both parents.

d. Recombinant offspring outnumber parental-type offspring when two genes are 50 map units apart on a chromosome.

e. The number of recombinant offspring is proportional to the distance between two gene loci on a chromosome.

Chapter 16

The Molecular Basis of Inheritance

Key Concepts

16.1 DNA is the genetic material

16.2 Many proteins work together in DNA replication and repair

16.3 A chromosome consists of a DNA molecule packed together with proteins

Framework

This chapter outlines the key evidence that was gathered to establish DNA as the molecule of inheritance. Watson and Crick's double helix, with its rungs of specifically paired nitrogenous bases and twisting phosphate-sugar side ropes, provided the three-dimensional model that explained DNA's ability to encode a great variety of information and produce exact copies of itself through semiconservative replication. The replication of DNA is an extremely fast and accurate process involving many enzymes and proteins. Eukaryotic chromosomes are complexes of DNA and protein that exhibit varying levels of packing during the cell cycle.

Chapter Review

Deoxyribonucleic acid, or DNA, is the genetic material. Nucleic acids' ability to direct their own replication allows for the precise copying and transmission of DNA to all the cells in the body and from one generation to the next. DNA encodes the blueprints that direct and control the biochemical, anatomical, physiological, and behavioral traits of organisms.

16.1 DNA is the genetic material

The Search for the Genetic Material: Scientific Inquiry
Chromosomes were shown to carry hereditary information and to consist of proteins and DNA. Until the 1940s, most scientists believed that proteins were the genetic material because of their specificity and heterogeneity. The role of DNA in heredity was first established through work with microorganisms—bacteria and viruses.

In 1928, F. Griffith was working with two strains of *Streptococcus pneumoniae.* When he mixed the remains of heat-killed pathogenic bacteria with harmless bacteria, some bacteria were changed into disease-causing bacteria. These bacteria incorporated external genetic material in a process called **transformation,** which results in a change in genotype and phenotype.

O. Avery worked to identify the transforming agent by testing the DNA, RNA, and proteins from heat-killed pathogenic cells. In 1944, Avery, McCarty, and MacLeod announced that DNA was the molecule that transformed bacteria.

Viruses consist of little more than DNA, or sometimes RNA, contained in a protein coat. They reproduce by infecting a cell and commandeering that cell's metabolic machinery. **Bacteriophages,** or **phages,** are viruses that infect bacteria. In 1952, A. Hershey and M. Chase showed that DNA was the genetic material of a phage known as T2 that infects the bacterium *E. coli.*

In 1950, E. Chargaff noted that the ratio of nitrogenous bases in the DNA from various organisms was species specific. Chargaff also determined that the number of adenines and thymines was approximately equal, and the number of guanines and cytosines was also equal. The A=T and G=C properties of DNA became known as *Chargaff's rules.*

INTERACTIVE QUESTION 16.1

Hershey and Chase devised an experiment using radioactive isotopes to determine whether the phage's DNA or protein entered the bacteria and was the genetic material of T2 phage.

a. How did they label phage protein?

b. How did they label phage DNA?

Separate samples of *E. coli* were infected with the differently labeled T2 cells, then blended and centrifuged to isolate the bacterial cells from the lighter viral particles.

c. Where was the radioactivity found in the samples with labeled phage protein?

d. Where was the radioactivity found in the samples with labeled phage DNA?

e. What did Hershey and Chase conclude from these results?

Building a Structural Model of DNA: Scientific Inquiry
By the early 1950s, the arrangement of covalent bonds in a nucleic acid polymer was established, but the three-dimensional structure of DNA was yet to be determined.

In X-ray crystallography, an X-ray beam passed through a substance produces an X-ray diffraction photo with a pattern of spots that a crystallographer interprets as information about three-dimensional molecular structure. J. Watson saw an X-ray photo produced by R. Franklin that indicated the helical shape of DNA. The width of the helix suggested that it consisted of two strands, thus the term **double helix.**

Watson and F. Crick constructed wire models to build a double helix that would conform to the X-ray measurements and the known chemistry of DNA. They arrived at a model that paired the nitrogenous bases on the inside of the helix with the sugar-phosphate chains on the outside. The helix makes one full turn every 3.4 nm; ten layers of nucleotide pairs, stacked 0.34 nm apart, are present in each turn of the helix.

To produce the molecule's uniform 2-nm width, a purine base must pair with a pyrimidine. The side groups of the bases permit two hydrogen bonds to form between adenine and thymine and three hydrogen bonds between guanine and cytosine. This complementary pairing explains Chargaff's rules. Van der Waals attractions between the closely stacked bases help to hold the molecule together.

In 1953, Watson and Crick published a paper in *Nature* reporting the double helix as the molecular model for DNA.

INTERACTIVE QUESTION   16.2

Review the structure of DNA by labeling the following diagrams.

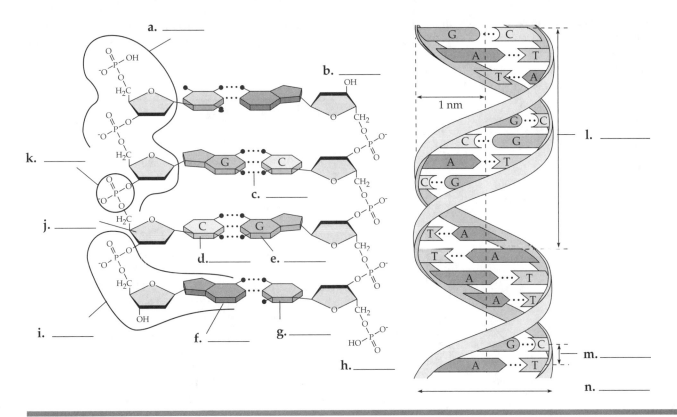

16.2   Many proteins work together in DNA replication and repair

The Basic Principle: Base Pairing to a Template Strand   Watson and Crick noted that the base-pairing rule of DNA sets up a mechanism for its replication. Each side of the double helix is an exact complement to the other. When the two sides of a DNA molecule separate, each strand serves as a template for rebuilding a double-stranded molecule.

The **semiconservative model** of DNA replication predicts that the two daughter DNA molecules each have one parental strand and one newly formed strand. In contrast, a conservative model predicts that the parent double helix re-forms and the duplicated molecule is totally new, whereas a dispersive model predicts that all four strands of the two DNA molecules are a mixture of parental and new DNA.

M. Meselson and F. Stahl tested these models by growing *E. coli* in a medium with ^{15}N, a heavy isotope that the bacteria incorporated into their nitrogenous bases. Cells with labeled DNA were transferred to a medium with the lighter isotope, ^{14}N. Samples were removed after one and two generations of bacterial growth, and DNA was extracted and centrifuged. The locations of the density bands in the centrifuge tubes confirmed the semiconservative model of DNA replication.

INTERACTIVE QUESTION   16.3

Using different colors for heavy (parental) and light (new) strands of DNA, sketch the DNA molecules formed in two replication cycles after *E. coli* were moved from medium containing ^{15}N to medium containing ^{14}N. Show the resulting density bands in the centrifuge tubes.

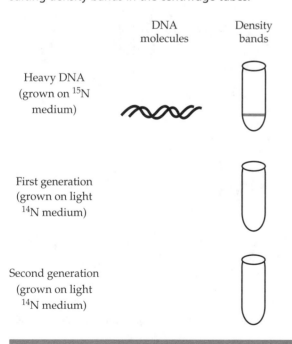

DNA Replication: A Closer Look   Replication begins at special sites, called **origins of replication,** where proteins that initiate replication bind to a specific sequence of nucleotides and separate the two strands to form a replication "bubble." Replication proceeds in both directions in the two Y-shaped **replication forks.**

An enzyme called **helicase** unwinds the helix and separates the parent strands at replication forks. **Single-strand binding proteins** keep the separated strands apart while they serve as templates. **Topoisomerase** helps relieve the strain from the twisting of DNA strands in front of helicase. An enzyme called **primase** joins about 10 RNA nucleotides paired to the parental strand to form the **primer** needed to start the new DNA strand.

Enzymes called **DNA polymerases** connect nucleotides to the growing end of a new DNA strand. DNA polymerase III and I are involved in replication in *E. coli*; at least 11 different DNA polymerases have been discovered so far in eukaryotes. A nucleoside triphosphate lines up with its complementary base on the template strand; it loses two phosphate groups and the hydrolysis of this pyrophosphate to two inorganic phosphates (Ⓟ$_i$) provides the energy for polymerization.

The two strands of a DNA molecule are antiparallel; their sugar-phosphate backbones run in opposite directions. The deoxyribose sugar of each nucleotide is connected to its own phosphate group at its 5' carbon and connects to the phosphate group of the adjacent nucleotide by its 3' carbon. Thus, a strand of DNA has polarity, with a 5' end where the first nucleotide's phosphate group is exposed, and a 3' end where the nucleotide at the other end has a hydroxyl group attached to its 3' carbon.

INTERACTIVE QUESTION   16.4

Look back to Interactive Question 16.2 and label the 5' and 3' ends of both strands of the DNA molecule.

DNA polymerases add nucleotides only to the free 3' end of a primer or growing DNA strand; DNA is replicated in a 5' → 3' direction. The **leading strand** is the new continuous strand being formed along one template strand by DNA polymerase III (DNA pol III). A "sliding clamp" protein moves DNA pol III along the template strand in the progressing replication fork.

The **lagging strand** is created as a series of short segments, called **Okazaki fragments,** that are formed in the 5' → 3' direction away from the replication fork. Each fragment on the lagging strand requires a primer. A continuous strand of DNA is produced after DNA polmerase I (DNA pol I) replaces the RNA primer with DNA nucleotides and an enzyme called **DNA ligase** joins the sugar-phosphate backbones of the fragments.

The various proteins that function in replication form a large DNA replication complex. In eukaryotic cells, many such complexes may anchor to the nuclear matrix and the DNA polymerase molecules may pull the parental DNA strands through them.

INTERACTIVE QUESTION   16.5

In this diagram showing bacterial DNA replication, label the following items: leading and lagging strands, Okazaki fragment, DNA pol III, DNA pol I, DNA ligase, helicase, single-strand binding proteins, primase, RNA primer, and 5′ and 3′ ends of parental DNA.

Proofreading and Repairing DNA Initial pairing errors in nucleotide placement may occur as often as 1 per 100,000 base pairs. The amazing accuracy of DNA replication (one error in ten billion nucleotides) is achieved as DNA polymerases check each newly added nucleotide against its template and remove incorrect nucleotides. Other enzymes also fix incorrectly paired nucleotides, called **mismatch repair.**

Reactive chemicals, radioactive emissions, X-rays, and UV light may alter DNA molecules. A large number of different types of DNA repair enzymes correct these changes. In **nucleotide excision repair,** the damaged strand is cut out by a **nuclease** and the gap is correctly filled through the action of DNA polymerase and ligase. In skin cells, nucleotide excision repair frequently corrects thymine dimers caused by ultraviolet rays of sunlight.

Replicating the Ends of DNA Molecules Because DNA polymerase cannot attach nucleotides to the 5′ end of a daughter DNA strand, repeated replications cause a progressive shortening of linear DNA molecules.

Multiple repetitions of a short nucleotide sequence at the ends of chromosomes, called **telomeres,** protect an organism's genes from being shortened during successive DNA replications.

The shortening of telomeres may limit cell division. In eukaryotic germ cells, however, the enzyme **telomerase** lengthens telomeres. Some somatic cancer cells and "immortal" strains of cultured cells produce telomerase and are thus capable of unregulated cell division.

INTERACTIVE QUESTION   16.6

Draw the last Okazaki fragment being formed on the lagging strand of a linear DNA molecule. Indicate how this results in a shortening of the end of the DNA molecule.

16.3   A chromosome consists of a DNA molecule packed together with proteins

The double-stranded, circular DNA molecule of a bacterial chromosome is supercoiled and tightly packed into a region of the cell called the **nucleoid.**

In eukaryotes, each chromosome consists of a single, extremely long DNA double helix precisely associated with a large amount of protein, forming a complex called **chromatin.** Eukaryotic cells have 1,000 times more DNA than a typical bacterium. This DNA fits into the nucleus through a multilevel system of coiling and folding.

Histones are small, positively charged proteins that bind tightly to the negatively charged DNA. Unfolded chromatin appears as a string of beads, each bead a **nucleosome** consisting of the DNA helix wound around a protein core of four pairs of different histones. The "string" between beads is called *linker DNA.* The amino end (histone tail) of each of the eight histones extends outward. Also called the 10-nm fiber because of its diameter, a nucleosome is the basic unit of DNA packing. A fifth histone, H1, attaches to the DNA near the bead.

The 30-nm fiber is a tightly coiled fiber of nucleosomes organized with the aid of histone H1, histone tails, and linker DNA. The looped domain is a loop of the 30-nm chromatin fiber attached to a protein scaffold, forming a 300-nm fiber. In a metaphase chromosome, looped domains coil and fold, further compacting the chromatin.

In interphase, chromatin is extended in the nucleus, whereas discrete, condensed chromosomes appear during mitosis. Some DNA packing is evident in interphase chromosomes, and their looped domains appear to be attached to discrete locations of the nuclear lamina inside the nuclear envelope, perhaps helping to organize areas of gene expression. Certain regions of chromatin, called **heterochromatin,** are in a highly condensed state during interphase. The more open form of interphase chromatin, called **euchromatin,** is available for transcription and thus gene expression.

Chemical modifications of histones, such as phosphorylation of amino acids on a histone tail, are involved in the changing structure of chromatin during mitosis, meiosis, and gene expression.

INTERACTIVE QUESTION   16.7

List the multiple levels of packing in a metaphase chromosome in order of increasing complexity.

Word Roots

helic- = a spiral (*helicase:* an enzyme that untwists the double helix of DNA at the replication forks)

liga- = bound or tied (*DNA ligase:* a linking enzyme for DNA replication)

-phage = to eat (*bacteriophages:* viruses that infect bacteria)

semi- = half (*semiconservative model:* type of DNA replication in which the replicated double helix consists of one old strand, derived from the parent molecule, and one newly made strand)

telos- = an end (*telomere:* the protective structure at each end of a eukaryotic chromosome)

trans- = across (*transformation:* a phenomenon in which external DNA is assimilated by a cell)

Structure Your Knowledge

1. Summarize the evidence and techniques Watson and Crick used to deduce the double-helix structure of DNA.

2. Review your understanding of DNA replication by describing the key enzymes and proteins (in the order of their functioning) that direct replication.

Test Your Knowledge

MULTIPLE CHOICE: *Choose the one best answer.*

1. One of the reasons most scientists believed proteins were the carriers of genetic information was that
   a. proteins were more heat stable than nucleic acids.
   b. the protein content of duplicating cells always doubled prior to division.
   c. proteins were much more complex and heterogeneous molecules than nucleic acids.
   d. early experimental evidence pointed to proteins as the hereditary material.
   e. proteins were found in DNA.

2. Transformation involves
   a. the uptake of external genetic material, often from one bacterial strain to another.
   b. the creation of a strand of RNA from a DNA molecule.
   c. the infection of bacterial cells by phage.
   d. the type of semiconservative replication shown by DNA.
   e. the replication of DNA along the lagging strand.

3. The DNA of an organism has thymine as 20% of its bases. What percentage of its bases would be guanine?

   a. 20%   **c.** 40%   **e.** 80%

   b. 30%   **d.** 60%

4. In his work with pneumonia-causing bacteria, Griffith found that

   a. DNA was the transforming agent.

   b. the pathogenic and harmless strains mated.

   c. heat-killed harmless cells could cause pneumonia when mixed with heat-killed pathogenic cells.

   d. some heat-stable chemical was transferred to harmless cells to transform them into pathogenic cells.

   e. a T2 phage transformed harmless cells to pathogenic cells.

5. T2 phage is grown in *E. coli* with radioactive phosphorus and then allowed to infect other *E. coli*. The culture is blended to separate the viral coats from the bacterial cells and centrifuged. Which of the following best describes the expected results of such an experiment?

   a. Both viral and bacterial DNA are labeled; radioactivity is found in the supernatant.

   b. Viral DNA is labeled; radioactivity is found in the pellet.

   c. Viral proteins are labeled; radioactivity is found in the supernatant but not in the pellet.

   d. Both viral and bacterial proteins are labeled; radioactivity is present in both the supernatant and the pellet.

   e. The virus destroyed the bacteria; no pellet is formed.

6. In their classic experiment, Meselson and Stahl

   a. provided evidence for the semiconservative model of DNA replication.

   c. were able to separate phage protein coats from *E. coli* by using a blender.

   c. found that DNA labeled with ^{15}N was of intermediate density.

   d. grew *E. coli* on labeled phosphorus and sulfur.

   e. found that DNA composition was species specific.

7. Watson and Crick concluded that each base could not pair with itself because

   a. there would not be room for the helix to make a full turn every 3.4 nm.

   b. the uniform width of 2 nm would not permit two purines or two pyrimidines to pair together.

   c. the bases could not be stacked 0.34 nm apart.

   d. identical bases could not hydrogen-bond together.

   e. they would be on antiparallel strands.

8. The joining of nucleotides in the polymerization of DNA requires energy from

   a. DNA polymerase.

   b. the hydrolysis of the terminal phosphate group of ATP.

   c. RNA nucleotides.

   d. the phosphate groups of the sugar-phosphate backbone.

   e. the hydrolysis of the pyrophosphates removed from nucleoside triphosphates.

9. The continuous elongation of a new DNA strand along one strand of DNA

   a. requires the action of DNA ligase as well as polymerase.

   b. occurs because DNA ligase can only elongate in the 5' → 3' direction.

   c. occurs on the leading strand.

   d. occurs on the lagging strand.

   e. a, b, and c are correct.

10. Which of the following statements about DNA polymerase is *incorrect*?

    a. It forms the bonds between complementary base pairs.

    b. It is able to proofread and correct for errors in base pairing.

    c. It is unable to initiate synthesis; it requires an RNA primer.

    d. It only works in the 5' → 3' direction.

    e. It is found in eukaryotes and prokaryotes.

11. Thymine dimers—covalent links between adjacent thymine bases in DNA—may be induced by UV light. When they occur, they are repaired by

    a. excision enzymes (nucleases).

    b. DNA polymerase.

    c. ligase.

    d. primase.

    e. a, b, and c are all needed.

12. How does DNA synthesis along the lagging strand differ from that on the leading strand?

    a. Nucleotides are added to the 5' end instead of the 3' end.

    b. Ligase is the enzyme that polymerizes DNA on the lagging strand.

    c. An RNA primer is needed on the lagging strand but not on the leading strand.

    d. Okazaki fragments, which each grow 5' → 3', must be joined along the lagging strand.

    e. Helicase synthesizes Okazaki fragments, which are then joined by ligase.

13. Which of the following enzymes or proteins is paired with an incorrect or inaccurate function?
    a. Helicase—unwind and separate parental double helix
    b. Telomerase—add telomere repetitions to ends of chromosomes
    c. Single-strand binding protein—hold strands of unwound DNA apart and straight
    d. Nuclease—cut out (excise) damaged DNA strand
    e. Primase—form DNA primer to start replication

Use the following diagram to answer Questions 14 through 17.

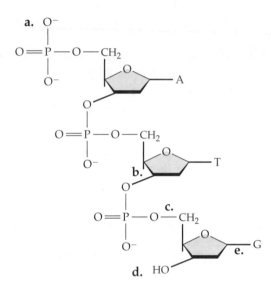

14. Which letter indicates the 5′ end of this single DNA strand?
    a.   b.   c.   d.   e.

15. At which letter would the next nucleotide be added?
    a.   b.   c.   d.   e.

16. Which letter indicates a phosphodiester bond formed by DNA polymerase?
    a.   b.   c.   d.   e.

17. The base sequence of the DNA strand made from this template would be (from top to bottom)
    a. A T C.   c. T A C.   e. A T G.
    b. C G A.   d. U A C.

18. What are telomeres, and what do they do?
    a. ever-shortening tips of chromosomes that may signal cells to stop dividing at maturity
    b. highly repetitive sequences at tips of chromosomes that protect the lagging strand during replication
    c. repetitive sequences of nucleotides at the centromere region of a chromosome

    d. enzymes that are present in germ cells that allow these cells to undergo repeated divisions
    e. Both a and b are correct.

19. You are trying to support your hypothesis that DNA replication is *conservative* (i.e., parental strands separate; complementary strands are made, but these new strands join together to make a new DNA molecule, and the parental strands rejoin.) You take *E. coli* that had grown in a medium containing only heavy nitrogen (^{15}N) and transfer a sample to a medium containing light nitrogen (^{14}N). After allowing time for only one DNA replication, you centrifuge a sample and compare the density band(s) formed with control bands for bacteria grown on either normal ^{14}N or ^{15}N medium. Which band location would support your hypothesis of *conservative* DNA replication?

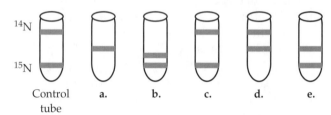

Bands after one replication on ^{14}N

20. Using the experiment explained in question 19, which centrifuge tube would represent the band distribution obtained after one replication showing the DNA replication is *semiconservative*?

21. Place the following structures in order from smallest to largest. Then choose the structure that is in the middle of that size range.
    a. looped domain
    b. histone H1
    c. nucleosome
    d. 30-nm fiber
    e. metaphase chromosome

22. Which of the following would *not* be involved in the proper behavior of chromosomes in meiosis?
    a. detachment of looped domains from the nuclear lamina
    b. phosphorylation of particular amino acids in histone tails
    c. condensing of centromeres and telomeres, converting them from euchromatin to heterochromatin
    d. coiling and folding of looped domains
    e. All of the above are involved in the behavior of chromosomes in meiosis.

Chapter 17

From Gene to Protein

Key Concepts

17.1 **Genes specify proteins via transcription and translation**

17.2 **Transcription is the DNA-directed synthesis of RNA:** *A Closer Look*

17.3 **Eukaryotic cells modify RNA after transcription**

17.4 **Translation is the RNA-directed synthesis of a polypeptide:** *A Closer Look*

17.5 **Point mutations can affect protein structure and function**

17.6 **Despite differences in gene expression among the domains of life, the concept of a gene is universal**

Framework

This chapter deals with the pathway from DNA to RNA to proteins. The instructions of DNA are transcribed to a sequence of codons in mRNA. In eukaryotes, mRNA is processed before it leaves the nucleus. Complexed with ribosomes, mRNA is translated as a sequence of amino acids in a polypeptide as tRNAs match their anticodons to the mRNA codons. Mutations, which alter the base pairs in DNA, may alter the protein product. A gene may be defined as a DNA sequence whose final product is either a polypeptide or an RNA molecule.

Chapter Review

The DNA-directed synthesis of proteins (or RNAs) is called **gene expression.**

17.1 Genes specify proteins via transcription and translation

Evidence from the Study of Metabolic Defects In 1909, A. Garrod first suggested that genes determine phenotype through the action of enzymes, reasoning that inherited diseases were caused by an inability to make certain enzymes.

G. Beadle and E. Tatum, working with mutants of a bread mold, *Neurospora crassa*, demonstrated the relationship between genes and enzymes. By growing nutritional mutants on complete growth medium and then transferring samples to various combinations of minimal medium and one added nutrient, they were able to identify the specific metabolic defect for each mutant. Three classes of *Neurospora* mutants that were unable to synthesize arginine were identified by supplementing different precursors of the pathway. Beadle and Tatum reasoned that the metabolic pathway of each class was blocked at a different enzymatic step. They formulated the *one gene–one enzyme hypothesis.*

Biologists revised this hypothesis to one gene–one protein. Because many proteins consist of more than one polypeptide chain, the axiom changed to one gene–one polypeptide. (Exceptions include genes that code for more than one protein due to alternative splicing and that code for RNA molecules.)

Basic Principles of Transcription and Translation RNA is the link between a gene and the protein for which it codes. RNA differs from DNA in three ways: The sugar component of its nucleotides is ribose, rather than deoxyribose; uracil (U) replaces thymine as one of its nitrogenous bases; and RNA is usually single stranded.

Transcription is the transfer of information from DNA to **messenger RNA (mRNA)** or another type of RNA, using the "language" of nucleic acids. **Translation** transfers information from mRNA to a polypeptide, changing from the language of nucleotides to that of amino acids. **Ribosomes** are the sites of translation—the synthesis of a polypeptide.

In prokaryotes, which lack a nucleus, transcription and translation can occur simultaneously. In eukaryotes, the mRNA must exit the nucleus before translation can begin. **RNA processing,** the modification of *pre-mRNA* or of the **primary transcript** of any gene (such as one coding for RNA) within the nucleus, occurs only in eukaryotes.

INTERACTIVE QUESTION   17.1

Fill in the sequence in the flow of genetic information, often called the *central dogma*. Put the name of the process above each arrow.

_____ $\longrightarrow$ _____ $\longrightarrow$ _____

The Genetic Code   A sequence of three nucleotides provides 4^3, or 64, possible unique sequences of nucleotides, more than enough to code for the 20 amino acids. The translation of nucleotides into amino acids uses a **triplet code** to specify each amino acid.

The base triplets along the **template strand** of a gene are transcribed into mRNA **codons.** The same strand of a DNA molecule can be the template strand for one gene and the complementary strand for another. The mRNA is complementary to the DNA template since its bases follow the same base-pairing rules, with the exception that uracil substitutes for thymine in RNA. Codons are read in the $5' \to 3'$ direction.

In the early 1960s, M. Nirenberg added artificial "poly U" mRNA to a test tube containing all the biochemical ingredients necessary for protein synthesis and obtained a polypeptide containing a single amino acid. Molecular biologists had deciphered all 64 codons by the mid-1960s. Three codons function as stop signals, or termination codons. The codon AUG both codes for methionine and functions as an initiation codon, a start signal for translation.

The code is often redundant, meaning that more than one codon may specify a single amino acid. The code is never ambiguous: No codon specifies two different amino acids.

The nucleotide sequence on mRNA is read in the correct **reading frame,** starting at a start codon and reading each triplet sequentially.

INTERACTIVE QUESTION   17.2

Practice using the dictionary of the genetic code in your textbook. Determine the amino acid sequence for a polypeptide coded for by the following mRNA transcript (written $5' \to 3'$):

AUGCCUGACUUUAAGUAG

The genetic code of codons and their corresponding amino acids is almost universal. A bacterial cell can translate the genetic messages of human cells. The near universality of a common genetic language provides compelling evidence of the antiquity of the code and the evolutionary connection of all living organisms.

17.2   Transcription is the DNA-directed synthesis of RNA: *A Closer Look*

Molecular Components of Transcription   The **promoter** is the DNA sequence where **RNA polymerase** attaches and initiates transcription. In bacteria, the **terminator** is the sequence that signals the end of transcription. A **transcription unit** is the sequence of DNA that is transcribed into one RNA molecule.

Bacteria have one type of RNA polymerase. Eukaryotes have three types; the one that synthesizes mRNA is called RNA polymerase II.

Synthesis of an RNA Transcript   The promoter is the binding site for RNA polymerase and determines where transcription starts and which DNA strand is used as the template. It includes recognition sequences, such as the **TATA box** common in eukaryotes, upstream from the start point. In eukaryotes, **transcription factors** must first recognize and bind to the promoter before RNA polymerase II can attach, at which point the assembly is called the **transcription initiation complex.**

RNA polymerase untwists the double helix, exposing DNA nucleotides for base pairing with RNA nucleotides, and joins the nucleotides to the 3' end of the growing polymer. The new RNA peels away from the DNA template and the DNA double helix re-forms. Several molecules of RNA polymerase may be transcribing simultaneously along a gene, enabling a cell to produce large quantities of mRNA.

In prokaryotes, transcription ends after RNA polymerase transcribes the terminator sequence. In eukaryotes, polymerase continues past a polyadenylation signal sequence (AAUAAA), proteins cut loose the pre-mRNA, and polymerase eventually falls off the DNA after transcribing hundreds more nucleotides.

Review the key steps of transcription in eukaryotes:

a.

b.

c.

17.3 Eukaryotic cells modify RNA after transcription

Alteration of mRNA Ends A modified guanine nucleotide is attached to the 5′ end of a pre-mRNA, and a string of adenine nucleotides, called a **poly-A tail,** is added to the 3′ end. The **5′ cap** and poly-A tail may facilitate transport of mRNA from the nucleus, aid ribosome attachment, and protect the ends of mRNA from hydrolytic enzymes. The cap and tail are attached to the untranslated regions (UTRs) at the 5′ and 3′ ends.

Split Genes and RNA Splicing Long segments of noncoding base sequences, known as **introns** or intervening sequences, occur within the boundaries of eukaryotic genes. The remaining coding regions are called **exons,** since they are expressed in protein synthesis (except that the 5′ and 3′ UTRs are not translated). A primary transcript is made of the gene—but introns are removed and exons joined before the mRNA leaves the nucleus—in a process called **RNA splicing.**

Signals for RNA splicing are sets of a few nucleotides at either end of each intron. *Small nuclear ribonucleoproteins (snRNPs),* composed of proteins and *small nuclear RNA (snRNA),* are components of a molecular complex called a **spliceosome.** The spliceosome snips an intron out of the RNA transcript and connects the adjoining exons. In addition to splice-site recognition and spliceosome assembly, a function of snRNA may be catalytic in intron removal.

RNA molecules that act as enzymes are called **ribozymes.** In some cases of RNA splicing, intron RNA catalyzes its own removal. The ability to function as enzymes relates to three properties of RNA: It is single-stranded and can base-pair with itself, forming a specific three-dimensional structure; some of its bases contain functional groups that can participate in catalysis; and it can hydrogen-bond with other nucleic acid molecules, allowing it to precisely locate slicing regions.

Some introns are involved in regulating gene activity, and splicing is necessary for the export of mRNA from the nucleus. **Alternative RNA splicing** allows some genes to produce different polypeptides. Exons may code for polypeptide **domains,** functional segments of a protein, such as binding and active sites. Introns may facilitate recombination of exons between different alleles or even between different genes. Such *exon shuffling* can result in novel proteins.

How does the mRNA that leaves the nucleus differ from the primary transcript pre-mRNA?

17.4 Translation is the RNA-directed synthesis of a polypeptide: *A Closer Look*

Molecular Components of Translation **Transfer RNA (tRNA)** molecules carry specific amino acids to ribosomes. They each have a base triplet, called an **anticodon,** that base-pairs with a complementary codon on mRNA, thus assuring that amino acids are arranged in the sequence prescribed by the transcription from DNA.

Transfer RNA is transcribed in the nucleus of a eukaryote and moves into the cytoplasm where it can be used repeatedly. These single-stranded, short RNA molecules are arranged into a cloverleaf shape by hydrogen bonding between complementary base sequences and fold into a three-dimensional, roughly L-shaped structure. The anticodon is at one end of the L; the 3′ end is the attachment site for its amino acid.

Each amino acid has a specific **aminoacyl-tRNA synthetase** that attaches it to its appropriate tRNA molecules to create an aminoacyl tRNA. The hydrolysis of ATP drives this process.

Sixty-one codons for amino acids can be read from mRNA, but there are only about 45 different tRNA molecules. A phenomenon known as **wobble** enables the third nucleotide of some tRNA anticodons to pair with more than one kind of base in the codon. Thus, one tRNA can recognize more than one mRNA codon, all of which code for the same amino acid carried by that tRNA.

INTERACTIVE QUESTION   17.5

Using some of the codons and the amino acids you identified in Interactive Question 17.2, fill in the following table.

DNA Triplet $3' \to 5'$	mRNA Codon $5' \to 3'$	Anticodon $3' \to 5'$	Amino Acid
			methionine
		GGA	
TTC			
	UAG		

Ribosomes facilitate the specific pairing of tRNA anticodons with mRNA codons during protein synthesis. They consist of a large and a small subunit, each composed of proteins and a form of RNA called **ribosomal RNA (rRNA).** Subunits are constructed in the nucleolus in eukaryotes. Prokaryotic ribosomes are smaller and differ enough in molecular composition that some antibiotics can inhibit them without affecting eukaryotic ribosomes.

A large and small subunit join to form a ribosome when they attach to an mRNA molecule. Ribosomes have a binding site for mRNA, a **P site** (peptidyl-tRNA site) that holds the tRNA carrying the growing polypeptide chain, an **A site** (aminoacyl-tRNA site) that holds the tRNA carrying the next amino acid, and an **E site** (exit site) from which discharged tRNAs leave the ribosome.

The protein components are mostly on the outside and the rRNA is in the interior at the interface between the subunits and at the A and P sites, supporting the hypothesis that rRNA performs a ribosome's catalytic functions.

Building a Polypeptide   The three stages of protein synthesis—initiation, elongation, and termination—all require the aid of protein "factors." The first two stages also require energy provided by the hydrolysis of GTP (guanosine triphosphate).

The initiation stage begins as the small subunit of the ribosome binds to an mRNA and an initiator tRNA carrying methionine, which attaches to the start codon AUG on the mRNA. With the aid of proteins called *initiation factors* and the expenditure of a GTP, the large subunit of the ribosome attaches to the small one, forming a *translation initiation complex.* The initiator tRNA fits into the P site.

The addition of amino acids in the elongation stage involves several proteins called elongation factors and occurs in a three-step cycle.

In codon recognition, an aminoacyl-tRNA base-pairs with the mRNA codon in the A binding site. This step requires energy from the hydrolysis of 2 GTP.

In the peptide bond formation step, an RNA molecule of the large subunit catalyzes the formation of a peptide bond between the carboxyl end of the polypeptide held in the P site and the amino acid in the A site. The polypeptide is now held by the amino acid in the A site.

In translocation, the tRNA carrying the growing polypeptide is translocated to the P site, a process requiring energy from the hydrolysis of a GTP molecule. The empty tRNA from the P site is moved to the E site and released. The next mRNA codon moves into the A site as the mRNA moves through the ribosome.

Termination occurs when a stop codon—UAA, UAG, or UGA—reaches the A site of the ribosome. A release factor binds to the stop codon and hydrolyzes the bond between the polypeptide and the tRNA in the P site. The completed polypeptide leaves through the *exit tunnel* of the large subunit. The two ribosomal subunits and other components dissociate.

An mRNA may be translated simultaneously by several ribosomes in strings called **polyribosomes** (or **polysomes**).

INTERACTIVE QUESTION   17.6

In the following diagrams of polypeptide synthesis, name the stages (1–4), identify the components (a–l), and then briefly describe what happens in each stage.

Completing and Targeting the Functional Protein
During and following translation, a polypeptide folds spontaneously into its secondary and tertiary structures. Chaperone proteins often facilitate the correct folding.

The protein may need to undergo *post-translational modifications:* Amino acids may be chemically modified; one or more amino acids at the beginning of the chain may be enzymatically removed; segments of the polypeptide may be excised; or several polypeptides may associate into a quaternary structure.

All ribosomes are identical, whether they are free ribosomes that synthesize cytosolic proteins or ER-bound ribosomes that make membrane and secretory proteins. Polypeptide synthesis begins in the cytoplasm. If a protein is destined for the endomembrane system or for secretion, its polypeptide chain will begin with a **signal peptide** that is recognized by a protein-RNA complex called a **signal-recognition particle (SRP),** which attaches the ribosome to a receptor protein that is part of a multiprotein translocation complex on the ER membrane. As the growing polypeptide threads into the ER, the signal peptide is usually removed.

Other signal peptides direct some proteins made in the cytosol to specific sites such as mitochondria, chloroplasts, or the interior of the nucleus.

INTERACTIVE QUESTION   17.7

What determines if a ribosome becomes bound to the ER?

17.5   Point mutations can affect protein structure and function

Mutations, changes in the genetic information of a cell (or virus), may involve large portions of a chromosome or affect just one base pair of nucleotides, as in a **point mutation.** If the mutation is in a cell that gives rise to a gamete, it may be passed on to offspring.

Types of Point Mutations   A **base-pair substitution** replaces one nucleotide and its complementary partner with another pair of nucleotides. Due to the redundancy of the genetic code, some base-pair substitutions do not change the amino acid translation and are called *silent mutations.* A **missense mutation**, which results in the insertion of a different amino acid, may not alter the character of the protein if the new amino acid has similar properties or is not located in a region crucial to that protein's function.

A base-pair substitution that results in a different amino acid in a critical portion of a protein, such as the active site of an enzyme, may significantly impair protein function.

Nonsense mutations occur when the point mutation changes an amino acid codon into a stop codon, prematurely halting the translation of the polypeptide chain and usually creating a nonfunctional protein.

Base-pair **insertions** or **deletions** that are not in multiples of three nucleotides alter the reading frame. All nucleotides downstream from the mutation will be improperly grouped into codons, creating extensive missense and usually ending in nonsense. These **frameshift mutations** almost always produce nonfunctional proteins.

Mutagens   Mutations can occur in a number of ways. *Spontaneous mutations* include base-pair substitutions, insertions, deletions, and longer mutations that occur during DNA replication, repair, or recombination. Physical agents, such as X-rays and UV light, and various chemical agents that cause mutations are called **mutagens.** Chemical mutagens include base analogs that substitute for normal bases and then pair incorrectly in DNA synthesis, chemicals that insert into and distort the double helix, and other agents that chemically change DNA bases. Tests can measure the mutagenic effects of chemicals and thus their potential carcinogenic risk.

INTERACTIVE QUESTION   17.8

Define the following, and explain what type of point mutation could cause each of these mutations.

a. silent mutation

b. missense mutation

c. nonsense mutation

d. frameshift mutation

17.6   Despite differences in gene expression among the domains of life, the concept of a gene is universal

Comparing Gene Expression in Bacteria, Archaea, and Eukarya   Archaeal RNA polymerase resembles that of eukaryotes; both depend on transcription factors. Termination of transcription in archaea may resemble that of eukaryotes. Archaeal ribosomes are the same size as bacterial ribosomes, but their sensitivity to chemical inhibitors is more similar to eukaryotic ribosomes. The initiation of translation is most similar between archaea and bacteria.

In bacteria and probably archaea, transcription and translation occur almost simultaneously, whereas in eukaryotes, transcription is physically separated from translation by the nuclear envelope, allowing extensive RNA processing to occur before RNA leaves the nucleus. Eukaryotes also target proteins to particular cell locations.

What Is a gene? Revisiting the Question   Our definition of a gene has evolved from Mendel's heritable factors, to Morgan's loci along chromosomes, to one gene–one polypeptide. Research continually refines our understanding of the structural and functional aspects of genes, which now include introns, promoters, and other regulatory regions. The best working definition of a gene is that it is a region of DNA whose final product is either a polypeptide or an RNA molecule.

Word Roots

anti- = opposite (*anticodon:* a specialized base triplet on one end of a tRNA molecule that recognizes a particular complementary codon on an mRNA molecule)

exo- = out, outside, without (*exon:* a coding region of a eukaryotic gene that is expressed)

intro- = within (*intron:* a noncoding, intervening sequence within a eukaryotic gene)

muta- = change; **-gen** = producing (*mutagen:* a physical or chemical agent that causes mutations)

poly- = many (*poly-A tail:* the modified end of the 3' end of an mRNA molecule consisting of the addition of some 50 to 250 adenine nucleotides)

trans- = across; **-script** = write (*transcription:* the synthesis of RNA on a DNA template)

Structure Your Knowledge

1. You have been introduced to several types of RNA in this chapter. List four of these types and their functions. What explains the functional versatility of RNA molecules?

2. Make sure you understand and can explain the processes of transcription and translation. You may find that filling in the table on the next page (for eukaryotic gene expression) in a study group helps you to review these processes.

	Transcription	Translation
Template		
Location		
Molecules involved		
Enzymes involved		
Control—start and stop		
Product		
Product processing		
Energy source		

3. What is the genetic code? Explain redundancy and the wobble phenomenon. What is the significance of the fact that the genetic code is nearly universal?

4. Prepare a concept map showing the types and consequences of point mutations.

Test Your Knowledge

MULTIPLE CHOICE: *Choose the one best answer.*

1. In Beadle and Tatum's study of *Neurospora*, they were able to identify three classes of mutants that needed arginine added to minimal media in order to grow. The production of arginine includes the following steps: precursor → ornithine → citrulline → arginine. What nutrient(s) had to be supplied for the mutants with a defective enzyme for the ornithine → citrulline step to grow?

   a. the precursor
   b. ornithine
   c. citrulline
   d. either ornithine or citrulline
   e. the precursor, ornithine, and citrulline

2. Transcription involves the transfer of information from
   a. DNA to RNA.
   b. RNA to DNA.
   c. mRNA to an amino acid sequence.
   d. DNA to an amino acid sequence.
   e. the nucleus to the cytoplasm.

3. If the 5' → 3' nucleotide sequence on the complementary (noncoding) DNA strand is CAT, what is the corresponding codon on mRNA?

   a. UAC   c. GUA   e. CAT
   b. CAU   d. GTA

4. RNA polymerase
   a. is the protein responsible for the production of ribonucleotides.
   b. is the enzyme that creates hydrogen bonds between nucleotides on the DNA template strand and their complementary RNA nucleotides.
   c. is the enzyme that transcribes exons but does not transcribe introns.
   d. is a ribozyme composed of snRNPs.
   e. moves along the template strand of DNA, elongating an RNA molecule in a 5' → 3' direction.

5. How is the template strand for a particular gene determined?
   a. It is the DNA strand that runs from the 5' → 3' direction.
   b. It is the DNA strand that runs from the 3' → 5' direction.
   c. It is established within the promoter.
   d. It doesn't matter which strand is the template because they are complementary and will produce the same mRNA.
   e. It is signaled by a polyadenylation signal sequence.

6. Which enzyme synthesizes tRNA?
   a. RNA replicase
   b. RNA polymerase
   c. aminoacyl-tRNA synthetase
   d. ribosomal enzymes
   e. ribozymes

7. Which of the following is true of RNA processing?
   a. Exons are excised before the mRNA is translated.
   b. The RNA transcript that leaves the nucleus may be much longer than the original transcript.
   c. Assemblies of protein and snRNPs, called spliceosomes, may catalyze splicing.
   d. Large quantities of rRNA are assembled into ribosomes.
   e. Signal peptides are added to the 5′ end of the transcript.

8. Which of the following is *not* involved in the formation of a eukaryotic transcription initiation complex?
   a. TATA box
   b. transcription factors
   c. snRNA
   d. RNA polymerase II
   e. promoter

9. A prokaryotic gene 600 nucleotides long can code for a polypeptide chain of how many amino acids (at most)?
   a. 100
   b. 200
   c. 300
   d. 600
   e. 1,800

10. All of the following are transcribed from DNA *except*
    a. exons.
    b. introns.
    c. tRNA.
    d. rRNA.
    e. promoter.

11. What might introns have to do with the evolution of new proteins?
    a. The excised introns are transcribed and translated as new proteins by themselves.
    b. Introns are more likely to accumulate mutations than exons, and these mutations then result in the production of novel proteins.
    c. Introns that are self-excising may also function as hydrolytic enzymes for other processes.
    d. Introns provide more area where crossing over may occur (without interfering with the coding sequences) and thus increase the probability of exon shuffling between alleles.
    e. Introns often correspond to domains in proteins that fold independently and have specific functions. Switching domains between nonallelic genes could produce novel proteins.

12. A ribozyme is
    a. an exception to the one gene–one RNA molecule axiom.
    b. an enzyme that adds the 5′ cap and poly-A tail to mRNA.
    c. an example of rearrangement of protein domains caused by RNA splicing.
    d. an RNA molecule that functions as an enzyme.
    e. an enzyme that produces both small and large ribosomal subunits.

13. All of the following would be found in a prokaryotic cell *except*
    a. mRNA.
    b. rRNA.
    c. simultaneous transcription and translation.
    d. snRNA.
    e. RNA polymerase.

14. Which of the following is transcribed and then translated to form a protein product?
    a. gene for tRNA
    b. intron
    c. gene for a transcription factor
    d. 5′ and 3′ UTRs
    e. gene for rRNA

15. Transfer RNA
    a. forms hydrogen bonds between its codon and the anticodon of an mRNA in the A site of a ribosome.
    b. binds to its specific amino acid in the active site of an aminoacyl-tRNA synthetase.
    c. has catalytic activity and is thus a ribozyme.
    d. is translated from mRNA.
    e. is produced in the nucleolus.

16. Place the following events in the synthesis of a polypeptide in the proper order.
    1. A peptide bond forms.
    2. An aminoacyl tRNA matches its anticodon to the codon in the A site.
    3. A tRNA translocates from the A to the P site, and an unattached tRNA leaves the ribosome from the E site.
    4. The large subunit attaches to the small subunit, with the initiator tRNA in the P site.
    5. A small subunit binds to an mRNA and an initiator tRNA.
       a. 4-5-3-2-1
       b. 4-5-2-1-3
       c. 5-4-3-2-1
       d. 5-4-1-2-3
       e. 5-4-2-1-3

17. Translocation in the process of translation involves
    a. the hydrolysis of a GTP molecule.
    b. the movement of the tRNA in the A site to the P site.
    c. the movement of the mRNA strand one triplet length.
    d. the release of the unattached tRNA from the E site.
    e. all of the above.

18. Which of the following type of molecule catalyzes the formation of a peptide bond?
    a. RNA polymerase
    b. rRNA
    c. mRNA
    d. aminoacyl-tRNA synthetase
    e. proteinase

19. Which of the following is *not* true of an anticodon?
    a. It consists of three nucleotides.
    b. It lines up in the $5' \rightarrow 3'$ direction along the $5' \rightarrow 3'$ mRNA strand.
    c. It extends from one loop of a tRNA molecule.
    d. It may pair with more than one codon.
    e. Its base uracil base-pairs with adenine.

20. Changes in a polypeptide following translation may involve
    a. the addition of sugars or lipids to certain amino acids.
    b. the action of enzymes to add amino acids at the beginning of the chain.
    c. the removal of poly-A from the end of the chain.
    d. the addition of a $5'$ cap of a modified guanosine residue.
    e. all of the above.

21. Several proteins may be produced at the same time from a single mRNA by
    a. the action of several ribosomes in a string, called a polyribosome.
    b. several RNA polymerase molecules working sequentially.
    c. signal peptides that associate ribosomes with rough ER.
    d. containing several promoter regions.
    e. the involvement of multiple spliceosomes.

22. A signal peptide
    a. is most likely to be found on cytosolic proteins produced by bacterial cells.
    b. directs an mRNA molecule into the lumen of the ER.
    c. is a sign to help bind the small ribosomal unit at the initiation codon.
    d. would be the first 20 or so amino acids of a protein destined for a membrane location or secretion from the cell.
    e. is part of the $5'$ cap.

23. A base deletion early in the coding sequence of a gene would most likely result in
    a. a nonsense mutation.
    b. a frameshift mutation.
    c. multiple missense mutations.
    d. a nonfunctional protein.
    e. all of the above.

24. Which of the following best characterizes gene expression in archaea?
    a. Their molecular machinery and processes are exactly like that of bacteria.
    b. Their molecular machinery and processes are exactly like that of eukaryotes.
    c. Their processes of transcription and translation differ substantially from both bacteria and eukaryotes.
    d. Their mechanisms of gene expression are most similar to eukaryotes, although they share some similarities with bacteria.
    e. Their processes of transcription and translation have not yet been studied.

Chapter 18

Regulation of Gene Expression

Key Concepts

18.1 Bacteria often respond to environmental change by regulating transcription

18.2 Eukaryotic gene expression can be regulated at any stage

18.3 Noncoding RNAs play multiple roles in controlling gene expression

18.4 A program of differential gene expression leads to the different cell types in a multicellular organism

18.5 Cancer results from genetic changes that affect cell cycle control

Framework

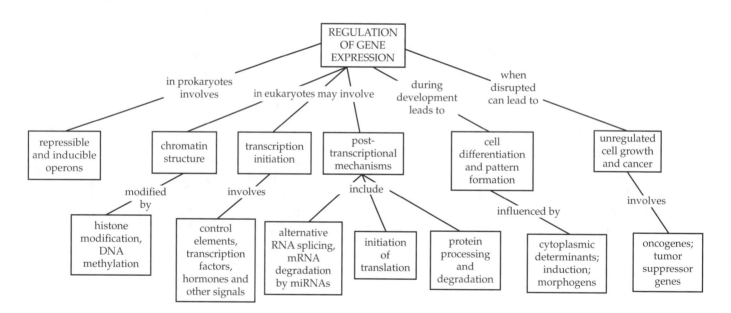

Chapter Review

18.1 Bacteria often respond to environmental change by regulating transcription

Bacteria can respond to short-term environmental fluctuations by regulating enzyme activity through *feedback inhibition*. Regulation of gene transcription controls enzyme production as metabolic needs change. F. Jacob and J. Monod described the *operon model* for gene regulation in 1961.

Operons: The Basic Concept In bacteria, genes for the different enzymes of a single metabolic pathway may be grouped together into one transcription unit and served by a single promoter. Thus, these genes are under *coordinate control.* An **operator** is a segment of DNA within or near the promoter that controls the access of RNA polymerase to the genes. An **operon** is the DNA segment that includes the clustered genes, the promoter, and the operator.

A **repressor** is a protein that binds to a specific operator, blocking attachment of RNA polymerase and thus turning an operon "off." **Regulatory genes** code for repressor proteins. These allosteric proteins, which may assume active or inactive shapes, are usually produced at a slow but continuous rate. The activity of the repressor protein may be determined by the presence or absence of a **corepressor.**

In the *trp* operon, tryptophan is the corepressor that binds to the repressor protein, changing it into its active shape, which has a high affinity for the operator and switches off the *trp* operon. Should the tryptophan concentration of the cell fall, repressor proteins are no longer bound with tryptophan; the operator is no longer repressed; RNA polymerase attaches to the promoter; and mRNA for the enzymes needed for tryptophan synthesis is produced.

Repressible and Inducible Operons: Two Types of Negative Gene Regulation The transcription of a *repressible operon*, such as the *trp* operon, is inhibited when a specific small molecule binds to a regulatory protein. The transcription of an *inducible operon* is stimulated when a specific small molecule interacts with a regulatory protein.

The *lac* operon, controlling lactose metabolism in *E. coli*, is an inducible operon that contains three genes. The *lac* repressor, coded for by the regulatory gene, *lacI*, is innately active, binding to the *lac* operator and switching off the operon. Allolactose, an isomer of lactose, acts as an **inducer,** a small molecule that binds to and inactivates the repressor protein, so that the operon can be transcribed.

INTERACTIVE QUESTION   18.1

In the following diagram of the *lac* operon, an operon for inducible enzymes, identify components a through i.

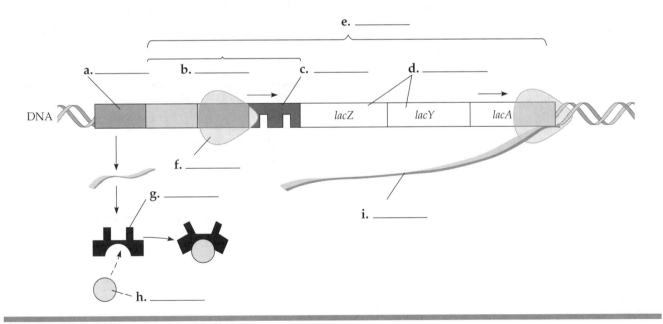

INTERACTIVE QUESTION 18.2

a. *Repressible enzymes* usually function in _____ pathways. The pathway's product serves as a _____ to activate the repressor and turn off enzyme synthesis and prevent overproduction of the product of the pathway. Genes for repressible enzymes are usually switched _____ and the repressor is synthesized in an _____ form.

b. *Inducible enzymes* usually function in _____ pathways. Nutrient molecules serve as _____ to stimulate production of the enzymes necessary for their breakdown. Genes for inducible enzymes are usually switched _____ and the repressor is synthesized in an _____ form.

Positive Gene Regulation   *E. coli* cells preferentially use glucose. Should glucose levels fall, transcription of operons for other catabolic pathways can be increased through the action of the allosteric regulatory protein *catabolite activator protein (CAP)*, which is an **activator. Cyclic AMP (cAMP)** accumulates in the cell when glucose is scarce and binds with CAP, changing it to its active shape. The active CAP attaches above the promoter region and stimulates transcription by facilitating the binding of RNA polymerase.

The regulation of the *lac* operon includes negative control by the repressor protein that is inactivated by the presence of lactose, and positive control by CAP when complexed with cAMP.

18.2   Eukaryotic gene expression can be regulated at any stage

All cells control the expression of their genes in response to their external and internal environments. Eukaryotic cells in multicellular organisms also regulate gene expression to create specialized cells.

Differential Gene Expression   Differences in cells with the same genome are the result of **differential gene expression.** Only a small portion of a cell's DNA codes for proteins; some codes for RNA; and most is noncoding DNA. Gene expression can be regulated at various stages, although, as in prokaryotes, it is most commonly regulated at transcription.

Regulation of Chromatin Structure   DNA packing and the location of a gene's promoter relative to nucleosomes and to the attachment to the chromosome scaffold may help control which genes are available for transcription. Genes within highly condensed heterochromatin are usually not expressed.

The attachment of acetyl groups ($-COCH_3$) to lysines in histone tails, called **histone acetylation,** interrupts the binding to neighboring nucleosomes such that transcription proteins may more easily access genes. Some enzymes that acetylate or deacetylate histones are associated with or are part of transcription factors. The addition of methyl groups ($-CH_3$) to histone tails leads to condensation of the chromatin. Phosphorylation of an amino acid next to a methylated amino acid undoes that effect. According to the *histone code hypothesis,* the specific patterns of modifications determine chromatin shape and influence transcription.

In a process called DNA methylation, methyl groups are added to DNA bases (usually cytosine). Genes are generally more heavily methylated in cells in which they are not expressed. Some proteins that bind to methylated DNA also interact with histone deacetylation enzymes, reinforcing the transcription repression.

DNA methylation may reinforce gene regulatory decisions of early development as enzymes act during DNA replication to methylate daughter DNA strands and pass on methylation records. Such methylation patterns account for **genomic imprinting** in mammals in which either a maternal or paternal allele of certain genes is permanently regulated.

Epigenic inheritance is the transmission of traits by mechanisms that do not involve changes in DNA sequences, such as chromatin modifications that affect gene expression in future generations of cells.

INTERACTIVE QUESTION 18.3

a. Give an example of highly methylated and inactive DNA common in mammalian cells.

b. Would histone tail deacetylation increase or decrease the transcription of a gene located in that nucleosome?

Regulation of Transcription Initiation   A typical gene consists of a promoter sequence, where a transcription initiation complex that includes RNA polymerase II attaches, and a sequence of introns interspersed among the coding exons. After transcription, RNA processing of the pre-mRNA removes the introns and adds a 5' cap and a poly-A tail at the 3' end. **Control elements** are noncoding sequences to which transcription factors bind that help regulate transcription.

General transcription factors bind to RNA polymerase and each other to initiate the transcription of all protein-coding genes. The binding of *specific transcription factors* to control elements, however, greatly increases transcription rates of particular genes.

Proximal control elements are located close to the promoter. **Enhancers** are groups of *distal control elements* located far upstream or downstream of a gene or in an intron. According to a current model, a protein-mediated bend in the DNA brings activators bound to enhancers into contact with *mediator proteins*, which interact with proteins at the promoter. These protein-protein interactions assemble the initiation complex. Hundreds of transcription activators have been identified. Most have a DNA-binding domain and one or more activation domains that bind other regulatory proteins.

Some specific transcription factors act as repressors, which can inhibit gene expression in several ways, such as blocking the binding of activators or binding to specific control elements in an enhancer. Some activators and repressors influence chromatin structure by recruiting proteins that either acetylate or deacetylate histones. Most eukaryotic gene repression may occur by this *silencing* at the level of chromatin modification.

Only a dozen or so short nucleotide sequences have been found in control elements, but each enhancer contains about 10 different control elements. The *combination* of these elements ensures that specific activator proteins must be present to activate transcription. Thus, the specific activators and repressors made in a cell determine which genes are expressed.

Prokaryotic genes are often organized into operons and are transcribed together. In a few cases, coexpressed eukaryotic genes (each with its own promoter) are clustered together and their regulation involves changes in chromatin structure, making them available for transcription. Most eukaryotic genes coding for enzymes in the same metabolic pathway are scattered on the chromosomes. The integrated control of these genes may involve a specific collection of control elements associated with each related gene and requiring the same activators.

For example, steroid hormones may activate multiple genes when they bind to specific receptor proteins within a cell that function as transcription activators for several genes. Other signal molecules bind to cell surface receptors and initiate signal transduction pathways that activate particular transcription activators or repressors. The same chemical signal will activate genes with the same control elements and thus coordinate gene expression.

INTERACTIVE QUESTION 18.4

Label the components of this diagram that shows how enhancers, mediator proteins, and transcription factors facilitate formation of a transcription initiation complex.

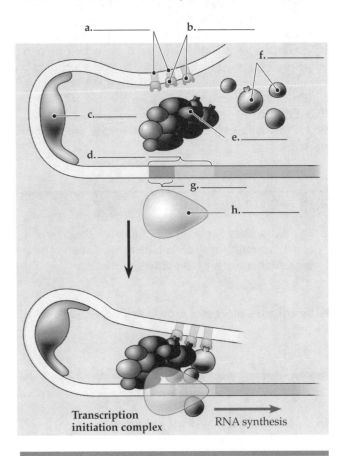

Transcription initiation complex   RNA synthesis →

Mechanisms of Post-Transcriptional Regulation Gene expression, measured by the amount of functional protein that is made, can be regulated at any post-transcriptional step. At the level of RNA processing, **alternative RNA splicing** may produce different mRNA molecules when regulatory proteins control which segments of the primary transcript are chosen as introns and which are chosen as exons.

In contrast to prokaryotic mRNA that is degraded after a few minutes, eukaryotic mRNA can last hours or even weeks. The hydrolysis of mRNA by nucleases is usually preceded by the enzymatic shortening of the poly-A tail and removal of the 5' cap. The untranslated region (UTR) at the 3' end of mRNA may contain nucleotide sequences that affect how long an mRNA lasts.

The translation of eukaryotic mRNA can be delayed by the binding of regulatory proteins to the 5' UTR that prevent ribosome binding. A great deal of mRNA is synthesized and stored in egg cells; translation may begin when enzymes add more A nucleotides to poly-A tails or when translation initiation factors are activated following fertilization.

Following translation, polypeptides are often cleaved or chemical groups added to yield an active protein. Some proteins must be transported to target locations.

Selective degradation of proteins may serve as a control mechanism in the cell. Molecules of the protein ubiquitin are added to mark proteins for destruction. **Proteasomes** recognize and degrade the marked proteins.

INTERACTIVE QUESTION   18.5

a. The untranslated regions (UTR) at both the 5' and 3' ends of an mRNA may contribute to regulation of gene expression. Describe their different effects.

b. Describe the process of protein degradation in a cell.

18.3   Noncoding RNAs play multiple roles in controlling gene expression

Mounting evidence indicates that a large amount of the genome may be transcribed into non-protein-coding RNAs (*or noncoding RNAs*). A large population of small RNA molecules may be involved in regulating gene expression.

Effects on mRNAs by MicroRNAs and Small Interfering RNAs   Long RNA precursors fold on themselves forming hairpin structures. These hairpins are cut off and an enzyme called Dicer trims the ends. One strand of the fragment becomes a **microRNA (miRNA).** These small, single-stranded RNA molecules associate with a complex of proteins and then base-pair with complementary sequences on target mRNA. This miRNA-protein complex either blocks translation or degrades the mRNA. An estimated one-third of human genes may be regulated by miRNAs.

This inhibition of gene expression by RNA molecules was first observed experimentally and called **RNA interference (RNAi).** Injection of double-stranded RNA into a cell turned off genes with matching sequences. It is now known that cells cut these longer double-stranded RNA molecules into many short RNA molecules called **small interfering RNAs (siRNAs).** SiRNAs appear to inhibit the expression of genes by the same mechanism as miRNAs.

Chromatin Remodeling and Silencing of Transcription by Small RNAs   Small RNAs are also involved in the formation of heterochromatin. Evidence from yeast experiments indicates that a double-stranded RNA produced from DNA in the centromere region is complexed with proteins and targets the centromere region, where it initiates chromatin remodeling into heterochromatin. Similar mechanisms may also block transcription of specific genes.

INTERACTIVE QUESTION   18.6

a. Describe the sequence from the formation of a primary miRNA transcript to the blocking of transcription of an mRNA.

b. What is the difference between microRNAs and small interfering RNAs?

c. Do small RNAs regulate gene expression through affecting translation or transcription?

18.4   A program of differential gene expression leads to the different cell types in a multicellular organism

A Genetic Program for Embryonic Development   The three key processes of embryonic development are cell division, the production of large numbers of cells; **cell differentiation,** the formation of cells specialized in structure and function; and **morphogenesis,** the physical processes that produce body shape. All three processes are based on differential gene expression resulting from differences in gene regulation in cells.

Cytoplasmic Determinants and Cell-Cell Signals   The cytoplasm of an unfertilized egg cell contains maternal mRNA, proteins, and other substances that are unevenly distributed, and the first few mitotic divisions separate these components and expose the nuclei in these new cells to different environments. These maternal components of the egg cell that influence early development by regulating gene expression are called **cytoplasmic determinants.**

The other important source of developmental control comes from signals received from other embryonic cells. Change in the gene expression of target cells resulting from signals from other cells is called **induction.**

Sequential Regulation of Gene Expression During Cellular Differentiation   A cell's developmental history leads to its eventual differentiation as a cell with a specific structure and function. **Determination** is a term used to describe the condition when a cell is irreversibly committed to its fate. When a cell becomes differentiated, it expresses genes for *tissue-specific proteins*, and that expression is usually controlled at the level of transcription.

The embryonic precursor cells from which muscle cells arise have the potential to develop into a number of different cell types. Once they become committed to becoming muscle cells, they are called *myoblasts.* Researchers have identified several "master regulatory genes" that cause myoblast determination. One of these is *myoD*, which codes for a transcription factor named MyoD that binds to control elements and initiates transcription of other muscle-specific transcription factors. These secondary transcription factors then activate muscle-protein genes. MyoD also turns on genes that block the cell cycle and stop cell division, and it activates its own transcription, thus maintaining the cell's differentiated state.

INTERACTIVE QUESTION   18.7

a. What is the difference between determination and differentiation?

b. The MyoD protein has been shown to be able to transform some, but not all, differentiated cells into muscle cells. Why doesn't it work on all kinds of cells?

Pattern Formation: Setting Up the Body Plan   **Pattern formation** is the ordering of cells and tissues into their characteristic structures and locations. An animal's three major body axes are laid out early in development. Cells and their progeny develop in response to molecular cues called **positional information** that tell a cell where it is located relative to the body axes and neighboring cells.

The genetic study of *Drosophila* development has led to the discovery of some common developmental principles. A fruit fly's body consists of a series of segments grouped into the head, thorax, and abdomen. The anterior-posterior and dorsal-ventral axes are determined by positional information provided by cytoplasmic determinants localized in the unfertilized egg. Each egg cell is surrounded by nurse cells and follicle cells that supply nutrients and other molecules needed for development. Eggs are laid following fertilization, and a wormlike larva develops. Following three larval stages, the larva becomes a pupa. Metamorphosis produces an adult fly.

In the 1940s, E. B. Lewis studied developmental mutants and was able to map certain mutations that control pattern formation to specific genes. In the 1970s, C. Nüsslein-Volhard and E. Wieschaus undertook a search for the genes that control segment formation. They studied mutations that were **embryonic lethals,** which prevented the development of viable larvae. They exposed flies to a chemical mutagen and then performed many thousands of crosses to detect recessive mutations causing the death of embryos or resulting in larvae with abnormal segmentation. They identified 120 genes involved in pattern formation leading to normal segmentation.

Maternal effect genes are genes of the mother that code for proteins or mRNA that are deposited in the unfertilized egg. These genes are also called **egg-polarity genes** because they determine the anterior-posterior and dorsal-ventral axes of the egg and consequently the embryo.

One egg-polarity gene is **bicoid.** The product of the *bicoid* gene is concentrated at one end of the embryo and responsible for determining its anterior end. Offspring of a mother defective for this gene have two tail regions and lack the front half of the body. Researchers located bicoid mRNA concentrated in the most anterior end of egg cells. Following fertilization, the mRNA is translated into Bicoid protein, which diffuses posteriorly, forming a gradient in the early embryo. Such substances, whose gradients establish an embryo's axes or other features, are called **morphogens.** Proteins whose gradients determine the posterior end and establish the dorsal-ventral axis have also been identified. Positional information then establishes the proper number of segments and finally triggers the formation of each segment's characteristic structures.

INTERACTIVE QUESTION   18.8

What type of evidence established that Bicoid protein is a morphogen that determines the anterior end of a fruit fly?

18.5   Cancer results from genetic changes that affect cell cycle control

Types of Genes Associated with Cancer Chemical carcinogens, X-rays, or certain viruses most often cause the changes in the genes that regulate cell growth and division that lead to cancer. Viruses that can cause cancer in animals are called *tumor viruses.*

Oncogenes, or cancer-causing genes, were first found in certain retroviruses. Similar genes were later recognized in the genomes of humans and other animals. Cellular **proto-oncogenes,** which code for proteins that stimulate cell growth and division, may become oncogenes by several mechanisms that result in an overproduction of growth-stimulating protein.

Mutations in **tumor-suppressor genes** can contribute to the onset of cancer when they result in a decrease in the activity of proteins that prevent uncontrolled cell growth.

INTERACTIVE QUESTION   18.9

a. List three genetic changes by which a proto-oncogene can become an oncogene.

b. List three possible functions of tumor-suppressor proteins.

Interference with Normal Cell-Signaling Pathways In about 30% of human cancers, the *ras* proto-oncogene is mutated. The *ras* **gene** codes for a G protein that connects a growth-factor receptor on the plasma membrane to a cascade of protein kinases that leads to the production of a cell cycle stimulating protein. A mutation may create a hyperactive version of the Ras protein that relays a signal without binding of a growth factor.

The *p53* **gene** is mutated in about 50% of human cancers. It codes for a tumor-suppressor protein that is a transcription factor involved in the synthesis of several growth-inhibiting proteins. Such proteins may activate the *p21* gene, whose product binds to cyclin-dependent kinases, halting the cell cycle and allowing time for the cell to repair damaged DNA. The p53 protein can also activate genes involved in DNA repair. Should DNA damage be irreparable, p53 activates "suicide genes" that initiate apoptosis.

The Multistep Model of Cancer Development More than one somatic mutation appears to be needed to produce a cancerous cell. Mutation of a single proto-oncogene can stimulate cell division, but usually both alleles for several tumor-suppressor genes must be defective to allow uncontrolled cell growth. In many malignant tumors, the gene for telomerase becomes activated and cells become able to divide indefinitely.

Inherited Predisposition and Other Factors Contributing to Cancer A genetic predisposition to certain cancers may involve the inheritance of an oncogene or a recessive mutant allele for a tumor-suppressor gene. Approximately 15% of colorectal cancers involve inherited mutations, often in the tumor-suppressor gene *APC*, which regulates cell migration and adhesion.

About 5–10% of breast cancer cases are linked to an inherited mutant allele for either *BRCA1* or *BRCA2*, both of which appear to be tumor-suppressor genes.

The ultraviolet radiation in sunlight and chemicals in cigarette smoke may contribute to cancer through their DNA damaging effects. Viruses can cause cancer when the insertion of viral DNA introduces an oncogene or affects a proto-oncogene or tumor-suppressor gene. Viral proteins may also inactivate p53 or other tumor-suppressor proteins.

Word Roots

morph- = form; **-gen** = produce (*morphogen:* a substance that provides positional information in the form of a concentration gradient along an embryonic axis)

proto- = first, original; **onco-** = tumor (*proto-oncogene:* a normal cellular gene corresponding to an oncogene)

Structure Your Knowledge

1. Complete this concept map to help you review the mechanisms by which bacteria regulate their gene expression in response to varying metabolic needs.

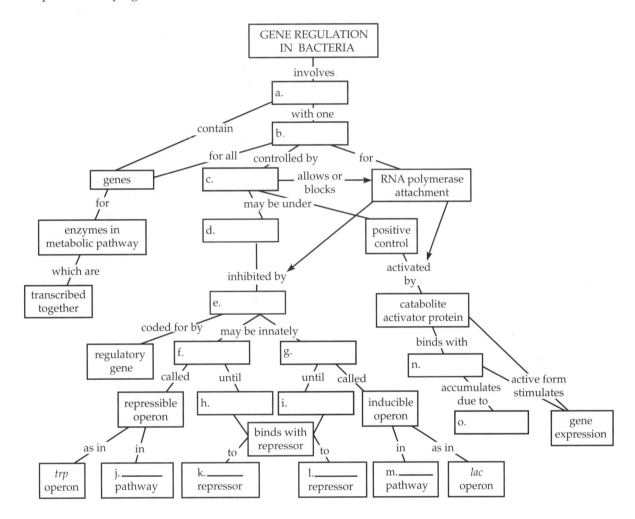

2. Fill in the following table to help you organize the major mechanisms that can regulate the expression of eukaryotic genes.

Level of Control	Examples
Chromatin structure	**a.**
Transcriptional regulation	**b.**
Post-transcriptional regulation	**c.**
Transcriptional regulation	**d.**
Post-transcriptional regulation	**e.**

3. How might the mechanism of transcriptional regulation differ for cytoplasmic determinants and for the cell-cell signaling involved in induction?

Test Your Knowledge

MULTIPLE CHOICE: *Choose the one best answer.*

1. Inducible enzymes
   a. are usually involved in anabolic pathways.
   b. are produced when a small molecule inactivates the repressor protein.
   c. are produced when an activator molecule enhances the attachment of RNA polymerase with the operator.
   d. are regulated by inherently inactive repressor molecules.
   e. are regulated almost entirely by feedback inhibition.

2. In *E. coli,* tryptophan switches off the *trp* operon by
   a. inactivating the repressor protein.
   b. inactivating the gene for the first enzyme in the pathway by feedback inhibition.
   c. binding to the repressor and increasing the latter's affinity for the operator.
   d. binding to the operator.
   e. binding to the promoter.

3. In the control of gene expression in bacteria, a regulatory gene
   a. has its own promoter.
   b. is transcribed continuously.
   c. is not contained in the operon it controls.
   d. codes for repressor proteins.
   e. is or does all of the above.

4. A mutation that renders nonfunctional the product of a regulatory gene for a repressible operon would result in
   a. continuous transcription of the genes of the operon.
   b. complete blocking of the attachment of RNA polymerase to the promoter.
   c. irreversible binding of the repressor to the operator.
   d. no difference in transcription rate when an activator protein was present.
   e. negative control of transcription.

5. The control of gene expression is more complex in eukaryotic cells because
   a. DNA is associated with protein.
   b. gene expression differentiates specialized cells.
   c. the chromosomes are linear and more numerous.

   d. operons are controlled by more than one promoter region.
   e. inhibitory or activating molecules may help regulate transcription.

6. DNA methylation of cytosine residues
   a. initiates the acetylation of histones.
   b. may be a mechanism of epigenic inheritance when methylation patterns are repeated in daughter cells.
   c. occurs in the promoter region and enhances binding of RNA polymerase.
   d. is a signal for proteasomes to degrade a protein.
   e. may be related to the transformation of proto-oncogenes to oncogenes.

7. Which of the following is *not* true of enhancers?
   a. They may be located thousands of nucleotides upstream from the genes they affect.
   b. When bound with activators, they interact with the promoter region and other transcription factors to produce an initiation complex.
   c. They may complex with steroid-activated receptor proteins, which selectively activate specific genes.
   d. They may coordinate the transcription of enzymes involved in the same metabolic pathway when they contain the same combination of control elements.
   e. Each gene may have several enhancers, and each enhancer may be associated with and regulate several genes.

8. Which of the following is *not* an example of the control of gene expression that occurs after transcription?
   a. mRNA stored in the cytoplasm needing activation of translation initiation factors
   b. the length of time mRNA lasts before it is degraded
   c. rRNA genes amplified in multiple copies in the genome
   d. alternative RNA splicing before mRNA exits from the nucleus
   e. splicing or modification of a polypeptide

9. Which of the following might a proto-oncogene code for?
   a. DNA polymerase
   b. RNA polymerase
   c. receptor proteins for growth factors
   d. an enhancer
   e. transcription factors that inhibit cell division genes

10. A gene can develop into an oncogene when
    a. it is present in more copies than normal.
    b. it undergoes a translocation that removes it from its normal control region.
    c. a mutation results in a more active or resistant protein.
    d. a mutation in a control element increases expression.
    e. any of the above occur.

11. A tumor-suppressor gene could cause the onset of cancer if
    a. both alleles have mutations that decrease the activity of the gene product.
    b. only one allele has a mutation that alters the gene product.
    c. it is inherited in mutated form from a parent.
    d. a proto-oncogene has also become an oncogene.
    e. both a and d have happened.

12. A eukaryotic gene typically has all of the following associated with it *except*
    a. a promoter.
    b. an operator.
    c. enhancers.
    d. introns and exons.
    e. control elements.

13. What is apoptosis?
    a. a cell suicide program that may be initiated by p53 protein in response to DNA damage
    b. metastasis, or the spread of cancer cells to a new location in the body
    c. the transformation of a normal cell to a cancer cell
    d. the mutation of a G protein into a hyperactive form
    e. the transformation of a proto-oncogene to an oncogene by a point mutation

14. Which of the following would you expect to find as part of a receptor protein that binds with a steroid hormone?
    a. a TATA box
    b. a domain that binds to DNA and protein-binding domains
    c. an activated operator region that allows attachment of RNA polymerase
    d. an enhancer sequence located at some distance upstream or downstream from the promoter
    e. transmembrane domains that facilitate its localization in a plasma membrane

15. What are proteasomes?
    a. complexes of proteins that excise introns
    b. single-stranded RNA molecules complexed with proteins that block translation of or degrade mRNA
    c. small, positively charged proteins that form the core of nucleosomes
    d. enormous protein complexes that degrade proteins marked with ubiquitin
    e. complexes of transcription factors whose protein-protein interactions are required for enhancing gene transcription

16. A larger portion of the DNA in a eukaryotic cell is transcribed than would be predicted by the proteins made by the cell. What is being transcribed and what is its function?
    a. Multiple enhancer regions are being transcribed to amplify the transcription of protein-coding genes.
    b. These transcriptions are of noncoding "junk" DNA that is a remnant of mutated protein-coding segments. The transcripts are degraded by nuclear enzymes.
    c. These transcriptions produce double-stranded small interfering RNAs that regulate the transcription of other genes.
    d. The additional DNA that is transcribed is introns that are excised from the primary transcript in the production of mRNA.
    e. Many RNA coding genes are transcribed. Precursor RNAs fold into hairpin structures, which are cut and processed into miRNAs that regulate translation of mRNAs.

17. Cytoplasmic determinants are
    a. unevenly distributed cytoplasmic components of an unfertilized egg.
    b. often involved in transcriptional regulation.
    c. usually separated in the first few mitotic divisions following fertilization.
    d. maternal contributions that help to direct the initial stages of development.
    e. all of the above.

18. Pattern formation in animals is based on
    a. positional information a cell receives from gradients of morphogens.
    b. the induction of cells by the nurse cells in the mother's ovary.
    c. the packing of chromatin in the nucleus.
    d. differentiation of cells, which then migrate together to form tissues and organs.
    e. the first few mitotic divisions.

19. What would be the fate of a *Drosophila* larva that inherits two copies of a mutant *bicoid* gene (one mutant allele from each heterozygous parent)?
    a. It develops two heads, one at each end of the larva.
    b. It develops two tails, one at each end of the larva.
    c. It develops normally but, if female, produces mutant larvae that have two tail regions.
    d. It develops into an adult with legs growing out of its head.
    e. It receives no *bicoid* mRNA from the nurse cells of its mother.

20. In this hypothetical embryo, a high concentration of a morphogen called morpho is needed to activate gene *P*; gene *Q* is active at medium concentrations of morpho or above; and gene *R* is expressed as long as there is any quantity of morpho present. A different morphogen called phogen has the following effects: activates gene *S* and inactivates gene *Q* when at medium to high concentrations. If morpho and phogen are diffusing from where they are produced at the opposite ends of the embryo, which genes will be expressed in region 2 of this embryo? (Assume diffusion through the three regions from high at source to medium in the middle to low concentration at the opposite end.)

Morpho
produced here

Phogen
produced here

    a. genes *P*, *Q*, *R*, and *S*
    b. genes *P*, *Q*, and *R*
    c. genes *Q* and *R*
    d. genes *R* and *S*
    e. gene *R*

21. Which of the following would most likely account for a family history of colorectal cancer?
    a. a family diet that is low in fats and high in fiber
    b. inheritance of one mutated *APC* allele that regulates cell adhesion and migration
    c. a family history of breast cancer
    d. inheritance of the *ras* oncogene that locks the G protein in an active configuration
    e. inheritance of a proto-oncogene

22. How is the coordinated transcription of genes involved in the same pathway regulated?
    a. The genes are transcribed in one transcription unit, although each gene has its own promoter.
    b. The genes are located in the same region of the chromosome, and enzymes deacetylate the entire region so that transcription may begin.
    c. The genes all respond to the same general transcription factors, although they may respond to different specific transcription factors.
    d. A steroid hormone selectively binds to the promoters for all the genes.
    e. The genes have the same combination of control elements in the enhancer.

Viruses

Key Concepts

19.1 A virus consists of a nucleic acid surrounded by a protein coat

19.2 Viruses reproduce only in host cells

19.3 Viruses, viroids, and prions are formidable pathogens in animals and plants

Framework

A virus is an infectious particle consisting of a genome of single- or double-stranded DNA or RNA enclosed in a protein capsid and sometimes a membrane envelope derived from the host. Viruses replicate using the metabolic machinery of their bacterial, animal, or plant host. Viral infections may destroy the host cell and cause diseases within the host organism. Viruses may have evolved from plasmids or transposons.

Chapter Review

19.1 A virus consists of a nucleic acid surrounded by a protein coat

The Discovery of Viruses: **Scientific Inquiry** The search for the cause of tobacco mosaic disease led to the discovery of viruses. The infectious agent could not be removed from infected sap by passing it through a filter designed to remove bacteria. Unlike bacteria, the infectious agent could not be cultivated on nutrient media, but it was able to reproduce within plants. In 1935, W. Stanley crystallized the infectious particle, now known as tobacco mosaic virus (TMV). Since that time, many viruses have been seen with the electron microscope.

Structure of Viruses Viral genomes may be single- or double-stranded DNA or single- or double-stranded RNA. Viral genes are usually contained on a single linear or circular nucleic acid molecule.

The **capsid,** or protein shell, is built from a large number of often identical protein subunits *(capsomeres)* and may be rod-shaped (helical viruses), polyhedral (icosahedral viruses), or more complex in shape. **Viral envelopes,** derived from membranes of the host cell but also including viral proteins and glycoproteins, may cloak the capsids of viruses found in animals. Some viruses also contain a few viral enzymes.

Complex capsids are found among **bacteriophages,** or **phages,** the viruses that infect bacteria. Of the phages that infect the bacterium *E. coli,* T2, T4, and T6 have similar capsid structures consisting of a polyhedral head and a protein tail piece with tail fibers for attaching to a bacterium.

19.2 Viruses reproduce only in host cells

Viruses are obligate intracellular parasites that lack metabolic enzymes and other equipment needed to reproduce. Each virus type has a limited **host range** due to proteins on the outside of the virus that recognize only specific receptor molecules on the host cell surface.

General Features of Viral Reproductive Cycles Once the viral genome enters the host cell, the cell's enzymes, nucleotides, amino acids, ribosomes, ATP, and other resources are used to replicate the viral genome and produce capsid proteins. DNA viruses use host DNA polymerases to copy their genome, whereas RNA viruses use virus-encoded polymerases for replicating their RNA genome.

After replication, viral nucleic acid and capsid proteins spontaneously assemble to form new viruses within the host cell, a process called self-assembly. Hundreds or thousands of newly formed virus particles are released, often destroying the host cell in the process.

Reproductive Cycles of Phages   A reproductive cycle of a phage that culminates in lysis of the host cell and release of newly produced phages is known as a **lytic cycle. Virulent phages** reproduce only by a lytic cycle.

The T4 phage uses its tail fibers to stick to a receptor site on the surface of an *E. coli* cell. The sheath of the tail contracts and thrusts its viral DNA into the cell, leaving the empty capsid behind. The *E. coli* cell begins to transcribe and translate phage genes, one of which codes for an enzyme that chops up host cell DNA. Nucleotides from the degraded bacterial DNA are used to create viral DNA. Capsid proteins are made and assembled into phage tails, tail fibers, and heads. The viral components assemble into 100 to 200 phage particles that are released after an enzyme is manufactured that damages the bacterial cell wall.

Mutations that change their receptor sites and **restriction enzymes** that chop up viral DNA once it enters the cell help to defend bacteria against viral infection.

In a **lysogenic cycle,** a virus reproduces its genome without killing its host. **Temperate phages** can reproduce by the lytic and lysogenic cycles.

When the phage lambda (λ) injects its DNA into an *E. coli* cell, it can begin a lytic cycle, or its DNA may be incorporated into the host cell's chromosome and begin a lysogenic cycle as a **prophage.** Most of the genes of the inserted phage genome are repressed by a protein coded for by a prophage gene. Reproduction of the host cell replicates the phage DNA along with the bacterial DNA. The prophage may exit the bacterial chromosome, usually in response to environmental stimuli, and start the lytic cycle.

Expression of prophage genes may cause a change in the bacterial phenotype. Several disease-causing bacteria would be harmless except for the expression of prophage genes that code for toxins.

INTERACTIVE QUESTION   19.1

In this diagram of a lytic and lysogenic cycle, describe steps 1–8 and label structures a–e.

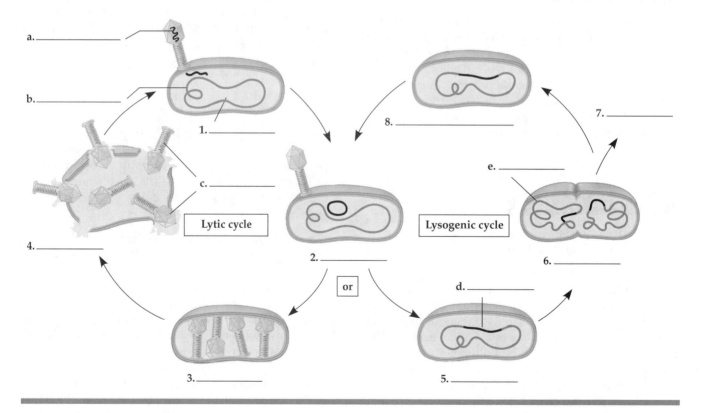

Reproductive Cycles of Animal Viruses Animal viruses are classified based on their type of genome. A viral envelope surrounds the capsid of almost all animal viruses that have RNA genomes. Glycoproteins extending from the viral membrane attach to receptor sites on a host cell plasma membrane. Either the two membranes fuse, or endocytosis transports the capsid into the cell. The viral genome replicates and directs the synthesis of viral proteins. Glycoproteins are produced and embedded in the ER membrane, and then transported to the plasma membrane. New viruses bud off within an envelope derived from the host's plasma membrane bearing viral glycoproteins.

Herpesviruses reproduce within the host cell nucleus and are temporarily cloaked in host cell nuclear membrane. The herpesvirus' double-stranded DNA can remain latent as a minichromosome in the host cell nucleus until it initiates herpes infections in times of stress.

The single-stranded RNA of animal class IV viruses can serve directly as mRNA. The RNA genome of class V viruses must first be transcribed into a strand of complementary RNA (using a viral enzyme packaged inside the capsid) that then serves as mRNA and a template for making genome RNA.

In the complicated reproductive cycle of **retroviruses** (class VI), the viral RNA genome is transcribed into double-stranded DNA by a viral enzyme, **reverse transcriptase.** This viral DNA is then integrated into a chromosome, where it is transcribed by the host cell into viral RNA, which acts both as new viral genome and as mRNA for viral proteins. **HIV (human immunodeficiency virus)** is a retrovirus that causes **AIDS (acquired immunodeficiency syndrome).** The integrated viral DNA remains as a **provirus** within the host cell DNA. New viruses, assembled with two copies of both the RNA genome and reverse transcriptase within a capsid, bud off covered in host cell plasma membrane studded with viral glycoproteins.

INTERACTIVE QUESTION   19.2

Summarize the flow of genetic information during replication of a retrovirus. Indicate the enzymes that catalyze this flow.

$$\rightarrow \qquad \rightarrow$$

Enzymes:

Evolution of Viruses Viruses may have evolved from fragments of cellular nucleic acids that moved from one cell to another and eventually evolved special packaging. Sources of viral genomes may have been plasmids, self-replicating circles of DNA found in bacteria and yeast, and transposons, segments of DNA that can change locations within a cell's genome.

The genomes of viruses are often more similar to those of their host cells than to the genomes of viruses infecting other hosts. Recent sequencing of viral genomes, however, has found genetic similarities among viruses that otherwise seem distantly related.

Some researchers propose that the recently discovered mimivirus, with its large size and genome coding for many cellular enzymes, most likely evolved before the first cells. According to this hypothesis, the viral genome became reduced after the viruses began exploiting their cellular hosts.

19.3   Viruses, viroids, and prions are formidable pathogens in animals and plants

Viral Diseases in Animals The symptoms of a viral infection may be caused by viral-programmed toxins produced by infected cells, cells killed or damaged by the virus, or the body's defense mechanisms fighting the infection.

Vaccines are variants or derivatives of pathogens that induce the immune system to react against the actual disease agent. Vaccinations have greatly reduced the incidence of many viral diseases.

Unlike bacteria, viruses use the host's cellular machinery to replicate, and few drugs have been found to treat or cure viral infections. Some antiviral drugs resemble nucleosides and interfere with viral nucleic acid synthesis.

Emerging Viruses The emergence of "new" viral diseases may be linked to the mutation of an existing virus (as in influenza viruses), the dissemination of an existing virus to a more widespread population (as in HIV), or the spread from one host species to another (as in the SARS virus in which bats are a natural reservoir). Many new human diseases are estimated to originate by the third mechanism.

A general outbreak of a disease is called an **epidemic.** Flu epidemics may be caused by new strains of viruses to which people have little immunity. **Pandemics** (global epidemics) have been caused by influenza A strains that infect a wide range of animals. If different strains of virus undergo genetic recombination within an animal's cells and accumulate mutations that allow the virus to infect human cells, the recombinant virus may be highly pathogenic. The H5N1 virus (named for the form of the viral surface proteins hemagglutinin and neuraminidase) is called "avian flu" and has killed many domestic birds and 160 people as of 2006. Its human mortality rate is greater than 50%. Millions of birds have been killed or vaccinated in the attempt to prevent its emergence as a virus that is transmitted easily between humans.

Viral Diseases in Plants   Most plant viruses are RNA viruses. Plant viral diseases may spread through *vertical transmission* from a parent plant or through *horizontal transmission* from an external source. Plant injuries increase susceptibility to viral infections, and insects can act as carriers of viruses.

INTERACTIVE QUESTION   19.3

How does a virus spread throughout a plant? Are there cures for viral plant diseases?

Viroids and Prions: The Simplest Infectious Agents   **Viroids** are very small infectious molecules of circular RNA that can replicate in host plant cells, causing abnormal development and stunted growth.

Prions are protein infectious agents that may be linked to several degenerative brain diseases, such as mad cow disease and Creutzfeldt-Jacob disease in humans. Prions apparently spread disease by converting normal cellular proteins into the defective form of the prion.

Word Roots

capsa- = a box (*capsid:* the protein shell that encloses the viral genome)

lyto- = loosen (*lytic cycle:* a type of viral replication cycle resulting in the release of new phages by death or lysis of the host cell)

-phage = to eat (*bacteriophages:* viruses that infect bacteria)

pro- = before (*provirus:* viral DNA that inserts into a host genome)

retro- = backward (*retrovirus:* an RNA virus that reproduces by transcribing its RNA into DNA and then inserting the DNA into a cellular chromosome)

virul- = poisonous (*virulent virus:* a virus that reproduces only by a lytic cycle)

Structure Your Knowledge

1. Create a concept map that describes the lytic and lysogenic cycles of a phage.

2. Complete the concept map below to help organize your understanding of viruses.

Test Your Knowledge

MULTIPLE CHOICE: *Choose the one best answer.*

1. The study of viruses has done all of the following *except*
   a. provide information on the molecular biology of all organisms.
   b. illuminate the sexual reproductive cycles of viruses.
   c. develop new techniques for manipulating genes.
   d. develop an understanding of the causes of diseases.
   e. show the role of mutation in the relationship between host and virus.

2. Beijerinck concluded that the cause of tobacco mosaic disease was not a filterable toxin because
   a. the infectious agent could not be cultivated on nutrient media.
   b. a plant sprayed with filtered sap would develop the disease.
   c. the infectious agent could be crystallized.
   d. the infectious agent reproduced and could be passed on from a plant infected with filtered sap.
   e. the filtered sap was infectious even though microbes could not be found in it.

3. Viral genomes may be any of the following *except*
   a. single-stranded DNA.
   b. double-stranded RNA.
   c. misfolded infectious proteins.
   d. a linear single-stranded RNA molecule.
   e. a circular double-stranded DNA molecule.

4. Retroviruses have a gene for reverse transcriptase that
   a. uses viral RNA as a template for making complementary RNA strands.
   b. protects viral DNA from degradation by restriction enzymes.
   c. destroys the host cell DNA.
   d. translates RNA into proteins.
   e. uses viral RNA as a template for DNA synthesis.

5. Virus particles are formed from capsid proteins and nucleic acid molecules
   a. by spontaneous self-assembly.
   b. at the direction of viral enzymes.
   c. by using host cell enzymes.
   d. using energy from ATP stored in the tail piece.
   e. by both b and d.

6. A virus has a base ratio of (A + G)/(U + C) = 1. What type of virus is this?
   a. a single-stranded DNA virus
   b. a single-stranded RNA virus
   c. a double-stranded DNA virus
   d. a double-stranded RNA virus
   e. a retrovirus

7. Vertical transmission of a plant viral disease may involve
   a. the movement of viral particles through the plasmodesmata.
   b. the inheritance of an infection from a parent plant.
   c. a bacteriophage transmitting viral particles.
   d. insects carrying viral particles between plants.
   e. the transfer of filtered sap.

8. Bacteria defend against viral infection through the action of
   a. antibiotics that they produce.
   b. restriction enzymes that chop up foreign DNA.
   c. their R plasmids.
   d. reverse transcriptase.
   e. incorporating viral DNA into the bacterial chromosome.

9. Drugs that are effective in treating viral infections
   a. induce the body to produce antibodies.
   b. inhibit the action of viral ribosomes.
   c. interfere with the synthesis of viral nucleic acid.
   d. change the cell-recognition sites on the host cell.
   e. are vaccines that stimulate the immune system to create immunity.

10. Which of the following is true of prions?
    a. They are emerging viruses.
    b. They are fast-acting infectious agents.
    c. They probably evolved from transposons.
    d. They are infectious proteins that may convert brain proteins into misfolded forms.
    e. They may be transferred between animals by sexual contact.

11. The replication of the genome of an RNA virus uses
    a. DNA polymerase from the host.
    b. RNA replicating enzymes coded for by viral genes or carried in the viral capsid.
    c. reverse transcriptase to synthesize RNA.
    d. RNA polymerase from the host.
    e. restriction enzymes from the host.

12. The replication of the genome of a DNA virus uses
    a. DNA polymerase from the host.
    b. RNA replicating enzymes coded for by viral genes or carried in the viral capsid.
    c. reverse transcriptase to make an RNA copy from the DNA.
    d. RNA polymerase from the host.
    e. restriction enzymes from the host.

Chapter 20

Biotechnology

Key Concepts

20.1 DNA cloning yields multiple copies of a gene or other DNA segment

20.2 DNA technology allows us to study the sequence, expression, and function of a gene

20.3 Cloning organisms may lead to production of stem cells for research and other applications

20.4 The practical applications of DNA technology affect our lives in many ways

Framework

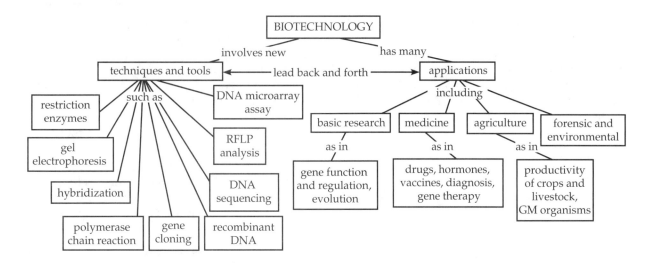

Chapter Review

DNA technology began with techniques for making **recombinant DNA,** which combines nucleotide sequences from different organisms or species into the same DNA molecule. **Genetic engineering,** the direct manipulation of genetic material for practical purposes, has begun a revolution in **biotechnology.** This use of living organisms or their components to manufacture desirable products dates back centuries, but advances in DNA technology have resulted in hundreds of new products and applications.

20.1   DNA cloning yields multiple copies of a gene or other DNA segment

DNA Cloning and Its Applications: A Preview   One of the approaches used to clone pieces of DNA makes use of the **plasmids** of bacterial cells. Recombinant DNA may be made by inserting foreign DNA into plasmids. These plasmids are put back into bacterial cells where they will replicate as the *recombinant bacteria* reproduce to form clones of identical cells. Such **gene cloning** provides multiple copies of the gene and may also be used to produce protein coded for by the foreign DNA.

Using Restriction Enzymes to Make Recombinant DNA   **Restriction enzymes** protect bacteria from phages or other bacteria by cutting up foreign DNA. Most restriction enzymes recognize short nucleotide sequences, called **restriction sites,** and cut at specific points within them. The cell protects its own DNA from restriction by methylating nucleotide bases within its own restriction sequences.

A restriction site is usually symmetrical, the same sequence of four to eight nucleotides running in opposite directions on the two strands. The most useful restriction enzymes cut the sugar-phosphate backbone in a staggered way, leaving **sticky ends** of short single-stranded sequences on both sides of the resulting **restriction fragment.**

DNA from different sources can be combined in the laboratory when the DNA is cut by the same restriction enzyme and the complementary bases on the sticky ends of the restriction fragments form hydrogen-bonded base pairs. **DNA ligase** is used to seal the strands together.

INTERACTIVE QUESTION   20.1

Which of these DNA sequences would most likely function as a restriction site for a restriction enzyme? Why?

··CAGCAG··   ··GTGCTG··   ··GAATTC··

··GTCGTC··   ··CACGAC··   ··CTTAAG··

Cloning a Eukaryotic Gene in a Bacterial Plasmid   **Cloning vectors** are DNA molecules that can move foreign DNA into a cell and replicate there. Recombinant plasmids returned to bacterial cells will replicate the foreign DNA as the bacteria reproduce.

The plasmid method of gene cloning involves treating plasmids containing an antibiotic-resistant gene with a restriction enzyme that cuts the DNA ring at a single restriction site and disrupts a gene whose activity is easily determined, such as *lacZ*, the gene for β-galactosidase. The clipped plasmids are mixed with DNA containing the gene of interest, which has also been treated with the same restriction enzyme, producing many different fragments. The sticky ends form hydrogen bonds with each other, and DNA ligase seals the recombinant molecules.

The plasmids are introduced by transformation into bacterial cells that have a mutation in their *lacZ* gene, and thus are unable to produce β-galactosidase. Bacteria are plated onto a medium containing ampicillin and X-gal, a compound that is cleaved by β-galactosidase and yields a blue product. Colonies that are able to grow on the medium (because they contain a plasmid with the *amp^R* gene) and are not blue (because of the foreign DNA inserted in the middle of the β-galactosidase gene) are carrying a recombinant plasmid.

An organism's DNA can be cut into thousands of pieces with restriction enzymes and inserted into plasmids. The collection of the thousands of clones of bacteria containing recombinant plasmids derived from this "shotgun" approach is called a **genomic library.**

Bacteriophages are also used as vectors for creating genomic libraries. DNA fragments are spliced into phage DNA, which is packaged into capsids and used to infect bacteria. The production of new phage clones the foreign DNA.

Cloning vectors also include **bacterial artificial chromosomes (BACs),** large plasmids that carry longer inserts than do standard plasmid or phage vectors. BAC clones are often stored in multiwelled plastic plates.

A partial genomic library can be produced using the mRNA molecules isolated from a cell. Using reverse transcriptase from retroviruses, mRNA is used as a template to produce **complementary DNA (cDNA).** Restriction sites are added to the ends and the cDNA is inserted into vector DNA, which is used to create a **cDNA library.**

INTERACTIVE QUESTION   20.2

Compare genomic and cDNA libraries with regard to their advantages and disadvantages.

The colonies with recombinant plasmids may be tested to find the ones that contain the gene of interest by **nucleic acid hybridization,** using a **nucleic acid probe** that has complementary sequences to segments of the gene. Cells from each clone are applied to a nylon membrane to create an *arrayed library.* The membrane is treated to break open the cells and denature the DNA. The probe can then hybridize to the single-stranded DNA and be located by its radioactively labeled molecules or fluorescent tag. Once the desired clone is identified, it can be grown in liquid culture and the gene of interest isolated in large quantities.

INTERACTIVE QUESTION   20.3

This schematic diagram shows the steps in plasmid cloning of a human gene. Identify components a–j. Briefly describe the five steps of the process. How are bacterial clones that have picked up the recombinant plasmid with the human gene of interest identified?

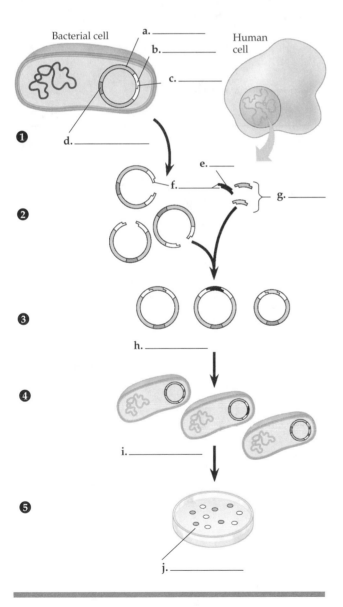

Expressing Cloned Eukaryotic Genes Differences between prokaryotic and eukaryotic mechanisms for gene expression can be overcome by using an **expression vector,** a cloning vector that has an active prokaryotic promoter just upstream from the eukaryotic gene insertion site. Using a cDNA gene removes the problem of long introns, especially since bacteria do not have RNA-splicing machinery.

Yeast cells are eukaryotic hosts used to clone and express eukaryotic genes; they are easy to grow and have plasmids that serve as vectors. **Yeast artificial chromosomes (YACs)** have also been constructed that carry foreign DNA and contain an origin for DNA replication, a centromere, and two telomeres. These vectors behave normally in mitosis and are able to clone large pieces of DNA.

Plant and animal cells in culture can serve as hosts and may be necessary when a protein must be modified following translation.

More efficient means for introducing DNA into eukaryotic cells include **electroporation,** in which an electric pulse briefly opens holes in the plasma membrane through which DNA can enter, injection into cells using microscopically thin needles, or using *Agrobacterium* to introduce DNA into plant cells.

Amplifying DNA in Vitro: The Polymerase Chain Reaction (PCR) The **polymerase chain reaction (PCR)** can produce billions of copies of a section of DNA in only a few hours. DNA containing the region of interest is incubated with the four nucleotides, a special heat-resistant type of DNA polymerase, and specially synthesized primers that bind upstream from the target sequence on both DNA strands. The solution is heated to separate the DNA strands, then cooled so the primers can anneal (hydrogen-bond) to complementary sequences. DNA polymerase adds nucleotides to the 3′ ends of the primers. The solution is heated again and the process repeated. The desired DNA segment does not need to be purified from the starting material, and very small samples can be used.

INTERACTIVE QUESTION   20.4

a. Why is PCR often used prior to cloning a gene in cells?

b. Why even bother cloning genes in cells, since PCR produces so many copies so fast?

20.2   DNA technology allows us to study the sequence, expression, and function of a gene

Gel Electrophoresis and Southern Blotting   Many of the methods for analyzing and comparing DNA make use of **gel electrophoresis,** a technique that separates nucleic acids and proteins on the basis of their size and electrical charge. Due to the negative charge of their phosphate groups, DNA molecules migrate through the electric field produced in a thin slab of gel toward the positive electrode. Linear molecules of DNA move at a rate inversely proportional to their length, producing band patterns in the gel of fragments of decreasing size.

Cutting a long DNA molecule with a particular restriction enzyme and separating the resulting restriction fragments by gel electrophoresis produces a characteristic pattern of bands, a process called *restriction fragment analysis*. Pure samples of such bands can be recovered from the gel and retain their biological activity.

Two DNA samples, such as alleles of a gene, will produce different patterns of bands when differences in their DNA sequences add or delete restriction sites. The addition of nucleic acid hybridization with a probe allows two or more unpurified samples of DNA (such as the entire genome) to be compared for the presence and band location of a particular DNA sequence. In a technique known as **Southern blotting,** DNA is treated with a restriction enzyme; the fragments are separated on a gel and then transferred to nitrocellulose paper by blotting; labeled probes of single-stranded DNA are added and hybridize with complementary DNA sequences; and the restriction fragment bands of interest are identified by autoradiography.

INTERACTIVE QUESTION   20.5

A bloody crime has occurred. Police have collected blood samples from the victim, two suspects, and blood found at the scene. Briefly list the steps the lab went through to produce the following autoradiograph (Southern blotting).

a.

b.

c.

d.

e.

Which suspect would you charge with the crime?

DNA Sequencing In the *dideoxyribonucleotide chain-termination method* of DNA sequencing, developed by F. Sanger, a sample of a denatured DNA fragment is incubated with a primer, DNA polymerase, the four deoxyribonucleotides, and the four modified nucleotides (dideoxyribonucleotides) that randomly block further synthesis when they are incorporated into a growing DNA strand. The sets of fluorescently tagged strands of varying lengths are separated through a polyacrylamide gel in a capillary tube. A fluorescence detector reads the sequence of their colors, which corresponds to the nucleotide sequence complementary to the DNA fragment.

Analyzing Expression of Genes Labeled nucleic acid probes that hybridize with mRNAs can identify the location and timing of the expression of a gene in an organism. **Northern blotting,** which separates samples of mRNA from different developmental stages, identifies when a gene is expressed.

With **reverse transcriptase-polymerase chain reaction (RT-PCR),** reverse transcriptase makes cDNA from mRNAs isolated from different samples. The cDNA is amplified by PCR using primers specific for that gene, and gel electrophoresis reveals bands of amplified DNA only in samples in which the gene has been transcribed to mRNA.

In *in situ* hybridization, fluorescently labeled probes can identify the location of specific mRNAs in an intact organism.

In order to study patterns of gene expression, researchers isolate the mRNA made in different cells, create a cDNA library using reverse transcriptase, and then use the cDNA as probes to explore collections of genomic DNA. Using this cDNA in **DNA microarray assays,** scientists can test all the genes expressed in a tissue for hybridization with short, single-stranded DNA fragments from thousands of genes arrayed on a grid (called a DNA chip). Gene expression in different tissues and at different stages of development can be compared.

Determining Gene Function The function of unknown genes can be studied using *in vitro* mutagenesis, in which changes are made to a cloned gene, the gene is returned to the cell, and it "knocks out" the normal gene. Changes in physiology or developmental patterns that result from the missing gene product are monitored. A new way to stop the expression of selected genes in cells is called **RNA interference (RNAi).** Synthetic double-stranded RNA molecules that match a gene sequence trigger the breakdown or block translation of that gene's mRNA.

INTERACTIVE QUESTION 20.6

a. What are some of the benefits of determining the nucleotide sequence of a gene of unknown function?

b. Describe a recent research method that you could use to determine in which tissues a particular gene is expressed.

c. Describe a research method that would allow you to compare both which genes are expressed and what their relative rates of expression are in several tissues.

d. In one study RNAi was used to prevent the expression, one gene at a time, of 86% of the genes in early nematode embryos. What did this allow researchers to do?

20.3 Cloning organisms may lead to production of stem cells for research and other applications

Organismal cloning involves producing genetically identical individuals (clones) from a single somatic cell of a multicellular organism.

Cloning of Plants: Single Cell Cultures   F. C. Steward demonstrated *genomic equivalence* in plants by growing new carrot plants from differentiated root cells. Most plant cells remain **totipotent,** retaining the ability to give rise to a complete new organism.

Cloning of Animals: Nuclear Transplantation   Early evidence of genomic equivalence in animals was provided by the work of Briggs, King, and Gurdon, who transplanted nuclei from embryonic and tadpole cells into enucleated frog egg cells, a method called *nuclear transplantation.* The ability of the transplanted nucleus to direct normal development was inversely related to its developmental age.

In 1997, Scottish researchers reported cloning an adult sheep by transplanting a nucleus from a fully differentiated mammary cell into an unfertilized enucleated egg cell, and then implanting the resulting early embryo into a surrogate mother. The mammary cell was induced to dedifferentiate by culturing in a nutrient-poor medium.

The *reproductive cloning* of numerous mammals has shown that cloned animals do not always look and behave identically. Environmental influences and random events play a role in development.

INTERACTIVE QUESTION   20.7

Although numerous mammals have now been cloned successfully, most cloned embryos fail to develop normally, and many cloned animals have various defects. What is a likely cause of these developmental failures?

Stem Cells of Animals   **Stem cells** are relatively unspecialized cells that continue to reproduce themselves and can, under proper conditions, differentiate into one or more types of cells. *Embryonic stem cells* taken from early embryos can be cultured indefinitely and can differentiate into all cell types. *Adult stem cells* have been isolated from various tissues and grown in culture. Such cells are capable of producing multiple but not all types of cells. Both types of stem cells can be induced to differentiate into specialized cells. Stem cell research has the potential to provide cells to repair organs that are damaged or diseased. *Therapeutic cloning* of embryonic stem cells, although different from reproductive cloning of humans, still raises ethical debates.

20.4   The practical applications of DNA technology affect our lives in many ways

Medical Applications   DNA technology is identifying genes responsible for genetic diseases; it is hoped this will lead to new ways to diagnose, treat, and even prevent those disorders. DNA microarray assays allow comparisons of gene expression in healthy and diseased tissues.

PCR, labeled DNA probes, and RT-PCR are being used to identify pathogens and to diagnose infectious diseases and genetic disorders. Even if a gene has not yet been cloned, a disease allele may be diagnosed when closely linked with a *genetic marker*, a sequence variation, or *polymorphism*, found among individuals. **Single nucleotide polymorphisms (SNPs)** are variations that occur in at least 1% of the human population. When such variations change a restriction site, they result in **restriction fragment length polymorphisms (RFLPs).** These genetic markers can now be detected by microarray analysis or PCR.

Gene therapy may provide the means for correcting genetic disorders in individuals by replacing or supplementing defective genes. New genes would be introduced into somatic cells of the affected tissue. For the correction to be permanent, the cells must be types that actively reproduce within the body, such as bone marrow cells. To date, a few trials have used retroviral vectors to carry a normal allele into bone marrow cells, but results have been mixed.

INTERACTIVE QUESTION   20.8

a. Why is it easier to perform a test for Huntington's disease now that the gene has been cloned?

b. What are some of the practical and ethical considerations in human gene therapy?

New approaches to cancer treatments include genetically engineered proteins that block a protein necessary for a tumor cell's survival. Many pharmaceutical proteins are produced using biotechnology. Engineering host cells so that they secrete a protein as it is made simplifies its purification.

Recombinant DNA techniques can been used to make vaccines that consist of surface molecules on pathogens, or to modify the genome of a pathogen so as to attenuate it (make it nonpathogenic) so it can be used as a vaccine.

Transgenic animals are produced by injecting a foreign gene into egg cells fertilized *in vitro*. The eggs are then transplanted into surrogate mothers. "Pharm" animals have been engineered to produce large quantities of a pharmaceutical protein, often by secretion in the animal's milk. "Pharm" plants are also being created to produce human and viral proteins for medical uses and vaccines.

Forensic Evidence and Genetic Profiles RFLP analysis has been used in criminal cases to compare the **genetic profile** of a suspect with that of crime DNA samples. Variations in the number of **short tandem repeats (STRs)** found in various loci are now commonly used in DNA analysis. PCR can amplify specific STRs from small samples, which are then separated and their length (corresponding to the number of repeats) determined by electrophoresis. Forensic tests are able to provide a high statistical probability that matching genetic profiles come from the same individual.

Environmental Cleanup Genetically engineered microorganisms that are able to extract heavy metals, such as copper, lead, and nickel, may become important in mining and cleaning up mining waste. Engineering organisms to degrade chlorinated hydrocarbons and other toxic compounds is an area of active research. *Biofuels* make use of microorganisms to ferment plant sugars to "bioethanol" or plant oils to "biodiesel."

Agricultural Applications A potential use of transgenic animals is to increase productivity or beneficial traits.

The ability to regenerate plants from single cells growing in tissue culture has made plant cells easier to genetically manipulate than animal cells. The most commonly used vector is the **Ti plasmid** from the bacterium *Agrobacterium tumefaciens,* which integrates a segment of its DNA into plant chromosomes. Using DNA technology, foreign genes are inserted into Ti plasmids, and the recombinant plasmids are introduced into plant cells growing in culture. When these cells regenerate whole plants, the foreign gene is included in the plant genome.

Many crops have been engineered with bacterial genes for herbicide resistance. Crop plants are being engineered to be resistant to infectious pathogens and insects and to grow in high salinity. Research efforts are also focusing on producing more nutritional transgenic crops, as in the "golden" rice enriched with beta-carotene.

Safety and Ethical Questions Raised by DNA Technology Safety regulations have focused on potential hazards of engineered microbes. Most public concern, however, now centers on agricultural **genetically modified (GM) organisms,** which contain artificially acquired genes from a different variety or different species. One environmental risk of transgenic plants is that their herbicide- or insect-resistant genes might pass to wild plants, creating "superweeds." Some fear that GM foodstuffs may be hazardous to human health. An international Biosafety Protocol requires exporters to label GM organisms in bulk food shipments so that importing countries can decide on potential environmental or health risks. In the United States, several federal agencies evaluate and regulate new products and procedures.

Ethical questions about human genetic information include who should have access to information about a person's genome and how that information should be used. Potential ethical, environmental, and health issues must be considered in the development of these powerful genetic techniques and remarkable products of biotechnology.

Word Roots

electro- = electricity (*electroporation:* a technique to introduce recombinant DNA into cells by applying a brief electrical pulse to a solution containing cells)

liga- = bound, tied (*DNA ligase:* a linking enzyme essential for DNA replication)

muta- = change; **-genesis** = origin, birth (*in vitro mutagenesis:* a technique to discover the function of a gene by introducing specific changes into the sequence of a cloned gene, reinserting the mutated gene into a cell, and studying the phenotype of the mutant)

poly- = many; **morph-** = form (*single nucleotide polymorphisms:* one-base-pair variations in the genome sequence)

Structure Your Knowledge

1. Describe several examples of the many possible applications for DNA technology in agriculture and in medicine.

2. Fill in the following table on the basic tools of gene manipulation used in DNA technology.

Technique or Tool	Brief Description	Some Uses in DNA Technology
Restriction enzymes	a.	
Gel electrophoresis	b.	
cDNA	c.	
Labeled probe	d.	
Southern blotting	e.	
DNA sequencing	f.	
PCR	g.	
RFLP analysis	h.	
DNA microarray assay	i.	
RT-PCR	j.	
in vitro mutagenesis	k.	

Test Your Knowledge

MULTIPLE CHOICE: *Choose the one best answer.*

1. The role of restriction enzymes in DNA technology is to

   a. provide a vector for the transfer of recombinant DNA.

   b. produce cDNA from mRNA.

   c. produce a cut (usually staggered) at specific restriction sites on DNA.

   d. reseal "sticky ends" after base pairing of complementary bases.

   e. denature DNA into single strands that can hybridize with complementary sequences.

2. The segment of DNA shown at right has restriction sites I and II, which create restriction fragments a, b, and c. Which of the following gels produced by electrophoresis would represent the separation and identity of these fragments?

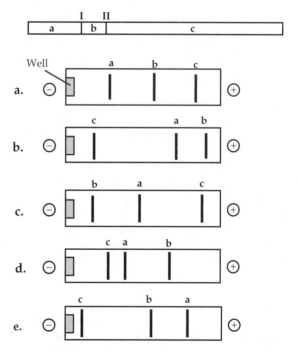

3. Yeast has become important in genetic engineering because it

   a. has RNA splicing machinery.

   b. has plasmids that can be genetically engineered.

   c. allows the study of eukaryotic gene regulation and expression.

   d. grows readily and rapidly in the laboratory.

   e. does all of the above.

4. Which of the following DNA sequences would most likely be a restriction site?
   a. AACCGG
      TTGGCC
   b. GGTTGG
      CCAACC
   c. AAGG
      TTCC
   d. AATTCCGG
      TTAAGGCC
   e. GAATTC
      CTTAAG

5. Which of the following is *not* true of adult stem cells?
   a. They have been found not only in bone marrow, but also in other tissues, including the adult brain.
   b. Although more difficult to grow than embryonic stem cells, they have been successfully grown in culture and made to differentiate into specialized cells.
   c. They are differentiated cells that can be induced to dedifferentiate and become totipotent.
   d. They are capable of developing into several but not all types of cells under appropriate conditions.
   e. These relatively unspecialized cells continually reproduce themselves.

6. Petroleum-lysing bacteria are being engineered for the treatment of oil spills. What is the most realistic danger of these bacteria to the environment?
   a. mutations leading to the production of a strain pathogenic to humans
   b. extinction of natural microbes due to the competitive advantage of the "petro-bacterium"
   c. destruction of natural oil deposits
   d. poisoning of the food chain
   e. contamination of the water

7. You are attempting to introduce a gene that imparts larval moth resistance to bean plants. Which of the following vectors are you most likely to use?
   a. phage DNA
   b. *E. coli* plasmid
   c. Ti plasmid
   d. yeast plasmid
   e. bacterial artificial chromosome

8. An attenuated virus
   a. is a virus that is nonpathogenic.
   b. is an elongated viral particle.
   c. can transfer recombinant DNA to other viruses.
   d. will not produce an immune response.
   e. is made with cDNA.

9. Which of the following is a difficulty in getting prokaryotic cells to express eukaryotic genes?
   a. The signals that control gene expression are different and prokaryotic promoter regions must be added to the vector.
   b. The genetic code differs because prokaryotes substitute the base uracil for thymine.
   c. Prokaryotic cells cannot transcribe introns because their genes do not have them.
   d. The ribosomes of prokaryotes are not large enough to handle long eukaryotic genes.
   e. The RNA splicing enzymes of bacteria work differently from those of eukaryotes.

10. Complementary DNA does not create as complete a library of genes as the shotgun approach because
    a. it has eliminated introns from the genes.
    b. a cell produces mRNA for only a small portion of its genes.
    c. the shotgun approach produces more restriction fragments.
    d. cDNA is not as easily integrated into plasmids.
    e. reverse transcriptase cannot transcribe introns.

11. Which of the following is not true of restriction sites?
    a. Modification by methylation of bases within them prevents restriction of bacterial DNA.
    b. They are usually symmetrical sequences of four to eight nucleotides.
    c. They signal the attachment of RNA polymerase.
    d. Each restriction site is cut by a specific restriction enzyme.
    e. Cutting a restriction site in the middle of a functional and identifiable gene is used to screen clones that have taken up foreign DNA.

Use the following choices to answer questions 12–14.
    a. restriction enzyme
    b. reverse transcriptase
    c. ligase
    d. DNA polymerase
    e. RNA polymerase

12. Which is the first enzyme used in the production of cDNA?

13. Which enzyme is used in the polymerase chain reaction?

14. Which is the first enzyme used in the production of RFLPs?

15. A plasmid has two antibiotic resistance genes, one for ampicillin and one for tetracycline. It is treated with a restriction enzyme that cuts in the middle of the ampicillin gene. DNA fragments containing a human globin gene were cut with the same enzyme. The plasmids and fragments are mixed, treated with ligase, and used to transform bacterial cells. Clones that have taken up the recombinant DNA are the ones that
 a. are blue and can grow on plates with both antibiotics.
 b. can grow on plates with ampicillin but not with tetracycline.
 c. can grow on plates with tetracycline but not with ampicillin.
 d. cannot grow with any antibiotics.
 e. can grow on plates with tetracycline and are blue.

16. STRs (short tandem repeats) are a valuable tool for
 a. DNA microarray assays.
 b. infecting plant cells with recombinant DNA.
 c. acting as probes in Southern blots.
 d. genetic profiling.
 e. PCR to produce multiple copies of a DNA segment.

17. You have affixed the chromosomes from a cell onto a microscope slide. Which of the following would not make a good radioactively labeled probe to help map a particular gene to one of those chromosomes? (Assume DNA of chromosomes and probes is single stranded.)
 a. cDNA made from the mRNA transcribed from the gene
 b. a portion of the amino acid sequence of that protein
 c. mRNA transcribed from the gene
 d. a piece of the restriction fragment on which the gene is located
 e. a sequence of nucleotide bases determined from the genetic code needed to produce a known sequence of amino acids found in the protein product of the gene

18. If the first three nucleotides in a six-nucleotide restriction site are CTG, what would the next three nucleotides most likely be?
 a. AGG   **c.** CTG   **e.** GAC
 b. GTC   **d.** CAG

19. What is a "pharm" animal?
 a. a transgenic animal that produces large quantities of a pharmaceutical product
 b. an animal used by the pharmaceutical industry to test new medical treatments

 c. a cloned animal that was produced from an adult cell nucleus inserted into an ovum
 d. a transgenic animal that produces more meat
 e. a genetically modified organism whose production is permitted in the United States but not in the European Union

20. Which of the following processes or procedures does not involve any nucleic acid hybridization?
 a. separation of fragments by gel electrophoresis
 b. Southern blotting
 c. polymerase chain reaction
 d. DNA profiling
 e. DNA microarray assay

21. Which of the following would be useful in identifying disease-causing alleles?
 a. RNA interference
 b. *in situ* hybridization
 c. short tandem repeats
 d. single nucleotide polymorphisms
 e. RT-PCR

22. This restriction fragment contains a gene whose recessive allele is lethal. The normal allele has restriction sites for the restriction enzyme PSTI at sites I and II. The recessive allele lacks restriction site I. An individual who had a sister with the lethal trait is being tested to determine if he is a carrier of that allele. Indicate which of these band patterns would be produced on a gel if he is a carrier (heterozygous for the gene)?

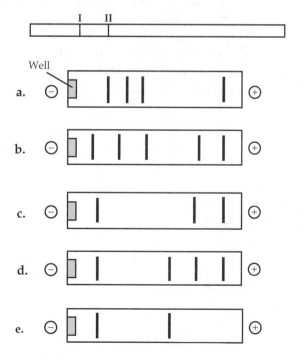

Chapter 21

Genomes and Their Evolution

Key Concepts

21.1 New approaches have accelerated the pace of genome sequencing

21.2 Scientists use bioinformatics to analyze genomes and their functions

21.3 Genomes vary in size, number of genes, and gene density

21.4 Multicellular eukaryotes have much noncoding DNA and many multigene families

21.5 Duplication, rearrangement, and mutation of DNA contribute to genome evolution

21.6 Comparing genome sequences provides clues to evolution and development

Framework

This chapter introduces genomics and bioinformatics, new approaches to analyzing and comparing the genomes of life's diverse organisms. Genomics is allowing researchers to address questions about genome organization, gene expression, growth and development, and evolution. The various types of noncoding DNA in the human genome are described. The chapter also covers the genetic processes that contribute to genome evolution.

Chapter Review

The complete genome sequences of humans, chimpanzees, and several other model organisms are enabling the study of whole sets of genes and their interactions, called **genomics.** The new field of **bioinformatics** is applying computational methods to the analysis of the ever-growing volumes of biological data.

21.1 New approaches have accelerated the pace of genome sequencing

The international effort to map the human genome, called the **Human Genome Project,** was begun in 1990 and largely completed in 2003.

Three-Stage Approach to Genome Sequencing Cytogenetic maps based on banding patterns and fluorescence *in situ* hybridization (FISH) provided the basis for mapping chromosomes. The first stage in mapping the human genome was to develop a **linkage map** of several thousand genetic markers, either genes or other identifiable sequences such as RFLPs or simple tandem repeats (STRs). Determining recombination frequencies with these known markers enabled researchers to order genes and locate other markers.

In a **physical map,** the actual distances between markers are determined. The DNA of each chromosome is cut into restriction fragments, and their order on the chromosome is determined. First large fragments are cut, cloned, and ordered. Then those fragments are cut and their fragments ordered. Cloning vectors include yeast artificial chromosomes, which can carry long fragments, and bacterial artificial chromosomes (BACs), which carry shorter inserts. Finally small fragments are cloned in plasmids or phages.

The nucleotide sequences of the short fragments were determined with sequencing machines using the dideoxy chain termination method. The development of "high-throughput" sequencing machines speeded the process.

Whole-Genome Shotgun Approach to Genome Sequencing J. C. Venter, founder of the company Celera Genomics, developed a *whole-genome shotgun approach* that relies on powerful computer programs to order the large number of sequenced short fragments. The whole-genome shotgun approach is now widely used, although it can miss some duplicated sequences.

INTERACTIVE QUESTION   21.1

Compare the strategies of the public consortium of the Human Genome Project and of the whole-genome shotgun approach for mapping a genome.

INTERACTIVE QUESTION   21.2

The software program BLAST, available on the NCBI website, allows a researcher to compare a DNA sequence to every sequence in Genbank to identify similar regions. What are some uses of this function?

21.2   Scientists use bioinformatics to analyze genomes and their functions

Centralized Resources for Analyzing Genome Sequences   The National Center for Biotechnology Information (NCBI) and other genome centers maintain websites with databases of DNA sequences and protein sequences and structures, as well as software for analyzing and comparing sequences and structures. These sites make bioinformatics resources available to researchers worldwide.

Identifying Protein-Coding Genes within DNA Sequences   Computer software can scan DNA sequences for signs of genes, such as start and stop signals; RNA-splicing sites; and *expressed sequence tags*, or *ESTs* (sequences similar to those of known mRNAs). The sequences of new gene candidates are compared with sequences of known genes of other species to look for similarities that might indicate the function of the new gene. A combination of biochemical techniques (three-dimensional structure and binding sites) and functional studies (blocking the gene with RNAi to determine effect on phenotype) help to reveal protein function.

Understanding Genes and Their Products at the Systems Level   **Proteomics** is the identification and study of entire protein sets *(proteomes)* coded for by a genome. The interaction of proteins is the basis for the functioning of cells and organisms.

With lists compiled of DNA sequences and proteins, researchers are now looking at the functional integration of these parts in biological systems. This systems biology approach is attempting to define gene circuits and protein interaction networks, using computers, mathematics, and new software to process and integrate huge amounts of data.

The Cancer Genome Atlas is taking a systems biology approach to analyzing changes in genes and patterns of gene expression in cancer cells. Silicon "chips" holding arrays of most known human genes are used to analyze gene expression in patients with various diseases. Such approaches may help to match treatments with a person's unique genetic makeup.

21.3   Genomes vary in size, number of genes, and gene density

Genome Size   Of the 600 genomes that had been sequenced by mid-2007, most are of bacteria, whose genomes have between 1 and 6 million base pairs (Mb). Archaeal genomes appear to be similar; whereas most multicellular eukaryotes have genomes of at least 100 Mb. There does not appear to be a correlation between genome size and an organism's phenotype.

Number of Genes   Bacteria and archaea have from 1,500 to 7,500 genes; eukaryotes range from about 5,000 genes for unicellular fungi to 40,000 for multicellular organisms. The human genome is estimated to contain 20,488 genes. Alternate splicing of the multiple exons in most human genes may yield about 75,000 different polypeptides.

Gene Density and Noncoding DNA   Eukaryotes have fewer genes per million base pairs than bacteria or archaea. Mammals appear to have the lowest gene density. Most of the DNA of eukaryotic genomes is noncoding DNA both within and between genes and includes introns and complex regulatory sequences.

INTERACTIVE QUESTION   21.3

Look at organisms listed in Table 21.1 in your text to answer these questions.

a. Which organism has the highest gene density? _____ the lowest gene density? _____

b. Which organism has the largest number of genes? _____ the smallest number? _____

c. Which organism has the largest haploid genome size? _____ the smallest genome size? _____

d. What is the estimated number of genes in the human genome? _____ Explain the fact that there are many more different human polypeptides than human genes.

21.4 Multicellular eukaryotes have much noncoding DNA and many multigene families

Noncoding DNA, which was previously referred to as "junk DNA," may turn out to have important functions. Almost 500 identical regions of noncoding DNA have been identified in humans, rats, and mice, a higher level of sequence conservation than for protein-coding regions in these species.

About 1.5% of the human genome consists of exons that code for proteins, rRNA, or tRNA. The rest includes regulatory sequences, introns, nonfunctional former genes called **pseudogenes,** and sequences present in many copies, called **repetitive DNA.** More than half of this repetitive DNA (44% of the human genome) is either made up of or related to transposable elements.

Transposable Elements and Related Sequences
Stretches of DNA that can move about within a genome through a process called *transposition* are called *transposable genetic elements* or **transposable elements.**

Transposons move about a genome as a DNA intermediate, either by a "cut-and-paste" or a "copy-and-paste" mechanism. **Retrotransposons,** which make up the majority of transposable elements, are first transcribed into an RNA intermediate. This RNA transcript is converted back to DNA by reverse transcriptase, coded for by the retrotransposon itself.

Transposable elements may be represented as multiple (although not identical) copies of transposons or as related sequences that have lost the ability to move. In primates, perhaps 10% of the genome is made up of *Alu elements,* 300-nucleotide-long sequences, many of which are transcribed into RNA but are of unknown function.

About 17% of the human genome consists of *LINE-1,* or *L1,* retrotransposons. The introns of about 80% of analyzed human genes contain L1 sequences, suggesting that L1 may help regulate gene expression.

INTERACTIVE QUESTION   21.4

Why do retrotransposons always move by the "copy-and-paste" mechanism?

Other Repetitive DNA, Including Simple Sequence DNA About 15% of the human genome is repetitive DNA that appears to have arisen by mistakes in DNA replication. Scattered large-segment duplications account for 5% of the genome. **Simple-sequence DNA,** on the other hand, makes up 3% and consists of multiple copies of tandemly repeated sequences. When the repeat consists of two to five nucleotides, the unit is called a **short tandem repeat,** or **STR.** The variation in repeat numbers between genomes is the basis of genetic profiles. Much of this so-called *satellite DNA* is located at centromeres, where it functions in cell division and chromatin organization, and telomeres, which protect the tips of chromosomes.

Genes and Multigene Families Sequences coding for proteins and structural RNAs make up 1.5% of the human genome. Including introns and regulatory sequences, the amount of gene-related DNA is 25%. More than half of the coding DNA occurs in **multigene families,** collections of similar or identical genes.

With the exception of the genes for histone proteins, identical multigene families code for RNA products. The genes coding for the three largest rRNA molecules are arranged in a single transcription unit repeated in huge tandem arrays, enabling cells to produce the millions of ribosomes needed for protein synthesis.

Examples of multigene families of nonidentical genes are the two families of genes that code for globins, including the α and β polypeptide chains of hemoglobin. Different versions of each globin subunit are clustered together on two different chromosomes and are expressed at the appropriate time during development. (See Interactive Question 21.5 on the following page.)

21.5 Duplication, rearrangement, and mutation of DNA contribute to genome evolution

Duplication of Chromosome Sets Extra sets of chromosomes may arise by accidents in meiosis. The resulting extra genes might diverge through mutation, leading to genes with novel functions. Polyploidy is more common in plants.

Alterations of Chromosome Structure Using genomic sequence information, researchers can compare the locations of DNA sequences on chromosomes among different species and reconstruct the evolutionary history of chromosomal rearrangements. Duplications and inversions of chromosomes are thought to contribute to speciation. The breakage points associated with these alterations appear to be recombination "hotspots." Such sites are associated with some human congenital diseases.

Duplication and Divergence of Gene-Sized Regions of DNA Errors such as unequal crossing over during meiosis (as may occur between copies of a transposable

INTERACTIVE QUESTION   21.5

Match the letter of the description and fill in the percentage (listed below) for each type of DNA sequence found in the human genome.

Types of DNA	Description	%
1. Exons and RNA-coding	_____	_____
2. Introns and regulatory sequences	_____	_____
3. Transposable elements and related sequences	_____	_____
4. Simple sequence DNA	_____	_____
5. Large-segment duplications	_____	_____
6. Unique noncoding DNA	_____	_____
7. *Alu* elements	_____	_____
8. L1 sequences	_____	_____

Descriptions

A. satellite DNA in centromeres and telomeres

B. multiple copies of movable sequences

C. gene fragments and pseudogenes

D. protein and RNA-coding sequences

E. family of short (300 nucleotide) transposable elements

F. multiple copies of large sequences

G. retrotransposons found in introns of most genes

H. DNA related to gene expression

Percentages to choose from: 1.5, 3, 5, 10, 15, 17, 24, and 44 (These percentages do not add up to 100 because some of these types of DNA are subsets of other categories and some types are not listed.)

element on misaligned nonsister chromatids) and slippage of template strands during DNA replication might lead to the duplication of individual genes.

The α-globin and β-globin gene families appear to have evolved from a common ancestral globin gene, which was duplicated and then diverged. Multiple duplications and mutations within each family have led to the current family of genes with related functions along with several intervening pseudogenes.

In other cases, mutations of a duplicated gene may lead to a protein product with a new function.

INTERACTIVE QUESTION   21.6

Lysozyme and α-lactalbumin have similar sequences but different functions. Both genes are found in mammals, but birds have only the gene for lysozyme. What does this observation suggest about the evolution of these genes?

Rearrangements of Parts of Genes: Exon Duplication and Exon Shuffling   Unequal crossing over can lead to a gene with a duplicated exon. Exons often code for domains of a protein, and their duplication could provide a protein with enhanced structure and function. Errors in meiotic recombination could also lead to *exon shuffling* within a gene or between nonallelic genes. For instance, the gene for TPA has three different exons that may have originated in the genes for epidermal growth factor, fibronectin, and plasminogen.

How Transposable Elements Contribute to Genome Evolution   Recombination events can take place between homologous transposable element sequences that are scattered throughout the genome, causing chromosomal mutations that may occasionally be beneficial to the organism. Transposable elements that insert within a gene may disrupt its functioning; those that insert within regulatory sequences may increase or decrease gene expression. A transposable element can also move a copy of a gene or an exon to a new location. The transposition of an *Alu* element into introns may provide an alternative splice site in the RNA transcript, producing a new portion of a protein that may have an altered function. The increased genetic diversity provided by these mechanisms provides raw material for natural selection.

INTERACTIVE QUESTION   21.7

a. Explain two ways in which exon shuffling could occur.

b. What is a potential benefit of exon shuffling?

21.6   Comparing genome sequences provides clues to evolution and development

Comparing Genomes   Comparisons of *highly conserved* genes illuminate the evolutionary relationships among species that are distantly related. Such analyses support the theory that bacteria, archaea, and eukaryotes represent the three domains of life and also reinforce the relevance of using model organisms to study both basic biological processes and human biology.

The similarity between genomes of closely related species allows researchers to use one genome sequence as a framework to map the second genome. The small number of differences between genomes translates into the phenotypic divergence of species. The human and chimpanzee genomes differ in single base substitutions by only 1.2%. Insertions or deletions of larger regions in the genome result in an additional 2.7% difference. There are more *Alu* elements in the human genome, and a third of the human duplications are not in the chimpanzee genome.

Comparisons of genetic changes since species diverged show that some genes are changing faster in humans than in the chimpanzee or mouse. Many of these faster evolving genes code for transcription factors, such as the *FOXP2* gene that appears to function in vocalization in vertebrates and in human speech and language. Mutations in this gene cause verbal impairment in humans; it is expressed in the brains of songbirds during the period that they are learning their songs; and knock-out experiments with mice have shown that homozygous mutant mice had malformed brains and did not produce their normal vocalizations. The human and chimpanzee FOXP2 proteins differ by two amino acids; the mouse protein differs by a third amino acid.

Comparisons of human genomes have revealed several million single nucleotide polymorphism (SNP) sites, as well as inversions, deletions, and duplications. These genetic markers will contribute to the study of human evolution and the identification of disease-related genes.

Comparing Developmental Processes Biologists in the field of evolutionary developmental biology (**evo-devo**) compare developmental processes to understand how they have evolved and how minor changes in gene sequence or regulation may lead to diverse forms of life.

A sequence of 180 nucleotides called a **homeobox**, which codes for a *homeodomain*, has been found in *Drosophila* homeotic genes. The same or very similar homeobox nucleotide sequences have been identified in homeotic genes of many animals. Homeotic genes in the fruit fly and mouse are found in the same linear sequence on chromosomes. Related sequences are found in regulatory genes of yeast and plants. These similarities indicate that the homeobox sequence must have arisen early and been conserved through evolution as part of genes involved in regulation of gene expression and development.

Homeotic genes are often called *Hox* genes in mammals. Proteins with homeodomains probably coordinate the transcription of groups of developmental genes and thus control pattern formation.

Many other genes involved in development, such as those coding for components of signaling pathways, are highly conserved. The differing patterns of expression of these genes in different body areas may explain the development of animals with different body plans.

INTERACTIVE QUESTION 21.8

If all *Hox* genes contain the same or very similar homeobox, how can they control different developmental sequences?

The processes of development evolved independently in plants and animals due to their ancient divergence. The rigid cell walls of plants restrict movement, and morphogenesis relies more on the orientation of cell divisions and cell enlargement. Some similarities persist from their common ancestral single-celled eukaryote. In both, development involves a cascade of transcriptional regulators, although the master control genes are different in the two groups. Although both *Hox* genes and *Mads-box* genes are found in animals and plants, the former act as master regulatory switches in animals, whereas the latter do so in plants.

Word Roots

pseudo- – false (*pseudogenes:* DNA segments very similar to real genes but which do not yield functional products)

retro- = backward (*retrotransposons:* transposable elements that move in a genome as an RNA intermediate, a transcript of the retrotransposon DNA)

Structure Your Knowledge

1. About 25% of the human genome relates to the production of proteins or RNA products (exons, introns, regulatory sequences). Is the remaining 75% just junk? Describe the following types of noncoding DNA and include some possible functions.
   a. transposable elements
   b. *Alu* elements
   c. L1 sequences
   d. simple sequence DNA
   e. pseudogenes

2. Describe some of the processes that contribute to genome evolution.

Test Your Knowledge

MULTIPLE CHOICE: *Choose the one best answer.*

1. Computer software is used to identify putative genes in the nucleotide sequences in data banks. Which of the following is the software probably *not* scanning for?
   a. promoters
   b. RNA-splicing sites
   c. STRs (simple tandem repeats)
   d. ESTs (expressed sequence tags)
   e. stop signals for transcription

2. The human genome appears to have only one-third more genes than the simple nematode, *C. elegans*. Which of the following best explains how a more complex human can have relatively few genes?
   a. The unusually long introns in human genes are involved in regulation of gene expression.
   b. More than one polypeptide can be produced from a gene by alternative splicing.
   c. Human genes code for many more types of domains.
   d. The human genome has a high proportion of noncoding DNA.
   e. The large number of SNPs (single nucleotide polymorphisms) in the human genome provides for a great deal of genetic variability.

3. Tissue plasminogen activator (TPA) is a protein with three types of domains. One of each of these types is found in the protein's epidermal growth factor, fibronectin, and plasminogen. What is a likely explanation for this?
   a. All four genes are members of a multigene family involved in cell signaling.

   b. TPA was the first gene to evolve; the other three genes each lost two domains.
   c. The gene for TPA arose by several instances of exon shuffling from the other three genes.
   d. TPA must have many *Alu* elements that allow for alternative splice sites to incorporate these exons.
   e. Several duplication events led to the evolution of the TPA gene.

4. Which of the following best describes what pseudogenes and introns have in common?
   a. They are not expressed—they do not produce a functional product.
   b. They are DNA segments that lack a promoter but do have other control regions.
   c. They are RNA molecules that are not translated into proteins.
   d. They code for RNA products, not proteins.
   e. They appear to have arisen from retrotransposons.

5. Bacterial genes have an average length of 1,000 base pairs; human genes average about 27,000 base pairs. Which of the following is the best explanation for that difference?
   a. Prokaryotes have smaller, but many more, individual genes.
   b. Prokaryotes are more ancient organisms; longer genes arose later in evolution.
   c. Prokaryotes are unicellular; humans have many types of differentiated cells.
   d. Prokaryotic genes do not have introns; human genes have multiple introns.
   e. Prokaryotic proteins are not as large and complex as human proteins.

6. Which of the following techniques can be used to determine the function of a newly identified gene?
   a. comparisons with genes of known functions that have similar sequences and whose products have similar domains
   b. blocking gene function with RNAi to see the effect on the phenotype
   c. searching for *Alu* sequences that may indicate alternate splicing sites
   d. both a and b
   e. a, b, and c

7. Why is the whole-genome shotgun approach more efficient than the three-stage approach to genome sequencing?
   a. It uses more powerful high-throughput sequencing machines.

b. It makes use of already existing sequenced genomes to provide scaffolding for its fragments.

c. It does not need to take time to clone the gene fragments that it sequences.

d. It uses only one, very efficient restriction enzyme to create fragments of optimal size.

e. It does not produce a linkage map or a physical map, but uses powerful computer programs to assemble overlapping fragments.

8. Genes that are highly conserved are useful for

   a. tracing the evolution of genes that have led to new species.

   b. searching data banks to determine the function of newly found genes.

   c. establishing the sequence of divergence of closely related species.

   d. tracing the relationships of groups that diverged early in the evolution of life.

   e. both a and c.

9. Why is proteomics important in the systems biology approach?

   a. The interactions of networks of proteins are central to the functioning of cells and organisms.

   b. This bioinformatics field allows for the mathematical modeling of biological systems.

   c. Determining the proteins expressed in a cell identifies the genes more accurately than can be done through genomics.

   d. By determining the proteins involved in cellular development, researchers can predict the proteins involved in the development of systems.

   e. Comparing the proteins produced by a normal gene and the gene associated with a human disease will facilitate the development of improved treatment systems.

10. A highly conserved nucleotide sequence that has been found in developmental regulatory genes in many diverse organisms is called a

    a. homeodomain.

    b. homeobox.

    c. retrotransposon.

    d. homeotic gene.

    e. L1 sequence.

11. Which of the following has probably *not* contributed to genome evolution?

    a. duplications and inversions of parts of chromosomes

    b. exon duplication and shuffling

    c. recombination between transposable elements on different chromosomes

    d. duplications of genes that then diverged as mutations accumulated

    e. All of the above have contributed.

12. What is meant by saying that the *FOXP2* gene has been evolving faster in the human lineage?

    a. It is part of a multigene family, whose genes are expressed at different times during development.

    b. It has mutated and now resembles the *FOXP2* gene found in songbirds but not in other mammals.

    c. It has more *Alu* sequences in its introns, indicating that it undergoes alternative splicing.

    d. It has two amino acid differences that occurred in the 6 million years since the human lineage split from other primates; whereas there is only one amino acid difference between other primates and mammals such as mice.

    e. It is linked with language development in humans, which differs from the vocalizations of other mammals.

UNIT 4 — Mechanisms of Evolution

Chapter 22

Descent with Modification: A Darwinian View of Life

Key Concepts

22.1 The Darwinian revolution challenged traditional views of a young Earth inhabited by unchanging species

22.2 Descent with modification by natural selection explains the adaptations of organisms and the unity and diversity of life

22.3 Evolution is supported by an overwhelming amount of scientific evidence

Framework

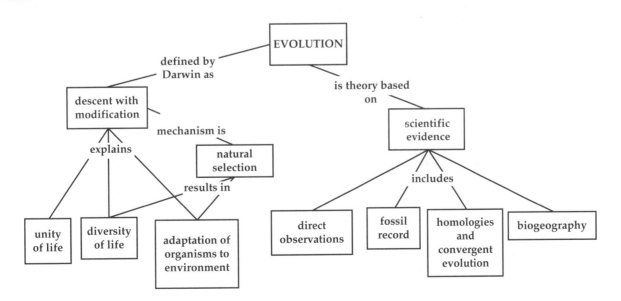

Chapter Review

Charles Darwin presented his scientific explanation for the adaptations of organisms to their environments, the diversity of life, and the unity of life in his book *On the Origin of Species*. **Evolution** may be defined in Darwin's terms as *descent with modification*, or more narrowly as the changes in a population's genetic composition over time.

Evolution can be viewed as both the *pattern* of evolutionary change observable in the natural world and the *process* or mechanisms underlying those changes.

22.1 The Darwinian revolution challenged traditional views of a young Earth inhabited by unchanging species

Scala Naturae and Classification of Species The Greek philosopher Aristotle proposed that all living forms could be arranged on a "scale of nature" of increasing complexity and that each group of organisms was permanent and perfect.

The Old Testament account of creation supported the idea of perfect and fixed species. In the 1700s, Linnaeus developed both a binomial system for naming organisms according to their genus and species, and a hierarchy of groups organizing his named species.

Ideas About Change Over Time **Fossils** are remnants or traces of past organisms, usually found in sedimentary rocks formed through the compression of layers of sand and mud into superimposed layers called **strata.** Cuvier developed **paleontology,** the study of fossils. Advocating **catastrophism,** he maintained that the extinctions and differences in the fossils found in different strata are the result of local sudden catastrophic events and not indicative of evolution.

Hutton proposed that immense changes in Earth's geology are the cumulative result of slow but continuous processes. Lyell, a contemporary of Darwin, developed a theory of **uniformitarianism,** stating that the rates of geologic processes have remained the same through Earth's history and continue in the present.

Darwin took two ideas from the observations of Hutton and Lyell: Earth must be very old, and very slow processes can produce substantial change, perhaps even in living species.

Lamarck's Hypothesis of Evolution Lamarck's hypothesis, published in 1809, explained the mechanism of evolution with two principles: The *use and disuse* of body parts leading to their development or deterioration, and *the inheritance of acquired characteristics*. Although present genetic knowledge rejects his

mechanism, Lamarck proposed several key evolutionary ideas: that evolution is the best explanation for the fossil record and that adaptation to the environment is the main result of evolution.

INTERACTIVE QUESTION   22.1

a. Match the theory or philosophy and its proponent(s) with the following descriptions.

A. catastrophism	**a.** Aristotle
B. inheritance of acquired characteristics	**b.** Cuvier
C. gradual geologic changes	**c.** Darwin
D. descent with modification	**d.** Hutton
E. classification system	**e.** Lamarck
F. scale of nature	**f.** Linnaeus
G. uniformitarianism	**g.** Lyell

Theory   Proponents

1. ——— ——— Describing the diversity of God's creations by naming and classifying species

2. ——— ——— History of Earth marked by sudden floods or droughts that resulted in extinctions

3. ——— ——— Proposed mechanism of evolution that modifications due to use or disuse are passed to offspring

4. ——— ——— Profound change is the cumulative product of slow but continuous processes

5. ——— ——— Fixed species on a continuum from simple to complex

6. ——— ——— All of life is related; present day species differ from ancestral species

7. ——— ——— Geologic processes have constant rates throughout time

b. Now place 1 through 7 in chronological order.

——— ——— ——— ——— ——— ——— ———

22.2 Descent with modification by natural selection explains the adaptations of organisms and the unity and diversity of life

Darwin's Research Darwin was 22 years old when he sailed from Great Britain on the HMS *Beagle*. He spent the voyage collecting thousands of specimens of the fauna and flora of South America, observing the various adaptations of organisms living in very diverse habitats, and making special note of the geographic distribution of the distinctly South American species. Darwin also read and was influenced by Lyell's *Principles of Geology*.

Adaptations are characteristics that help organisms survive and reproduce in specific environments. Darwin proposed that adaptations arise through **natural selection,** a process in which individuals with beneficial inherited characteristics produce more offspring than those without those characteristics. In 1844 he wrote an essay on the origin of species and natural selection but did not publish it. In 1858 Darwin received Wallace's manuscript describing an identical theory. Darwin then published *On the Origin of Species by Means of Natural Selection* in 1859.

The Origin of Species Darwin's book developed two main points: descent with modification as the basis of life's unity and diversity, and natural selection as the mechanism that matches organisms with their environment.

Darwin's concept of descent with modification explains that all organisms are related through descent from some unknown ancestor and develop increasing modifications as they adapt to various habitats. The history of life is analogous to a tree with a common ancestor at the fork of each new branch and present day species at the tips of the youngest twigs. The taxonomy developed by Linnaeus provided a hierarchical organization of groups that suggested to Darwin this branching tree of life.

Using the example of **artificial selection** in the breeding of domesticated plants and animals, Darwin presented evidence that selection among the variations present in a population can lead to substantial changes.

Darwin found support for the capacity of organisms to overproduce in the essay on human population growth published by Malthus in 1798.

Remember these three points about evolution by natural selection: Natural selection results in the evolution of populations, not individuals. Natural selection affects only those traits that are heritable—acquired characteristics cannot evolve. And natural selection depends on the specific environmental factors present in a region at a given time. If the environment changes, different adaptations will be favored.

INTERACTIVE QUESTION 22.2

List the four observations from which Darwin drew the two inferences that explain natural selection:

Observation 1:

Observation 2:

Observation 3:

Observation 4:

Inference 1: Individuals whose inherited characteristics give them a better chance of surviving and reproducing in a given environment are likely to leave more offspring.

Inference 2: Unequal survival and reproduction leads to an accumulation of favorable traits over generations.

22.3 Evolution is supported by an overwhelming amount of scientific evidence

Direct Observations of Evolutionary Change J. Endler, studying guppy populations in isolated pools in Trinidad over many years, observed variations in the color patterns of male guppies that correlate with the intensity of predation—the more predators, the less brightly colored the fish. He transplanted guppies from pools with active predators to pools with less active predators and observed, after 15 generations, an increase in the number and size of colored spots, documenting evolution in a natural setting over a relatively short period of time.

The evolution of drug resistance in HIV illustrates two facets of natural selection: It is an editing, not a creative, mechanism that selects for variations already present in a population. And it is regional and temporal, selecting for traits that are beneficial in the local environment at that current time.

INTERACTIVE QUESTION 22.3

Within a few weeks of treatment with the drug 3TC, a patient's HIV population consists entirely of 3TC-resistant HIV. Explain how this rapid evolution of drug resistance is an example of natural selection.

The Fossil Record The fossil record documents that present and past organisms differ, and that many species have become extinct. Fossils also trace the evolution of new groups, as in the origin of whales from land mammals.

The major branches of evolutionary descent established with evidence from anatomy and molecular data have been tested and supported by the sequence of fossil forms found in the fossil record.

Homology Homology is similarity resulting from common ancestry. The forelimbs of all mammals are **homologous structures,** containing the same skeletal elements regardless of function or external shape. Comparative anatomy and embryology illustrate that evolution is a remodeling process in which ancestral structures become modified for new functions.

Vestigial organs are rudimentary structures, of little or no value to the organism, that are historical remnants of ancestral structures.

Homologies can be seen on a molecular level. DNA, RNA, and an essentially universal genetic code, which have been passed along through all branches of evolution, are important evidence that all forms of life descended from the earliest organisms and are thus related.

Homology is evident on different hierarchical levels, reflecting evolutionary history in a nested pattern of descent. An **evolutionary tree** diagrams ancestral relationships. Each branch point represents the common ancestor of all the species beyond that point. Evolutionary trees are hypotheses based on available data that represent patterns of evolutionary descent.

Distantly related organisms may appear similar as a result of **convergent evolution,** the independent evolution of similar characteristics. These **analogous** structures arise as a result of evolutionary adaptation to similar environments.

Biogeography The geographic distribution of species, or **biogeography,** has been influenced by **continental drift,** the slow movements of Earth's continents. The single land mass called **Pangaea** formed 250 million years ago and began to break apart 200 million years ago. The gradual separation of continents helps to explain and predict where fossils of different groups and their descendants are found.

Islands often have **endemic** species, found nowhere else and usually closely related to species on the nearest island or mainland. Widely separated areas having similar environments are not likely to be populated by closely related species. Rather, each area is more likely to have species that are taxonomically related to those of their region, regardless of environment.

INTERACTIVE QUESTION 22.4

Complete the following concept map that summarizes the main sources of evidence for evolution.

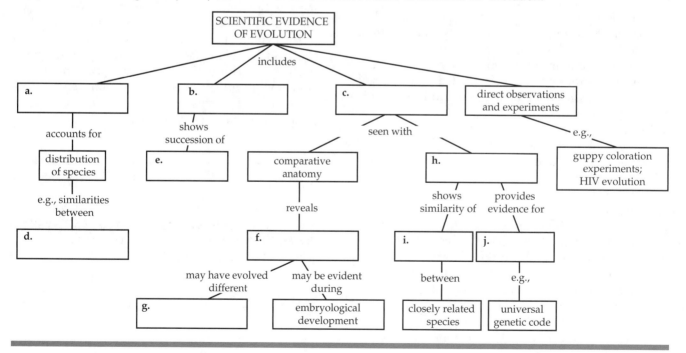

What Is Theoretical About the Darwinian View of Life? A scientific "theory" is a unifying concept with broad explanatory power and with predictions that have been and continue to be tested by experiments and observations.

Word Roots

bio- = life; **geo-** = the Earth (*biogeography:* the study of the past and present distribution of species)

end- − within (*endemic:* a type of species that is found only in one region and nowhere else in the world)

homo- = like, resembling (*homology:* similarity in characteristics resulting from a shared ancestry)

paleo- = ancient (*paleontology:* the scientific study of fossils)

vestigi- = trace (*vestigial organs:* structures of marginal, if any, importance to an organism, historical remnants of structures that had important functions in ancestors)

Structure Your Knowledge

1. Explain in your own words the main components of Darwin's theory of evolution.

Test Your Knowledge

MULTIPLE CHOICE: *Choose the one best answer.*

1. The best description of natural selection is
   a. the survival of the fittest.
   b. the struggle for existence.
   c. the reproductive success of the members of a population best adapted to the environment.
   d. the overproduction of offspring in environments with limited natural resources.
   e. a change in the proportion of alleles within a population.

2. To Cuvier, the differences in fossils from different strata were evidence for
   a. changes occurring as a result of cumulative but gradual processes.
   b. divine creation.
   c. evolution by natural selection.
   d. continental drift.
   e. local catastrophic events such as droughts or floods.

3. Darwin proposed that new species evolve from ancestral forms by
   a. the gradual accumulation of adaptations to changing or different environments.
   b. the inheritance of acquired adaptations to the environment.
   c. the struggle for limited resources.
   d. the accumulation of mutations.
   e. the excessive reproduction of species.

4. All of the following influenced Darwin as he synthesized the theory of evolution by natural selection *except*
   a. the biogeographic distribution of species such as the mockingbirds on the Galápagos Islands.
   b. Lyell's book, *Principles of Geology,* on the gradualness of geologic changes.
   c. Linnaeus's hierarchical classification of species, which could be interpreted as evidence of evolutionary relationships.
   d. examples of artificial selection that produce rapid changes in domesticated species.
   e. Mendel's paper in which he described his "laws of inheritance."

5. What might you conclude from the observation that the bones in your arm and hand are similar to the bones that make up a bat's wing?
   a. The bones in the bat's wing are vestigial structures, no longer useful as "arm" bones.
   b. The bones in a bat's wing are homologous to your arm and hand bones.
   c. Bats and humans evolved in the same geographic area.
   d. Bats lost their opposable digits during the course of evolution.
   e. Our ancestors could fly.

6. The remnants of pelvic and leg bones in a snake
   a. are vestigial structures.
   b. show that lizards evolved from snakes.
   c. are homologous structures.
   d. provide evidence for inheritance of acquired characteristics.
   e. resulted from artificial selection.

7. The hypothesis that whales evolved from land-dwelling ancestors is supported by
   a. evidence from the biogeographic distribution of whales.
   b. molecular comparisons of whales, fish, and reptiles.
   c. historical accounts of walking whales.
   d. the ability of captive whales to be trained to walk.
   e. fossils of extinct whales that had small hind limbs.

8. Darwin's claim that all of life descended from a common ancestor may best be supported with evidence from
   a. the fossil record.
   b. comparative embryology.
   c. taxonomy.
   d. molecular biology.
   e. comparative anatomy.

9. The smallest unit that can evolve is
   a. a genome.
   b. an individual.
   c. a species.
   d. a population.
   e. a community.

10. Which of the following would *not* be considered part of the process of natural selection?
    a. Many of the variations among individuals in a population are heritable.
    b. More offspring are produced than are able to survive and reproduce.
    c. Individuals with traits best adapted to the environment are likely to leave more offspring.
    d. Many adaptive traits may be acquired during an individual's lifetime, helping that individual to evolve.
    e. Differential reproductive success leads to gradual change in a population.

11. When cytochrome *c* molecules are compared, yeasts and molds are found to differ by approximately 46 amino acids per 100 residues (amino acids making up a protein), whereas insects and vertebrates are found to differ by approximately 29 amino acids per 100 residues. What can one conclude from these data?
    a. Very little can be concluded unless the DNA sequences for the cytochrome *c* genes are compared.
    b. Yeasts evolved from molds, but vertebrates did not evolve from insects.

c. Insects and vertebrates diverged from a common ancestor more recently than did yeasts and molds.
d. Yeasts and molds diverged from a common ancestor more recently than did insects and vertebrates.
e. The evolution of cytochrome *c* occurred more rapidly in yeasts and molds than in insects and vertebrates.

Use this evolutionary tree representing the relationships among a group of vertebrates to answer questions 12 and 13. The letters at each branch point indicate the common ancestor for groups beyond that point.

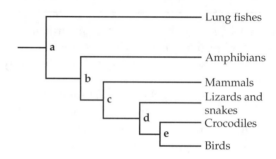

12. Which was the last common ancestor of crocodiles with lizards and snakes?

13. Which of the following can be concluded from this evolutionary tree?
    a. Mammals are more closely related to amphibians than to birds.
    b. Mammals are more closely related to lizards and snakes than to birds.
    c. Birds are more closely related to lizards and snakes than to mammals.
    d. Birds and mammals are more closely related than are birds to lizards and snakes.
    e. Lungfish are not related to any of the other groups.

14. Chimpanzees and humans share many of the same genes, indicating that most likely
    a. the two groups belong to the same species.
    b. the two groups belong to the same phylum.
    c. the two groups share a relatively recent common ancestor.
    d. humans evolved from chimpanzees.
    e. chimpanzees evolved from humans.

15. Which of the following is an example of convergent evolution?

   a. the increase in size and number of colored spots on guppies transplanted to pools with less active predators

   b. similarities between the marsupial sugar glider and the eutherian flying squirrel

   c. two very different plants that are found in different habitats, but evolved from a fairly recent common ancestor

   d. the remodeling of the vertebrate forelimb in the evolution of a bird wing

   e. the many different bill sizes and shapes of finches on the Galápagos Islands

The Evolution of Populations

23.1 Mutation and sexual reproduction produce the genetic variation that makes evolution possible

23.2 The Hardy-Weinberg equation can be used to test whether a population is evolving

23.3 Natural selection, genetic drift, and gene flow can alter allele frequencies in a population

23.4 Natural selection is the only mechanism that consistently causes adaptive evolution

Framework

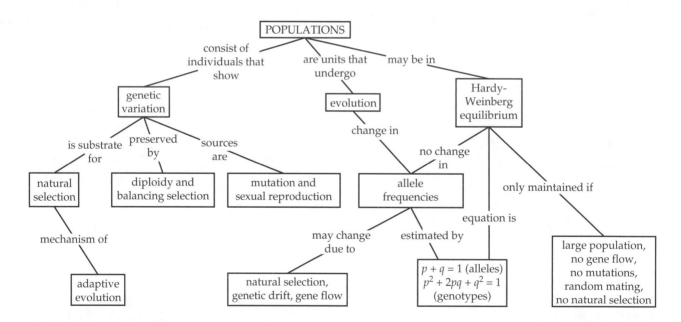

Chapter Review

Although it is individuals that are selected for or against by natural selection, it is populations that actually evolve. **Microevolution** is defined as changes in the allele frequencies of a population from generation to generation.

23.1 Mutation and sexual reproduction produce the genetic variation that makes evolution possible

Genetic Variation   Individual variation, the slight differences between individuals as a result of their unique genomes, is the raw material for natural selection. *Discrete characters* vary as distinct, either-or phenotypes and usually are determined by a single gene locus. *Quantitative characters,* those that vary along a continuum and are usually affected by two or more gene loci, represent most of the heritable variation within a population.

Measures of genetic variation look at both gene variability and nucleotide variability. The **average heterozygosity** of a population is the average percent of loci that are heterozygous. Nucleotide variability measures the average percent of differences in nucleotide sites between individuals of a population.

Geographic variations are regional genetic differences between populations. A **cline,** or graded variation in a character across a geographic area, may parallel an environmental gradient.

Mutation New alleles originate by **mutation,** changes in the sequence of nucleotides in DNA. Most mutations occur in somatic cells and cannot be passed on to the next generation.

Point mutations that occur in noncoding DNA or do not change the amino acid sequence of a protein are often harmless. Mutations that do alter the phenotype are usually harmful. Rarely, however, a mutant allele may enhance an individual's reproductive success.

Chromosomal mutations are most often deleterious. Duplication of small DNA segments, introduced by transposons or errors in cell division, may provide an expanded genome with extra loci that could eventually take on new functions by mutation.

Mutation rates in animals and plants average about one in every 100,000 genes per generation. Mutation produces genetic variation very rapidly in prokaryotes and viruses due to their short generation spans.

Sexual Reproduction In a sexually reproducing population, most of the genetic variation comes from the reshuffling of alleles into new combinations in each individual. In sexual reproduction, the processes of crossing over and independent assortment in meiosis and random combination of gametes in fertilization create unique genetic combinations in offspring.

INTERACTIVE QUESTION 23.1

a. What is a major source of genetic variation for prokaryotes and viruses?

b. What is the major source of genetic variation for plants and animals?

c. Explain why your answers to a and b are different.

23.2 The Hardy-Weinberg equation can be used to test whether a population is evolving

Gene Pools and Allele Frequencies A **population** is a localized group of individuals of a species that interbreed. The **gene pool** is the term for all the alleles at all the loci present in a population. If all individuals are homozygous for the same allele, the allele is said to be *fixed.* More often, two or more alleles are present in the gene pool in some relative proportion or frequency. In a case with two alleles at a particular gene locus, the letters p and q represent the frequencies of the two alleles within the population, and their combined frequencies must equal 1: $p + q = 1$.

INTERACTIVE QUESTION 23.2

In a population of 200 mice, 98 are homozygous dominant for brown coat color (*BB*), 84 are heterozygous (*Bb*), and 18 are homozygous recessive (*bb*).

a. The genotype frequencies of this population are

_____ *BB* _____ *Bb* _____ *bb*.

b. The allele frequencies of this population are

_____ B allele _____ b allele.

The Hardy-Weinberg Principle If only Mendelian segregation and recombination of alleles at fertilization are involved, the gene pool of a population will remain constant from one generation to the next, as described by the **Hardy-Weinberg principle.** Such a nonevolving gene pool is said to be in **Hardy-Weinberg equilibrium.**

The allele frequency within a population determines the proportion of gametes that will contain that allele. The random combination of gametes will yield offspring with genotypes that reflect and reconstitute the allele frequencies of the previous generation.

With the equation for Hardy-Weinberg equilibrium, the frequencies of genotypes in the next generation can be calculated from the probability of each combination of alleles. According to the rule of multiplication, the probability that two gametes containing the same allele will come together is equal to ($p \times p$) or p^2, or ($q \times q$) or q^2. A p and q allele can combine in two different ways, depending on which parent contributes which allele; therefore, the frequency of a heterozygous offspring is equal to $2pq$. The sum of the frequencies of all possible genotypes in the population adds up to 1: $p^2 + 2pq + q^2 = 1$.

INTERACTIVE QUESTION   23.3

Use the allele frequencies you determined in Interactive Question 23.2 to predict the genotype frequencies of the next generation.

Frequencies of

B (p) _____ b (q) _____

BB = p^2 = _____ Bb = 2pq = _____ bb = q^2 = _____

Hardy-Weinberg equilibrium is maintained only if all of the following five conditions are met: no mutations, random mating (nonrandom mating changes genotype frequencies), no natural selection (no differential survival and reproductive success), an extremely large population to offset chance fluctuations (genetic drift), and no gene flow or movement of alleles into or out of populations.

If the frequency of homozygous recessive individuals is known (q^2), then the frequency of q may be estimated as the square root of q^2 (assuming the population is in Hardy-Weinberg equilibrium for that gene). If the frequency of individuals with a recessively inherited disease is known (q^2), then the frequency of carriers of that recessive allele in the population can be calculated.

INTERACTIVE QUESTION   23.4

Practice using the Hardy-Weinberg equation so that you can easily determine genotype frequencies from allele frequencies and vice versa.

a. The allele frequencies in a population are A = 0.6 and a = 0.4. Predict the genotype frequencies for the next generation.

   AA _____ Aa _____ aa _____

b. What would the allele frequencies be for the generation you predicted in part a?

   A _____ a _____

23.3 Natural selection, genetic drift, and gene flow can alter allele frequencies in a population

Natural Selection   Individuals that have traits that are better suited to their environment tend to be more successful in producing viable, fertile offspring and pass their alleles to the next generation in disproportionate numbers, resulting in adaptive evolution.

Genetic Drift   Chance deviations from expected results are more likely to occur in a small sample. Chance fluctuations in a population's allele frequencies from one generation to the next are called **genetic drift.**

Genetic drift that occurs when only a few individuals colonize a new area is known as the **founder effect.** Allele frequencies in the small sample are unlikely to be representative of the parent population.

The **bottleneck effect** occurs when some disaster or other factor reduces the population size dramatically, and the few surviving individuals are unlikely to represent the genetic makeup of the original population. Genetic drift will remain a factor until the population grows large enough for chance events to be less significant.

Genetic drift has a larger effect on small populations, causes gene frequencies to randomly fluctuate over time, can lead to the loss of alleles within populations, and can cause fixation of harmful alleles.

Gene Flow   **Gene flow,** the migration of individuals or the transfer of gametes between populations, may change allele frequencies. Differences in allele frequencies between populations tend to be reduced by gene flow.

INTERACTIVE QUESTION   23.5

Fill in the following concept map that summarizes three causes of microevolution. Even better, create your own concept map to help you review the ways in which a population's genetic composition may be altered.

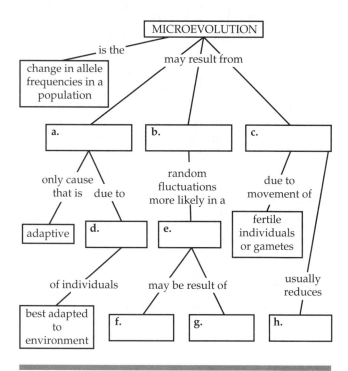

23.4 Natural selection is the only mechanism that consistently causes adaptive evolution

A Closer Look at Natural Selection   **Relative fitness** is a measure of an individual's contribution to the gene pool of the next generation, relative to the contributions of others. Natural selection acts on the phenotype of the whole organism. The relative fitness of an allele depends on the entire genetic and environmental context in which it is expressed.

The frequency distribution of a trait may be affected by three modes of selection. **Directional selection** occurs most frequently during periods of environmental change, when individuals on one end of a phenotypic range may be favored. **Disruptive selection** occurs when the environment favors individuals on both extremes of a phenotypic range. **Stabilizing selection** acts against extreme phenotypes and favors more intermediate forms, tending to reduce phenotypic variation.

The Key Role of Natural Selection in Adaptive Evolution   As natural selection increases the frequencies of alleles that enhance survival and reproduction, the match between organisms and their environment increases. Natural selection is the agent of adaptive evolution.

Sexual Selection   **Sexual selection** is the selection for traits that enhance an individual's chances of obtaining mates. Sexual selection can lead to **sexual dimorphism,** the distinction between males and females on the basis of secondary sexual characteristics. Sexual selection may involve **intrasexual selection,** in which individuals of the same sex compete for mates, or **intersexual selection,** in which individuals of one sex (usually female) discriminate in choosing a mate. Also called *mate choice,* intersexual selection may be based on showy traits that reflect the general health of the male and, thus, the quality of his genes.

The Preservation of Genetic Variation   The diploidy of most eukaryotes maintains genetic variation by hiding recessive alleles in heterozygotes, enabling these alleles to persist in the population and be selected for, should the environment change.

Balancing selection can maintain stable frequencies of two of more phenotypes in a population. **Heterozygote advantage** tends to maintain two or more alleles at a locus when heterozygotes have survival and reproductive advantages. In **frequency-dependent selection,** a phenotype's reproductive success declines if it becomes too common in the population.

Some of the genetic variation seen in populations may be **neutral variations** that do not confer a selective advantage or disadvantage. Noncoding regions of DNA or alleles that do not affect fitness may change randomly by genetic drift.

INTERACTIVE QUESTION   23.6

a. Why hasn't the highly deleterious sickle-cell allele been selected against and eliminated from the gene pool of the U.S. population?

b. Why is this allele at such a relatively high frequency in the gene pool of some African populations?

Why Natural Selection Cannot Fashion Perfect Organisms   Natural selection can act only on variations that are available; new alleles do not arise as they are needed. Each species has evolved from a long line of ancestral forms, many of whose structures have been coopted for new situations. Adaptations are often compromises between the need to do several different things, such as swim and walk. And finally, chance events affect a population's evolutionary history.

Word Roots

inter- = between (*intersexual selection:* individuals of one sex are choosy in selecting their mates from individuals of the other sex, also called mate choice)

intra- = within (*intrasexual selection:* a direct competition among individuals of one sex for mates of the opposite sex)

micro- = small (*microevolution:* a change in the gene pool of a population over a succession of generations)

muta- = change (*mutation:* a change in the DNA of genes that ultimately creates genetic diversity)

Structure Your Knowledge

1. **a.** What is the Hardy-Weinberg principle?
   b. Define the variables of the equation for Hardy-Weinberg equilibrium. Make sure you can use this equation to determine allele frequencies and predict genotype frequencies.

2. It seems that natural selection would work toward genetic unity; the genotypes that are most fit produce the most offspring, increasing the frequency of adaptive alleles and eliminating less beneficial alleles from the population. Yet there remains a great deal of variability within populations of a species. Describe some of the factors that contribute to this genetic variability.

Test Your Knowledge

MULTIPLE CHOICE: *Choose the one best answer.*

1. Which of the following describes an organism's relative fitness?
   a. survival
   b. number of matings
   c. adaptation to the environment
   d. successful competition for resources
   e. relative number of viable offspring

2. According to the Hardy-Weinberg principle,
   a. the allele frequencies of a population should remain constant from one generation to the next if the population is large and only sexual reproduction is involved.
   b. only natural selection, resulting in unequal reproductive success, will cause evolution.
   c. the square root of the frequency of individuals showing the dominant trait will equal the frequency of p.
   d. p and q can only be determined for a population that is not evolving.
   e. all of the above are correct.

3. If a population has the following genotype frequencies, $AA = 0.42$, $Aa = 0.46$, and $aa = 0.12$, what are the allele frequencies?
   a. $A = 0.42$ $a = 0.12$ d. $A = 0.76$ $a = 0.24$
   b. $A = 0.6$ $a = 0.4$ e. $A = 0.88$ $a = 0.12$
   c. $A = 0.65$ $a = 0.35$

4. In a population with two alleles, B and b, the allele frequency of b is 0.4. What would be the frequency of heterozygotes if the population is in Hardy-Weinberg equilibrium?
   a. 0.16
   b. 0.24
   c. 0.48
   d. 0.6
   e. You cannot tell from this information.

5. In a population that is in Hardy-Weinberg equilibrium for two alleles, C and c, 16% of the population show a recessive trait. Assuming C is dominant to c, what percent show the dominant trait?
   a. 36%   d. 84%
   b. 48%   e. 96%
   c. 60%

6. Genetic drift is likely to be seen in a population
   a. that has a high migration rate.
   b. that has a low mutation rate.
   c. in which natural selection is occurring.
   d. that is very small.
   e. for which environmental conditions are changing.

7. Gene flow often results in
   a. populations that move to better environments.
   b. an increase in the adaptive evolution of the next generation.
   c. stabilizing selection.
   d. the fixation of alleles.
   e. a reduction of the allele frequency differences between populations.

8. All of the following tend to maintain two or more alleles at a particular locus in a population *except*
   a. balancing selection.
   b. disruptive selection.
   c. heterozygote advantage.
   d. directional selection.
   e. frequency-dependent selection.

9. The average heterozygosity of *Drosophila* is estimated to be about 14%, which means that
   a. 86% of fruit fly genes are identical.
   b. on average, 14% of a fruit fly's gene loci are heterozygous.
   c. 14% of nucleotide sites differ between individuals.
   d. nucleotide variability must be very great between individuals.
   e. the fruit fly population never experienced a bottleneck effect.

10. Mutations are rarely a direct source for microevolution in eukaryotes because
    a. they are most often harmful and do not get passed on.
    b. they do not directly produce most of the genetic variation present in a diploid population.
    c. they occur very rarely.
    d. they are only passed on when they occur in gametes.
    e. all of the above.

11. In a study of a population of field mice, you find that 48% of the mice have a coat color that indicates that they are heterozygous for a particular gene. What would be the frequency of the dominant allele in this population?
    a. 0.24
    b. 0.48
    c. 0.50
    d. 0.60
    e. You cannot estimate allele frequency from this information.

12. In a random sample of a population of shorthorn cattle, 73 animals were red ($C^R C^R$); 63 were roan, a mixture of red and white ($C^R C^r$); and 13 were white ($C^r C^r$). Estimate the allele frequencies of C^R and C^r, and determine whether the population is in Hardy-Weinberg equilibrium.
    a. $C^R = 0.64$, $C^r = 0.36$; because the population is large and a random sample was chosen, the population is in equilibrium.
    b. $C^R = 0.7$, $C^r = 0.3$; the genotype ratio is not what would be predicted from these frequencies and the population is not in equilibrium.
    c. $C^R = 0.7$, $C^r = 0.3$; the genotype ratio is close to what would be predicted from these frequencies and the population is in equilibrium.
    d. $C^R = 1.04$, $C^r = 0.44$; the allele frequencies add up to greater than 1 and the population is not in equilibrium.
    e. You cannot estimate allele frequency from this information.

13. A scientist observes that the height of a certain species of asters decreases as the altitude on a mountainside increases. She gathers seeds from samples at various altitudes, plants them in a uniform environment, and measures the height of the new plants. All of her experimental asters grow to approximately the same height. From this she concludes that
    a. height is not a quantitative trait.
    b. the cline she observed was due to genetic variations.
    c. the differences in the parent plants' heights were due to directional selection.
    d. the height variation she initially observed was an example of nongenetic environmental influence.
    e. stabilizing selection was responsible for height differences in the parent plants.

14. Sexual selection will
    a. select for traits that enhance an individual's chance of mating.
    b. increase the size of individuals.
    c. result in individuals better adapted to the environment.
    d. produce more offspring.
    e. select for traits that increase fertility.

15. The greatest source of genetic variation in plant and animal populations is from
    a. mutations.
    b. sexual reproduction.

c. selection.
d. geographic variation.
e. recessive masking in heterozygotes.

16. A plant population is found in an area that is becoming more arid. The average surface area of leaves has been decreasing over the generations. This trend is an example of
    a. a cline.
    b. directional selection.
    c. disruptive selection.
    d. gene flow.
    e. genetic drift.

17. Mice that are homozygous for a lethal recessive allele die shortly after birth. In a large breeding colony of mice, you find that a surprising 5% of all newborns die from this trait. In checking lab records, you discover that the same proportion of offspring have been dying from this trait in this colony for the past three years. (Mice breed several times a year and have large litters.) How might you explain the persistence of this lethal allele at such a high frequency?
    a. Homozygous recessive mice have a reproductive advantage.
    b. A large mutation rate keeps producing this lethal allele.
    c. There is some sort of heterozygote advantage and perhaps selection against the homozygous dominant trait.
    d. Genetic drift has kept the recessive allele at this high frequency in the population.
    e. Since this is a diploid species, the recessive allele cannot be selected against when it is in the heterozygote.

18. Humans have an estimated 1,000 olfactory receptor genes. This is most likely an example of
    a. gene flow.
    b. gene duplication.
    c. frequency-dependent selection.
    d. neutral variation.
    e. a quantitative character.

19. Genetic analysis of a large population of mink inhabiting an island in Michigan revealed an unusual number of loci where one allele was fixed. Which of the following is the most probable explanation for this genetic homogeneity?

    a. The population exhibited nonrandom mating, producing a high proportion of homozygous genotypes.

    b. A very small number of mink may have colonized this island, and this founder effect and subsequent genetic drift could have fixed many alleles.

    c. The gene pool of this population never experienced gene flow.

    d. Natural selection has selected for and fixed the best adapted alleles at these loci.

    e. The colonizing population may have had much more genetic diversity, but very recent genetic drift may have fixed these alleles by chance.

20. If an allele is recessive and lethal in homozygotes shortly after birth,

    a. the allele is present in the population at a frequency of 0.001.

    b. the allele will be removed from the population by natural selection in approximately 1,000 years.

    c. the relative fitness of the homozygous recessive genotype is 0.

    d. the allele will most likely remain in the population at a low frequency because it cannot be selected against when in a heterozygote.

    e. Both c and d are correct.

21. Which of these types of selection is *mismatched* with its example?

    a. disruptive—a population of black-bellied seedcrackers consists of birds with either small bills (more effective at eating soft seeds) or large bills (able to crack hard seeds)

    b. intrasexual—elephant seal males are more than four times larger than females; males fight over areas of beach where females congregate during breeding season

    c. intersexual—female gray tree frogs choose mates that give long mating calls

    d. stabilizing—the frequency of A, B, AB, and O blood groups remains constant in a population

    e. frequency-dependent—as fish of one coloration become more numerous, predators form a "search image" for that coloration and preferentially feed on them

Chapter 24

The Origin of Species

Framework

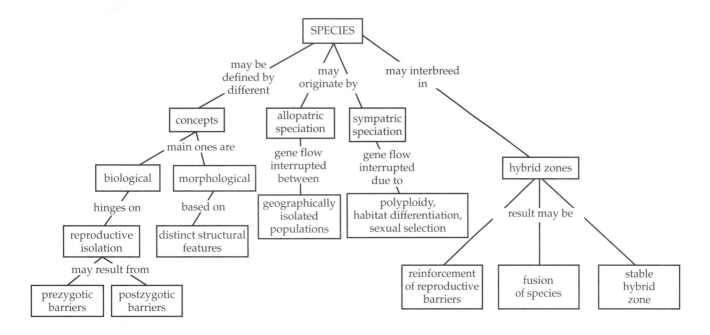

Chapter Review

Speciation, the origin of new species from an ancestral species, creates the diversity of life while maintaining the relatedness or unity of life. **Microevolution** explains evolutionary changes within the gene pool of a population. **Macroevolution** considers broader changes, such as the origin of new taxonomic groups above the species level.

24.1 The biological species concept emphasizes reproductive isolation

Different kinds of organisms are most often characterized by their physical form or morphology, although differences in physiology, biochemistry, and DNA sequences also support the existence of distinct species.

The Biological Species Concept According to the **biological species concept,** a **species** is a group of populations of individuals that have the potential to interbreed in nature and produce viable, fertile offspring, but which do not successfully interbreed with members of other species. Gene flow between populations tends to hold the gene pool of a species together.

The **reproductive isolation** that is necessary for the formation of a new species results from barriers that prevent individuals of different species from producing viable, fertile **hybrids. Prezygotic barriers** function before the formation of a zygote by preventing mating between species or successful fertilization should gametes meet. Should a hybrid zygote form, **postzygotic barriers** prevent it from developing into a viable, fertile adult.

The biological species concept does not work for species that are asexual, such as prokaryotes. Extinct species also cannot be grouped based on the criterion of interbreeding. And gene flow does occur between some species that remain morphologically and ecologically different.

Other Defnitions of Species Most species have been identifed on the basis of physical characteristics, an approach called the **morphological species concept.** The **ecological species concept** defines species on the basis of their ecological niche, the role they play and resources they use in the specific environments in which they are found. In the **phylogenetic species concept,** a species is an evolutionary lineage that represents one branch on the tree of life and has distinct morphology or molecular sequences.

INTERACTIVE QUESTION   24.1

Name the type of reproductive isolation and whether it is a pre- or postzygotic barrier for the following examples.

Type of Isolation	Pre- or Post-	Example
a.	b.	Two species of frogs mate in a laboratory setup and produce viable, but sterile, offspring.
c.	d.	Two species of sea urchins release gametes at the same time, but the sperm fail to fuse with eggs of a different species.
e.	f.	The genital openings of two species of land snails cannot line up because their shells spiral in opposite directions.
g.	h.	Two species of short-lived mayflies emerge during different weeks in spring.
i.	j.	Two species of salamanders mate and produce offspring, but the hybrid's offspring are sterile.
k.	l.	Two similar species of birds have different mating rituals.
m.	n.	Embryos of two species of mice bred in the lab usually abort.
o.	p.	Peepers breed in woodland ponds; leopard frogs breed in swamps.

INTERACTIVE QUESTION   24.2

Fill in the following table to review four of the approaches that biologists have proposed for conceptualizing a species.

Concept	Emphasis
biological	a.
b.	anatomical differences, most commonly used
c.	unique roles in specific environments
phylogenetic	d.

24.2   Speciation can take place with or without geographic separation

Allopatric ("Other Country") Speciation **Allopatric speciation** occurs when geographic isolation interrupts gene flow between two subpopulations. Geographic separation alone is not a reproductive barrier in the biological sense. Intrinsic reproductive barriers may arise coincidentally as allopatric populations go down separate evolutionary paths due to mutation, genetic drift and natural selection.

Sympatric ("Same Country") Speciation In **sympatric speciation,** reproductive barriers prevent gene flow between populations that share the same area.

Mistakes during cell division may lead to **polyploidy,** the presence of extra sets of chromosomes. An **autopolyploid** has more than two sets of chromosomes that have all come from the same species. Failures in cell division can produce tetraploids (4*n*), which can fertilize themselves or other tetraploids, but cannot reproduce with

diploids from the parent population, resulting in reproductive isolation in just one generation.

Polyploid species may also arise through **allopolyploidy.** Interspecific hybrids may propagate asexually, but are usually sterile due to difficulties in the meiotic production of gametes. Future mitotic or meiotic mistakes can result in the production of a fertile polyploid.

Polyploid speciation has been frequent and important in plant evolution. Researchers now hybridize plants by inducing meiotic and mitotic errors to create new species.

INTERACTIVE QUESTION 24.3

a. A new plant species B forms by autopolyploidy from species A, which has a chromosome number of $2n = 10$. How many chromosomes would species B have?

b. If species A were to hybridize by allopolyploidy with species C ($2n = 14$) and produce a new, fertile species, D, how many chromosomes would species D have?

Sympatric speciation in animals may involve isolation within the geographic range of the parent population based on different resource usage. Sexual selection may also lead to sympatric speciation. In a laboratory study of two sympatric species of cichlids, mate choice based on coloration was shown to be the reproductive barrier that normally separates the two species.

INTERACTIVE QUESTION 24.4

a. Differentiate between allopatric and sympatric speciation.

b. How might reproductive barriers arise in each type of speciation?

24.3 Hybrid zones provide opportunities to study factors that cause reproductive isolation

Patterns Within Hybrid Zones Areas where diverging allopatric populations come back into contact and members of the different species interbreed are called **hybrid zones.** In the hybrid zone found between the yellow-bellied toad and the fire-bellied toad, the frequencies of alleles specific to one species range from 100% near that species' edge of the hybrid zone to 0% at

the edge near the other species. What keeps these hybrids from introducing the alleles of each other's species into the parent populations? Their poor survival and reproduction mean that these hybrids produce few viable offspring if they mate with the parent species.

Hybrid Zones Over Time In a process called **reinforcement,** natural selection may strengthen prezygotic barriers to reproduction when hybrids are less fit than individuals of the parent species. In other cases, weak reproductive barriers may allow sufficient gene flow that barriers break down even more, and the two hybridizing species fuse into a single species. Many hybrid zones appear to be stable, with hybrids continuing to be formed, but with the two parent gene pools remaining separate.

INTERACTIVE QUESTION 24.5

Use the following diagram to explain the three possible outcomes for populations in a hybrid zone.

a.

b.

c.

24.4 Speciation can occur rapidly or slowly and can result from changes in few or many genes

The Time Course of Speciation In the fossil record, new forms often appear rather suddenly, persist unchanged for a long time, and then disappear. **Punctuated equilibria** is the term used to describe these long periods of stasis punctuated by episodes of relatively rapid speciation and change. The fossil record for other species indicates a much more gradual divergence.

A number of studies suggest that once divergence begins, the process of speciation may be relatively rapid. For example, the chromosomes of experimentally produced hybrid sunflowers came to resemble those of the wild sunflower hybrid species in only five generations of breeding in the lab.

INTERACTIVE QUESTION 24.6

If the process of speciation appears to occur relatively rapidly, why don't we see new species evolving all the time? And what does your answer have to do with concern about the current high rate of extinction of Earth's species?

Studying the Genetics of Speciation Researchers have been able to identify the genes that played a key role in speciation in some organisms. For example, reproductive isolation in two species of monkeyflowers has been linked to a few genes, including a flower color gene that affects pollinator preference.

From Speciation to Macroevolution As species diverge and diversify, some differences may accumulate that lead to the formation of whole new groups of organisms, becoming part of the large-scale evolutionary changes of macroevolution.

Word Roots

auto- = self; **poly-** = many (*autopolyploid:* a type of polyploid species resulting from one species doubling its chromosome number to become tetraploid)

macro- = large (*macroevolution:* long-term evolutionary change that produces new groups of organisms)

post- = after (*postzygotic barrier:* any of several species-isolating mechanisms that prevent hybrids produced by two different species from developing into viable, fertile adults)

sym- = together; **-patri** = father (*sympatric speciation:* a mode of speciation occurring as a result of a radical change in the genome that produces a reproductively isolated subpopulation in the midst of its parent population)

Structure Your Knowledge

1. How are speciation and microevolution different?
2. In which type of speciation, sympatric or allopatric, would you expect reproductive barriers to arise more easily? Explain your answer.
3. What does the term *punctuated equilibria* describe?

Test Your Knowledge

MULTIPLE CHOICE: *Choose the one best answer.*

1. Which of the following is *not* a type of *intrinsic* reproductive isolation?
   a. mechanical isolation
   b. behavioral isolation
   c. geographic isolation
   d. gametic isolation
   e. temporal isolation

2. Two species of frogs occasionally mate, but the embryos do not complete development. What type of barrier isolates these gene pools?
   a. gametic isolation
   b. prezygotic barrier
   c. hybrid breakdown
   d. reduced hybrid viability
   e. reduced hybrid fertility

3. Which concept of species would be most useful to a field biologist identifying new species in a tropical rain forest?
   a. biological
   b. ecological
   c. phylogenetic
   d. morphological
   e. both a and b

4. For which of the following is the biological species concept least appropriate?
   a. plants
   b. animals
   c. prokaryotes
   d. fossils
   e. both c and d

5. A horse ($2n = 64$) and a donkey ($2n = 62$) can mate and produce a mule. How many chromosomes would there be in a mule's cells?
   a. 31   c. 63   e. 126
   b. 62   d. 64

6. Allopatric speciation is more likely to occur when an isolated population
   a. is large and thus has more genetic variation.
   b. is reintroduced to its original homeland.
   c. is small and exposed to different selection pressures in its new habitat.
   d. inhabits an island close to its parent species' mainland.
   e. All of the above contribute to allopatric speciation.

7. A tetraploid plant species (with four identical sets of chromosomes) is probably the result of
   a. allopolyploidy.
   b. autopolyploidy.
   c. hybridization and nondisjunction.
   d. allopatric speciation.
   e. a and c.

8. A botanist identifies a new species of plant that has 32 chromosomes. It grows in the same habitat with three similar species: species A ($2n = 14$), species B ($2n = 16$), and species C ($2n = 18$). Suggest a possible speciation mechanism for the new species.
   a. allopatric divergence by development of a reproductive isolating mechanism
   b. change in a key developmental gene that causes the plants to flower at different times
   c. autopolyploidy, perhaps due to a nondisjunction in the formation of gametes of species B
   d. allopolyploidy, a hybrid formed from species A and C
   e. Either answer c or d could account for the formation of this new plant species.

9. Which of the following would *not* contribute to allopatric speciation?
   a. geographic separation
   b. genetic drift
   c. gene flow
   d. different selection pressures
   e. founder effect

10. Which of the following is the best description of punctuated equilibria?
    a. Long periods of stasis are punctuated by episodes of relatively rapid speciation and change.
    b. The equilibrium of separate species may be punctuated by gene flow within hybrid zones.
    c. Most rapid speciation events involve polyploidy in plants; speciation in animals is a much more gradual process.
    d. Rapid environment changes produce rapid speciation events. Gradual environmental changes result in gradual speciation events. Speciation does not occur if the environment remains constant.
    e. In the framework of geologic time periods, speciation events occur rapidly and the equilibrium of species is punctuated by frequent extinctions.

11. When hybrids formed in a hybrid zone can breed with themselves and both parent species, and they have equal fitness to the parent species, one would predict that
    a. the hybrid zone would be stable.

    b. allopatric speciation would occur.
    c. reinforcement of reproductive barriers would keep the parent species separate.
    d. reproductive barriers would lessen and the two parent species would fuse.
    e. a new hybrid species would form by sympatric speciation.

12. This chapter introduced several research studies that illustrate various aspects of speciation. Which of the following is *not* an accurate description of a conclusion based on one of these studies?
    a. After several generations of being raised on different food sources, both "maltose flies" and "starch flies" tended to mate with like-adapted partners. These mating preferences appear to be the beginning of reproductive isolation by way of a behavioral prezygotic barrier between the two populations.
    b. Two closely related species of monkeyflowers have different pollinators (bumblebees and hummingbirds). Transferring an allele for flower color between the two species resulted in both types of pollinators visiting the newly formed hybrids. Thus, this reproductive barrier based on pollinator choice may have been influenced by a change in a single gene locus.
    c. In allopatric populations of the pied flycatcher and the collared flycatcher, males of the two species look very similar. In sympatric populations, the males have quite different colorations. Reinforcement of reproductive barriers is the best explanation for these observed differences.
    d. The distribution of two species of ground crickets is strongly influenced by temperature. Hybrids formed in the many hybrid zones between the two species are often as fit as the parent species. Changes in local temperatures may cause one or the other parent species to become locally extinct. Thus, hybrid zones and the hybrids formed in them may not last long enough for reproductive barriers to break down and fusion of the gene pools to occur.
    e. In the many species of cichlids found in Lake Victoria, the reproductive barriers between species as a result of female mate choice based on male coloration appear to be breaking down because the introduced Nile perch is a strong predator of cichlids.

The History of Life on Earth

Key Concepts

25.1 Conditions on early Earth made the origin of life possible

25.2 The fossil record documents the history of life

25.3 Key events in life's history include the origins of single-celled and multicelled organisms and the colonization of land

25.4 The rise and fall of dominant groups reflect continental drift, mass extinctions, and adaptive radiations

25.5 Major changes in body form can result from changes in the sequence and regulation of developmental genes

25.6 Evolution is not goal oriented

Framework

Macroevolution describes the long evolutionary history of life on Earth chronicled in the fossil record. Starting with hypotheses on how life may have originated, this chapter describes some of the major events in life's history, such as the rise of photosynthetic prokaryotes, the origin of eukaryotic cells and multicellularity, and the colonization of land. Divisions between eras in the geologic record are marked by major biological transitions, linked with continental drift, mass extinctions, and adaptive radiations. The mechanisms underlying the major changes seen in the fossil record involve changes in and control of developmental genes.

Chapter Review

25.1 Conditions on early Earth made the origin of life possible

Synthesis of Organic Compounds on Early Earth Earth formed about 4.6 billion years ago and was bombarded by huge rocks and ice until about 3.9 billion years ago. As Earth cooled, water vapor condensed into oceans, and

the atmosphere probably contained nitrogen and its oxides, carbon dioxide, methane, ammonia, and hydrogen sulfide.

In the 1920s, A. Oparin and J. Haldane independently hypothesized that conditions on early Earth, in particular the reducing atmosphere, lightning, and intense ultraviolet radiation, favored the synthesis of organic compounds.

In 1953, S. Miller and H. Urey experimentally supported the Oparin-Haldane hypothesis using an apparatus that simulated the hypothetical conditions of early Earth. Numerous laboratory replications using various combinations of atmospheric gases have also produced organic compounds. The atmosphere may not have been reducing enough to play a significant role in the production of organic molecules.

Submerged volcanoes and deep-sea vents may have provided the environment in which the first abiotic synthesis of organic molecules occurred. Carbonaceous chondrites found in meteorites contain carbon compounds, including many amino acids, providing additional evidence that such abiotic synthesis is possible.

Abiotic Synthesis of Macromolecules Researchers have created amino acid polymers by dripping solutions of organic monomers onto hot sand, clay, or rock. Such polymers may have served as weak catalysts on early Earth.

Protobionts Life requires accurate replication and metabolism, and the two are interdependent. **Protobionts** are collections of abiotically produced organic molecules surrounded by a membrane. Laboratory experiments in which mixtures of organic ingredients assemble into liposomes indicate that protobionts could have formed spontaneously.

Self-Replicating RNA and the Dawn of Natural Selection RNA probably functioned as the first hereditary material. **Ribozymes** are enzyme-like RNA catalysts. Some ribozymes can make complementary copies of parts of their sequence. Protobionts with self-replicating, catalytic RNA could grow, split, pass some of their RNA to daughters, and be acted on by natural selection. As the more successful offspring continued to reproduce, metabolic and hereditary improvements may have accumulated. DNA, a more stable genetic molecule, eventually replaced RNA as the carrier of genetic information. RNA may then have taken on its current role in the translation of genetic information.

INTERACTIVE QUESTION 25.1

Why do we say that, for life to have begun, both accurate replication and metabolism are required?

25.2 The fossil record documents the history of life

The Fossil Record The fossil record, based on the sequence of fossils found in the strata of sedimentary rocks, provides evidence of great changes in the organisms that have lived on Earth. The record is incomplete, however, because large numbers of species that lived probably left no fossils; geologic processes destroy many fossils; and only a fraction of existing fossils have been found.

How Rocks and Fossils Are Dated The order in which fossils appear in the strata of sedimentary rocks indicates their relative age.

Radiometric dating is used to determine the actual ages of rocks and fossils. Each radioactive isotope has a fixed rate of decay—its **half-life,** which is the number of years it takes for 50% of the "parent" isotope to decay to its "daughter" isotope. During an organism's lifetime, it accumulates isotopes in proportions equal to their relative abundance in the environment. After the organism dies, its radioactive isotopes decay at their fixed rate. Carbon-14 has a half-life of 5,730 years. Isotopes that decay more slowly can be used to date older fossils. The age of volcanic rock can be used to infer the age of fossils associated with such layers of rock.

Patterns of magnetic reversals, when the orientation of Earth's magnetic field has reversed, can also be used to date rocks.

INTERACTIVE QUESTION 25.2

A fossil has one-eighth of the atmospheric ratio of C-14 to C-12. Estimate the age of this fossil.

The Origin of New Groups of Organisms The fossil record can detail the origin of new groups of organisms,

as in the gradual evolution of mammals from a group of tetrapods called synapsids.

25.3 Key events in life's history include the origins of single-celled and multicelled organisms and the colonization of land

Geologists have established a **geologic record** of Earth's history. The Archaean and the Proterozoic eons lasted approximately 4 billion years. The Phanerozoic eon covers the last half billion years and is divided into three eras: the Paleozoic, Mesozoic, and Cenozoic. These eras are delineated by major extinction events.

The First Single-Celled Organisms The oldest known fossils are 3.5-billion-year-old fossil **stromatolites,** which are rocklike layers of prokaryotes and sediment. It is possible that single-celled organisms originated as early as 3.9 billion years ago.

Banded iron formations in marine sediments and the rusting of iron in terrestrial rocks that began about 2.7 billion years ago provide evidence of oxygen accumulation from the photosynthesis of ancient cyanobacteria. The relatively rapid accumulation of atmospheric oxygen around 2.2 billion years ago caused the extinction of many prokaryotic groups and relegated others to anaerobic habitats.

Fossils that are generally accepted as eukaryotic date from 2.1 billion years ago. The cytoskeleton of eukaryotic cells allows them to change shape and engulf other cells. According to the **endosymbiosis** hypothesis, the mitochondria and plastids of eukaryotes may have originated as small prokaryotes that were undigested prey or internal parasites of larger cells.

An aerobic heterotrophic prokaryote, living as an endosymbiont within a host cell, may have evolved into a mitochondrion. Plastids may have originated from photosynthetic endosymbionts. According to the hypothesis of **serial endosymbiosis,** mitochondria probably evolved first because they or their genetic remnants are present in all eukaryotes.

INTERACTIVE QUESTION 25.3

List some of the evidence supporting an endosymbiotic origin of mitochondria and plastids.

The Origin of Multicellularity A great diversity of single-celled eukaryotes evolved from the structurally complex first eukaryotic cells. The first multicellular eukaryotes gave rise to some algae, plants, fungi, and animals.

Molecular evidence suggests that the common multicellular ancestor arose 1.5 billion years ago. The oldest known fossils are small algae from 1.2 billion years ago.

According to the "snowball Earth" hypothesis, severe ice ages that occurred from 750 to 580 million years ago accounted for the limited diversity and distribution of multicellular eukaryotes, with the first major diversification occurring after Earth thawed. Fossils of these larger and more diverse organisms, known as the Ediacaran biota, lived from 575 to 545 million years ago.

Fossils of Porifera, Cnidaria, and Mollusca date from the late Proterozoic. The appearance of many of the major phyla of animals between 535 and 525 million years ago is referred to as the **Cambrian explosion.** But molecular evidence indicates that many animal phyla began to diverge perhaps as early as 700 million to 1 billion years ago. And a diversity of algae and animal forms (including embryos) are represented in Chinese fossils dating back 570 million years ago.

INTERACTIVE QUESTION   25.4

What lifestyle change may be related to the diversity of forms that evolved in the 10 million years known as the Cambrian explosion?

The Colonization of Land   Cyanobacteria coated damp terrestrial surfaces more than a billion years ago, but plants, fungi, and animals began to move onto land only about 500 million years ago. This colonization was associated with adaptations that helped prevent dehydration and permitted reproduction on land. Plants and fungi appear to have colonized the land together in symbiotic associations. Arthropods and tetrapod vertebrates are the most widespread and diverse land animals.

25.4 The rise and fall of dominant groups reflect continental drift, mass extinctions, and adaptive radiations

Continental Drift   The continents rest on great plates that float on Earth's molten mantle. Earthquakes occur and islands and mountains are formed in regions where these shifting plates abut. In the past 1.1 billion years, **continental drift** has brought all the landmasses together to form a supercontinent three times. The formation of the supercontinent **Pangaea** about 250 million years ago destroyed and altered habitats, causing many species to become extinct. The shifting of landmasses north or south greatly changes their climate. The separation and drifting apart of continents

creates huge geographic isolation events, promoting allopatric speciation. Continental drift helps explain the past and current distribution of organisms.

INTERACTIVE QUESTION   25.5

Marsupials probably evolved in what is now North America and Asia, yet their greatest diversity is found in Australia. How can you account for this biogeographic distribution?

Mass Extinctions   A species may become extinct due to a change in its physical or biological environment. **Mass extinctions** have occurred during periods of major environmental change. There have been five mass extinctions over the past 500 million years.

The Permian mass extinction, occurring at the boundary between the Paleozoic and Mesozoic eras about 250 mya, claimed about 96% of marine animal species and many terrestrial ones. These extinctions may be related to massive volcanic eruptions in Siberia, which probably warmed the global climate and slowed the mixing of oceans, reducing the oxygen supply to marine organisms.

The Cretaceous mass extinction, which marks the boundary between the Mesozoic and Cenozoic eras about 65.5 mya, claimed more than one-half of the marine species, many families of terrestrial plants and animals, and most of the dinosaurs. The thin layer of clay rich in iridium (an element common in meteorites) may have been the fallout from a huge cloud of dust created when an asteroid or comet collided with Earth. This cloud would have blocked sunlight and severely affected weather.

The current rate of extinction may be 100 to 1,000 times the typical rate seen in the fossil record. A sixth mass extinction caused by human destruction of habitat may occur in the next few centuries or millennia.

The fossil record indicates it takes 5 to 10 million years for the diversity of life to recover following a mass extinction. Once an evolutionary lineage becomes extinct, it cannot reappear. Mass extinctions may alter the types of organisms making up an ecological community.

Adaptive Radiations   **Adaptive radiations** are periods in which numerous, variously adapted species evolve and fill new ecological niches. Adaptive radiations have occurred following mass extinctions, the evolution of major adaptations that allow organisms to exploit new ecological roles, and the colonization of new regions.

a. Mammals originated 180 million years ago but did not change much until their adaptive radiation 65 mya. Explain this observation.

b. Give some examples of adaptive radiations following major evolutionary innovations that opened new ecological roles.

c. What factors have contributed to the adaptive radiation of the thousands of endemic species of the Hawaiian Archipelago?

25.5   Major changes in body form can result from changes in the sequence and regulation of developmental genes

Evolutionary Effects of Developmental Genes The combination of evolutionary and developmental biology, called "evo-devo," explores how slight changes in developmental genes can result in major morphological differences between species.

Heterochrony is an evolutionary change in the rate or timing of development. A minor genetic alteration that affects the relative growth rates of body parts can produce a very differently proportioned adult form.

Paedomorphosis is the retention in the adult of juvenile traits of an ancestral species and can occur when the timing of reproductive development accelerates.

Homeotic genes control the spatial arrangement of body parts. Mutations in homeotic genes called *Hox* genes, whose products provide positional information in embryos, can drastically alter body form. Two duplications of *Hox* genes may have been central to the evolution of vertebrates.

The Evolution of Development Changes in the number and sequences of developmental genes as well as changes in gene regulation may produce new body forms.

a. Researchers identified the key difference in the *Hox* gene of *Artemia* (brine shrimp) and *Drosophila* (fruit fly) that is responsible for the suppression of legs in *Drosophila*. What was the significance of this study?

b. Marine populations of threespine stickleback fish have protective spines on their ventral surface; lake populations of these fish do not. How did researchers determine that the loss or reduction of these spines was caused by a difference in the regulation of a developmental gene and not in the gene itself?

25.6   Evolution is not goal oriented

Evolutionary Novelties Often very complex organs, such as the eyes of vertebrates and some molluscs, have evolved gradually from simpler structures that served similar needs in ancestral species.

Evolutionary novelties may also evolve by the gradual modification of existing structures for new functions. *Exaptation* is the term for structures that evolved and functioned in one setting and were then co-opted for a new function.

Evolutionary Trends *Equus*, the modern horse, descended from its much smaller, browsing, multitoed ancestor, *Hyracotherium*, through a series of speciation episodes that produced many different species and diverging trends, as documented in the fossil record.

a. According to S. Stanley's model of *species selection*, "differential speciation success" plays a role in macroevolution similar to the role of differential reproductive success in microevolution. Explain.

b. Do evolutionary trends indicate that evolution is goal directed?

Word Roots

hetero = different (*heterochrony*: evolutionary changes in the timing or rate of development)

macro- = large (*macroevolution*: long-term evolutionary change including the origin of novel designs, adaptive radiations, and mass extinctions)

paedo- = child (*paedomorphosis*: the retention in the adult organism of the juvenile features of its evolutionary ancestors)

proto- = first (*protobionts*: aggregates of abiotically produced molecules)

stromato- = something spread out; **-lite** = a stone (*stromatolite*: layered rocks made by the actions of prokaryotes that bind sediment, in which are found the most ancient forms of life)

Structure Your Knowledge

1. List the sequence of four stages that could have led to the origin of the first living cells.

2. Label the eons, eras, and the key events shown on this clock analogy of Earth's history.

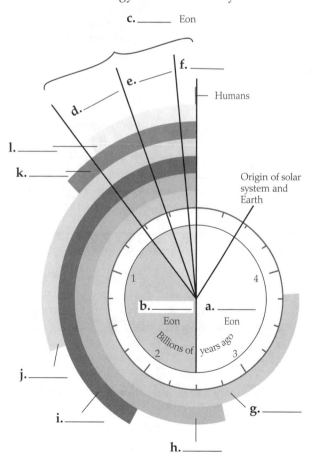

3. Describe three major processes that have influenced the changing diversity of life on Earth.

4. What genetic changes would most likely lead to the evolution of new morphological forms?

Test Your Knowledge

MULTIPLE CHOICE: *Choose the one best answer.*

1. The primitive atmosphere of Earth may have favored the abiotic synthesis of organic molecules because it
   a. was highly oxidative.
   b. was reducing (near volcanic openings) and had energy sources in the form of lightning and UV radiation.
   c. had a great deal of methane and organic fuels.
   d. had plenty of water vapor, carbon, oxygen, and nitrogen, providing the C, H, O, and N needed for organic molecules.
   e. consisted almost entirely of hydrogen gas, creating a very reducing environment.

2. The half-life of carbon-14 is 5,730 years. A fossil that is 22,920 years old would have what amount of the normal proportion of C-14 to C-12?
   a. $\frac{1}{2}$ c. $\frac{1}{6}$ e. $\frac{1}{16}$
   b. $\frac{1}{4}$ d. $\frac{1}{8}$

3. The Permian mass extinction between the Paleozoic and Mesozoic eras
   a. affected a huge percentage of marine and many terrestrial species.
   b. coincided with the breaking apart of Pangaea.
   c. made way for the adaptive radiation of mammals, birds, and pollinating insects.
   d. appears to have been caused by a large asteroid striking Earth.
   e. involved all of the above.

4. Which of the following characteristics would increase the likelihood that organisms would be well represented in the fossil record?
   a. abundant and widespread
   b. exist over a long time period
   c. have hard shells or skeletons
   d. live in regions where geology favors fossil formation
   e. All of the above contribute.

5. Life on Earth is thought to have begun
   a. 540 million years ago, at the beginning of the Paleozoic era.

b. 635 million years ago, at the start of the Ediacaran.

c. 2.7 billion years ago, when oxygen began to accumulate in the atmosphere.

d. between 3.5 and 3.9 billion years ago, after Earth cooled and rocks began to form.

e. 4.5 billion years ago, shortly after Earth first formed.

6. Stromatolites are

a. early prokaryotic fossils found in sediments around hydrothermal vents.

b. protobionts that form when lipids assemble into a bilayer surrounding organic molecules.

c. layers of rusted terrestrial rocks formed when O_2 produced by early photosynthetic prokaryotes entered the atmosphere.

d. fossils appearing to be eukaryotes that are about twice the size of prokaryotes.

e. layered communities of prokaryotes, fossils of which represent the oldest known prokaryotes.

7. Liposomes are

a. spontaneously forming droplets surrounded by a lipid membrane.

b. prokaryotes that lived as endosymbionts in larger prokaryotic cells.

c. amino acid polymers formed in the laboratory by dripping organic monomers onto hot rocks.

d. abiotically produced lipids.

e. RNA self-replicating molecules.

8. Which of the following is a proposed hypothesis for the origin of genetic information?

a. Early DNA molecules coded for RNA, which then catalyzed the production of proteins.

b. Early polypeptides were the first "genes," followed by RNA and then DNA.

c. Short RNA strands were capable of self-replication and evolved by natural selection.

d. As protobionts grew and split, they distributed copies of their molecules to their offspring.

e. Early RNA molecules coded for the order of amino acids in a polypeptide.

9. Which of the following provides the least evidence that life may have formed spontaneously?

a. the discovery of ribozymes, showing that prebiotic RNA molecules may have been autocatalytic.

b. the laboratory synthesis of liposomes.

c. the fossil record.

d. the abiotic synthesis of polymers from monomers dripped onto hot rocks.

e. the production of organic compounds in the Miller-Urey experimental apparatus.

10. Banded iron formations in marine sediments provide evidence of

a. the first prokaryotes around 3.5 billion years ago.

b. oxidized iron layers in terrestrial rocks.

c. the accumulation of oxygen in the seas from the photosynthesis of ancient cyanobacteria.

d. the evolution of photosynthetic archaea near deep-sea vents.

e. the crashing of meteorites onto Earth, possibly transporting abiotically produced organic molecules from space.

11. Diverse soft-bodied organisms, known as the Ediacaran biota, appear in the fossil record around 575 million years ago. This time period may correspond to

a. the thawing of snowball Earth.

b. the first appearance of multicellular eukaryotes in the fossil record.

c. the Cambrian explosion of animal forms.

d. the colonization of land by plants and fungi, providing the food source for animals to follow.

e. the rapid increase in atmospheric oxygen that allowed for the active lifestyle of animals.

12. The more than 500 species of fruit fly on the various Hawaiian Islands, all apparently descended from a single ancestral species, are an excellent example of

a. continental drift.

b. adaptive radiation.

c. biogeography.

d. exaptation.

e. an evolutionary trend.

13. According to the hypothesis of endosymbiosis,

a. the first primitive cells may have originated when a lipid bilayer spontaneously formed around abiotically produced polymers.

b. plants were able to colonize land with fungi as symbionts with their roots.

c. the mitochondria and plastids of eukaryotic cells were originally prokaryotic endosymbionts that became permanent parts of the host cell.

d. the infoldings and specializations of the plasma membrane led to the evolution of the endomembrane system.

e. the nuclear membrane evolved first, followed by plastids, and then mitochondria.

14. What is meant by the concept of species selection?
    a. Reproductive isolating mechanisms are responsible for maintaining the integrity of individual species.
    b. Characteristics that increase the probability that a species will separate into two or more species will accumulate in the most successful species.
    c. Natural selection can act on the species level as well as the population level.
    d. The species that last the longest and speciate the most often influence the direction of evolutionary trends.
    e. The colonization of new and diverse habitats can lead to the adaptive radiation of numerous species.

15. The evolution of the middle ear bones from bones of the jaw hinge of cynodonts is an example of
    a. heterochrony.
    b. exaptation.
    c. paedomorphosis.
    d. changes in homeotic gene expression.
    e. an evolutionary trend.

Key Concepts

26.1 Phylogenies show evolutionary relationships

26.2 Phylogenies are inferred from morphological and molecular data

26.3 Shared characters are used to construct phylogenetic trees

26.4 An organism's evolutionary history is documented in its genome

26.5 Molecular clocks help track evolutionary time

26.6 New information continues to revise our understanding of the tree of life

Framework

A goal of systematics is to reconstruct the phylogenetic history of species. Cladistics identifies monophyletic taxa based on shared derived characters. The best phylogenetic hypotheses are parsimonious trees based on molecular homologies as well as on morphological and fossil evidence. Current hypotheses for the comprehensive tree of life suggest that eukaryotes resulted from a fusion of a bacterium and archaean, and that horizontal gene transfers were common in the early history of life.

Chapter Review

Phylogeny is the evolutionary history of a species or group. **Systematics** is the discipline focused on classifying organisms and determining evolutionary relationships.

26.1 Phylogenies show evolutionary relationships

Binomial Nomenclature As begun by Linnaeus, each species is assigned a two-part latinized name—a **binomial**—consisting of the name of the **genus** and the specific epithet, designating one species in that genus.

Hierarchical Classification **Taxonomy** is concerned with naming and classifying species. In the Linnaean taxonomic system, genera are grouped into **families,** which are then grouped into **orders, classes, phyla,** and **kingdoms,** and more recently into **domains.** A taxonomic unit at any level is called a **taxon.**

Linking Classification and Phylogeny   A **phylogenetic tree** represents hypotheses about evolutionary relationships among groups, with branching patterns that may reflect their Linnaean classification. The "phylocode" system of classification is based totally on evolutionary relationships, only naming groups that include a common ancestor and all descendants.

Each dichotomous **branch point** on a phylogenetic tree represents the divergence of two taxa from a common ancestor. **Sister taxa** share the most recent common ancestor. Trees are **rooted,** with the oldest branch to the left. A **polytomy** is a branch point involving more than two descendants, indicating that these evolutionary relationships are not yet clear.

What We Can and Cannot Learn from Phylogenetic Trees   Phylogenetic trees show patterns of descent but do not usually indicate when species evolved or the amount of genetic change in each lineage. Also, species next to each other on the tree indicate that they shared a common ancestor, not that one evolved from the other.

Applying Phylogenies   The information contained in phylogenetic trees has practical applications such as molecular forensics or improvement of crops.

INTERACTIVE QUESTION   26.1

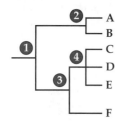

In this hypothetical phylogenetic tree, which number represents the common ancestor of all the taxa?
Which branch point is a polytomy?   Which taxa are sister taxa?

26.2   Phylogenies are inferred from morphological and molecular data

Morphological and Molecular Homologies   Homologies are similarities due to shared ancestry. The more similar morphologies or DNA sequences are, the more likely that organisms are closely related.

Sorting Homology from Analogy   **Analogy** is similarity due to convergent evolution, in which unrelated species develop similar features because natural selection has led to similar adaptations. Analogous structures are also called **homoplasies.** In general, when two

complex structures share many similar features, they more likely were inherited from a common ancestor. In molecular comparisons, the more nucleotide sequences shared between comparable genes, the more likely the genes are to be homologous.

INTERACTIVE QUESTION 26.2

What two complications may make it difficult to determine phylogenetic relationships based on morphological similarities between species? Give examples.

Evaluating Molecular Homologies   Molecular comparisons are complicated by insertion or deletion mutations that change the lengths of homologous regions of DNA. Computer programs identify similar sequences and then insert gaps in order to align homologous DNA segments properly for nucleotide comparisons. The matching bases between organisms that are not closely related may be molecular homoplasies. Mathematical tools have been developed to identify "distant" homologies between extremely divergent sequences.

Molecular systematics uses DNA and other molecular comparisons to infer evolutionary relationships.

INTERACTIVE QUESTION   26.3

Base deletions have changed the lengths and alignments of these two homologous regions of DNA. Determine the best possible fit between these two DNA sequences. How many deletions and base changes have occurred in these DNA segments?

ACGTGCACG

AGTGAGG

26.3   Shared characters are used to construct phylogenetic trees

Cladistics   Classification based on common descent forms the basis of **cladistics.** A **clade** consists of an ancestral species and all of its descendant species. Such a clade is **monophyletic.** A **paraphyletic** group excludes some species that share a common ancestor with other species in the group, and a **polyphyletic** group includes several groups with different ancestors.

Shared ancestral characters are found in a particular clade but evolved in an ancestor that is not part of that clade. **Shared derived characters** are evolutionary novelties that are unique to a particular clade.

To determine the branching sequence of a group of related species, the group is compared to an **outgroup,** a species or group of species that diverged before the group being studied (the **ingroup**). A comparison of the numbers of characters that are present in each taxon of the ingroup indicates the sequence in which shared derived characters evolved and determines the branch points used to produce a phylogenetic tree.

Place the taxa (outgroup, A, B, C, and D) on the phylogenetic tree based on the presence or absence of the characters 1–4 as shown in this table. Indicate before each branch point the number for the shared derived character that evolved in the ancestor of the clade.

	Outgroup O	A	Taxa B	C	D
1	0	1	1	1	1
2	0	0	1	0	1
3	0	1	1	0	1
4	0	0	1	0	0

(Characters label runs vertically along the left side of rows 1–4.)

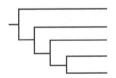

Phylogenetic Trees with Proportional Branch Lengths The branching pattern of most phylogenetic trees is relative rather than absolute, indicating only the order in which members of each clade last shared a common ancestor. The branch length of some trees can be scaled to reflect rates of evolutionary change (for instance, number of changes in DNA sequence) or time (using the dates of branch points as indicated in the fossil record).

Maximum Parsimony and Maximum Likelihood Systematists use morphological characters or molecular comparisons to choose among many possible phylogenetic trees using the principle of **maximum parsimony**— the smallest number of evolutionary changes is the simplest explanation and thus the best hypothesis to consider first. According to the principle of **maximum likelihood,** a tree can be found that represents the most likely sequence of events, given a certain set of assumptions

(such as equal rates of DNA changes). Computer programs search for the most parsimonious and most likely trees.

Phylogenetic Trees as Hypotheses The most parsimonious tree based on the fewest number of evolutionary changes represents the best hypothesis of the relationships among a set of species. A hypothesis of phylogenetic relationships becomes more reliable with the more data that can be compared.

Phylogenetic hypotheses can be used to make and test predictions. In **phylogenetic bracketing,** features shared by two closely related organisms are predicted to be found in their common ancestor and all its descendants. Fossil discoveries support the prediction that the ancestor of birds and crocodiles, and therefore also dinosaurs, built nests and brooded their eggs.

According to the principle of parsimony, the evolution of the four-chambered heart should place birds and mammals in the same clade. Why does the most accepted evolutionary tree show them as separate branches from the reptilian line?

26.4 An organism's evolutionary history is documented in its genome

Molecular systematics uses nucleic acid or other molecular comparisons to uncover phylogenetic relationships, making it possible both to compare genetic divergence among groups within a species and to reconstruct phylogeny among very distantly related species. The ability to span this length of time depends on genes that evolve either very slowly or very rapidly.

Gene Duplications and Gene Families Gene duplications are important evolutionary mutations, increasing the number of genes in the genome. **Orthologous genes** are homologous genes found in different species, which then diverge following speciation. **Paralogous genes** result from gene duplication in the same genome, and as they diverge, they provide new opportunities for evolutionary change within a species.

Genome Evolution Widespread orthologous genes indicate that all organisms share many biochemical and developmental pathways. There is a relatively small difference in the number of genes in organisms of very different complexity. Genes of more complex organisms may be more versatile.

INTERACTIVE QUESTION   26.6

a. Give an example of genes that would be compared to uncover phylogenetic relationships among the earliest branches on the tree of life.

b. Give an example of genes that evolve very rapidly and are used to discriminate among closely related species or even among populations of the same species.

c. Give an example of paralogous genes.

26.5   Molecular clocks help track evolutionary time

Molecular Clocks   Some regions of DNA appear to evolve at constant rates, and comparisons of the number of nucleotide substitutions in orthologous genes can serve as **molecular clocks** to estimate the time since species branched from their common ancestor. The number of differences in paralogous genes is proportional to the time since the gene was duplicated.

Some genes appear to have a reliable average rate of evolution. Graphs that plot nucleotide or amino acid differences against times for known evolutionary branch points can be used to estimate phylogenetic branchings that are not evident from the fossil record.

According to the **neutral theory,** much of the evolutionary change in genes does not affect fitness. Harmful mutations are removed quickly from the gene pool, but neutral mutational changes that have little effect on fitness should occur at a constant rate.

INTERACTIVE QUESTION   26.7

Using the neutral theory of molecular evolution, explain why different genes might have a different molecular clock rate.

Natural selection, which favors some DNA changes over others, should disrupt the smooth running of the molecular clock. Evolutionists disagree about the extent of neutral mutations, and thus about whether molecular clocks are reliable for timing evolution. Many are skeptical when molecular clocks are used to date evolutionary divergences that occurred billions of years ago.

When molecular clocks are calibrated using many genes, fluctuations due to natural selection may average out. Molecular and fossil-based estimates of divergence times in vertebrate evolution agree closely.

Applying a Molecular Clock: The Origin of HIV   By comparing sequences of HIV viruses from samples taken at various times during the epidemic, researchers have observed a remarkably consistent rate of evolution. They estimate that the HIV-1 M strain first infected humans in the 1930s.

26.6   New information continues to revise our understanding of the tree of life

From Two Kingdoms to Three Domains   Historically, taxonomists divided the diversity of life into two kingdoms—plants and animals. In the five-kingdom system, the prokaryotes were set apart from the eukaryotes and placed in the kingdom Monera. The kingdom Protista contained mostly unicellular eukaryotes, while kingdoms Plantae, Fungi, and Animalia consisted of multicellular eukaryotes.

The current three-domain system creates a taxon above the kingdom level. The domains Bacteria and Archaea, although both consisting of single-celled prokaryotes, differ in many characteristics. The domain Eukarya includes all the eukaryotes: plants, fungi, animals, and many groups of single-celled organisms.

A Simple Tree of All Life   The rRNA genes have evolved so slowly that they have been used as the basis for constructing a basic tree of life. The tree shows the first major split as the divergence of the bacteria from the other two domains. Analysis of other genes, however, suggests that eukaryotes share a more recent common ancestor with bacteria than with archaea.

Is the Tree of Life Really a Ring?   Genome comparisons from the three domains indicate that **horizontal gene transfer,** perhaps through transposable elements and plasmid exchange, or even through fusions of different organisms, occurred during the early history of life. Some scientists suggest that the early history of life is best represented by a ring. If eukaryotes arose as an endosymbiosis between a bacterium and an archaean, then eukaryotes are most closely related to both.

Word Roots

analog- = proportion (*analogy:* similarity due to convergence)

bi- = two; **nom-** = name (*binomial:* a two-part latinized name of a species)

clado- = branch (*cladogram:* a dichotomous phylogenetic tree that branches repeatedly)

homo- = like, resembling (*homology:* similarity in characteristics resulting from a shared ancestry)

mono- = one (*monophyletic:* pertaining to a taxon derived from a single ancestral species that gave rise to no species in any other taxa)

parsi- = few (*principle of parsimony:* the premise that a theory about nature should be the simplest explanation that is consistent with the facts)

phylo- = tribe; **-geny** = origin (*phylogeny:* the evolutionary history of a taxon)

Structure Your Knowledge

1. Draw a phylogenetic tree that best represents the relationships among taxa A–E described below:

   Taxon A was the first to diverge from the other evolutionary lines—it shares its last common ancestor with B, C, D, and E. B and C are sister taxa. They share a more recent common ancestor with D than with E.

2. This comprehensive tree of life represents current hypotheses on the origin of the three domains. Identify the domains. Number one represents the last common ancestor of all living things. Describe the two major episodes of horizontal gene transfer illustrated by #2 and #3.

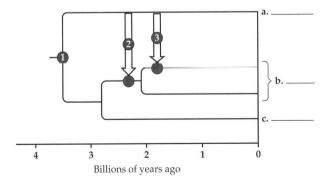

Billions of years ago

Test Your Knowledge

MULTIPLE CHOICE: *Choose the one best answer.*

1. Related families are grouped into the next-highest taxon called a
   a. class. c. order. e. kingdom.
   b. phylum. d. genus.

2. Convergent evolution may occur
   a. when ancestral structures are co-opted for new functions.
   b. when homologous structures are adapted for different functions.
   c. from adaptive radiation.
   d. when species are widely separated geographically.
   e. when species have similar ecological roles.

3. Which of these taxonomic units are most equivalent across all eukaryotes?
   a. family
   b. species
   c. phyla
   d. kingdoms
   e. All of these units are determined by set criteria, and all are comparable across different lineages of eukaryotes.

4. Which of these is a shared derived character for monotreme, marsupial, and eutherian mammals?
   a. parental care
   b. internal fertilization
   c. amniotic egg
   d. production of milk for young
   e. complete embryonic development inside a uterus

5. Which of the following are the best examples of homoplasies?
   a. forelimbs of bat and mole
   b. silversword plants of Hawaii
   c. four-chambered hearts of birds and bats
   d. skulls of apes and humans
   e. hindlegs of Australian and North American moles

6. Four of these trees describe the same phylogenetic relationships among taxa A, B, C, D, E, and F. Which tree shows a different phylogeny?

a.

b.

c.

d.

e.

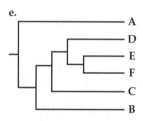

7. Shared derived characters are
   a. those that characterize all the species on a branch of a phylogenetic tree.
   b. determined through a computer comparison of paralogous genes in a group of species.
   c. homologous structures that develop during adaptive radiation.
   d. characters found in the outgroup but not in the species under consideration.
   e. used to identify species but not higher taxa.

8. Which of the following is the best description of our current hypothesis of the tree of life?
   a. The tree of life consists of three domains: Bacteria, Archaea, and Eukarya.
   b. The base of the tree of life is still uncertain because the molecular clock is not accurate for evolutionary events that occurred that long ago.
   c. The domain Archaea is known to be the first branch; domains Bacteria and Eukarya are more closely related to each other.
   d. There was substantial horizontal gene transfer and perhaps even fusion of different organisms during the early history of life.
   e. Both a and d represent our current hypothesis.

9. How could a molecular clock be useful for studying orthologous genes?
   a. It could help determine when the original gene duplication occurred.
   b. It could help date the divergence of different lineages that all contain that homologous gene.
   c. Because orthologous genes do not have neutral mutations, the molecular clock could not be used to study them.
   d. It could date when the gene first arose.
   e. Because the evolution of orthologous genes does not occur at a consistent rate, a molecular clock would not be useful.

10. A comparative study of which of the following would provide the best data on the early evolution of fungi and plants?
    a. mitochondrial DNA
    b. DNA from a chloroplast
    c. the amino acid sequence of chlorophyll
    d. DNA for ribosomal RNA
    e. the morphology of present-day specimens

11. A comparative study of which of the following would provide the best data on the ancestry of people from Germany, Italy, and Spain?
    a. mitochondrial DNA
    b. orthologous genes
    c. ribosomal RNA
    d. paralogous genes
    e. the fossil record

12. A taxon such as the class Reptilia, which does not include its relatives, the birds, is
    a. really an order.
    b. a clade.
    c. monophyletic.
    d. polyphyletic.
    e. paraphyletic.

13.

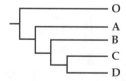

According to this phylogenetic tree, C and D are most closely related because they
   a. do not share a common ancestor with O, A, or B.
   b. are monophyletic.
   c. evolved from a common ancestor a long time ago.
   d. have the most shared derived characters in common.
   e. have the youngest species.

14. A biologist is studying the evolution of four similar species of birds. Which approach would allow her to choose the best phylogenetic tree from all possible phylogenies?

a. Draw the simplest tree and choose that one.

b. From a comparison of nucleotide sequences, determine the number of evolutionary events required for each tree and choose the most parsimonious tree.

c. Compare the entire genomes of each species; the two most similar genomes are the two species that are most closely related.

d. Determine which species can interbreed; those that can evolved from a common ancestor most recently.

e. Choose the tree that has the most evolutionary changes required as the most likely explanation for how these very similar birds have evolved into four distinct species.

Chapter 27

Bacteria and Archaea

Key Concepts

27.1 Structural and functional adaptations contribute to prokaryotic success

27.2 Rapid reproduction, mutation, and genetic recombination promote genetic diversity in prokaryotes

27.3 Diverse nutritional and metabolic adaptations have evolved in prokaryotes

27.4 Molecular systematics is illuminating prokaryotic phylogeny

27.5 Prokaryotes play crucial roles in the biosphere

27.6 Prokaryotes have both harmful and beneficial impacts on humans

Framework

This chapter presents the morphology, reproduction, phylogeny, and ecology of prokaryotes. These generally single-celled organisms greatly outnumber and outweigh all eukaryotes combined and flourish in all habitats, including ones that are too harsh for any other forms of life. The collective impact of these microscopic organisms is huge.

Chapter Review

27.1 Structural and functional adaptations contribute to prokaryotic success

Prokaryotic cells are usually 0.5–5 μm in diameter. The three most common cell shapes are spheres (cocci), rods (bacilli), and spirals.

Cell-Surface Structures Prokaryotic cell walls maintain cell shape and protect the cell from bursting in a hypotonic surrounding. Most prokaryotes plasmolyze in a hypertonic environment. The cell walls of bacteria (but not archaea) contain **peptidoglycan,** a matrix composed of modified-sugar polymers cross-linked by short polypeptides.

The **Gram stain** is an important tool for identifying bacteria as **gram-positive** (bacteria with walls containing a thicker layer of peptidoglycan) or **gram-negative** (bacteria with more complex walls including an outer lipopolysaccharide membrane).

Many prokaryotes secrete a sticky **capsule** outside the cell wall that serves as protection from attack by a host's immune system and as glue for adhering to a substrate or other prokaryotes. Some bacteria may also attach by means of surface hairlike appendages called **fimbriae.** Fimbriae are shorter and more numerous than **sex pili,** which are specialized for the exchange of DNA between prokaryotes.

Motility Many prokaryotes are equipped with flagella, either scattered over the cell surface or concentrated at one or both ends of the cell. These flagella differ from eukaryotic flagella in their lack of a plasma membrane cover, structure, and function. ATP-driven pumps and the diffusion of H^+ back into the cell power the basal apparatus, embedded in the cell wall, which turns a hook attached to the protein filament of the flagellum.

Many motile prokaryotes exhibit **taxis,** an oriented movement in response to chemical, light, or other stimuli.

Internal and Genomic Organization Extensive internal compartmentalization is not found in prokaryotic cells, although some may have membranes, usually infoldings of the plasma membrane that function in respiration or photosynthesis.

The circular DNA chromosome, which has relatively little protein associated with it, is concentrated in a region called the **nucleoid.** Smaller rings of DNA, called **plasmids,** may carry a few genes.

The ribosomes of prokaryotes are smaller than eukaryotic ribosomes and differ in their protein and RNA content. Some antibiotics work by binding to prokaryotic ribosomes and blocking protein synthesis.

Reproduction and Adaptation   With generation times from 20 minutes to 3 hours, prokaryotes have a huge reproductive potential. The growth of colonies usually stops due to the exhaustion of nutrients or the toxic accumulation of wastes.

Some bacteria produce **endospores,** which are tough-walled dormant cells formed in response to a lack of nutrients. Endospores are so durable that microbiologists must use steam at high pressure to sterilize laboratory equipment and media.

Due to short generation times, favorable mutations are rapidly propagated, facilitating adaptive evolution to environmental changes. Scientists have studied such evolution in colonies of *E.coli,* documenting rapid adaptation to a challenging new, low-glucose environment in 5,000 generations, with a slower increase in fitness for the next 15,000 generations.

INTERACTIVE QUESTION 27.1

Fill in the table with a brief description of the characteristics of prokaryotic cells.

Property	Description
Cell shape	a.
Cell size	b.
Cell surface	c.
Motility	d.
Internal membranes	e.
Genome	f.
Reproduction and growth	g.

27.2 Rapid reproduction, mutation, and genetic recombination promote genetic diversity in prokaryotes

Rapid Reproduction and Mutation   Mutations, although statistically rare, produce a great deal of genetic diversity because prokaryotes reproduce so rapidly and have such large population sizes.

Genetic Recombination   The mechanisms of genetic recombination in bacteria are different from the eukaryotic mechanisms of meiosis and fertilization.

Prokaryotes can take up DNA from the environment in a process called **transformation.** The foreign DNA is integrated into the bacterial chromosome by an exchange of homologous DNA segments. Many bacteria have surface proteins specialized for uptake of DNA from closely related species.

Bacteriophages can transfer genes from one bacterium to another by a process called **transduction.** A random piece of host DNA may be accidentally packaged within a phage and introduced into a new bacterium. Recombination may occur when the newly introduced DNA replaces the homologous region of a bacterial chromosome.

In **conjugation,** a bacterial cell temporarily joins to another cell by sex pili, and the donor cell transfers DNA to the recipient through a "mating bridge." The ability to form pili and donate DNA usually results from the presence of an **F factor,** a special piece of DNA that is either part of the chromosome or a plasmid.

Bacterial cells containing the F factor on the **F plasmid** are called F$^+$ cells. During conjugation, a single strand of the plasmid DNA is transferred through a mating bridge to the recipient cell. The single parental strands in both cells replicate, and the recipient cell is now also an F$^+$ cell.

Cells in which the F factor is inserted into the bacterial chromosome are called *Hfr cells*, for "high frequency of recombination." When these cells undergo conjugation, one parental strand of part of the F factor transfers an attached strand of the bacterial chromosome to the recipient cell. Movements disrupt the mating, usually resulting in a partial transfer of genes and F factor. Replication in both donor and recipient cells produces double-stranded DNA. Crossovers between the new DNA and the recipient cell's chromosome produce new genetic combinations in this recombinant F$^-$ bacterium. Such horizontal gene transfers are probably also important in archaea.

R plasmids carry genes that code for various mechanisms of resistance to antibiotics. R plasmids also have genes coding for sex pili and are transferred to nonresistant cells during conjugation, creating the medical problem of antibiotic-resistant pathogens.

INTERACTIVE QUESTION   27.2

Complete the following concept map that summarizes the genetic characteristics of bacteria.

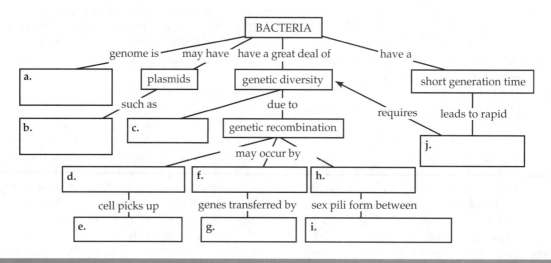

27.3   Diverse nutritional and metabolic adaptations have evolved in prokaryotes

Nutrition refers to how an organism obtains energy (photo- or chemotroph) and the carbon it uses for synthesizing organic compounds (auto- or heterotroph).

Thus, there are four major nutritional categories: (1) **photoautotrophs,** which use light energy and CO_2 to synthesize organic compounds; (2) **chemoautotrophs,** which obtain energy by oxidizing inorganic substances (such as H_2S, NH_3, or Fe^{2+}) and need only CO_2 as a carbon source; (3) **photoheterotrophs,** which use light energy but must obtain carbon in organic form; and (4) **chemoheterotrophs,** which use organic molecules as both an energy and a carbon source.

Metabolic Relationships to Oxygen   **Obligate aerobes** need O_2 for cellular respiration; **facultative anaerobes** can use O_2 but also can grow in anaerobic conditions using anaerobic respiration or fermentation;

obligate anaerobes are poisoned by O_2. Some obligate anaerobes use **anaerobic respiration** to break down nutrients, with inorganic molecules other than oxygen serving as the final electron acceptor.

Nitrogen Metabolism   Nitrogen is an essential component of proteins and nucleic acids. Some cyanobacteria and a few other prokaryotes obtain nitrogen through **nitrogen fixation,** converting atmospheric N_2 to ammonia (NH_3).

Metabolic Cooperation   The cyanobacterium *Anabaena* forms filaments in which most cells carry out photosynthesis, while a few cells called **heterocytes** perform nitrogen fixation. Members of the colony share nutrients through intercellular connections. **Biofilms** are surface-coating colonies characterized by intercellular signaling, proteins that adhere cells to each other and to the substrate, and channels in the colony for movement of nutrients and wastes.

INTERACTIVE QUESTION 27.3

a. Identify the large structure inside this *Bacillus anthracis* cell. What is its significance? What mode of nutrition do you think this bacterium uses?

Bacillus anthracis

1 μm

b. This is a micrograph of the cyanobacterium *Anabaena*. What mode of nutrition does this bacterium use? What is the function of the spherical cell indicated by the arrow?

Anabaena

5 μm

27.4 Molecular systematics is illuminating prokaryotic phylogeny

While shape, staining characteristics, nutritional mode, and motility may be helpful characters for clinically identifying bacteria, they have not provided a phylogenetic classification.

Lessons from Molecular Systematics Molecular comparisons of base sequences of SSU-rRNA indicate that domains Bacteria and Archaea diverged early in the history of life. Continuing analysis of prokaryotic genomes shows that the genetic diversity of prokaryotes is huge and that horizontal gene transfer has been and continues to be significant in the evolution of prokaryotes.

Archaea Archaea have some characteristics in common with bacteria, some with eukaryotes, and some unique characteristics. Many archaea are **extremophiles,** species that live in extreme habitats. **Extreme halophiles** (salt-lovers) live in extremely saline waters. **Extreme thermophiles** may be found in hot sulfur springs and near deep-sea hydrothermal vents.

Methanogens have a unique energy metabolism in which CO_2 is used to oxidize H_2, producing the waste product methane (CH_4). These strict anaerobes often live in swamps and marshes, are important decomposers in sewage treatment, and are gut inhabitants that contribute to the nutrition of cattle and other herbivores.

The methanogens and many halophiles are placed into the clade Euryarchaeota. Most thermophiles fit into the Crenarchaeota. Not all archaea are extremeophiles; numerous archaea have been found in more moderate habitats.

Prokaryote phylogeny continues to change as new groups are discovered. Examples include the oldest lineage, Korarchaeota, and the clade of tiny Nanoarchaeota.

Bacteria Bacteria show great diversity in their modes of nutrition and metabolism. Five major groups are described in the text.

Proteobacteria are a nutritionally diverse group of aerobic and anaerobic gram-negative bacteria. Its five subgroups are **alpha proteobacteria,** many of which, such as *Rhizobium* and *Agrobacterium,* are mutual symbionts or parasites of eukaryotes; **beta proteobacteria,** which include the important soil bacteria *Nitrosomonas;* **gamma proteobacteria,** which include photosynthetic sulfur bacteria and some serious pathogens and enterics (intestinal inhabitants such as *E. coli*); **delta proteobacteria,** including the colony-forming myxobacteria; and **epsilon proteobacteria,** the mostly pathogenic group that includes the stomach ulcer–causing *Helicobacter pylori.*

Chlamydias are obligate intracellular animal parasites. One of these gram-negative species is the most common cause of blindness and nongonococcal urethritis, a sexually transmitted disease.

Spirochetes are helical heterotrophs that move in a corkscrew fashion. Spirochetes cause syphilis and Lyme disease.

The **gram-positive bacteria** are a diverse group. The subgroup actinomycetes includes the species that cause tuberculosis and leprosy. Most actinomycetes are colonial soil bacteria, some of which are cultured to produce antibiotics. There are diverse solitary species of gram-positive bacteria, including those responsible for anthrax and botulism, and the species of *Staphylococcus* and *Streptococcus.* Mycoplasmas, the smallest of all cells, are the only bacteria that lack cell walls.

The **cyanobacteria** have plantlike, oxygenic photosynthesis and are important producers in freshwater and marine ecosystems. Some filamentous cyanobacteria have cells specialized for nitrogen fixation.

INTERACTIVE QUESTION   27.4

a. Evidence indicates that mitochondria evolved from aerobic alpha proteobacteria through endosymbiosis. From which bacterial group did chloroplasts likely evolve?

b. What is genetic prospecting? How does it contribute to prokaryotic phylogeny?

27.5   Prokaryotes play crucial roles in the biosphere

Chemical Cycling Prokaryotes are indispensable in recycling chemical elements. As **decomposers,** they return carbon, nitrogen, and other elements to the environment for assimilation into new living forms. Through photosynthesis and nitrogen fixation, prokaryotes convert inorganic compounds into forms that other organisms can use.

Ecological Interactions **Symbiosis** is an ecological relationship involving close contact between organisms of different species. The larger organism is called the **host,** and the smaller is known as the **symbiont.** In **mutualism,** both species benefit. Many of the 500 to 1,000 species of bacteria living in the human intestines are mutualists, digesting food and synthesizing important nutrients. In **commensalism,** one organism benefits while the other is neither harmed nor helped. In **parasitism,** a **parasite** eats portions of its host. **Pathogens** are parasites that cause disease.

INTERACTIVE QUESTION   27.5

Explain how cyanobacteria play an important role in chemical cycling.

27.6   Prokaryotes have both harmful and beneficial impacts on humans

Pathogenic Prokaryotes About one-half of all human diseases are caused by pathogenic prokaryotes. Pathogens most commonly cause disease by producing toxins. **Exotoxins** are proteins secreted by prokaryotes that cause such diseases as botulism and cholera. **Endotoxins,** which are lipopolysaccharides released from the outer membrane of gram-negative bacteria that have died, cause such diseases as typhoid fever and *Salmonella* food poisoning.

Improved sanitation and the development of antibiotics have decreased the incidence of bacterial disease in developed countries. The evolution of antibiotic-resistant strains of pathogenic bacteria, however, poses a serious health threat. Horizontal gene transfer also spreads genes connected with virulence, as in the emergence of the dangerous *E.coli* strain, O157:H7. Some pathogenic prokaryotes are potential bioterrorism weapons.

Prokaryotes in Research and Technology **Bioremediation** is the use of organisms to remove environmental pollutants. Prokaryotes are used to treat sewage and clean up oil spills. They are used in the mining industry to recover metals from ores. Prokaryotes have been engineered to make vitamins, antibiotics, and other products.

Word Roots

-gen = produce (*methanogen:* microorganisms that obtain energy by using carbon dioxide to oxidize hydrogen, producing methane as a waste product)

-oid = like, form (*nucleoid:* a dense region of DNA in a prokaryotic cell)

an- = without, not; **aero-** = the air (*anaerobic:* lacking oxygen; referring to an organism, environment, or cellular process that lacks oxygen and may be poisoned by it)

anti- = against; **-biot** = life (*antibiotic:* a chemical that kills bacteria or inhibits their growth)

bi- = two (*binary fission:* the type of cell division by which prokaryotes reproduce; each dividing daughter cell receives a copy of the single parental chromosome)

chemo- = chemical; **hetero-** = different (*chemoheterotroph:* an organism that must consume organic molecules for both energy and carbon)

endo- = inner, within (*endotoxin:* a component of the outer membranes of certain gram-negative bacteria responsible for generalized symptoms of fever and ache)

exo- = outside (*exotoxin:* a toxic protein secreted by a bacterial cell that produces specific symptoms even in the absence of the bacterium)

halo- = salt; **-philos** = loving (*halophile:* microorganisms that live in unusually highly saline environments such as the Great Salt Lake or the Dead Sea)

mutu- = reciprocal (*mutualism:* a symbiotic relationship in which both the host and the symbiont benefit)

photo- = light; **auto-** = self; **-troph** = food, nourish (*photoautotroph:* an organism that harnesses light energy to drive the synthesis of organic compounds from carbon dioxide)

sym- = with, together; **-bios** = life (*symbiosis:* an ecological relationship between organisms of two different species that live together in direct contact)

thermo- = temperature (*thermophiles:* microorganisms that thrive in hot environments, often 60–80°C)

Structure Your Knowledge

1. One might think of prokaryotes as primitive, simple single cells. This chapter, however, provides many examples of how diverse, highly adaptable, metabolically versatile and cooperative, and communal they can be. List some examples of these prokaryotic characteristics.

2. Describe four positive ways in which prokaryotes have an impact on our lives and on the world around us.

Test Your Knowledge

MULTIPLE CHOICE: *Choose the one best answer.*

1. Many prokaryotes secrete a sticky capsule outside the cell wall that
   a. allows them to glide along a slime thread.
   b. serves as protection from host defenses and glue for adherence.
   c. reacts with the Gram stain.
   d. is used for attaching cells during conjugation.
   e. is composed of peptidoglycan, polymers of modified sugars cross-linked by short polypeptides.

2. Chemoautotrophs
   a. are photosynthetic.
   b. use organic molecules for an energy and carbon source.
   c. oxidize inorganic substances for energy and use CO_2 as a carbon source.
   d. use light to generate ATP but need organic molecules for a carbon source.
   e. use light energy to extract electrons from H_2S.

3. Which of the following is *not* descriptive of conjugation between an Hfr and F⁻ bacterium?
   a. The Hfr cell has an F plasmid integrated into its chromosome.
   b. The Hfr cell forms sex pili and transfers one strand of its chromosome into an F⁻ cell.
   c. Random movements often break the conjugation bridge before the entire single strand of its chromosome and F factor is transferred.
   d. Hfr conjugation is the major mechanism that transfers antibiotic resistance genes between bacteria.
   e. Crossing over between homologous genes creates a recombinant cell.

4. Genetic variation in prokaryotes may be a result of
   a. horizontal gene transfer.
   b. mutation.
   c. conjugation.
   d. transformation and transduction.
   e. all of the above.

5. Facultative anaerobes
   a. can survive with or without oxygen.
   b. are poisoned by oxygen.
   c. are anaerobic chemoautotrophs.
   d. are able to fix atmospheric nitrogen to make NH_3.
   e. include the methanogens that reduce CO_2 to CH_4.

6. Some prokaryotes have specialized membranes that
   a. contain ribosomes and function in protein synthesis.
   b. arose from endosymbiosis of smaller prokaryotes, creating mitochondria and chloroplasts.
   c. form from infoldings of the plasma membrane and may function in cellular respiration or photosynthesis.
   d. are produced by the endoplasmic reticulum, but differ in composition from eukaryotic membranes.
   e. enclose the nucleoid region and separate plasmids from the single prokaryotic chromosome.

7. Archaea
   a. are thought to be more closely related to eukaryotes than to bacteria.
   b. have cell walls that lack peptidoglycan.
   c. are often found in harsh habitats, reminiscent of the environment of early Earth.
   d. include the methanogens, extreme halophiles, and extreme thermophiles, as well as groups found in more moderate habitats.
   e. are all of the above.

8. Organizing prokaryotes into major clades within the domains Archaea and Bacteria has been based on
   a. fossil records of prokaryotes that have been recently discovered.
   b. molecular comparisons.
   c. the Gram stain and colony characteristics when grown on solid media.
   d. shape and motility.
   e. nutritional modes and ecology.

9. Some cyanobacteria are capable of nitrogen fixation. This process
   a. oxidizes nitrogen-containing compounds to produce ATP.
   b. allows these bacteria to live in anaerobic environments.
   c. removes soil nitrogen and returns N_2 gas to the atmosphere.
   d. converts N_2 to ammonia, making nitrogen available to plants for incorporation into proteins and nucleic acids.
   e. is an essential part of the nitrogen cycle that rescues nitrogen trapped in mineral deposits and makes it available to organisms for their nitrogen metabolism.

10. Which of the following groups is matched with an incorrect description?
    a. Proteobacteria—diverse gram-negative bacteria including pathogens such as *Salmonella* and *Helicobacter pylori,* as well as beneficial species such as *Rhizobium*
    b. Chlamydias—intracellular parasites, including a species that causes blindness and nongonococcal urethritis
    c. Spirochetes—helical heterotrophs, including pathogens causing syphilis and Lyme disease
    d. Gram-positive bacteria—diverse group that includes actinomycetes, mycoplasmas, and pathogens that cause anthrax, botulism, and tuberculosis
    e. Cyanobacteria—photosynthetic filamentous colonies; some species use the pigment bacteriorhodopsin and inhabit very saline waters

FILL IN THE BLANKS

1. _____ the name for spherical prokaryotes

2. _____ region in which the prokaryotic chromosome is found

3. _____ common laboratory technique that identifies two groups of bacteria

4. _____ surface appendages of prokaryotes used for adherence to substrate

5. _____ an oriented movement in response to light or chemical stimuli

6. _____ resistant cell that can survive harsh conditions

7. _____ surface-coating cooperative colonies of prokaryotes

8. _____ proteins that are secreted by pathogens and are potent poisons

9. _____ use of organisms to remove environmental pollutants

10. _____ type of respiration that uses molecules other than O_2 as final electron acceptor

Chapter 28

Protists

Key Concepts

28.1 Most eukaryotes are single-celled organisms

28.2 Excavates include protists with modified mitochondria and protists with unique flagella

28.3 Chromalveolates may have originated by secondary endosymbiosis

28.4 Rhizarians are a diverse group of protists defined by DNA similarities

28.5 Red algae and green algae are the closest relatives of land plants

28.6 Unikonts include protists that are closely related to fungi and animals

28.7 Protists play key roles in ecological relationships

Framework

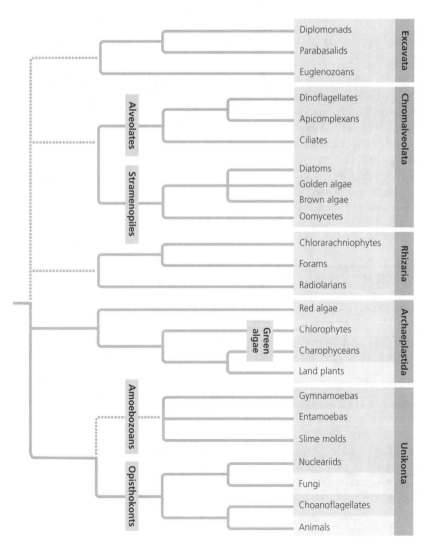

This chapter surveys the incredible diversity of protists—eukaryotes that are not fungi, plants, or animals. A tentative phylogeny of eukaryote groups, organized in five larger "supergroups," is on the preceding page.

Chapter Review

The traditional kingdom Protista has been abandoned; many lineages of this paraphyletic group are recognized as their own kingdoms. The informal term **protist** is still used for eukaryotes that are not plants, animals, or fungi.

28.1 Most eukaryotes are single-celled organisms

Structural and Functional Diversity in Protists Protists include eukaryotic, primarily unicellular organisms, with a few multicellular species. Nutritionally, protists can be photoautotrophs, heterotrophs, or **mixotrophs,** which are both photosynthetic and heterotrophic. These modes of nutrition are spread throughout the various protist lineages. Reproduction and life cycles are highly diverse.

Endosymbiosis in Eukaryotic Evolution Mitochondria evolved from endosymbiotic alpha proteobacteria within the earliest eukaryotes. All eukaryotes studied thus far have mitochondria, or signs of having had them in the past. Plastids evolved from a photosynthetic cyanobacterium that became an endosymbiont of a heterotrophic eukaryote. This lineage of cells eventually gave rise to red and green algae. In several instances of **secondary endosymbiosis,** a red or green alga was engulfed by a heterotrophic eukaryote, leading to new protist lineages.

INTERACTIVE QUESTION   28.1

The plastids of chlorarachniophytes are surrounded by four membranes and enclose a vestigial nucleus called a *nucleomorph.* What does this evidence suggest?

Five Supergroups of Eukaryotes A current hypothesis of the phylogeny of protists includes five supergroups: Excavata, Chromalveolata, Rhizaria, Archaeplastida, and Unikonta.

28.2   Excavates include protists with modified mitochondria and protists with unique flagella

The proposed clade **Excavata** is based on morphological studies of the cytoskeleton and an "excavated" feeding groove found in some members.

Diplomonads and Parabasalids Most diplomonads and parabasalids are found in anaerobic environments. They have modified mitochondria and do not have plastids.

The *mitosomes* of **diplomonads** lack functional electron transport chains. Diplomonads, such as the intestinal parasite *Giardia intestinalis,* have two nuclei and multiple flagella.

The reduced mitochondria of **parabasalids** are called *hydrogenosomes* and generate energy anaerobically, releasing H_2. *Trichomonas vaginalis* is a sexually transmitted parasite.

Euglenozoans The diverse clade of **euglenozoans** includes predatory heterotrophs, autotrophs, and pathogenic parasites, all of which have a spiral or crystalline rod inside their flagella.

A single large mitochondrion containing a mass of DNA called a kinetoplast is characteristic of the group named **kinetoplastids.** These protists include free-living heterotrophs as well as a number of parasites of plants, animals, and other protists.

The **euglenids** are characterized by one or two flagella that emerge from the anterior end. Many species of the photosynthetic *Euglena* switch to heterotrophy in the absence of sunlight.

INTERACTIVE QUESTION   28.2

Sleeping sickness is caused by the kinetoplastid *Trypanosoma* and transmitted by the African tsetse fly. Why is it so difficult for the immune system to attack this parasite?

28.3   Chromalveolates may have originated by secondary endosymbiosis

The large, diverse clade **Chromalveolata** is proposed based on some DNA data and some evidence that the chromalveolates originated by secondary endosymbiosis when their ancestor engulfed a red alga.

Alveolates Membrane-bounded alveoli under the plasma membrane are characteristic of the **alveolates.** These sacs may help to stabilize the cell surface or to osmoregulate. As supported by molecular systematics, the alveolates include the dinoflagellates, apicomplexans, and ciliates.

Most species of **dinoflagellates** are unicellular. The beating of two flagella in perpendicular grooves on the cell surface between the cellulose plates of the cell produces a characteristic spinning movement.

Dinoflagellates make up a large proportion of marine and freshwater phytoplankton. Blooms of some populations are responsible for the often-harmful "red tides." Half of all dinoflagellates are heterotrophic.

Nearly all **apicomplexans** are animal parasites, and most have complex life cycles that often include sexual and asexual stages and several host species. The parasites spread by tiny infectious cells called *sporozoites.* They have an apical complex of organelles specialized for invading host cells and a nonphotosynthetic plastid called an apicoplast.

The control of malaria, caused by *Plasmodium,* is complicated by multiple factors: development of insecticide resistance in *Anopheles* mosquitoes, which spread the disease; drug resistance in *Plasmodium;* the sequestering of the parasite within human liver and blood cells; and the ability of the parasite to change its surface proteins. The genome of *Plasmodium* has been sequenced and the expression of most genes has been tracked. This research may help the development of vaccines or other drugs.

Ciliates are characterized by the cilia they use to move and feed. Their numerous cilia may be widespread or clumped. Ciliates have two types of nuclei: Large macronuclei control everyday functions of the cell and have multiple copies of the genome. Ciliates reproduce asexually by binary fission. Small micronuclei are exchanged in a process called conjugation, which provides for genetic recombination.

INTERACTIVE QUESTION 28.3

Label the indicated structures in this diagram of a *Paramecium.* To which subgroup, clade, and supergroup does this organism belong?

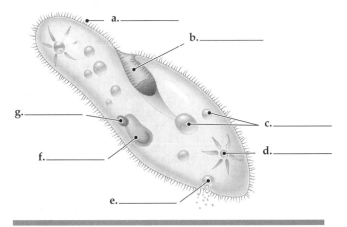

a._____
b._____
g._____
c._____
f._____
d._____
e._____

Stramenopiles The **stramenopiles** include several heterotrophic clades and various important groups of algae, all characterized by hairlike projections on a flagellum. In most cases, a shorter "smooth" flagellum is paired with the "hairy" one.

Diatoms are a major component of marine and freshwater phytoplankton with unique, protective, boxlike silica walls. These unicellular protists usually reproduce asexually. Massive quantities of fossilized diatom walls make up diatomaceous earth. Diatoms may make a contribution to the field of nanotechnology as nano-engineers study the self-assembly of their intricate shells. Their population explosions (blooms) are part of a biological carbon pump that sequesters carbon in ocean deposits.

The color of **golden algae** results from yellow and brown carotenoids. Most golden algae are unicellular, biflagellated, and found among marine and freshwater plankton. Some species are mixotrophic.

The mostly marine **brown algae** include some of the largest and most complex algae. Large marine brown algae (and some red and green algae) are called seaweeds. Some seaweeds have tissues and organs that are analogous to those found in plants. Seaweeds may have a plantlike **thallus,** or body, consisting of a rootlike **holdfast** and a stemlike **stipe** that supports leaflike **blades.** Some brown algae have floats, which keep the photosynthesizing blades near the surface. The giant, fast-growing brown algal kelps live in deeper waters.

Some multicellular algae have an **alternation of generations.** The multicellular, haploid gametophyte produces gametes. After syngamy, the diploid zygote grows into the multicellular sporophyte, which then produces spores by meiosis. The gametophyte and sporophyte may be similar in appearance **(isomorphic)** or distinct **(heteromorphic).**

INTERACTIVE QUESTION 28.4

What adaptations help brown algal seaweeds inhabit intertidal zones?

Oomycetes include water molds, white rusts, and downy mildews. Unlike their plastid-bearing ancestors, oomycetes do not have plastids. Although many resemble fungi in appearance, oomycetes have cell walls made of cellulose, and molecular systematics shows they are not closely related to fungi.

The large egg cell of a water mold is fertilized by a smaller sperm nucleus. Zygotes, following dormancy, germinate to form hyphae, which are tipped by zoosporangia that asexually produce biflagellated zoospores. Water molds are important decomposers in aquatic ecosystems. White rusts and downy mildews can be destructive parasites of land plants. Genetic studies of *Phytophthora infestans,* the oomycete that causes potato late blight, indicate that it has acquired genes in recent decades that make it more aggressive and resistant to pesticides.

28.4   Rhizarians are a diverse group of protists defined by DNA similarities

Molecular systematics indicates that the clade **Rhizaria** is a monophyletic yet diverse group that includes chlorarachniophytes, forams, and radiolarians.

Amoebas, originally defined on the basis of their cellular extensions called **pseudopodia,** are spread across many eukaryotic taxa. The amoebas in clade Rhizaria have threadlike pseudopodia.

Forams   **Foraminiferans,** or **forams,** are known for their porous shells, called **tests,** made of organic material and calcium carbonate. Pseudopodia extending through the pores function in swimming, test formation, and feeding. Many forams obtain nourishment from symbiotic algae.

Radiolarians   Slender pseudopodia help these organisms phagocytize microscopic food organisms. **Radiolarians** are primarily marine and have delicate silica shells.

INTERACTIVE QUESTION   28.5

How are foram fossils used?

28.5   Red algae and green algae are the closest relatives of land plants

Descendants of the heterotrophic protist that acquired a cyanobacterium endosymbiont evolved into red algae and green algae more than a billion years ago. Land plants arose from the green algal lineage at least 475 million years ago. The supergroup **Archaeplastida** is a monophyletic group that includes red algae, green algae, and land plants.

Red Algae   The accessory pigment phycoerythrin produces the color of **red algae** and allows them to absorb the wavelengths of light that penetrate into deep water.

Most red algae species are marine and multicellular, often with filamentous, delicately branched thalli. Alternation of generations is common, although life cycles are diverse and they have no flagellated stages in their life cycle.

Green Algae   The chloroplasts of **green algae** resemble those of plants, and molecular systematics indicates that green algae and plants are closely related. Some systematists favor including green algae in an extended kingdom, Viridiplantae. The two main green algal groups are chlorophytes and charophyceans. The charophyceans are very closely related to land plants.

Most chlorophytes live in fresh water. Unicellular forms may be planktonic, inhabitants of damp soil, or symbionts in other eukaryotes.

Size and complexity have increased in the green algae in three ways: the formation of colonies; repeated nuclear divisions to produce multinucleated filaments; and cell division and differentiation to produce true multicellular forms.

The life cycles of most chlorophytes include sexual (with biflagellated gametes) and asexual stages. Some multicellular green algae have an alternation of generations, as in the seaweed *Ulva.*

28.6   Unikonts include protists that are closely related to fungi and animals

The recently proposed supergroup **Unikonta** includes animals, fungi, and some protists. Molecular systematics supports two major clades: the amoebozoans and the opisthokonts (animals, fungi, and closely related protists). Some studies support the close relationship of these two clades; other studies do not.

Research has yet to establish the root of the eukaryotic tree. A. Stechmann and T. Cavalier-Smith propose that the unikonts were the first eukaryotes to diverge. They base this hypothesis on the DHFR and TS genes, which remain separate in bacteria and the unikont groups (considered the ancestral character) and are fused (considered the derived character) in members of the Excavata, Chromalveolata, Rhizaria, and Archaeplastida.

Amoebozoans   Slime molds, gymnamoebas, and entamoebas are grouped as **amoebozoans,** a clade supported by molecular data. The resemblance of the slime molds (or mycetozoans) to fungi is the result of convergent evolution to similar lifestyles.

Plasmodial slime molds engulf food particles by phagocytosis as they grow through leaf litter or rotting logs. This multinucleate mass called a **plasmodium** is the feeding stage. Under harsh conditions, sporangia on erect stalks produce resistant spores by meiosis. Spores germinate, the amoeboid or flagellated haploid cells fuse, and the diploid nucleus repeatedly divides to form a new plasmodium.

During the feeding stage of the life cycle, **cellular slime molds** consist of haploid solitary amoeboid cells. As food is depleted, the individual cells congregate into a mass. Asexual fruiting bodies produce resistant spores. In sexual reproduction, two amoebas fuse to form a zygote that develops into a giant cell enclosed in a protective wall. Following meiosis and mitosis, haploid amoebas are released.

The cellular slime mold *Dictyostelium discoideum* is an experimental model for studying the evolution of multicellularity. Research indicates that a recognition system allows noncheater cells to aggregate differentially from "cheater" cells (mutant cells that always become spore cells).

Most amoebas of the diverse group gymnamoebas are found free-living in freshwater, marine, or soil habitats, where they consume bacteria and other protists.

Amoebas of the genus *Entamoeba* are parasites of vertebrates and some invertebrates. *E. histolytica* causes amebic dysentery in humans.

Opisthokonts Of the protists in this diverse clade that includes animals and fungi, nucleariids are more closely related to fungi, and choanoflagellates are more closely related to animals.

INTERACTIVE QUESTION 28.6

Compare the following aspects of the life cycles of plasmodial slime molds and cellular slime molds.

Group	Haploid or Diploid Feeding Stage	Body Form of Feeding Stage	Gametes or Sexual Cells
Plasmodial slime molds	a.	b.	c.
Cellular slime molds	d.	e.	f.

28.7 Protists play key roles in ecological relationships

Protists abound almost anywhere there is water—they occupy freshwater, marine, and moist terrestrial habitats or live symbiotically within the bodies of hosts. They form an important part of plankton, the drifting community of mostly microscopic organisms found in bodies of water.

Symbiotic Protists Dinoflagellates are important photosynthetic symbionts in coral. Other protists are destructive pathogens of plants and animals.

Photosynthetic Protists Many protists are **producers.** Most aquatic food webs are based on photosynthetic protists and bacteria.

Word Roots

-phyte = plant (*gametophyte:* the multicellular haploid form in organisms undergoing alternation of generations)

con- = with, together (*conjugation:* in ciliates, the transfer of micronuclei between two cells that are temporarily joined)

hetero- = different; **-morph** = form (*heteromorphic:* a condition in the life cycle of all modern plants in which the sporophyte and gametophyte generations differ in morphology)

iso- = same (*isomorphic:* alternating generations in which the sporophytes and gametophytes look alike, although they differ in chromosome number)

pseudo- = false; **-podium** = foot (*pseudopodium:* a cellular extension of amoeboid cells used in moving and feeding)

thallos- = sprout (*thallus:* a seaweed body that is plant-like but lacks true roots, stems, and leaves)

Structure Your Knowledge

1. Evidence indicates that all plastids evolved from a cyanobacterium that was engulfed by an ancestral heterotrophic eukaryote. Fill in the blanks in the figure on the next page, which diagrams the diversification of this ancestral eukaryote into red algae and green algae as well as several instances of secondary endosymbiosis that led to various protist groups.

Heterotrophic
eukaryote

Chlorarachniophytes

Test Your Knowledge

MATCHING: *Match the protist clades with their characteristics and examples.*

_____ **1.** modified mitochondria; *Giardia, Trichomonas*

_____ **2.** plant-type chloroplasts; *Chlamydomonas, Ulva*

_____ **3.** lobe-shaped pseudopodia; free-living amoeba, parasites, slime molds

_____ **4.** phycoerythrin, no flagellated stage; *Porphyra*

_____ **5.** hairy and smooth flagella; water molds, diatoms, golden and brown algae

_____ **6.** amoebas with threadlike pseudopodia; *Globigerian*

_____ **7.** subsurface sacs; dinoflagellates, apicomplexans, ciliates

_____ **8.** flagella with spiral or crystalline rod; *Trypanosoma, Euglena*

A. Alveolates
B. Amoebozoans
C. Diplomonads and parabasalids
D. Euglenozoans
E. Forams and Radiolarians
F. Green algae
G. Red algae
H. Stramenopiles

MULTIPLE CHOICE: *Choose the one best answer.*

1. Mixotrophs are
   a. plasmodial slime molds.
   b. cellular slime molds.
   c. gametophyte and sporophyte generations that differ in morphology.
   d. organisms that can be both heterotrophic and autotrophic.
   e. organisms that can be both parasitic and free-living.

2. Phytoplankton
   a. form the basis of most marine food chains.
   b. include the multicellular green, red, and brown algae.
   c. are mutualistic symbionts that provide food for coral reef communities.
   d. are unicellular heterotrophs that float near the ocean surface.
   e. Both a and d are correct.

3. According to the theory of secondary endosymbiosis,
   a. multicellularity evolved when primitive cells incorporated prokaryotic cells that then took on specialized functions.
   b. the symbiotic associations found in coral result from the incorporation of dinoflagellates into coral animals.
   c. cells that had obtained their plastids through endosymbiosis were engulfed and themselves became plastids in heterotrophic eukaryotic cells.
   d. the infoldings and specializations of the plasma membrane led to the evolution of the endomembrane system.
   e. the nuclear membrane evolved first, then mitochondria, and then plastids.

4. The diplomonads and parabasalids are unique among eukaryotes in that
   a. they lack nuclear membranes, even though diplomonads have two nuclei.
   b. they lack ribosomes.
   c. they lack plastids, and all other protist lineages have at least some members that are autotrophic.
   d. they have modified mitochondria and cannot perform aerobic respiration.
   e. they were an early branch from bacteria rather than from archaea.

5. Genetic variation is generated in the ciliate *Paramecium* when
   a. a micronucleus replicates its genome many times and becomes a macronucleus.
   b. zoospores, which are gametes, fuse.
   c. micronuclei are exchanged in conjugation.
   d. mutations occur in the many copies of the genome contained in the macronucleus and are passed on to offspring.
   e. plasmids are exchanged in conjugation.

6. The Chlorophyta, or green algae, are believed to share a common ancestor with plants because
   a. they are the only multicellular algal protists.
   b. they do not have flagellated gametes.
   c. they are the only protists whose plastids evolved from cyanobacteria.
   d. their chloroplasts are similar in ultrastructure and pigment composition to those of plants.
   e. they are the only algae that exhibit alternation of generations.

7. Which of the following is *not* true of seaweeds?
   a. They are found in the red, green, and brown algal groups.
   b. They have true roots that anchor them tightly to withstand the turbulence of waves.
   c. They are a source of food and commercial products.
   d. The gel-forming polysaccharides in their cell walls protect them from the abrasive action of waves.
   e. They are multicellular photoautotrophs.

8. Which protists are in the same eukaryotic supergroup as animals?
   a. dinoflagellates
   b. forams
   c. parabasalids
   d. stramenopiles
   e. amoebozoans

9. Which eukaryotic supergroup includes the land plants?
   a. Unikonta
   b. Archaeplastida
   c. Excavata
   d. Chromalveolata
   e. Rhizaria

Chapter 29

Plant Diversity I: How Plants Colonized Land

Key Concepts

29.1 Land plants evolved from green algae

29.2 Mosses and other nonvascular plants have life cycles dominated by gametophytes

29.3 Ferns and other seedless vascular plants were the first plants to grow tall

Framework

The evolution of land plants involved adaptations to terrestrial habitats. The kingdom Plantae includes multicellular, photoautotrophic eukaryotes that develop from embryos nourished by the parent plant. Plants exhibit an alternation of generations in which the diploid sporophyte is the more conspicuous stage in all groups except the bryophytes. This chapter outlines the major periods of plant evolution in which bryophytes and seedless vascular plants appeared and radiated.

Chapter Review

For more than the first 3 billion years of Earth's history, life was confined to aquatic environments. Plants began to move onto land about 500 million years ago, and have now diversified into 290,000 living species.

29.1 Land plants evolved from green algae

Morphological and Biochemical Evidence Land plants have chloroplasts containing chlorophylls *a* and *b* and cell walls of cellulose. These characteristics, however, are found in several algal groups and cannot be used to distinguish plants from algae. Four traits that plants and the green algal charophytes share are **rosette-shaped cellulose-synthesizing complexes,** which produce the cellulose microfibrils of their cell walls; enzymes within peroxisomes that help minimize photorespiration losses; similarities in sperm cells (in those land plants that have flagellated sperm); and the formation of a **phragmoplast** in mitosis. This alignment of cytoskeletal elements and vesicles in the synthesis of a cell plate is found only in plants and certain charophytes, such as *Chara* and *Coleochaete.*

Results from the analysis of nuclear and chloroplast genes confirm that charophytes are the closest living relatives of land plants.

Adaptations Enabling the Move to Land The tough polymer **sporopollenin** protects charophyte zygotes during fluctuations in water levels. With the accumulation of traits such as sporopollenin, ancient charophytes living along the edges of ponds and lakes may have given rise to the first plants to colonize land.

INTERACTIVE QUESTION 29.1

List some of the benefits and challenges of a terrestrial habitat for the first land plants.

Plant biologists are debating three different ways in which to establish the boundaries of the plant kingdom: the traditional kingdom Plantae containing only embryophytes (plants with embryos); the kingdom Streptophyta, which includes the charophytes; or the kingdom Viridiplantae, which includes the chlorophytes as well as the charophytes with the embryophytes. This textbook uses the traditional embryophyte definition of kingdom Plantae.

Derived Traits of Plants The following derived traits distinguish land plants: apical meristems, alternation of generations (along with a multicellular, dependent embryo), walled spores produced in sporangia, and multicellular gametangia.

Regions of cell division at the tips of roots and shoots, called **apical meristems,** produce linear growth and cells that differentiate into a protective epidermis and internal plant tissues. The elongation of roots and shoots provides increasing access to environmental resources.

Alternation of generations is found in all land plants. In this life cycle, the multicellular haploid **gametophyte** alternates with the multicellular diploid **sporophyte.** The sporophyte produces haploid **spores,** which are reproductive cells that directly develop into new organisms. Plant embryos, which are retained in the female gametophyte plant, have **placental transfer cells** that facilitate the transfer of nutrients from parental tissues. This derived trait is the basis for referring to land plants as **embryophytes.**

INTERACTIVE QUESTION 29.2

The gametophyte produces **a.** _____ by **b.** _____. Following fertilization, the **c.** _____ divides by **d.** _____ to develop into the **e.** _____ . The sporophyte produces **f.** _____ by **g.** _____. Spores germinate and develop into the **h.** _____.

The tough polymer sporopollenin protects spores as they disperse on land. Spores are produced and protected within multicellular **sporangia** on the sporophyte plant. Diploid cells called **sporocytes,** or spore mother cells, undergo meiosis and produce haploid spores in sporangia.

Gametes are produced within multicellular **gametangia.** Sperm are produced in **antheridia.** In many groups, sperm are flagellated and swim to the eggs. A single egg cell is produced and fertilized within an **archegonium,** where the zygote develops into an embryo.

Many land plants have a waxy **cuticle** coating their leaves and stems, which protects from water loss and microbial attack. Early plant fossils show symbiotic associations between plants and fungi. Such mycorrhizae facilitate nutrient absorption by plants. Many plants produce *secondary compounds* through side branches of their main metabolic pathways. These compounds may discourage herbivores and parasites, protect from UV radiation, or serve as signals.

The Origin and Diversification of Plants Fossil spores embedded in plant tissue have been found that date back 475 million years.

Vascular plants have **vascular tissue,** a transport system composed of cells joined into tubes. Liverworts, hornworts, and mosses do not have an extensive transport system and are referred to as nonvascular plants and informally called **bryophytes.** Whether or not the bryophytes form a clade is still debated.

Vascular plants, about 93% of plant species, do form a clade, and are divided into smaller clades. The **lycophytes** (club mosses and their relatives) and the **pterophytes** (ferns and their relatives) are informally called **seedless vascular plants.** The paraphyletic seedless vascular plants may be referred to as a **grade,** a group of organisms that shares the same level of biological organization.

A **seed** is an embryo enclosed with a supply of nutrients in a protective coat. The seed plants form a clade that contains two groups. **Gymnosperms** are called "naked seed" plants because their seeds are not enclosed. **Angiosperms** are flowering plants, in which seeds develop inside chambers called ovaries in a flower. Taxonomists recognize 10 phyla of extant plants.

INTERACTIVE QUESTION 29.3

Why are the seedless vascular plants considered a grade and not a clade?

29.2 Mosses and other nonvascular plants have life cycles dominated by gametophytes

The three separate phyla of nonvascular plants are **liverworts** (phylum Hepatophyta), **hornworts** (phylum Anthocerophyta), and **mosses** (phylum Bryophyta). The evolutionary relationships of these groups continue to be debated.

Bryophyte Gametophytes In the bryophytes, the gametophyte is the prevalent generation; sporophytes are present only part of the time. When moss spores germinate in a moist habitat, they grow into a mass of green filaments called a **protonema.** Meristem-containing buds grow into gamete-producing upright "stems" called **gametophores.** Bryophyte gametophytes are generally only a few cells thick and low growing. They are anchored by long, tubular cells or filaments of cells called **rhizoids.**

Some mosses have conducting tissues in their "stems." Plant biologists have yet to determine whether these tissues are homologous or analogous with vascular tissue.

Gametangia are enclosed in jackets of protective tissue. Sperm are produced in antheridia, from which they swim to eggs retained in archegonia. Zygotes develop into embryos, which rely on nutrients transported by placental transfer cells to develop into sporophytes. Many bryophyte species also reproduce asexually.

Bryophyte Sporophytes Although they are usually photosynthetic when young, bryophyte sporophytes remain attached to and dependent on their parental gametophytes. A sporophyte has a **foot** that obtains nutrients from the gametophyte, a stalk (**seta**) that conducts materials and may elongate for spore dispersal, and a sporangium or **capsule** in which millions of spores are produced by meiosis. The specialized "toothed" **peristome** gradually releases spores to be dispersed by wind currents.

The sporophytes of hornworts and mosses have **stomata,** pores through which gases are exchanged and water evaporates. There are three possible evolutionary origins of stomata: in the ancestor of all three groups and later lost in the liverwort lineage; in the ancestor of the branch leading to hornworts, mosses, and vascular plants; or in the ancestor of mosses and vascular plants, with an independent origin in hornworts.

INTERACTIVE QUESTION   29.4

Review the life cycle of a typical moss plant by filling in the following blanks.

The dominant generation is the **a.** _____. Female gametophytes produce eggs in **b.** _____. Male gametophytes produce sperm in **c.** _____. Sperm **d.** _____ through the damp environment to fertilize the egg. The zygote remains in the archegonium and grows into the **e.** _____, still attached to the female gametophyte. Spores are formed by the process of **f.** _____ in the **g.** _____. When shed, spores develop into the **h.** _____.

Ecological and Economic Importance of Mosses Bryophytes are widely distributed and common in moist habitats. Some are adapted to very cold or dry habitats due to their ability to dehydrate and rehydrate without dying. Research has shown that moss reduce the rate at which nitrogen is lost by water erosion in a sandy ecosystem.

Sphagnum, or peat moss, is an abundant wetland moss that forms undecayed organic deposits known as **peat.** The decay-resistant compounds in peat bogs can preserve organisms for thousands of years. Peat serves as huge stores of organic carbon and helps to stabilize atmospheric carbon dioxide concentrations. Peat is harvested for use as fuel, a soil conditioner, and plant packing material.

29.3   Ferns and other seedless vascular plants were the first plants to grow tall

Origins and Traits of Vascular Plants Fossils of ancient relatives of modern vascular plants date back 420 million years. These tiny plants had branched, independent sporophytes and terminal sporangia. The branching made possible more complex bodies and multiple sporangia. They lacked other derived traits of vascular plants.

In modern vascular plants, the diploid sporophyte is the larger and more complex plant. Seedless vascular plants remain limited to damp habitats because of their swimming sperm and fragile gametophytes, as shown in the life cycle of a fern in Interactive Question 29.5.

INTERACTIVE QUESTION 29.5

In the following diagram of the life cycle of a fern, label the processes (in the boxes) and structures (on the lines). Indicate which portion of the life cycle is haploid and which is diploid.

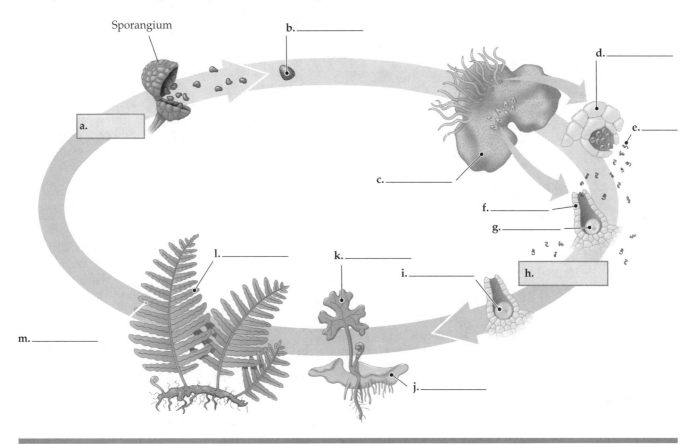

Water and minerals are conducted up from roots in **xylem.** The xylem of all vascular plants includes tube-shaped cells called **tracheids** with cell walls strengthened by **lignin.** Vascular plants are also called tracheophytes. **Phloem** transports sugars and other organic nutrients through living cells arranged into tubes.

Unlike the rhizoids of bryophytes, **roots** both anchor plants and absorb water and nutrients. Roots may have evolved from subterranean stems, either once in a common ancestor or independently in different lineages of vascular plants.

Leaves increase a plant's photosynthetic surface area. The single-veined leaves of lycophytes, known as **microphylls,** are usually small and spine-shaped. The branching vascular systems of **megaphylls,** typical of most other vascular plants, support larger leaves with greater photosynthetic capacity. Megaphylls appear in the fossil record at the end of the Devonian period, 40 million years after the appearance of microphylls.

Microphylls may have originated from sporangia along the stem, each supported by a single vascular tissue strand. Megaphylls may have evolved from closely lying branches when one branch came to *overtop* lower branches that then formed webbing to join them together.

Sporophylls are modified leaves bearing sporangia. In ferns, clusters of sporangia called **sori** are produced on the undersides of sporophylls. Cones, or **strobili,** are formed from groups of sporophylls in gymnosperms and many lycophytes. Most seedless vascular plants are **homosporous,** producing only one kind of spore, which develops into bisexual gametophytes. **Heterosporous** plants include all seed and some seedless vascular plants and produce two kinds of spores: **megaspores** that develop into female gametophytes and **microspores** that develop into male gametophytes.

INTERACTIVE QUESTION   29.6

Describe the traits of vascular plants that allow them to grow tall.

Classification of Seedless Vascular Plants   The two clades of living seedless vascular plant are classified into two phyla: phylum Lycophyta and phylum Pterophyta. Some systematists still retain three separate phyla in the pterophyte clade: Pterophyta (ferns), Sphenophyta (horsetails), and Psilophyta (whisk ferns).

Lycophytes were a major part of the landscape during the Carboniferous period. One evolutionary line, the giant lycophytes, became extinct when the climate became drier. The other line of small lycophytes is represented today by the club mosses, spike mosses, and quillworts. Many tropical species grow on trees as epiphytes—plants that anchor to other organisms but are not parasites. Sporophytes grow upright with many small leaves; horizontal stems grow along the surface and produce roots.

Ferns were also found in the great forests of the Carboniferous period and are the most numerous seedless vascular plants in the modern flora. The large fern leaves (megaphylls), or fronds, often grow from horizontal stems. Most fern species are homosporous; gametophytes are small.

Horsetails grew as tall plants during the Carboniferous period. A few species of the genus *Equisetum* are the only representatives found today. These small, upright plants grow in damp locations and many have cones at the tips of some of the green, jointed stems.

In spite of their dichotomous branching and lack of true leaves and roots, molecular evidence indicates that *Psilotum*, the whisk fern, is closely related to ferns.

The Significance of Seedless Vascular Plants   The seedless vascular plants of the Carboniferous forests decreased CO_2 levels in the atmosphere, perhaps contributing to the global cooling at the end of the Carboniferous. Dead plant material did not completely decay in the stagnant swamp waters, and great accumulations of peat developed. When the sea later covered the swamps and marine sediments piled on top, heat and pressure converted the peat to coal.

Word Roots

-angio = vessel (*gametangia:* the reproductive organ of bryophytes, consisting of the male antheridium and female archegonium; a multichambered jacket of sterile cells in which gametes are formed)

-phore = bearer (*gametophore:* the mature gamete-producing structure of a gametophyte body of a moss)

bryo- = moss; **-phyte** = plant (*bryophytes:* the mosses, liverworts, and hornworts; a group of nonvascular plants that inhabit the land but lack many of the terrestrial adaptations of vascular plants)

gymno- = naked; **-sperm** = seed (*gymnosperm:* a vascular plant that bears naked seeds not enclosed in any specialized chambers)

hetero- = different; **-sporo** = a seed (*heterosporous:* referring to plants in which the sporophyte produces two kinds of spores that develop into unisexual gametophytes, either female or male)

homo- = like (*homosporous:* referring to plants in which a single type of spore develops into a bisexual gametophyte having both male and female sex organs)

mega- = large (*megaspores:* a spore from a heterosporous plant that develops into a female gametophyte bearing archegonia)

micro- = small; **-phyll** = leaf (*microphylls:* the small leaves of lycophytes that have only a single, unbranched vein)

peri- = around; **-stoma** = mouth (*peristome:* the upper part of the moss capsule often specialized for gradual spore discharge)

phragmo- = a partition; **-plast** = formed, molded (*phragmoplast:* an alignment of cytoskeletal elements and Golgi-derived vesicles across the midline of a dividing plant cell)

proto- = first; **-nema** = thread (*protonema:* a mass of green, branched, one-cell-thick filaments produced by germinating moss spores)

pter- = fern (*pterophytes:* seedless plants with true roots and lignified vascular tissue; the group includes ferns, whisk ferns, and horsetails)

rhizo- = root; **-oid** = like, form (*rhizoids:* long tubular single cells or filaments of cells that anchor bryophytes to the ground)

Structure Your Knowledge

1. The evolution of land plants involved adaptations to the terrestrial habitat. List some of these adaptations that evolved in the bryophytes and the seedless vascular plants.

2. The following diagram presents a widely held view of plant phylogeny. Fill in blanks a–e with the informal names for the major groups and blanks f–k with the common names for the modern plant groups (list the three groups included in clade j).

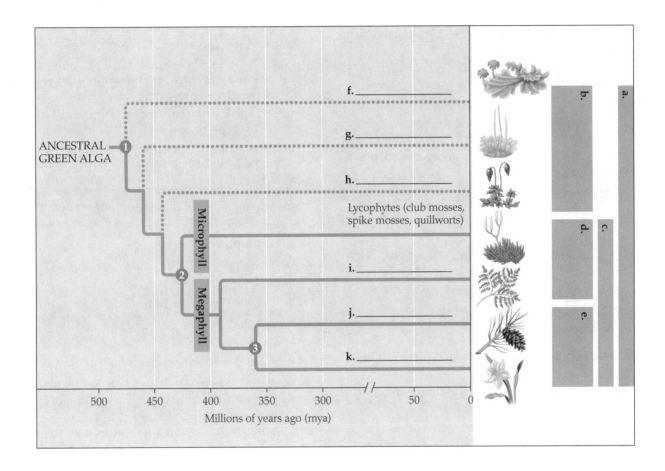

f. _____

g. _____

ANCESTRAL GREEN ALGA

h. _____

Lycophytes (club mosses, spike mosses, quillworts)

Microphyll

Megaphyll

i. _____

j. _____

k. _____

500   450   400   350   300   //   50   0

Millions of years ago (mya)

a.

b.

c.

d.

e.

Test Your Knowledge

MULTIPLE CHOICE: *Choose the one best answer.*

1. Adaptations for terrestrial life seen in all plants are
   a. chlorophylls *a* and *b*.
   b. cell walls of cellulose and lignin.
   c. sporopollenin, protection and nourishment of embryo by gametophyte.
   d. vascular tissue.
   e. cuticle and stomata.

2. Plants are thought to be most closely related to charophytes based on
   a. peroxisome enzymes that minimize photorespiration.
   b. rosette cellulose-synthesizing complexes in their plasma membranes.
   c. phragmoplast formation of cell plates in mitosis.
   d. similarities in sperm cells (in those land plants with flagellated sperm).
   e. all of the above.

3. Megaphylls
   a. are large leaves.
   b. are gametophyte plants that develop from megaspores.
   c. are leaves specialized for reproduction.
   d. are leaves with branching vascular systems.
   e. were a dominant group of the great coal forests.

4. Bryophytes differ from all other land plant groups because
   a. their gametophyte generation is dominant.
   b. they are lacking gametangia.
   c. they have flagellated sperm.
   d. they are not embryophytes.
   e. of all of the above.

5. Which of the following plant groups is *incorrectly* paired with its gametophyte generation?
   a. charophyte—no alternation of generations
   b. lycophyte—either green or nonphotosynthetic small plants
   c. moss—green matlike plant
   d. fern—frond growing from horizontal stem
   e. liverwort—flattened, low-growing green plant

6. Which of the following is most likely the closest relative of seed plants?
   a. charophytes
   b. pterophytes
   c. liverworts
   d. lycophytes
   e. psilophytes

7. The evolution of sporopollenin was important to the movement of plants onto land because it
   a. provided a mechanism for the dispersal of spores and pollen.
   b. provided the structural support necessary to withstand gravity.
   c. enclosed developing gametes and embryos in maternal tissue and prevented desiccation.
   d. provided a tough coating for spores so they could disperse on land.
   e. initiated the alternation of generations that is characteristic of all plants.

8. If a plant's life cycle includes both a male and female gametophyte, the sporophyte plant must be
   a. heterosporous.
   b. homosporous.
   c. homologous.
   d. analogous.
   e. megasporous.

9. Xylem and phloem are found in
   a. all plants.
   b. bryophytes, ferns, conifers, and angiosperms.
   c. only the gametophytes of vascular plants.
   d. the vascular plants, which include ferns, conifers, and flowering plants.
   e. only the vascular plants with seeds.

10. Which of the following functions may secondary compounds serve?
    a. protect plant from ultraviolet radiation
    b. inhibit microbes
    c. discourage herbivores and parasites
    d. function as signaling molecules
    e. all of the above

11. If you could take a time machine back to the Carboniferous period, which of the following scenarios would you most likely encounter?
    a. creeping mats of low-growing bryophytes
    b. fields of tall grasses swaying in the wind
    c. swampy forests dominated by large lycophytes, horsetails, and ferns
    d. huge forests of naked-seed trees filling the air with pollen
    e. the dominance of flowering plants

12. Which of the following has been proposed by plant biologists as the clade that establishes the boundary of the plant kingdom?
    a. kingdom Plantae, which includes only the embryophyte clade
    b. kingdom Streptophyta, which also includes the charophytes
    c. kingdom Viridiplantae, which also includes all green algae (chlorophytes)
    d. kingdom Vasculata, which includes only the vascular plants
    e. a, b, and c have all been proposed as the clade that defines the plant kingdom

13. Large stores of organic carbon that help to stabilize atmospheric carbon dioxide concentrations are found in
    a. huge tropical swamps dominated by mosses and ferns.
    b. peat.
    c. coal deposits formed by Cambrian forests.
    d. the abundant epiphytic lycophytes found in tropical regions.
    e. large tracts of small seedless vascular plants found in boreal regions.

14. Bryophytes were the dominant plants in the first 100 million years of plant evolution. By the Carboniferous period, seedless vascular plants formed giant forests. Why were these plants able to outcompete bryophytes?
    a. Their protected embryos were better able to withstand dry conditions, providing a selective advantage in dominating terrestrial habitats.
    b. They did not require water for their sperm to swim to fertilize the eggs, allowing them to colonize dry habitats.
    c. Because they were diploid, they could grow faster and taller than haploid bryophytes.
    d. As heterosporous plants, their spores had more genetic variation, enabling natural selection to favor taller growth.
    e. Their vascular tissue enabled them to grow tall, competing for light and shading out other plants.

Chapter 30

Plant Diversity II: The Evolution of Seed Plants

Key Concepts

30.1 Seeds and pollen grains are key adaptations for life on land

30.2 Gymnosperms bear "naked" seeds, typically on cones

30.3 The reproductive adaptations of angiosperms include flowers and fruits

30.4 Human welfare depends greatly on seed plants

■ **Framework**

The reduction and protection of the gametophyte within the parent sporophyte, the use of seeds as a means of dispersal, and the development of pollen for the transfer of sperm are reproductive adaptations of seed plants that enhance their success on land.

Gymnosperms bear their seeds "naked" on modified sporophylls. Angiosperms are the most diverse and widespread of plants, owing much of their success to their efficient reproductive apparatus—the flower—and their dispersal mechanism—the fruit. Agriculture is based almost entirely on angiosperms.

■ **Chapter Review**

Seed plants originated about 360 million years ago. One of the key adaptations that enabled seed plants to become the dominant members of terrestrial ecosystems is the seed—a mobile embryo packaged with food in a protective coat. The cultivation of seed plants, which transformed human societies, began about 13,000 years ago.

30.1 Seeds and pollen grains are key adaptations for life on land

Advantages of Reduced Gametophytes The extremely reduced gametophytes of seed plants develop within the sporangium, and thus are protected from drying out and UV radiation. They are also nourished by the sporophyte plant.

Heterospory: The Rule Among Seed Plants All seed plants are heterosporous, with megaspores that give rise to female gametophytes, and microspores that give rise to male gametophytes. A megaspongium produces a single functional megaspore; a microsporangium produces many microspores.

Ovules and Production of Eggs The megasporangium is surrounded by protective **integuments** derived from sporophyte tissue. An unfertilized **ovule** consists of the integument(s), megasporangium, and megaspore. The female gametophyte develops from the megaspore and produces one or more eggs.

Pollen and Production of Sperm Microspores develop into male gametophytes contained in sporopollenin-protected **pollen grains.** The transfer of pollen to the part of a plant containing ovules is called **pollination.** When pollen grains germinate, they release sperm through a pollen tube to the female gametophyte.

INTERACTIVE QUESTION 30.1

Why is the evolution of pollen an important terrestrial adaptation?

The Evolutionary Advantage of Seeds   An ovule develops into a **seed,** consisting of a sporophyte embryo surrounded by a food supply and enclosed in a protective coat derived from the integument(s) of the ovule. Seeds enable seed plants to withstand harsh environmental conditions and to disperse offspring. Unicellular spores perform those functions for bryophytes and seedless vascular plants.

INTERACTIVE QUESTION 30.2

Label the parts in this generalized diagram of an unfertilized ovule and seed of a gymnosperm. Indicate whether structures are diploid or haploid tissues.

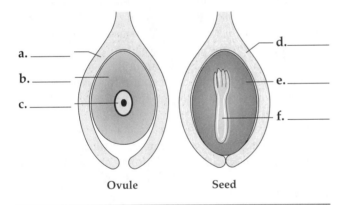

a. _____
b. _____
c. _____

d. _____
e. _____
f. _____

Ovule   Seed

30.2   Gymnosperms bear "naked" seeds, typically on cones

Gymnosperms are named for their naked seeds that develop on the surface of modified leaves, which usually form cones (strobili).

Gymnosperm Evolution   Transitional species of heterosporous, wood-producing, but seedless vascular plants are sometimes called **progymnosperms.** The first seed plants appear in the fossil record from around 360 million years ago. These early gymnosperm lineages became extinct. There are two surviving seed plant lineages, the gymnosperms and angiosperms. During the Carboniferous period, early gymnosperms lived in ecosystems dominated by seedless vascular plants. Drier conditions during the Permian gave gymnosperms a selective advantage. Gymnosperms supported the giant herbaceous dinosaurs of the Mesozoic era. With the environmental changes at the end of the Mesozoic, most of the dinosaurs became extinct, but many gymnosperms persisted.

There are four gymnosperm phyla. Cycadophyta includes cycads, with large cones and palmlike leaves, which thrived during the Mesozoic era. Ginkgophyta has only one extant species, the deciduous, ornamental ginkgo tree. Gnetophyta includes three genera: the bizarre *Welwitschia*, the tropical *Gnetum*, and the desert shrub *Ephedra*. The largest gymnosperm phylum is Coniferophyta. Most **conifers** (cone-bearing trees) are evergreens with needle-shaped leaves. Coniferous trees are some of the largest and oldest living organisms.

The Life Cycle of a Pine: A Closer Look   The pine tree is a heterosporous sporophyte. Pollen cones consist of many scalelike structures that bear sporangia. Meiosis gives rise to microspores that develop into pollen grains enclosing the male gametophytes. Scales of an ovulate cone hold two ovules, each of which contains a megasporangium. A megasporocyte undergoes meiosis, and one of the resulting megaspores undergoes repeated divisions to produce a female gametophyte in which a few archegonia develop.

Following pollination, a pollen tube grows and digests its way through the megasporangium. Fertilization may occur more than a year after pollination. The zygote develops into a sporophyte embryo, which is nourished by the remaining female gametophyte tissue and is enclosed in a seed coat derived from the integument of the parent sporophyte.

INTERACTIVE QUESTION 30.3

a. Describe a pollen cone and the formation of a male gametophyte.

b. Describe an ovulate cone and the formation of a female gametophyte.

INTERACTIVE QUESTION 30.4

a. Name the four floral organs that make up a flower.

b. What does a seed consist of?

c. What is a fruit?

30.3 The reproductive adaptations of angiosperms include flowers and fruits

Characteristics of Angiosperms The phylum Antho-phyta contains more than 90% of plants species—over 250,000 species.

The **flower,** the reproductive structure of an angiosperm, has up to four rings of modified leaves called floral organs. The outer **sepals** are usually green, whereas **petals** are brightly colored in most flowers that are pollinated by insects and birds. **Stamens,** the microsporophylls, produce microspores, which develop into pollen grains containing male gametophytes. A stamen has a stalk, called a filament, and a terminal **anther,** in which pollen is produced. **Carpels,** the megasporophylls, produce megaspores, which develop into female gametophytes. The carpel has a sticky **stigma,** which receives pollen, and a **style,** which leads to the **ovary.** The ovary contains ovules, which develop into seeds.

A **fruit** is a mature ovary that functions in the protection and dispersal of seeds. As seeds develop, the ovary enlarges, becoming the thickened wall (pericarp) of the fruit. Fruits may be fleshy or dry. The dry fruits of grasses are the major foods for humans. Fruits may be modified in various ways to disperse seeds, enlisting the aid of wind or animals.

In the angiosperm life cycle, the two highly reduced gametophytes are produced within the flower. Pollen grains, which develop from microspores within the anthers, contain the male gametophytes, consisting of a generative cell (that divides to form two sperm) and a tube cell. Ovules contain the female gametophyte, called an **embryo sac,** which consists of a few cells.

Most flowers have some mechanism to ensure **cross-pollination.** A pollen grain germinates on the stigma and extends a pollen tube down the style, through the **micropyle,** and into the ovule, where it releases two sperm cells into the embryo sac. In a process called **double fertilization,** one sperm unites with the egg to form the zygote and the other sperm fuses with the two polar nuclei in the large cell in the center of the female gametophyte.

The zygote divides and forms the sporophyte embryo, consisting of a rudimentary root and one or two seed leaves, called **cotyledons.** The **endosperm,** which develops from the triploid central cell, serves as a food reserve for the embryo.

INTERACTIVE QUESTION 30.5

In this diagram of the life cycle of an angiosperm, label the indicated structures and processes. Which structures represent the male and female gametophyte generations?

Angiosperm Evolution Angiosperms originated at least 140 million years ago. By 100 million years ago, angiosperms were becoming the dominant plants. The origin of angiosperms from gymnosperm ancestors is still a mystery.

A phylogenetic comparison of recently discovered fossils of 125-million-year-old angiosperms called *Archaefructus* have shown this group (which has both derived and primitive traits) to be most closely related to all living angiosperms. These fossils suggest that "proto-angiosperms" may have been herbaceous, aquatic plants, although some paleobotanists dispute this interpretation.

The monophyletic living gymnosperms appear to have diverged from the ancestors of angiosperms perhaps 280 million years ago. The extinct gymnosperm *Bennettitales*, with flower-like structures, may be a possible ancestor of angiosperms.

Studies of plant development and the genes that control flower development may help elucidate the origin of flowering plants.

Angiosperm Diversity The angiosperms were traditionally divided into two main groups, the **monocots** and **dicots.** Recent DNA comparisons reveal that, although the monocots are a monophyletic group, the dicots are not. The clade **eudicots** includes most previously classified dicots, but several other small lineages have been identified. The three oldest lineages are informally called **basal angiosperms.** The oldest lineage is represented today by one species, *Amborella trichopoda*, which lacks the xylem vessels found in more derived angiosperms. The next lineage to diverge includes water lilies, with the lineage of star anise and its relatives diverging later. A lineage known as the **magnoliids** is more closely related to monocots and eudicots than to the basal angiosperms.

INTERACTIVE QUESTION   30.6

List some of the differences between monocots and eudicots for the following characteristics.

Characteristic	Monocot	Eudicot
Number of cotyledons		
Leaf venation		
Vascular bundles in stems		
Root system		
Flower parts		
Openings in pollen grain		

Evolutionary Links Between Angiosperms and Animals   Animals influenced the evolution of plants and vice versa. Defenses against herbivores and herbivores that can overcome plant defenses would be favored by natural selection. Some animals became beneficial as pollinators and seed dispersers. Pollinators can enter bilateral flowers from only one direction and thus more specifically receive and disperse pollen to flowers of the same species. A research study compared the number of species in clades with bilaterally symmetrical flowers with the number of species in sister clades with radially symmetrical flowers. Clades with bilateral symmetry had more species, supporting the hypothesis that bilateral flower shape promotes speciation, perhaps by reducing gene flow between diverging populations.

30.4   Human welfare depends greatly on seed plants

Products from Seed Plants   Our fruit and vegetable crops are angiosperms, and angiosperms are used to feed livestock. The dramatic evolution of domesticated plants is a result of artificial selection in plant breeding. Seed plants also supply wood and medicines.

Threats to Plant Diversity   The growing human population with its demand for space, food, and natural resources is pushing hundreds of species toward extinction each year. Tropical rain forests, where plant diversity is greatest, are rapidly being destroyed, and humans are losing their potential supply of new food crops and medicines.

Word Roots

endo- = inner (*endosperm:* a nutrient-rich tissue formed by the union of a sperm cell with two polar nuclei during double fertilization, which provides nourishment to the developing embryo in angiosperm seeds)

peri- = around; **-carp** = fruit (*pericarp:* the thickened wall of a fruit)

pro- = before; **gymno-** = naked; **-sperm** = seed (*progymnosperm:* an extinct group of plants that is probably ancestral to gymnosperms and angiosperms)

Structure Your Knowledge

1. List the four gymnosperm phyla.

2. The angiosperms are grouped into a single phylum, Anthophyta, which traditionally contained the classes monocots and dicots. Morphological and molecular evidence now suggests six angiosperm lineages. Label those lineages on the following tree.

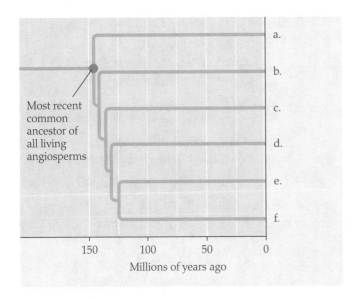

150   100   50   0
Millions of years ago

3. List the characteristics of seed plants that are evolutionary adaptations to their terrestrial habitat.

4. What adaptations helped angiosperms to become the most successful and widespread land plants?

Test Your Knowledge

MULTIPLE CHOICE: *Choose the one best answer.*

1. In which of the following groups do sperm no longer have to swim to reach the female gametophyte?
   a. bryophytes
   b. ferns
   c. most gymnosperms
   d. angiosperms
   e. Both c and d are correct.

2. What provides food for a developing sporophyte embryo in a gymnosperm seed?
   a. endosperm
   b. female gametophyte tissue
   c. female sporophyte tissue
   d. male gametophyte tissue
   e. pine cone

3. Which of the following is the correct path that a pollen tube takes to reach the female gametophyte in an angiosperm?
   a. stigma, style, ovary, ovule, embryo sac
   b. anther, stigma, filament, ovule, ovum
   c. stigma, filament, carpel, ovary, ovule
   d. carpel, megasporophyll, ovary, ovule, embryo sac
   e. stigma, style, sepal, ovule, ovary

4. Which correctly states the number and describes the generations that are represented in a pine seed?
   a. one: the new sporophyte generation
   b. two: seed coat and food supply from female gametophyte and sporophyte embryo
   c. two: seed coat from integument of parent sporophyte and new sporophyte embryo
   d. three: seed coat from parent sporophyte, food supply from gametophyte, and sporophyte embryo
   e. three: seed coat from female gametophyte, food supply from parent sporophyte, and sporophyte embryo

5. Gymnosperms rose to dominance during which of the following periods?
   a. Carboniferous, when they formed important components of the great "coal forests"
   b. Devonian, when they successfully competed with the short-statured bryophytes
   c. Permian, when they replaced the seedless vascular plants as the climate became drier
   d. Cretaceous, when cooler climates favored their growth over the ferns and other seedless vascular plants
   e. Cretaceous, when they replaced angiosperms as the largest and most widespread group of plants

6. Where would you find a microsporangium in the life cycle of a pine?
   a. within the embryo sac in an ovule
   b. in the pollen sacs in an anther
   c. on a sporophyll in a pollen cone
   d. on a scalelike sporophyll found in an ovulate cone
   e. forming a seed coat surrounding a pine seed

7. Which of the following is a key difference between seedless vascular plants and plants with seeds?

   a. The gametophyte generation is dominant in the seedless plants, whereas the sporophyte is dominant in the seed plants.

   b. The spore is the agent of dispersal in the first, whereas the seed functions in dispersal in the second.

   c. Seedless plants are heterosporous; seed plants are homosporous.

   d. The embryo is unprotected and totally independent in the seedless plants but retained within the female reproductive structure in the seed plants.

   e. Sporopollenin is not found in the seedless plants.

8. Which of these plants is believed to be the only survivor of the oldest branch of the angiosperm lineage?

   a. the ornamental ginkgo tree

   b. the water lilies

   c. the desert plant *Welwitschia*

   d. the star anise

   e. the small shrub *Amborella*

TRUE OR FALSE: *Indicate T or F, and then correct the false statements.*

_____ 1. All photoautotrophic, multicellular eukaryotes are plants.

_____ 2. Heterosporous plants produce male and female spores.

_____ 3. The gametophyte generation is most reduced in the gymnosperms.

_____ 4. The *Ginkgo,* cycads, and conifers are naked-seed plants.

_____ 5. A sporangium produces spores, no matter what group it is found in.

_____ 6. A fruit consists of an embryo, nutritive material, and a protective coat.

_____ 7. A stamen consists of a filament and anther in which microspores are produced, which give rise to pollen grains.

_____ 8. The female gametophyte in angiosperms consists of haploid cells in which a few archegonia develop.

Fungi

Key Concepts

31.1 Fungi are heterotrophs that feed by absorption

31.2 Fungi produce spores through sexual or asexual life cycles

31.3 Fungi descended from an aquatic, single-celled, flagellated protist

31.4 Fungi have radiated into a diverse set of lineages

31.5 Fungi play key roles in nutrient cycling, ecological interactions, and human welfare

Framework

This chapter describes the morphology, life cycles, evolutionary history, diversity, and economic and ecological importance of the kingdom Fungi. Fungi play an essential ecological role, both as decomposers and by their mycorrhizal association with plant roots. A flagellated protistan may have been the common ancestor to fungi and animals.

Chapter Review

31.1 Fungi are heterotrophs that feed by absorption

Nutrition and Ecology Fungi are heterotrophs that obtain their nutrients by absorption; they secrete digestive enzymes into the surrounding food and absorb the resulting small organic molecules. Fungi may be decomposers that break down nonliving organic material; parasites that absorb nutrients from living cells; or mutualists that feed on, but also benefit, their hosts.

Body Structure Most fungi are composed of multicellular filaments. Some are single-celled **yeasts,** which may inhabit liquid or moist habitats. Some species can grow both as filaments and yeasts.

A typical multicellular fungal body consists of a network of **hyphae,** which form a mass called a **mycelium.** This body form provides an extensive surface area for absorption. The tubular cell walls are composed of **chitin.** Although fungi are nonmotile, they rapidly enter into new food territory by the fast growth of their hyphae.

The hyphae of most fungi are divided into cells by cross-walls called **septa,** which usually have pores through which nutrients and cell organelles can pass. Some hyphae are aseptate, and these **coenocytic fungi** consist of a continuous cytoplasmic mass with many nuclei.

Symbiotic fungi may penetrate host cells with specialized hyphae called **haustoria.** Mutualistic relationships with plant roots are known as **mycorrhizae.** **Ectomycorrhizal fungi** form hyphal sheaths around a root and also grow into spaces in the root cortex. **Arbuscular mycorrhizal fungi** extend their hyphae through root cell walls, pushing in the plant cell membrane to form tubes within root cells.

INTERACTIVE QUESTION 31.1

What do mycorrhizal fungi do for plants? What do plants do for the fungi?

31.2 Fungi produce spores through sexual or asexual life cycles

Fungi release huge quantities of spores, produced either sexually or asexually, that aid in dispersal.

Sexual Reproduction Nuclei of hyphae and spores of most species are haploid. Sexual reproduction involves the release of sexual signaling **pheromones,** which may cause the signaling hypha of suitable mating types to grow toward and fuse with each other. Such cytoplasmic fusion is called **plasmogamy** and forms a mycelium with genetically different nuclei, or a **heterokaryon.** The haploid nuclei may mingle and exchange genes in a process similar to crossing over. Nuclei from the two parents may pair up in cells and divide in tandem, forming **dikaryotic** cells. After a period of time, the nuclei fuse (**karyogamy**), forming the only diploid stage in most fungi. Meiosis produces haploid spores.

Asexual Reproduction　Many fungi reproduce asexually by spores; some species reproduce only asexually.

Molds are rapidly growing mycelia that reproduce asexually by spores. Many molds can reproduce sexually if they encounter other mating types. Asexual reproduction in single-celled yeasts occurs by cell division or budding.

Molds and yeasts for which no sexual stage is known are traditionally called **deuteromycetes.** When mycologists discover a sexual stage in one of these fungi, they classify it accordingly. Genetic analysis is also used to classify fungi.

INTERACTIVE QUESTION　31.2

Briefly define the following terms that relate to the structure and reproduction of fungi:

a. mycelium

b. septa

c. coenocytic

d. heterokaryon

e. plasmogamy

f. dikaryon

g. karyogamy

31.3　Fungi descended from an aquatic, single-celled, flagellated protist

Fungi and Animalia are more closely related to each other than to plants or other eukaryotes.

The Origin of Fungi　Although most fungi lack flagella, phylogenetic systematics indicates that they evolved from a flagellated ancestor. The fungi, animals, and their protistan relatives are called **opisthokonts.** DNA data indicate that fungi are most closely related to

nucleariids, a unicellular group of protists. This evidence suggests that the ancestor of fungi was unicellular. Animals are more closely related to a different group of protists, suggesting that fungi and animals evolved multicellularity independently from different single-celled ancestors. The molecular clock estimates that the ancestors of animals and fungi diverged 1.5 billion years ago. The oldest undisputed fossils of fungi date back 460 million years.

Are Microsporidia Closely Related to Fungi? Microsporidia are unicellular parasites of animals and protists that do not have conventional mitochondria. Molecular evidence indicates that these highly derived parasites may be an early diverging lineage of fungi.

The Move to Land　The first vascular plant fossils have evidence of mycorrhizae. The adaptive radiation of fungi may have begun with this move to land.

INTERACTIVE QUESTION　31.3

Fill in the blanks in this phylogenetic tree of the opisthokonts: fungi, animals, and their protistan relatives.

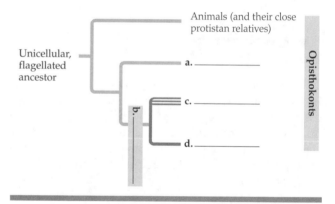

31.4　Fungi have radiated into a diverse set of lineages

Chytrids　The **chytrids,** phylum Chytridiomycota, are unique among fungi because of their flagellated **zoospores.** Molecular evidence indicates that this paraphyletic group diverged earliest in fungal evolution. Chytrids are primarily aquatic decomposers or parasites. Some form colonies with hyphae; others are single-celled. Some are mutualists in the digestive tracts of sheep and cattle.

Zygomycetes　**Zygomycetes** of phylum Zygomycota have diverse life histories and include molds, parasites, and commensal symbionts.

Rhizopus stolonifer, the black bread mold, is a common zygomycete that spreads horizontal coenocytic hyphae across its food substrate and erects hyphae with bulbous black sporangia containing hundreds of asexual haploid spores. Sexual reproduction between mycelia of opposite mating types produces a heterokaryotic **zygosporangium** with a tough protective coat that remains dormant until conditions are favorable. Karyogamy occurs between paired nuclei, followed by meiosis to produce haploid spores.

Glomeromycetes   Analysis of fungal genomes has identified the **glomeromycetes** as a separate clade (phylum Glomeromycota). All glomeromycetes form arbuscular mycorrhizae. Hyphae that invade plant roots branch into tiny treelike arbuscles. About 90% of all plants form mycorrhizae with glomeromycetes.

Ascomycetes   The **ascomycetes** (phylum Ascomycota), or **sac fungi,** are found in a wide variety of habitats. They produce sexual spores in saclike **asci,** usually contained in fruiting bodies called **ascocarps.**

Ascomycetes range in complexity from yeasts to elaborate cup fungi and morels. Some are serious plant pathogens; others are important decomposers. Many species are in symbiotic associations with algae or cyanobacteria called lichens and some form mycorrhizae with plants.

Asexual reproduction involves spores called **conidia,** which are produced in chains or clusters at the ends of hyphae called conidiophores. Plasmogamy between different mating types produces dikaryotic hyphae. Karyogamy occurs in terminal cells, which develop into asci. Meiosis yields haploid spores called ascospores.

The genome of *Neurospora crassa* has 10,000 genes with few stretches of noncoding DNA.

Basidiomycetes   The **basidiomycetes** (phylum Basidiomycota), which include the mushrooms and shelf fungi, produce a club-shaped **basidium** in which karyogamy followed by meisois occurs. These **club fungi** include important decomposers, mycorrhizae-forming mutualists, and the plant parasites rusts and smuts.

In response to environmental stimuli, an elaborate fruiting body called a **basidiocarp** is formed from the long-lived dikaryotic mycelium. Karyogamy and meiosis occur in numerous basidia, producing huge numbers of haploid sexual spores. As in the ascomycetes, many genetic recombinations from a single mating event result from the extended dikaryotic stage, in which many asci and basidia, respectively, are produced.

INTERACTIVE QUESTION   31.4

Indicate whether the following diagrams (1, 2, 3) are from a zygomycete, ascomycete, or basidiomycete life cycle. Identify the labeled structures.

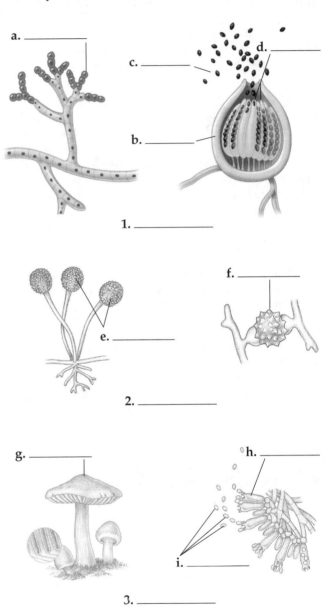

a. _____

c. _____

d. _____

b. _____

1. _____

f. _____

e. _____

2. _____

g. _____

h. _____

i. _____

3. _____

31.5   Fungi play key roles in nutrient cycling, ecological interactions, and human welfare

Fungi as Decomposers   Fungi and bacteria are decomposers of organic matter, making possible the essential recycling of chemical elements for plant growth.

Fungi as Mutualists   Mycorrhizae are very common and important in natural ecosystems and agriculture. All plant species studied thus far have symbiotic **endophytes** in their leaves or other plant parts, which benefit plants by producing toxins that deter herbivores, increasing plant tolerance for heat or drought, or playing a role in defending against pathogens.

Fungal inhabitants of the guts of grazing mammals help digest plant material. Farmer leaf-cutter ants raise fungal crops by gathering and feeding leaves to the fungus.

Lichens are symbiotic associations of millions of photosynthetic organisms in a lattice of fungal hyphae. The fungus is usually an ascomycete, and the partner is a unicellular or filamentous green algae or cyanobacteria. The alga provides the fungus with food and, in the case of cyanobacteria, nitrogen. The fungus creates most of the mass of the lichen, provides protection for the alga, and absorbs water and minerals.

Lichens reproduce asexually, either as fragments or as tiny clusters called **soredia.** In addition, it is common for the fungal component to reproduce sexually and for the algal component independently to reproduce asexually.

Lichens are important colonizers of bare rock and soil, but cannot tolerate air pollution.

Fungi as Pathogens   Fungal parasites of plants are common, killing chestnut and pine trees and spoiling grain and fruit crops. An ascomycete forms ergots on rye that can cause serious symptoms when accidentally milled into four. Lysergic acid, the raw material of LSD, is one of the toxins in the ergots.

Fungal infections in animals are called **mycoses.** Parasitic fungi cause ringworm, athlete's foot, vaginal yeast infections, and lung infections. Systemic mycoses are very serious. Many mycoses are opportunistic.

Practical Uses of Fungi   Commercially cultivated mushrooms are eaten, and fungi are used to ripen some cheeses. *Saccharomyces cerevisiae* is the yeast used in baking, brewing, and molecular research and biotechnology. Some fungi are the source of medical compounds, including antibiotics.

Word Roots

-osis = a condition of (*mycosis:* the general term for a fungal infection)

coeno- = common; **-cyto** = cell (*coenocytic:* referring to a multinucleated condition resulting from the repeated division of nuclei without cytoplasmic division)

di- = two; **-karyo** = nucleus (*dikaryotic:* a mycelium with two haploid nuclei per cell)

hetero- = different (*heterokaryon:* a mycelium formed by the fusion of two hyphae that have genetically different nuclei)

myco- = fungus; **rhizo-** = root (*mycorrhizae:* mutualistic associations of plant roots and fungi)

plasmo- = plasm; **-gamy** = marriage (*plasmogamy:* the fusion of the cytoplasm of cells from two individuals; occurs as one stage of syngamy)

Structure Your Knowledge

1. Fill in the following table that summarizes the characteristics of the five fungal phyla.

2. The kingdom Fungi contains members that are decomposers, parasites, and mutualists. How do these lifestyles relate to the ecological and economic importance of fungi?

Phylum	Examples	Key Features and Reproduction
Chytridiomycota	a.	b.
Zygomycota	c.	d.
Glomeromycota	e.	f.
Ascomycota	g.	h.
Basidiomycota	i.	j.

Test Your Knowledge

FILL IN THE BLANKS

_____ 1. division between cells in fungal hyphae

2. hyphae with nuclei from two parents

_____ 3. component of cell walls in most fungi

4. produced by fungi for absorptive nutrition

_____ 5. club-shaped reproductive structure found in mushrooms

6. sacs that contain sexual spores in sac fungi

_____ 7. mutualistic associations between plant roots and fungi

8. resistant structures produced by zygomycetes

_____ 9. hyphae with many nuclei

10. fungal group that diverged earliest

_____ 11. distinctive feature of glomeromycetes

12. clade that includes fungi, animals, and related flagellated protists

MULTIPLE CHOICE: *Choose the one best answer.*

1. The major difference between fungi and plants is that fungi
   a. have an absorptive form of nutrition.
   b. do not have a cell wall.
   c. are not eukaryotic.
   d. are multinucleate but not multicellular.
   e. reproduce by spores.

2. Chytrids had previously been classified with protists because they
   a. do not have chitin in their cell walls.
   b. do not have absorptive nutrition.
   c. have flagellated zoospores, whereas most fungi lack flagella.
   d. have metabolic pathways like those of protists.
   e. are aquatic, and fungi are terrestrial.

3. A fungus that is both a parasite and a decomposer is one that
   a. digests the nonliving portions of its host's body.
   b. lives off the sap within its host's body.
   c. first lives as a parasite but then consumes the host after it dies.
   d. lives as a mutualistic symbiont on its host.
   e. causes athlete's foot and vaginal infections.

4. The fact that karyogamy and meiosis do not immediately follow plasmogamy
   a. is necessary to create coenocytic hyphae.
   b. is characteristic of all fungi.
   c. allows fungi to reproduce asexually most often.
   d. results in heterokaryotic cells that may benefit from the variation present in two genomes.
   e. is characteristic of yeasts.

5. The traditional group deuteromycetes
   a. represent the most ancient lineage of fungi.
   b. include the fungal components of lichens.
   c. have abnormal forms of sexual reproduction.
   d. are fungi that are predatory.
   e. are fungi whose sexual stage is lacking or unknown.

6. In the Ascomycota,
   a. sexual reproduction occurs by conjugation.
   b. spores often line up in a sac in the order in which they were formed by meiosis.
   c. asexual spores form in sporangia on erect hyphae.
   d. most hyphae are dikaryotic.
   e. reproduction is always sexual.

7. Lichens are symbiotic associations that
   a. usually involve an ascomycete and a green alga or cyanobacterium.
   b. can reproduce sexually by forming soredia.
   c. require moist environments to grow.
   d. fix nitrogen for absorption by plant roots.
   e. are unusually resistant to air pollution.

8. The name given to three of the fungal phyla is based on
   a. the structure in which karyogamy occurs during sexual reproduction.
   b. the location of plasmogamy during sexual reproduction.
   c. the location of the heterokaryotic stage in the life cycle.
   d. the structure that produces asexual spores.
   e. their ancestral origin.

9. Fungi and animals appear to have evolved from a common ancestor
   a. because neither of them are photosynthetic.
   b. based on similarities in cell structure.
   c. based on molecular analyses.
   d. about the time that fungi and plants moved onto land.
   e. based on homologous ultrastructure of their flagella.

10. How can fungi be classified as opisthokonts, which means "posterior flagella," when most fungi lack flagella?
    a. Most fungal lineages appear to have lost their flagella during their evolution.
    b. The most primitive lineage of fungi, the chytrids, have flagellated zoospores.
    c. Fungi are in the same lineage as all flagellated protists.
    d. Fungi are closely related to animals, and since animals, which do have flagella, are opisthokonts, fungi must be too.
    e. Both a and b are the best reasons given here for their classification.

Chapter 32

An Introduction to Animal Diversity

Key Concepts

32.1 Animals are multicellular, heterotrophic eukaryotes with tissues that develop from embryonic layers

32.2 The history of animals spans more than half a billion years

32.3 Animals can be characterized by "body plans"

32.4 New views of animal phylogeny are emerging from molecular data

Framework

Animals are multicellular eukaryotic heterotrophs that ingest their food. Fossil evidence for the origin and rapid divergence of animals is limited. A colonial flagellated protist is the probable ancestor, and most animal phyla diversified during the 20 million years of the Cambrian explosion. The traditional phylogenetic tree of animals is based on body plan grades and includes such characteristics as symmetry, tissue layers, body cavities, and development. Molecular systematics splits the protostomes into clades Ecdysozoa and Lophotrochozoa.

Chapter Review

32.1 Animals are multicellular, heterotrophic eukaryotes with tissues that develop from embryonic layers

Nutritional Mode   Most animals ingest their food, eating either other organisms or nonliving organic matter.

Cell Structure and Specialization   Animals are multicellular, and their cells lack walls. Structural proteins,

such as collagen, hold adjacent cells together. Muscle and nervous tissues are unique to animals.

Reproduction and Development   The diploid stage is usually dominant, and reproduction is primarily sexual, with a flagellated sperm fertilizing a larger, nonmotile egg. The zygote undergoes a series of mitotic divisions, called **cleavage**, usually passing through a **blastula** stage during embryonic development. The process of **gastrulation** produces layers of embryonic tissues, resulting in the **gastrula** stage.

While all eukaryotes have regulatory genes, many of which contain DNA sequences called homeoboxes, animals share a unique homeobox-containing family of genes, called *Hox* genes, which are involved in regulating gene expression in embryonic development. In the ancestors of animals more complex than sponges, *Hox* genes underwent duplications, providing a "toolkit" for regulating development.

INTERACTIVE QUESTION   32.1

a. What is a **larva**?

b. How does a larva differ from a juvenile form of an animal?

c. By what process does a larva transform into a juvenile?

32.2 The history of animals spans more than half a billion years

The diversity of the animal kingdom includes the estimated 99% of species that are extinct, the 1.3 million living species, and the millions more that have yet to be identified. Some molecular clock calculations indicate that the ancestors of fungi and animals diverged about one billion years ago. The common ancestor of all living animals may have lived 675 to 875 million years ago and probably resembled choanoflagellates. These colonial, flagellated protists are stationary suspension feeders. Evidence that choanoflagellates are closely related to animals includes molecular data and morphological similarities to the collar cells found in sponges and some other animals.

Neoproterozoic Era (1 Billion–542 Million Years Ago) The earliest generally accepted fossils of animals, known as the **Ediacaran biota,** range from 565 to 550 million years old. Some are sponges; others may be related to cnidarians. What appear to be fossilized animal embryos date from 575 million years ago.

Paleozoic Era (542–251 Million Years Ago) Fossils from the **Cambrian explosion** (535 to 525 mya) represent about half of all extant phyla and include the first animals with hard skeletons. There are three main hypotheses to explain the Cambrian explosion: The emergence of new predator–prey relationships may have triggered various evolutionary adaptations. The accumulation of atmospheric oxygen may have supported the active metabolisms of mobile and larger animals. The evolution of the *Hox* complex of regulatory genes may have provided the developmental flexibility that could account for the diverse body plans that appear during the Cambrian explosion.

During the Ordovician, Silurian, and Devonian periods, animal diversity continued to increase. Fishes became the top predators of the seas. Arthropods appeared on land by 460 million years ago. Vertebrates began to adapt to land around 360 million years ago. Of the several lineages of tetrapods that diversified, amphibians and amniotes survive today.

Mesozoic Era (251–65.5 Million Years Ago) The first coral reefs formed; some reptiles returned to water; and flight appeared in pterosaurs and birds. Large predatory and herbivorous dinosaurs emerged, as did the first mammals.

Cenozoic Era (65.5 Million Years Ago to the Present) Mass extinctions marked the beginning of this period and included the nonflying dinosaurs and marine reptiles. Mammals began to exploit the vacated niches.

INTERACTIVE QUESTION   32.2

Review the three hypotheses for the rapid radiation of animal phyla during the Cambrian period.

32.3   Animals can be characterized by "body plans"

An animal's set of morphological and developmental traits is often referred to as a **body plan.** Experimental evidence suggests that β-catenin, one of the molecular controls necessary for gastrulation, evolved more than 500 million years ago. Other aspects of body plans may have changed independently in different lineages. A group of animals that share a similar body plan is called a *grade*.

Symmetry Sponges lack symmetry. The parts of an animal with **radial symmetry** radiate from the center. Animals with **bilateral symmetry** have distinct **anterior** (head) and **posterior** (tail) ends, and left and right sides. Bilateral animals also have **dorsal** (top) and **ventral** (bottom) sides. Bilateral symmetry is associated with **cephalization,** the concentration of sensory organs and a central nervous system in the head end, which is an adaptation facilitating unidirectional movement.

Tissues True tissues are groups of specialized cells separated from other tissues by membranous layers. Sponges lack true tissues. During gastrulation, an embryo develops concentric layers of cells called *germ layers:* **Ectoderm** develops into the outer body covering and, in some phyla, into the central nervous system; **endoderm** lines the developing digestive tube, or **archenteron,** and gives rise to the lining of the digestive tract and associated organs. Cnidarians and comb jellies are **diploblastic,** forming only these two germ layers. The bilateria animals are **triploblastic,** producing a middle layer, the **mesoderm,** from which arise muscles and most other organs.

Body Cavities A fluid- or air-filled **body cavity** separating the digestive tract from the outer body wall is called a **coelom.** A "true" coelom is completely lined by mesodermally derived tissue, which extends from the dorsal and ventral sides and suspends internal organs. Animals with a true coelom are called **coelomates. Pseudocoelomates** are animals whose body cavity, called a pseudocoelom, is formed from mesoderm and endoderm. Triploblastic animals that have solid bodies are called **acoelomates.**

A fluid-filled body cavity cushions internal organs, allows organs to grow and move independently of the outer body wall, and also functions as a skeleton in soft-bodied animals. During animal evolution, coeloms and pseudocoeloms have evolved and been lost many times and cannot be used to distinguish clades.

Protostome and Deuterostome Development Three features distinguish **protostome development** from **deuterostome development.**

Protostome development, typical of molluscs and annelids, is characterized by **spiral cleavage,** in which the planes of cell division are diagonal and newly formed cells fit in the grooves between cells of adjacent tiers. **Determinate cleavage** sets the developmental fate of each embryonic cell very early. Deuterostome development, typical of echinoderms and chordates, involves **radial cleavage,** in which parallel and perpendicular cleavage planes result in aligned tiers of cells. **Indeterminate cleavage,** present in most animals with deuterostome development, means that cells from early cleavage divisions retain the capacity to develop into complete embryos.

In protostome development, the coelom forms from splits within solid masses of mesoderm. In deuterostome development, the coelom forms from mesodermal outpocketings from the archenteron.

The **blastopore** is the opening during gastrulation leading to the developing archenteron. In protostome ("first mouth") development, the blastopore develops into the mouth, and a second opening forms at the end of the archenteron to produce an anus. In deuterostome development, the blastopore becomes the anus, and the second opening develops into the mouth.

INTERACTIVE QUESTION 32.3

Fill in this table to review some of the differences in the early embryological development of protostomes and deuterostomes.

	Protostomes	Deuterostomes
Cleavage	a.	b.
Coelom formation	c.	d.
Blastopore fate	e.	f.

32.4 New views of animal phylogeny are emerging from molecular data

The relationships of the 35 recognized animal phyla continue to be debated. Modern phylogenetic systematics is based on identifying a hierarchy of clades nested within larger clades. Molecular systematics, new studies of lesser-known phyla, and analyses of fossils are contributing to the identification of such clades, defined by shared derived characters unique to the members of the clade and their common ancestor. Two current phylogenetic hypotheses are based on systematic analyses of morphological characters or on results from recent molecular studies.

Points of Agreement There are five major points of agreement between the morphological and molecular trees: (1) All animals share a common ancestor and thus represent a clade called Metazoa. (2) Sponges are basal animals, branching from the base of both trees. (3) Eumetazoa is a clade of animals with true tissues. The **eumetazoans** include all animals except sponges and a few other groups. Basal members of the clade include the diploblastic, radially symmetrical phyla Cnidaria and Ctenophora (comb jellies). (4) Most animal phyla belong to the clade Bilateria. The **bilaterians** are defined by the shared derived characters of bilateral symmetry and three germ layers. (5) Vertebrates and some other phyla belong to the clade Deuterostomia.

Progress in Resolving Bilaterian Relationships The morphology-based tree divides the bilaterians into the deuterostomes and protostomes, assuming that these developmental differences reflect phylogeny. This tree also groups arthropods with annelids, both of which have segmented bodies. Recent molecular studies distinguish a group of acoelomate flatworms (the Acoela) that appear to be basal bilaterians and three bilaterian clades: deuterostomes; **ecdysozoans,** which include the arthropods and nematodes; and **lophotrochozoans,** which include the annelids and molluscs. The name *Ecdysozoa* refers to the shedding of an exoskeleton, called ecdysis, a trait shared by some ecdysozoan phyla. The name *Lophotrochozoa* is based on the feeding apparatus called a **lophophore,** found in animals such as brachiopods, and the larval stage called the **trochophore larva** shared by other members of the lophotrochozoans.

Future Directions in Animal Systematics Large-scale comparisons of multiple genes across many animal phyla currently being conducted may help systematists to test hypotheses about animal phylogeny.

Word Roots

a- = without; **-koilos** = a hollow (*acoelomate:* the condition of lacking a coelom)

arch- = ancient, beginning (*archenteron:* the endoderm-lined cavity, formed during the gastrulation process, that develops into the digestive tract of an animal)

bi- = two (*Bilateria:* the branch of eumetazoans possessing bilateral symmetry)

blast- = bud, sprout; **-pore** = a passage (*blastopore:* the opening of the archenteron in the gastrula that develops into the mouth in protostomes and the anus in deuterostomes)

cephal- = head (*cephalization:* an evolutionary trend toward the concentration of sensory equipment on the anterior end of the body)

deutero- = second (*deuterostome:* one of two lines of coelomates characterized by radial, indeterminate cleavage; formation of the coelom from outpockets of mesoderm; and development of the anus from the blastopore)

di- = two (*diploblastic:* having two germ layers)

ecdys- = an escape (*Ecdysozoa:* one of two proposed clades within the protostomes; it includes the arthropods)

ecto- = outside; **-derm** = skin (*ectoderm:* the outermost of the three primary germ layers in animal embryos)

endo- = within (*endoderm:* the innermost of the three primary germ layers in animal embryos)

gastro- = stomach, belly (*gastrulation:* the formation of a gastrula from a blastula)

in- = into; **-gest** = carried (*ingestion:* a heterotrophic mode of nutrition in which other organisms or detritus are eaten whole or in pieces)

lopho- = a crest, tuft; **-trocho** = a wheel (*Lophotrochozoa:* one of two proposed clades within the protostomes that includes annelids and molluscs)

meso- = middle (*mesoderm:* the middle primary germ layer of an early embryo)

meta- = boundary, turning point; **-morph** = form (*metamorphosis:* the resurgence of development in an animal larva that transforms it into a sexually mature adult)

proto- = first; **-stoma** = mouth (*protostomes:* a member of one of two distinct evolutionary lines of coelomates characterized by spiral, determinate cleavage, formation of the coelom as splits in solid masses of mesoderm, and development of the mouth from the blastopore)

pseudo- = false (*pseudocoelom:* a body cavity that is not completely lined by mesoderm)

radia- = a spoke, ray (*radial symmetry:* characterizing a body shaped like a pie or barrel, lacking a left and right side, but having a top and bottom)

tri- = three (*triploblastic:* having three germ layers)

Structure Your Knowledge

1. These two very abbreviated trees represent the morphological and molecular hypotheses of animal phylogeny. In the boxes, fill in the clades or body plan grades for these two trees. In the table that follows, list some of the shared derived characters that define each group.

Clade or Body Plan Grade	Shared Derived Characters
a.	
b.	
c.	
d.	
e.	
f.	
g.	
h.	
i.	

Test Your Knowledge

MULTIPLE CHOICE: *Choose the one best answer.*

1. Sponges differ from the rest of the animals because
   a. they are completely sessile.
   b. they have radial symmetry and are suspension feeders.
   c. their simple body structure has no true tissues, and they have no symmetry.
   d. they are not multicellular.
   e. they have no flagellated cells.

2. A fly larva
   a. is a miniature version of the adult.
   b. is transformed into an adult by molting.
   c. ensures more genetic variation in the insect life cycle.
   d. is a sexually immature organism morphologically different from the adult and specialized for eating and growth.
   e. is all of the above.

3. Cephalization
   a. is the development of bilateral symmetry.
   b. is the formation of a coelom by budding from the archenteron.
   c. is a diagnostic characteristic of deuterostomes.
   d. is common in radially symmetrical animals.
   e. is associated with motile animals that concentrate sensory organs in a head region.

4. A true coelom
   a. has mesoderm-derived tissues extending from the dorsal and ventral sides that support internal organs.
   b. allows organs to grow and move independently of the outer body wall.
   c. is a fluid-filled cavity completely lined by mesoderm.
   d. may be used as a skeleton by soft-bodied coelomates.
   e. is all of the above.

5. Which of the following is descriptive of protostome development?
   a. radial and determinate cleavage, blastopore becomes mouth
   b. spiral and indeterminate cleavage, coelom forms as split in solid mass of mesoderm
   c. spiral and determinate cleavage, blastopore becomes mouth, coelom from split in mass of mesoderm
   d. spiral and indeterminate cleavage, blastopore becomes mouth, coelom from mesoderm outpockets
   e. radial and determinate cleavage, coelom from mesoderm outpockets, blastopore becomes anus

6. Which of the following is *not* descriptive of a pseudocoelomate?
   a. a body cavity incompletely lined by mesoderm
   b. radial symmetry
   c. triploblastic
   d. true tissues
   e. formation of body cavity from split in mass of mesoderm

7. Which of the following characteristics is found only in animals?
   a. homeobox-containing genes
   b. flagellated sperm
   c. heterotrophic nutrition
   d. *Hox* genes
   e. All of the above are exclusive animal traits.

8. The morphology-based and molecular-based phylogenetic trees agree in which of the following ways?
   a. the acoel flatworms as basal bilaterians
   b. sponges as basal animals, Eumetazoa as animals with true tissues, Bilateria as the clade that includes most animals
   c. sponges as the probable ancestor of animals
   d. the grouping of the segmented annelids and arthropods
   e. Deuterostomia as a monophyletic clade that includes chordates, echinoderms, and brachiopods

Chapter 33

Invertebrates

Key Concepts

33.1 Sponges are basal animals that lack true tissues

33.2 Cnidarians are an ancient phylum of eumetazoans

33.3 Lophotrochozoans, a clade identified by molecular data, have the widest range of animal body forms

33.4 Ecdysozoans are the most species-rich animal group

33.5 Echinoderms and chordates are deuterostomes

Framework

This chapter surveys the amazing diversity of invertebrate animals. Characteristics and representatives of the major animal phyla are presented. The groups covered are the sponges, cnidarians, flatworms, rotifers, lophophorate phyla, molluscs, annelids, nematodes, arthropods, echinoderms, and chordates.

Chapter Summary

More than 95% of known animal species are **invertebrates,** animals that lack backbones.

33.1 Sponges are basal animals that lack true tissues

Sponges (of phyla Calcarea and Silicea) are sessile animals found in fresh and marine waters. Water is drawn through pores in the body wall of these saclike animals into a central cavity, the **spongocoel,** and flows out through the **osculum.** Sponges are **suspension feeders,** collecting food particles by the action of collared,

flagellated **choanocytes** lining the spongocoel. Choanocytes resemble the cells of choanoflagellates. Molecular evidence indicates that animals originated from a choanoflagellate-like ancestor.

Sponges lack true tissues. In the **mesohyl,** or gelatinous matrix between the two body-wall layers, are **amoebocytes.** These cells take up food, digest it, and carry nutrients to other cells. Amoebocytes also form skeletal fibers, which may be sharp spicules or flexible fibers.

Most sponges are **hermaphrodites,** producing both eggs and sperm, although usually sequentially functioning as one sex or the other. Sperm, carried out through the osculum, fertilize eggs retained in the mesohyl of neighboring sponges. Flagellated larvae disperse to a suitable substratum and develop into sessile adults. Sponges produce defensive compounds that may have pharmaceutical uses.

INTERACTIVE QUESTION 33.1

Give the locations and functions of the following:

a. choanocytes

b. amoebocytes

33.2 Cnidarians are an ancient phylum of eumetazoans

One of the oldest lineages of clade Eumetazoa, animals with true tissues, is the phylum Cnidaria. The cnidarians include hydras, jellies, anemones, and corals. Their simple anatomy consists of a sac with a central **gastrovascular cavity** and a single opening serving as both mouth and anus. This body plan has two forms: **polyp,** which is a sessile, cylindrical form with mouth and tentacles extending upward, and **medusa,** which is a flattened, mouth-down polyp that moves by passive drifting and weak body contractions. Both these body forms occur in the life histories of some cnidarians.

Cnidarians use their ring of tentacles, armed with **cnidocytes,** to capture prey. Cnidocytes contain cnidae, which are capsules that can evert and discharge long threads; stinging capsules are called **nematocysts.** A gelatinous mesoglea is sandwiched between the epidermis and gastrodermis. Cells of these layers have bundles of microfilaments arranged into contractile fibers acting as simple muscles. A nerve net is associated with simple sensory receptors and coordinates the contraction of cells against the hydrostatic skeleton of the gastrovascular cavity, producing movement.

Hydrozoans   Most hydrozoans alternate between an asexually reproducing polyp and a sexually reproducing medusa form. The common, freshwater hydras exist only in polyp form.

Scyphozoans   The medusa stage is more prevalent in the scyphozoans. The sessile polyp stage often does not occur in the jellies of the open ocean.

Cubozoans   Cubozoans have a box-shaped medusa stage and complex eyes in the fringe of their medusae. Many species, such as the sea wasp, have highly toxic cnidocytes.

Anthozoans   Sea anemones and corals occur only as polyps. Corals secrete calcified external skeletons, and the accumulation of such skeletons produces coral.

INTERACTIVE QUESTION   33.2

Name these two cnidarian body plans and identify the indicated structures.

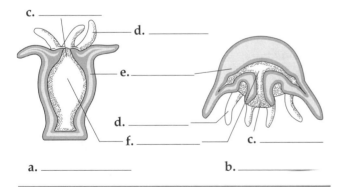

c. _____
d. _____
e. _____
d. _____
f. _____
a. _____
b. _____
c. _____

33.3 Lophotrochozoans, a clade identified by molecular data, have the widest range of animal body forms

Clade Bilateria consists of bilaterally symmetrical animals with triploblastic development. Most are coelomates. Molecular data points to three major clades of bilaterians: Lophotrochozoa, Ecdysozoa, and Deuterostomia. The clade **Lophotrochozoa** is named for the

lophophore, a feeding structure found in some members and the *trochophore larva* found in others. About 18 animal phyla are included in this morphologically diverse group.

Flatworms   Free-living flatworms live in marine, freshwater, and damp terrestrial habitats, and many are parasitic. They are triploblastic acoelomates. Most flatworms have a branching gastrovascular cavity, which functions in both digestion and distribution of food. Gas exchange and diffusion of nitrogenous wastes occur across the body wall. **Protonephridia** are networks of tubules with ciliated *flame bulbs* that pull fluid out of the body and function in osmoregulation. Phylum Platyhelminthes is divided into four classes.

Turbellarians include freshwater **planarians** and other, mostly marine, free-living flatworms. Planarians move using cilia to glide on secreted mucus or may use body undulations to swim. Eyespots on the head detect light, and lateral head flaps detect chemicals. Their nervous system consists of pairs of anterior ganglia and ventral nerve cords. Planarians reproduce asexually by regeneration or sexually by copulation between hermaphroditic worms.

Adapted to live as parasites in or on other animals, flukes (trematodes) have a tough outer covering, suckers, and extensive reproductive organs. The life cycles of trematodes are complex, usually including asexual and sexual stages and intermediate hosts in which larvae develop. Most monogeneans are external parasites of fish and have a simple life cycle with ciliated larvae dispersing to new hosts.

As parasites, mostly of vertebrates, tapeworms (class Cestoidea) consist of a scolex, with suckers and hooks for attaching to the host's intestinal lining, and a ribbon of proglottids packed with reproductive organs. Predigested food is absorbed from the host. The life cycles of tapeworms may include intermediate hosts.

INTERACTIVE QUESTION   33.3

a. Describe the digestive system of a planarian.

b. Why do tapeworms, which are also flatworms, lack a digestive system?

Rotifers   Rotifers are smaller than many protists but have an **alimentary canal,** with separate mouth and anus, and other organ systems. Fluid in the pseudocoelom functions as a hydrostatic skeleton and distributes nutrients. A crown of cilia draws microscopic food into the mouth to the pharynx with its grinding jaws. Some species reproduce totally asexually by **parthenogenesis,** in which female offspring develop from unfertilized eggs. In other species, two types of eggs develop parthenogenically: one type forming females and the other developing into simplified males that produce sperm. The resulting resistant zygotes survive harsh conditions in a dormant state.

Lophophorates: Ectoprocts and Brachiopods   These two phyla known as lophophorates have a lophophore, a crown of ciliated tentacles, which surrounds the mouth and functions in feeding. They have a true coelom.

Ectoprocts, commonly called bryozoans, are tiny, mostly marine animals living in colonies that are often encased in a hard **exoskeleton,** with pores through which their lophophores extend. Some species are important reef builders. **Brachiopods,** or lamp shells, attach to the seafloor by a stalk and open their hinged shell to allow water to flow through the lophophore.

Molluscs   Molluscs are soft-bodied, mostly marine animals, most of which are protected by a shell. The molluscan body plan has three main parts: a muscular **foot** used for movement, a **visceral mass** containing the internal organs, and a **mantle** that covers the visceral mass and may secrete a shell. In many molluscs, a **mantle cavity,** a water-filled chamber formed by the extension of the mantle, encloses the gills, anus, and excretory pores. A rasping **radula** is used for feeding by many molluscs. The nervous system includes a nerve ring and nerve cords. Nephridia remove wastes from the hemolymph.

Most molluscs have separate sexes. Many marine molluscs have a life cycle that includes a ciliated larva called the **trochophore.** Four of the eight molluscan classes are discussed in the text.

Chitons are oval marine animals with shells that are divided into eight dorsal plates. Chitons cling to and creep slowly over rocks, where they feed on algae.

Most gastropods are marine, although there are many freshwater species, and some snails and slugs are terrestrial. A distinctive feature of this class is **torsion,** the embryonic rotation of the visceral mass that results in the anus and mantle cavity being above the head. Most gastropods have single coiled shells. Many gastropods have distinct heads with eyes at the tips of tentacles. Moving by the rippling of the foot, most gastropods graze on plant material with their radula. Land snails lack gills; the lining of the mantle cavity functions as a lung.

Bivalves, such as clams, oysters, mussels, and scallops, have the two halves of their shell hinged at the mid-dorsal line. Most bivalves are suspension feeders; water flows into and out of the mantle cavity through siphons, and food particles are trapped in the mucus that coats the gills and then are swept to the mouth by cilia.

Cephalopods are rapid-moving carnivores. The mouth has beaklike jaws to bite prey and is surrounded by tentacles. The shell is reduced and internal in squids, absent in octopuses, and external only in the chambered nautilus. The foot has been modified to form parts of the tentacles and the muscular siphon, which animals may use to jet-propel themselves when water from the mantle cavity is forcibly expelled.

Cephalopods are the only molluscs with a closed circulatory system. They have a well-developed nervous system, sense organs, and a complex brain—important features for active predators. Shelled **ammonites** were the dominant invertebrate predators until their extinction at the end of the Cretaceous period.

INTERACTIVE QUESTION   33.4

a. Describe the three parts of the molluscan body plan.

1.

2.

3.

b. Compare the feeding behaviors and activity levels of snails, clams, and squid.

1. Snails:

2. Clams:

3. Squid:

Annelids   Annelids are segmented worms found in marine, freshwater, and damp soil habitats. This coelomate phylum is divided into three classes.

The oligochaetes include the earthworms and several aquatic species. Earthworm castings improve soil texture. The earthworm has a closed circulatory system, and respiration occurs across the moist, highly vascularized skin. Septa partition the coelom into segments, in each of which is found a pair of excretory metanephridia. The nervous system consists of cerebral ganglia and fused segmental ganglia along the ventral nerve cord. Earthworms are hermaphrodites; sperm are exchanged between worms during mating.

INTERACTIVE QUESTION 33.5

Identify the structures shown in this body segment of an earthworm.

g. _____

f. _____

e. _____

a. _____

b. _____

c. _____

d. _____

Polychaetes are mostly marine worms with parapodia on each segment that function in locomotion and often in gas exchange. Polychaetes may be planktonic, bottom burrowers, or tube dwellers.

Most leeches inhabit fresh water. Many feed on small invertebrates, whereas others are parasites that temporarily attach to animals, slit or digest a hole through the skin, and suck the blood of their host.

33.4 Ecdysozoans are the most species-rich animal group

Ecdysozoa includes animals that **molt,** shedding their external **cuticle** as they grow. The most abundant of animal groups, the nematodes and arthropods, are in the ecdysozoan clade.

Nematodes Nematodes, or roundworms, inhabit water, soil, and the bodies of plants and animals. These cylindrical worms have a tough cuticle, which is periodically shed as they grow. They have an alimentary canal. Fluid in the pseudocoelom circulates nutrients. Reproduction is usually sexual, fertilization is internal, and most zygotes form resistant cells.

Numerous nematode species are ecologically important decomposers. Other nematodes are serious agricultural pests and animal parasites. Parasitic nematodes can enlist the cells of their host to supply nutrients or to expand to house them.

INTERACTIVE QUESTION 33.6

Compare the locomotion of an earthworm and a nematode.

Arthropods In terms of species diversity, distribution, and vast numbers, **arthropods** are the most successful group of animals. Characteristics of arthropods include segmentation, which allows for regional specialization; a hard exoskeleton; and jointed appendages.

The earliest arthropods are found in the fossil record of the Cambrian explosion, which also includes a group called *lobopods.* The lobopods, with identical body segments, may have been the group from which arthropods evolved. Trilobites were early arthropods with fairly uniform appendages. Studies show that extant lobopods share all the arthropod *Hox* genes, including the unusual *Ubx* and *Abd-A* genes. Changes in the sequence or regulation of these genes may underlie the increased diversity in arthropod body segments.

Appendages have become modified for walking, feeding, sensing, mating, and defense. A cuticle of chitin and protein completely covers the body as an exoskeleton, providing protection and points of attachment for the muscles that move the appendages. To grow, an arthropod must molt, shedding its exoskeleton and secreting a larger one. The exoskeleton probably first evolved in the seas as protection and anchorage for muscles, but as arthropods diversified on land, its functions came to include protection from desiccation and support.

Arthropods have extensive cephalization and well-developed sensory organs. A heart pumps hemolymph through an **open circulatory system** consisting of short arteries and a network of sinuses known as the hemocoel. The embryonic coelom becomes reduced during development, and the hemocoel becomes the main body cavity. Gas exchange in most aquatic species occurs through gills, whereas most insects have tracheal systems of branching internal ducts.

Molecular and morphological data suggest that arthropods diverged into four evolutionary lineages: **cheliceriforms, myriapods, hexapods,** and **crustaceans.**

Cheliceriforms (horseshoe crabs, scorpions, spiders, mites, and ticks) have clawlike **chelicerae** that serve as pincers or fangs used for feeding. They have an anterior cephalothorax and an abdomen. The earliest cheliceriforms were large, predatory **eurypterids.** The horseshoe crab and sea spiders are the few marine cheliceriforms that survive today.

Most modern cheliceriforms are **arachnids.** Most ticks are bloodsucking parasites on reptiles or mammals. Many mites are parasites of plants and animals. In arachnids, the cephalothorax has six pairs of appendages: four pairs of walking legs, the chelicerae, and sensing or feeding appendages called *pedipalps.* In most spiders, **book lungs,** consisting of stacked internal plates, function in gas exchange. Many spiders spin characteristic webs of silk from special abdominal glands.

a. What are chelicerae?

b. How do spiders trap, kill, and eat their prey?

Millipedes and centipedes are in the subphylum Myriapoda. These terrestrial groups have a pair of antennae and three pairs of mouthparts, including jaw-like **mandibles.** Millipedes are wormlike vegetarians with two pairs of walking legs per segment. Centipedes are carnivores with one pair of legs per segment and poison claws on their front trunk segment.

Insects and their relatives in the subphylum Hexapoda have more known species than all other forms of life combined. In a typical insect, the head has fused segments and has one pair of antennae, a pair of compound eyes, and several pairs of mouthparts modified for various types of ingestion. The thorax and abdomen have obvious segments. The nervous system consists of a cerebral ganglion and paired ventral nerve cords with segmental ganglia. Malpighian tubules, outpocketings of the digestive tract, function in excretion. Trachael tubes form the respiratory system.

The oldest insect fossils are from the Devonian period (about 416 mya), but a major diversification occurred in the Carboniferous and Permian periods with the evolution of flight and modification of mouthparts for specialized feeding on plants. A diversification of insects appears to have accompanied the radiation of flowering plants.

Flight is a major key to the success of insects. Many insects have one or two pairs of wings that are extensions of the cuticle of the dorsal thorax. Dragonflies were among the first flying insects.

In **incomplete metamorphosis,** the young are smaller versions of the adult and pass through several molts before developing wings and becoming sexually mature. In **complete metamorphosis,** the larvae look entirely different from the adult. The larvae eat and grow; adults primarily reproduce and disperse. Metamorphosis occurs in a pupal stage. Reproduction is usually sexual; fertilization is usually internal.

Insects are classified into more than 30 orders. Insects affect humans as pollinators of crops, vectors of disease, and competitors for food.

The mostly aquatic crustaceans have two pairs of antennae; three or more pairs of mouthpart appendages, including mandibles; walking legs on the thorax; and appendages on the abdomen. Larger crustaceans have gills. Nitrogenous wastes pass by diffusion through thin areas of the cuticle, and a pair of glands regulates the salt balance of the hemolymph. Sexes usually are separate. One or more swimming larval stages occur in most aquatic crustaceans.

Isopods are mostly small aquatic crustaceans but include terrestrial pill bugs. Lobsters, crayfish, crabs, and shrimp are large crustaceans called **decapods.** Their cuticle is hardened by calcium carbonate, and a carapace covers the dorsal side of their cephalothorax. The small, very numerous **copepods** are important members of marine and freshwater plankton communities. Barnacles are sessile crustaceans that strain food from the water with their appendages.

Compare and contrast the following features for insects and crustaceans.

Feature	Insects	Crustaceans
Habitat	a.	b.
Locomotion	c.	d.
Respiration	e.	f.
Excretion	g.	h.
Antennae number	i.	j.
Appendages	k.	l.

33.5 Echinoderms and chordates are deuterostomes

The echinoderms and chordates are grouped together based on their common embryological traits of radial cleavage, coelom formation from the archenteron, and origin of the mouth opposite the blastopore. Molecular systematics supports Deuterostomia as a clade, while also indicating that some animals with deuterostome developmental features are not members.

Echinoderms Most **echinoderms** are sessile or slow-moving marine animals. They have a thin skin covering an endoskeleton of calcareous plates. A **water vascular system** with a network of hydraulic canals controls extensions called **tube feet** that function in locomotion, feeding, and gas exchange. Sexual reproduction usually involves separate sexes and external fertilization. Bilateral larvae metamorphose into adults that typically have five parts radiating from the center.

Sea stars (class Asteroidea) have multiple arms radiating from a central disk and use tube feet lining the undersurfaces of these arms to creep slowly and to grasp and open prey. The gripping of tube feet is the result of adhesive chemicals. Sea stars evert their stomach through their mouth and slip it into a slightly opened bivalve shell.

Brittle stars (class Ophiuroidea) have distinct central disks and move by lashing their long, flexible arms. They may be predators, scavengers, or suspension feeders.

Sea urchins and sand dollars, members of class Echinoidea, have no arms but are able to move slowly using their five rows of tube feet. Long spines also aid a sea urchin's movement. A sea urchin's mouth is ringed by complex jaw-like structures used to eat seaweeds and other foods.

Sea lilies live attached to the substrate by stalks; feather stars crawl about. Their long, flexible arms extend upward from around the mouth and are used in suspension feeding. Members of class Crinoidea have changed little in 500 million years of evolution.

Sea cucumbers (class Holothuroidea) are elongated animals that bear little resemblance to other echinoderms other than having five rows of tube feet.

Discovered in 1986, sea daisies (class Concentricycloidea) have an armless, disk-shaped body with a five-fold symmetry.

INTERACTIVE QUESTION 33.9

List the key characteristics that distinguish the phylum Echinodermata.

a.

b.

c.

Chordates Echinoderms and chordates have existed as separate phyla for at least 500 million years. Phylum Chordata contains two subphyla of invertebrates in addition to the hagfishes and vertebrates.

Word Roots

arthro- = jointed; **-pod** = foot (*Arthropoda:* segmented coelomates with exoskeletons and jointed appendages)

arachn- = spider (*Arachnida:* the arthropod group that includes scorpions, spiders, ticks, and mites)

brachio- = the arm (*brachiopod:* also called lamp shells, these animals superficially resemble clams and other bivalve molluscs, but the two halves of the brachiopod shell are dorsal and ventral to the animal rather than lateral, as in clams)

bryo- = moss; **-zoa** = animal (*bryozoan:* colonial animals (phylum Ectoprocta) that superficially resemble mosses)

cheli- = a claw (*chelicerae:* clawlike feeding appendages characteristic of the cheliceriform group)

choano- = a funnel; **-cyte** = cell (*choanocyte:* flagellated collar cells of a sponge)

cnido- = a nettle (*cnidocytes:* unique cells that function in defense and prey capture in cnidarians)

-coel = hollow (*spongocoel:* the central cavity of a sponge)

cope- = an oar (*copepods:* a group of small crustaceans that are important members of marine and freshwater plankton communities)

cuti- = the skin (*cuticle:* the exoskeleton of an arthropod)

deca- = ten (*decapod:* a large group of crustaceans that includes lobsters, crayfish, crabs, and shrimp)

diplo- = double (*Diplopoda:* the millipede class)

echino- = spiny; **-derm** = skin (*echinoderm:* sessile or slow-moving animals with a thin skin that covers an exoskeleton; the group includes sea stars, sea urchins, brittle stars, crinoids, sea cucumbers, and sea daisies)

eury- = broad, wide; **-pter** = a wing, a feather, a fin (*eurypterid:* mainly marine and freshwater, extinct cheliceriforms; these predators, also called water scorpions, ranged up to 3 meters long)

exo- = outside (*exoskeleton:* a hard encasement on the surface of an animal)

gastro- = stomach; **-vascula** = a little vessel (*gastrovascular cavity:* the central digestive compartment, usually with a single opening that functions as both mouth and anus)

hermaphrod- = with both male and female organs (*hermaphrodite:* an individual that functions as both male and female in sexual reproduction by producing both sperm and eggs)

in- = without (*invertebrates:* animals without a backbone)

iso- = equal (*isopods:* one of the largest groups of crustaceans, primarily marine, but including pill bugs common under logs and moist vegetation next to the ground)

lopho- = a crest, tuft; **-phora** = to carry (*lophophore:* a crown of ciliated tentacles that function in feeding and surround the mouth)

meso- = the middle; **-hyl** = matter (*mesohyl:* a gelatinous region between the two layers of cells of a sponge)

meta- = change; **-morph** = shape (*metamorphosis:* the resurgence of development in an animal larva that transforms it into a sexually mature adult)

nemato- = a thread; **-cyst** = a bag (*nematocysts:* the stinging capsules in cnidocytes, unique cells that func-
tion in defense and capture of prey)

nephri- = the kidney (*metanephridium:* in annelids, a
type of excretory tubule with internal openings
called nephrostomes that collect body fluids)

oscul- = a little mouth (*osculum:* a large opening in a
sponge that connects the spongocoel to the
environment)

partheno- = without fertilization; **-genesis** = producing
(*parthenogenesis:* a type of reproduction in which
females produce offspring from unfertilized eggs)

plan- = flat or wandering (*planarians:* flatworms that
prey on smaller animals or feed on dead animals)

tri- = three; **-lobi** = a lobe (*trilobite:* an extinct group of
arthropods with pronounced segmentation)

trocho- = a wheel (*trochophore:* distinctive larval stage
observed in certain invertebrates, including some
annelids and molluscs)

Structure Your Knowledge

1. Use the following simplified phylogenetic tree
(based on molecular evidence) to help you review
the diversity of animal phyla. Label the indicated
clades. Then identify each phylum and list some
key characteristics and common name examples
for each. The small sketches should help you identify the phyla. (Note that a sketch of a brachiopod
represents the two lophophore phyla.)

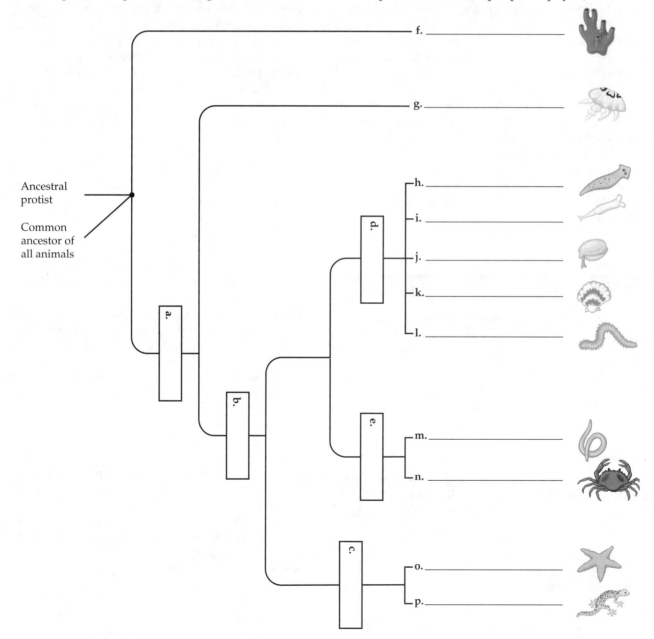

Test Your Knowledge

MATCHING: *Match the following organisms with their phylum and class or subphylum. Answers may be used more than once or not at all.*

Organism	Phylum	Class or Subphylum
1. jelly (jellyfish)	_____	_____
2. crayfish	_____	_____
3. snail	_____	_____
4. leech	_____	_____
5. tapeworm	_____	_____
6. cricket	_____	_____
7. scallop	_____	_____
8. tick	_____	_____
9. sea urchin	_____	_____
10. hydra	_____	_____
11. planaria	_____	_____
12. chambered nautilus	_____	_____

Phyla	Classes or Subphyla
A. Annelida	**a.** Arachnida
B. Arthropoda	**b.** Bivalvia
C. Cnidaria	**c.** Cephalopoda
D. Echinodermata	**d.** Cestoidea
E. Ectoprocta	**e.** Crustacea
F. Mollusca	**f.** Echinoidea
G. Nematoda	**g.** Gastropoda
H. Silicea	**h.** Hirudinea
I. Platyhelminthes	**i.** Hydrozoa
J. Rotifera	**j.** Insects (Hexopoda)
	k. Oligochaeta
	l. Scyphozoa
	m. Turbellaria

MULTIPLE CHOICE: *Choose the one best answer.*

1. Invertebrates include
   a. all animals except for the phylum Vertebrata.
   b. all animals without backbones.
   c. only animals that use hydrostatic skeletons.
   d. members of the protostomes, but not of the deuterostomes.
   e. all animals without an endoskeleton.

2. Which of the following is the best description of the sponges?
   a. no real symmetry, diploblastic, cnidocytes for capturing prey
   b. radial symmetry, triploblastic, nematocysts
   c. no real symmetry, without true tissues, choanocytes for trapping food particles
   d. bilateral symmetry, pseudocoel, flame bulbs for excretion
   e. bilateral symmetry, osculum and spongocoel for filtering water

3. Which of the following does *not* have a gastrovascular cavity for digestion?
   a. flatworm
   b. hydra
   c. polychaete worm
   d. sea anemone
   e. fluke

4. The basal group in clade Eumetazoa is
   a. Ectoprocta.
   b. Lophophorates.
   c. Platyhelminthes.
   d. Calcarea and Silicea.
   e. Cnidaria.

5. Hermaphrodites
   a. contain male and female sex organs but usually cross-fertilize.
   b. include sponges, earthworms, and most insects.
   c. are characteristically parthenogenic rotifers.
   d. are both a and b.
   e. are a, b, and c.

6. Which of the following is *not* true of cnidarians?
   a. An alternation of medusa and polyp stage is common in class Hydrozoa.
   b. They use a ring of tentacles armed with stinging cells to capture prey.
   c. They include hydras, jellies, sponges, and sea anemones.
   d. They have a nerve net that coordinates contraction of microfilaments for movement.
   e. They have a gastrovascular cavity.

7. Which of the following combinations of phylum and characteristics is *incorrect*?
   a. Brachiopoda—lophophore, stalked, marine animals with hinged shells
   b. Rotifera—parthenogenesis, crown of cilia, microscopic animals
   c. Nematoda—gastrovascular cavity, tough cuticle, ubiquitous
   d. Annelida—segmentation, closed circulatory system, hydrostatic skeleton
   e. Echinodermata—radial anatomy in adults, endoskeleton, water vascular system

8. Which of the following is an excretory or osmoregulatory structure that is *incorrectly* matched with its class?
   a. metanephridia—Oligochaeta
   b. Malpighian tubules—Echinoidea
   c. flame cells—Turbellaria
   d. thin region of cuticle—Crustacea
   e. diffusion across cell membranes—Hydrozoa

9. Torsion
   a. is embryonic rotation of the visceral mass that results in a U-shaped digestive tract in gastropods.
   b. is characteristic of molluscs.
   c. is responsible for the spiral growth of bivalve shells.
   d. describes the thrashing movement of nematodes.
   e. is responsible for the metamorphosis of insects.

10. Bivalves differ from other molluscs in that they
    a. are predaceous.
    b. have no heads and are suspension feeders.
    c. have shells.
    d. have open circulatory systems.
    e. use a radula to feed as they burrow through sand.

11. The exoskeleton of arthropods
    a. functions in protection and anchorage for muscles.
    b. is composed of chitin and cellulose.
    c. is absent in millipedes and centipedes.
    d. expands at the joints when the arthropod grows.
    e. functions in respiration and movement.

12. Which of the following does *not* function in suspension feeding?
    a. lophophore of ectoprocts
    b. radula of snails
    c. choanocytes of sponges
    d. mucus-coated gills of clams
    e. crown of cilia of rotifers

13. What do nematodes and arthropods have in common?
    a. They are both segmented.
    b. They are both pseudocoelomates.
    c. They include important members of plankton communities.
    d. They both have exoskeletons and undergo ecdysis (molting).
    e. Both a and d are correct.

14. Which of the following structures is *not* associated with prey capture?
    a. chaetae of earthworm
    b. mandibles of centipedes
    c. cnidocytes of hydra
    d. tentacles of squid
    e. tube feet of sea star

15. Many animals are parasitic. Which of the following is an *incorrect* description of one of these parasites?
    a. Ticks are bloodsucking parasites belonging to class Arachnida.
    b. Some roundworms (Nematoda) are internal parasites of humans.
    c. Lice are wingless ectoparasites in class Insecta.
    d. Flukes are flatworms and may have complex life cycles.
    e. Tapeworms are annelids that reproduce by shedding proglottids.

Chapter 34

Vertebrates

Key Concepts

34.1 Chordates have a notochord and a dorsal, hollow nerve cord

34.2 Craniates are chordates that have a head

34.3 Vertebrates are craniates that have a backbone

34.4 Gnathostomes are vertebrates that have jaws

34.5 Tetrapods are gnathostomes that have limbs

34.6 Amniotes are tetrapods that have a terrestrially adapted egg

34.7 Mammals are amniotes that have hair and produce milk

34.8 Humans are mammals that have a large brain and bipedal locomotion

Framework

This chapter focuses on the phylogeny of chordates, introducing the derived characters that define the major clades. Fill in the phylogenetic tree of chordates on p. 251 as you work through this chapter. Fossil evidence and hypotheses about human ancestry are also described.

Chapter Review

There are about 52,000 species of **vertebrates,** animals with a vertebral column, or backbone. This group is known for its *disparity*, its wide range of differences such as body mass.

34.1 Chordates have a notochord and a dorsal, hollow nerve cord

The deuterostome clade of the bilaterians includes the echinoderms and **chordates.** The chordates include the urochordates, cephalochordates, hagfishes, and vertebrates.

Derived Characters of Chordates Chordates have four distinguishing traits, which may appear only during the embryonic stage: (1) a flexible rod called a **notochord** that provides skeletal support along the length of the animal; (2) a dorsal, hollow nerve cord that develops into the brain and spinal cord; (3) **pharyngeal clefts,** or grooves that may develop into **pharyngeal slits** that open from the pharynx to the outside and function in suspension feeding or are modified for gas exchange; and (4) a post-anal tail containing skeletal elements and muscles. In **tetrapods** (vertebrates with limbs), the pharyngeal clefts contribute to the ear and other structures in the head and neck.

Lancelets Lancelets (Cephalochordata) are tiny marine animals that retain all four chordate characteristics into the adult stage. These suspension feeders burrow backward into the sand and filter food particles through a mucous net secreted across the pharyngeal slits. The lancelet swims by coordinated contractions of serial muscles that flex the notochord from side to side. Blocks of mesoderm, called *somites,* develop into these muscle segments.

Tunicates Recent molecular studies indicate that **tunicates** (Urochordata) are more closely related to other chordates than are lancelets. In this saclike, sessile animal, commonly called a sea squirt, water flows in the incurrent siphon, through pharyngeal slits where food is filtered by a mucous net, and out the excurrent siphon.

Early Chordate Evolution The ancestral chordate may have resembled a lancelet, the basal branch of chordates. The sequenced genome of tunicates indicates that, while genes associated with the vertebrate heart and thyroid gland are shared, genes involved with the long-range transmission of nerve impulses are found only in the vertebrate lineage. Studies have shown that several *Hox* genes that organize the major regions of the vertebrate brain are expressed in the anterior end of the lancelet dorsal nerve cord in the same anterior-to-posterior pattern.

INTERACTIVE QUESTION 34.1

What is the name for the animal shown in this diagram, and to which chordate group does it belong? Identify the labeled structures and indicate the four chordate characters.

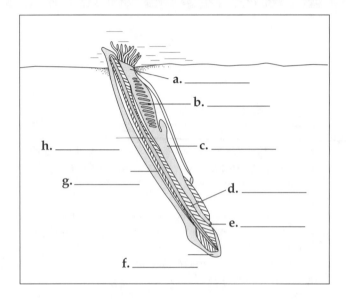

a. _____

b. _____

h. _____

c. _____

g. _____

d. _____

e. _____

f. _____

34.2 Craniates are chordates that have a head

The origin of a head with eyes and other sensory organs and a skull enclosing a brain was a major evolutionary transition seen in **craniates.**

Derived Characters of Craniates   Derived characters that distinguish craniates from other chordates include two clusters of *Hox* genes instead of only one, and duplicated families of genes for signaling molecules and transcription factors. A group of embryonic cells called the **neural crest** is found along the margin of the embryonic folds that form the nerve cord. These cells migrate in the embryo and contribute to various structures, including teeth and some of the bones and cartilage of the skull. The pharyngeal clefts of craniates evolved into gill slits with muscles and nerves that facilitate sucking in food and gas exchange. Craniates have a higher metabolism, more extensive muscles, a heart with at least two chambers, red blood cells, hemoglobin, and kidneys.

The Origin of Craniates   Newly discovered 530 million-year-old fossils (from the Cambrian explosion) appear to be transitional stages. *Haikouella* resembled lancelets, but has craniate characters such as eyes, hardened toothlike structures, and a small but well-formed brain. Another fossil from this period, *Myllokunmingia,* had a skull and is considered to be a true craniate.

Hagfishes   Hagfishes (clade Myxini) are the least derived surviving craniate lineage. Their cartilaginous skeleton includes a skull and a notochord; they lack vertebrae and jaws. The marine hagfishes are mostly bottom-dwelling scavengers that produce slime to repulse predators and competing scavengers.

INTERACTIVE QUESTION 34.2

List the derived characters of craniates.

34.3 Vertebrates are craniates that have a backbone

Derived Characters of Vertebrates   The gene duplication involving the *Dlx* family of transcription factor genes may be associated with nervous system innovations and the more extensive skull and backbone of vertebrates. In most vertebrates, the vertebrae enclose the spinal cord. The dorsal, ventral, and anal fins supported by fin rays of aquatic vertebrates provide thrust and steering control for swimming.

Lampreys   The jawless lampreys of clade Petromyzontida are the oldest living lineage of vertebrates. Lampreys are suspension feeders as larvae and blood-sucking parasites as adults. The notochord is the main axial skeleton and is surrounded by a cartilaginous pipe with pairs of projections that partially enclose the nerve cord.

Fossils of Early Vertebrates   **Conodants** were soft-bodied vertebrates with prominent eyes and barbed, mineralized hooks in their mouth, as well as dental elements in the pharynx for processing food. Conodonts appeared in the late Cambrian and were abundant for more than 300 million years.

During the Ordovician, Silurian, and Devonian periods, vertebrates with paired fins and an inner ear with semicircular canals emerged. These animals lacked jaws but had a muscular pharynx and were armored with mineralized bone. These armored jawless vertebrates disappeared by the end of the Devonian period.

Origins of Bone and Teeth   One hypothesis for mineralization in vertebrates is that it is associated with the switch from suspension-feeding to scavenging and predation. The earliest known mineralized structures are conodont dental elements. The armor of jawless vertebrates was derived from dental mineralization.

INTERACTIVE QUESTION 34.3

This phylogenetic hypothesis shows the major clades of chordates with some of the derived characters that define them. As you work through this chapter, fill in the eight large clades along the side and examples and names of each lineage.

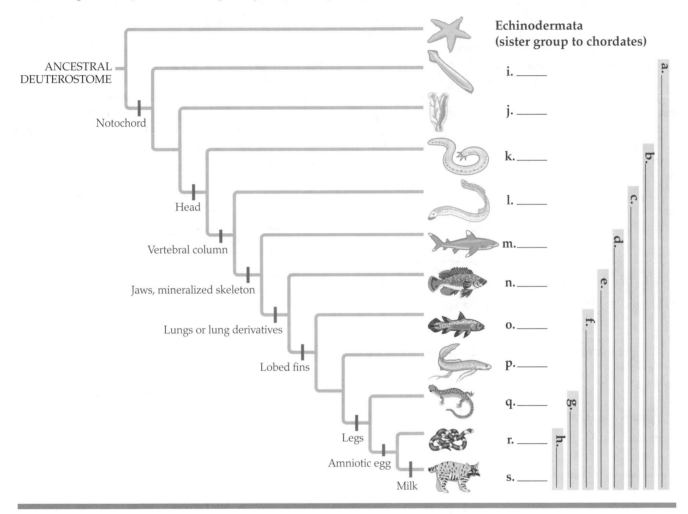

**Echinodermata
(sister group to chordates)**

i. ____

j. ____

k. ____

l. ____

m. ____

n. ____

o. ____

p. ____

q. ____

r. ____

s. ____

ANCESTRAL DEUTEROSTOME

Notochord

Head

Vertebral column

Jaws, mineralized skeleton

Lungs or lung derivatives

Lobed fins

Legs

Amniotic egg

Milk

a.
b.
c.
d.
e.
f.
g.
h.

INTERACTIVE QUESTION 34.4

List the derived characters of vertebrates.

34.4 Gnathostomes are vertebrates that have jaws

Derived Characters of Gnathostomes The hinged jaws of **gnathostomes** may have evolved from the skeletal rods of the anterior pharyngeal slits, with the remaining gill slits functioning in gas exchange.

Other derived characters of gnathostomes include an addition duplication of *Hox* genes, duplication of other gene clusters, an enlarged forebrain associated with enhanced senses of smell and vision, and a pressure-sensing lateral line system.

Fossil Gnathostomes Beginning in the mid-Ordovician period, about 470 mya, gnathostomes became increasingly diverse. Their paired fins, tail, and hinged jaws facilitated a predatory lifestyle. The earliest gnathostomes were the armored placoderms. Acanthodians radiated during the Devonian period.

Chondrichthyans (Sharks, Rays, and Their Relatives) **Chondrichthyans** have a skeleton made predominantly of cartilage, which appears to be a derived condition.

The largest group of chondrichthyans consists of sharks, rays, and skates. Sharks swim using powerful muscles in their trunk and caudal fin. The paired pectoral and pelvic fins provide lift. Swimming maintains buoyancy and moves water past the gills. Although most sharks are carnivores, the largest sharks and rays are suspension feeders.

Sharks have keen senses: sharp vision, nostrils, and skin regions in the head that can detect electric fields generated by muscle contractions of nearby animals. The shark's body transmits sound waves to the inner ear.

Following internal fertilization, **oviparous** species of sharks lay eggs that hatch outside the mother; **ovoviviparous** species retain the fertilized eggs until they hatch; and, in the few **viviparous** species, developing young are nourished by a yolk sac placenta, from nutrients in a uterine fluid, or by eating other eggs.

Rays, which propel themselves with their enlarged pectoral fins, are primarily bottom dwellers that feed on molluscs and crustaceans.

INTERACTIVE QUESTION   34.5

a. List the derived characters of gnathostomes.

b. What is the function of the **spiral valve** in a shark's intestine?

c. What is a **cloaca**?

Ray-Finned Fishes and Lobe-Fins   Nearly all vertebrates belong to a clade of gnathostomes called Osteichthyes, which historically included only the "bony fishes" but now includes the tetrapods. Almost all living **osteichthyans** have an ossified endoskeleton with a hard calcium phosphate matrix.

Aquatic osteichthyans (informally called fishes) move water through the mouth and out between the gills by the contraction of muscles in the gill chambers and movement of the protective bony flap called the **operculum.** Most aquatic osteichthyans also have a **swim bladder**, an air sac that controls buoyancy. It appears that lungs arose first as a supplement to gas exchange in the gills, and later evolved into swim bladders in some lineages.

The flattened, bony scales that cover the skin of most aquatic osteichthyans are coated with mucus to reduce drag. Fishes have a lateral line system, and most have external fertilization and are oviparous.

Ray-finned fishes (Actinopterygii) probably originated in fresh water during the Devonian period. They include most of the familiar modern fish, many of which spread to the seas during their evolution and some of which later returned to fresh water.

The **lobe-fins** (Sarcopterygii), known mainly from the fossil record, have paired muscular fins supported by a series of rod-shaped bones that may have enabled these lobe-fins to "walk." Only three lobe-fin lineages survive: Two populations of coelacanths (Actinistia) found in deep marine waters, three genera of lungfishes (Dipnoi) found in stagnant ponds and swamps in the Southern Hemisphere, and the tetrapods found on land.

INTERACTIVE QUESTION   34.6

Go back to the phylogenetic tree in Interactive Question 34.3 and make sure you have filled in the information up to letter *p*.

34.5   Tetrapods are gnathostomes that have limbs

About 360 million years ago, the fins of some lobe-fins evolved into the limbs and feet of tetrapods.

Derived Characters of Tetrapods   Tetrapods have limbs that support their weight on land and feet with digits. The neck has vertebrae (the atlas and axis) that allow the head greater movement. The pelvic girdle is fused to the backbone, transferring forces generated by the hind legs to the rest of the body. Pharyngeal clefts formed during development give rise to parts of the ears, glands, and other structures.

The Origin of Tetrapods   A wide range of lobe-fins evolved during the Devonian, probably using their stout fins to crawl through shallow waters and supplementing their gas exchange with their lungs. In one lineage, the fins became more limb-like. A diversity of tetrapods emerged during the Devonian and Carboniferous periods, most of which probably remained tied to the water. Many newly discovered fossils document the origin of tetrapods.

Amphibians **Amphibians** (class Amphibia) include salamanders (order Urodela), frogs (order Anura), and caecilians (order Apoda).

Salamanders, which may be aquatic or terrestrial, move with a lateral bending of the body. Frogs are more specialized for moving on land, using their powerful hind legs for hopping. Protective adaptations of frogs include camouflage coloring and distasteful or poisonous skin mucus that may be advertised by bright coloration. Tropical caecilians are burrowing, legless, nearly blind, wormlike animals.

Many frogs undergo a metamorphosis from a tadpole—the aquatic, herbivorous larval form with gills, a lateral line system, and a finned tail—to the carnivorous, terrestrial adult form with legs and lungs. Some amphibians are exclusively aquatic or terrestrial. Amphibians are most abundant in damp habitats, and some of their gas exchange occurs across their moist skin.

Fertilization is generally external. Oviparous species lay their eggs in aquatic or moist environments. Some species show varying amounts of parental care. During breeding season, some frogs communicate with vocalizations. Some species migrate to specific breeding sites, using various forms of communication and navigation.

INTERACTIVE QUESTION   34.7

a. What are the three lineages of lobe-fins?

b. Describe *Acanthostega*, a Devonian relative of tetrapods.

34.6 Amniotes are tetrapods that have a terrestrially adapted egg

Amniotes include reptiles (and birds) and mammals.

Derived Characters of Amniotes The **amniotic egg** contains *extraembryonic membranes* that protect and nourish the embryo. Most reptiles and some mammals have shelled amniotic eggs. The shells of bird eggs are calcareous; those of nonbird reptiles are leathery and flexible. Most mammals retain the embryo within the mother. Other terrestrial adaptations of the amniotes include a waterproof skin and increased use of the rib cage to ventilate the lungs.

INTERACTIVE QUESTION   34.8

Identify the four extraembryonic membranes in this sketch of an amniotic egg. List the function of each membrane.

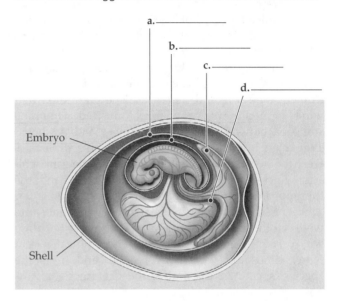

a.

b.

c.

d.

Early Amniotes The common ancestor of amphibians and amniotes likely lived about 370 million years ago. Early amniotes lived in drier environments than earlier tetrapods.

Reptiles The **reptile** clade includes a number of extinct groups (such as dinosaurs), and tuatara, lizards, snakes, turtles, crocodilians, and birds. We must infer the derived characters that distinguished early reptiles from other tetrapods. The skin of reptiles is covered with keratinized, waterproof scales. Most species lay shelled amniotic eggs, and fertilization is internal. Some species are viviparous, in which case the extraembryonic membranes form a placenta.

Many reptiles are **ectothermic.** They absorb external heat rather than generating their own, although they may regulate their body temperature through behavioral adaptations. Birds are **endothermic,** warming their bodies with metabolic heat.

The oldest reptilian fossils date from the late Carboniferous period, 310 million years ago. The first major branch to emerge was the **parareptiles,** large herbivores with dermal plates on their skin.

The **diapsids** diversified next. Their most obvious derived character is a pair of holes on each side of the skull. The two main lineages of diapsids are the **lepidosaurs,** which include tuataras, lizards, and snakes (and a number of now extinct huge marine reptiles) and the **archosaurs,** which produced the crocodilians and the extinct pterosaurs and dinosaurs. The **pterosaurs** had wings formed from a vascularized, muscled membrane stretched between an elongated finger and the body. Two main branches of the highly diverse **dinosaurs** were the herbivorous ornithischians and the saurischians, which gave rise to long-necked giants and the **theropods,** which included the ancestors of birds.

Scientists continue to debate whether dinosaurs were ectothermic or endothermic. The ancestor of birds was certainly endothermic. Evidence indicates that many dinosaurs were active, agile, and, in some species, social. At the end of the Cretaceous, all dinosaurs (except birds) became extinct.

The lizardlike tuatara represents one lineage of lepidosaurs, surviving only on islands off New Zealand. The other living lineage of lepidosaurs are the squamates. Lizards are the most numerous and diverse reptiles (aside from birds). Snakes are limbless, although some species have vestigial pelvic and limb bones. Snakes are carnivorous, with adaptations for locating (heat-detecting and olfactory organs), killing (sharp teeth, sometimes that inject toxins), and swallowing (loosely articulated jaws and elastic skin) their prey.

The earliest turtle fossils, from 220 mya, already had fully developed shells, so the origin of the turtle shell has not been established. This protective, usually hard shell has an upper and lower shield fused to the vertebrae, clavicles, and ribs. Molecular data suggest that turtles are related to crocodiles and Triassic archosaurs that had bony plates along their backbones.

Crocodiles and alligators, collectively called crocodilians, belong to an archosaur lineage. Living in warm regions, they are adapted to aquatic habitats.

INTERACTIVE QUESTION   34.9

The two main lineages of diapsids are archosaurs and lepidosaurs. List several representatives, both living and extinct, of these two lineages.

Birds are archosaurs, whose reptilian features have been modified as they became adapted to flight. Many of the derived characters of birds are adaptations that reduce weight. Feathers made of β-keratin are extremely light and strong, and shape the wing into an airfoil. Large pectoral muscles attached to a keel on the sternum flap the aerodynamic wings. The evolution of strong flying ability enhanced hunting, foraging, escape from predators, and migration to favorable habitats.

Birds are endothermic; feathers and a fat layer in some species help retain metabolic heat. The four-chambered heart and efficient respiratory system facilitate a high metabolic rate.

Birds have excellent vision. Their relatively large brains permit detailed visual processing and motor coordination. Birds display complex behaviors, particularly in courtship rituals and parenting. Eggs are brooded during development by the female and/or male bird.

Cladistic analysis indicates that birds are theropods, a group of bipedal saurischians. Feathered, but flightless theropod fossils have recently been found in Chinese sediments. Several species had feathers with vanes, others had filamentous feathers. Functions of early feathers may have been insulation, camouflage, and courtship displays.

Flight may have evolved as ground-running dinosaurs used feathers to gain extra lift, as small dinosaurs flapped their feathered fore limbs running up hills, or as tree-climbing dinosaurs used feathers to glide.

Feathered theropods had evolved into birds by 150 million years ago. *Archaeopteryx,* the oldest bird known, had feathered wings with clawed digits, teeth, and a long tail. Fossils from the Cretaceous period show a gradual loss of dinosaur features and acquisition of bird innovations.

The clade Neornithes, which includes the 28 orders of living birds, appeared about 65.5 million years ago. Most birds can fly. The **ratites** include the ostrich, kiwi, and emu. These flightless birds lack a keeled breastbone and enlarged pectoral muscles. The general body form of many flying birds is similar. The beak has assumed a great variety of shapes associated with different diets. Foot structure also shows variation.

INTERACTIVE QUESTION   34.10

List several adaptations for flight found in birds.

34.7 Mammals are amniotes that have hair and produce milk

Derived Characters of Mammals   **Mammals** are the other amniote lineage. Their derived characters include mammary glands, which produce milk to nourish the young, and hair, which (along with a layer of fat under the skin) helps to insulate these endothermic animals. Active metabolism is provided for by an efficient respiratory system that uses a diaphragm to help ventilate the lungs and a circulatory system with a four-chambered heart.

Mammals generally have large brains and many are capable of learning. Extended parental care of the young provides time to observe and learn survival skills.

Mammalian teeth come in a diverse assortment of shapes and sizes that are specialized for eating a variety of foods.

Early Evolution of Mammals   Mammals are part of the amniote group known as **synapsids,** distinguished by the temporal fenestra, a hole behind each eye socket. During the evolution of mammals from non-mammalian synapsids, the jaw was remodeled, with a stronger, single-bone lower jaw and the incorporation of two former jaw-joint bones into the middle ear.

Synapsids were the dominant tetrapods during a time in the Permian period, but many became extinct at the end of that period. Mammal-like synapsids emerged 200 million years ago at the end of the Triassic.

The first true mammals arose and diversified during the Jurassic period. The still small Mesozoic mammals, which were probably nocturnal and insectivorous, co-existed with dinosaurs. The three major lineages of living mammals emerged by the early Cretaceous. Mammals underwent an extensive radiation in the wake of the Cretaceous extinctions.

Monotremes   The platypus and echidnas (spiny anteaters) are the only extant **monotremes,** egg-laying mammals. Although egg-laying is a primitive character for amniotes, monotremes have hair and produce milk for their young. Monotremes appear to have diverged from other mammals about 180 million years ago.

Marsupials   Marsupials, including opossums, kangaroos, and koalas, share derived characters with eutherians including a higher metabolic rate, nipples that provide milk, and giving birth to live young. The embryo develops in the uterus. A **placenta,** formed from the uterine lining and extraembryonic membranes of the embryo, nourishes the developing embryo. A marsupial is born very early in development and completes its embryonic development attached to a nipple in a maternal pouch called a marsupium.

Australian marsupials have radiated and filled the niches occupied by eutherian mammals in other parts of the world. After the breakup of Pangaea, South America and Australia became island continents where marsupials diversified in isolation from eutherians. Eutherians reached South America in several migrations, whereas Australia has remained isolated since about 65 mya.

Eutherians (Placental Mammals)   **Eutherians** complete development attached to a placenta within the maternal uterus. Eutherians and marsupials may have diverged about 140 million years ago.

Molecular systematics is helping to establish the phylogeny of eutherian orders, although there is still no consensus. There appear to be at least four main clades. One clade evolved in Africa and includes the elephants, manatees, hyraxes, and aardvarks. A second clade radiated in South America and includes the sloths, anteaters, and armadillos. The third and largest clade includes rabbits, rodents, and primates. The fourth diverse clade includes many groups: carnivores, hoofed herbivores, bats, some shrews and moles, and cetaceans (dolphins and whales).

Primates include the lemurs, tarsiers, monkeys, and apes (including humans). Derived characters include grasping hands and feet, and flat nails on their digits. Compared to other mammals, primates have a large brain, short jaws, and forward-looking eyes with overlapping visual fields that enhance depth perception. They also exhibit complex social behavior and extensive parental care. Many of these traits were shaped by natural selection in the tree-dwelling early primates. All primates, except humans, have a wide separation between the big toe and the other toes, allowing them to grasp branches with their feet. Monkeys and apes have an **opposable thumb,** which functions in a grasping "power grip" in monkeys and apes other than humans, but is adapted for precise manipulation in humans.

There are three main groups of living primates: the lemurs, lorises, and pottos; the tarsiers; and the anthropoids. The lemurs, lorises, and pottos probably resemble early arboreal primates. The **anthropoids** include monkeys and apes. The oldest known anthropoid fossils are 45 million years old and point to tarsiers as the closest relative. Monkeys appeared in the New World during the Oligocene, after first evolving in the Old World. New World monkeys are strictly arboreal. Most monkeys in both groups are diurnal and live in social bands.

The primates informally called apes include gibbons, orangutans, gorillas, chimpanzees and bonobos, and humans. Apes diverged from Old World monkeys about 20–25 million years ago. Nonhuman apes live in the Old World tropics. All living apes have relatively long arms, short legs, and no tails. Apes have a larger brain and their behavior is more flexible than that of other primates. Gorillas and chimpanzees are highly social.

INTERACTIVE QUESTION   34.11

a. List the derived characteristics of mammals.

b. Go back to the phylogenetic tree in Interactive Question 34.3 and make sure you have filled in all the information.

34.8 Humans are mammals that have a large brain and bipedal locomotion

Derived Characters of Humans The species *Homo sapiens* is about 200,000 years old. Characters that distinguish humans from other apes are upright stance and bipedal locomotion, larger brain with the capacity for language and symbolic thought, and the manufacture and use of complex tools. Humans also have reduced jaws and a shorter digestive tract.

Although the genomes of humans and chimpanzees are 99% identical, recent studies show that they differ in the expression of regulatory genes, which may account for the many differences between the two groups.

The Earliest Hominins **Paleoanthropology** is the study of human origins. Approximately 20 extinct species of **hominins,** species that are more closely related to humans than to chimpanzees, have been identified. The oldest of these, *Sahelanthropus*, lived about 6 to 7 million years ago. These early hominins shared some derived characters of humans, such as reduced canine teeth, relatively flat faces, and upright stance, as indicated by the location of the foramen magnum underneath the skull. Different human features have evolved at different rates. Thus, for example, upright posture predates an enlarged brain.

These early hominins and chimpanzees represent divergent branches from a common ancestor. Human evolution has not occurred within an unbranched hominin line; there have been times when several different hominin species coexisted.

Australopiths Hominin diversity increased between 4 and 2 million years ago, and many of the hominins from this period are called australopiths, although their phylogeny is unresolved. The name came from the 1924 discovery of *Australopithecus africanus*. It appears that *A. africanus* lived between 3 and 2.4 million years ago and was an upright hominin, with human-like hands and teeth but a small brain.

In 1974, a 3.2-million-year-old *Australopithecus* skeleton was discovered in the Afar region of Ethiopia. "Lucy" and similar fossils have been designated *Australopithecus afarensis,* a species that appears to have existed for about 1 million years. Although bipedal, *A. afarensis* had a brain size similar to that of a chimpanzee, a long lower jaw, and long arms.

Another lineage from that time was the "robust" australopiths, such as *Paranthropus boisei,* which had sturdy skulls with powerful jaws and teeth.

Bipedalism By about 10 million years ago, the Himalayan range had formed, the climate became drier, and the forests of Africa and Asia contracted, presenting an increased savannah habitat that may have influenced hominin evolution. But all recently discovered fossils of early hominins show signs of bipedalism, and these hominins lived in mixed forests and open woodlands. About 1.9 million years ago hominins that lived in more arid environments began to walk long distances on two legs.

Tool Use The first evidence of complex tool use is 2.5-million-year-old cuts on animal bones, suggesting the use of stone tools. *Australopithecus garhi,* whose fossils were found nearby, had a relatively small brain.

Early Homo Larger-brained fossils dating from 2.4 to 1.5 million years old are the first to be placed in the genus *Homo*. Sometimes found with sharp stone tools, *Homo habilis*, or "handy man," had some derived hominin characters. Fossils from 1.9 to 1.6 years ago are recognized by a number of paleoanthropologists as the species *Homo ergaster*, which had a larger brain, long legs and hip joints adapted for long-distance walking, relatively short and straight fingers, smaller teeth, and more sophisticated stone tools.

The sexual dimorphism of size difference between the sexes was also less in *Homo ergaster* than in *Australopithecus*, a trend that continued in humans, with males now only about 1.2 times the weight of females. In living primates, reduced sexual dimorphism is associated with more pair-bonding. Thus, *H. ergaster* may have engaged in pair-bonding, perhaps associated with long-term biparental care of babies.

Homo erectus was the first hominin to migrate out of Africa. The oldest fossils of hominins outside of Africa date back 1.8 million years. Most paleoanthropologists conclude that *H. erectus* became extinct sometime after 200,000 years ago.

Neanderthals The Neanderthals appear to be descendants of *Homo heidelbergensis*, which originated about 600,000 years ago in Africa. Neanderthals lived in Europe and the Near East from about 200,000 to 28,000 years ago, at which point they became extinct.

They had large brains and made tools from stone and wood. Comparisons of mtDNA from two Neanderthal fossils with DNA of living humans show that Neanderthals form a clade distinct from the clade of all living humans.

Homo sapiens Evidence indicates that the ancestors of humans originated in Africa. The oldest known fossils of *H. sapiens* date from 195,000 years ago. These early humans lacked heavy brow ridges and were more slender than other hominins. DNA comparisons show that Europeans and Asians share a relatively recent common ancestor, with many African lineages branching off earlier. Comparisons of mitochondrial DNA and Y chromosomes of various populations support a common African ancestor of all *H. sapiens*.

The oldest fossils outside Africa come from the Middle East and date back about 115,000 years. One or more waves of humans appear to have first spread into Asia, then to Europe and Australia. The date of arrival in the New World may be 15,000 years ago. The 2004 discovery of a small hominin, named *Homo floresiensis*, illustrates how new information contributes to the ongoing inquiry into human origins.

The rapid expansion of our species may be tied to the evolution of human cognition. Symbolic thought may have emerged along with human language, enabling the construction of new tools and the ability to teach others to build them.

The gene *FOXP2* has been identified as essential for human language. Comparisons with homologous genes in other mammals indicate that the gene underwent intense natural selection within the past 200,000 years.

INTERACTIVE QUESTION 34.12

List some of the derived characters of humans.

Word Roots

arch- = ancient (*archosaurs:* the reptilian group that includes crocodiles, alligators, dinosaurs, and birds)

aktin- = a ray; **-pterygi** = a fin (*Actinopterygii:* the class of ray-finned fishes)

cephalo- = head (*cephalochordates:* a chordate without a backbone, represented by lancelets)

crani- = the skull (*craniata:* the chordate clade that possess a cranium)

crocodil- = a crocodile (*Crocodilia:* the reptile group that includes crocodiles and alligators)

di- = two (*diapsids:* a group of amniotes distinguished by a pair of holes on each side of the skull)

dino- = terrible; **-saur** = lizard (*dinosaurs:* an extremely diverse group of ancient reptiles varying in body shape, size, and habitat)

endo- = inner; **-therm** = heat (*endotherm:* an animal that uses metabolic energy to maintain a constant body temperature, such as a bird or mammal)

eu- = good (*eutherians:* placental mammals; those whose young complete their embryonic development within the uterus, joined to the mother by the placenta)

extra- = outside, more (*extraembryonic membranes:* four membranes that support the developing embryo in reptiles and mammals)

gnantho- = the jaw; **-stoma** = the mouth (*gnathostomes:* the vertebrate clade that possesses jaws)

homin- = man (*hominin:* a term that refers to mammals that are more closely related to humans than to any other living species)

lepido- = a scale (*lepidosaurs:* the reptilian group that includes lizards, snakes, and tuatara)

marsupi- = a bag, pouch (*marsupial:* a mammal, such as a koala, kangaroo, or opossum, whose young complete their embryonic development inside a maternal pouch called the marsupium)

mono- = one (*monotremes:* an egg-laying mammal, represented by the platypus and echidna)

neuro- = nerve (*neural crest:* a band of cells along the border where the neural tube pinches off from the ectoderm)

noto- = the back; **-chord** = a string (*notochord:* a longitudinal, flexible rod formed from dorsal mesoderm and located between the gut and the nerve cord in all chordate embryos)

opercul- = a covering, lid (*operculum:* a protective flap that covers the gills of fishes)

osteo- = bone; **-ichthy** = fish (*Osteichthyans:* the vertebrate clade that includes the ray-finned fishes and lobe-fins)

ovi- = an egg; **-parous** = bearing (*oviparous:* referring to a type of development in which young hatch from eggs laid outside the mother's body)

paedo- = a child; **-genic** = producing (*paedogenesis:* the precocious development of sexual maturity in a larva)

paleo- = ancient; **anthrop-** = man; **-ology** = the science of (*paleoanthropology:* the study of human origins and evolution)

placo- = a plate (*placoderm:* a member of an extinct group of gnathostomes that had jaws and were enclosed in a tough, outer armor)

ptero- = a wing (*pterosaurs:* winged reptiles that lived during the time of dinosaurs)

ratit- = flat-bottomed (*ratites:* the group of flightless birds)

soma- = body (*somites:* blocks of mesoderm that give rise to muscle segments in chordates)

syn- = together (*synapsids:* an amniote group distinguished by a single hole behind each eye socket)

tetra- = four; **-podi** = foot (*tetrapod:* a terrestrial lobe-fin, possessing two pairs of limbs, such as amphibians, reptiles, and mammals)

tunic- = a covering (*tunicates:* members of the subphylum Urochordata)

uro- = tail (*urochordate:* a chordate without a backbone, commonly called a tunicate)

uro- = (note) tail; **-del** = visible (*Urodela:* the order of salamanders that includes amphibians with tails)

vivi- = alive (*ovoviviparous:* referring to a type of development in which young hatch from eggs that are retained in the mother's uterus)

Structure Your Knowledge

1. Review the phylogenetic tree on page 251. Make note of the derived characters that define each clade.

2. Describe several examples from vertebrate evolution that illustrate the common evolutionary theme that new adaptations usually evolve from preexisting structures.

3. How do various vertebrate groups meet the challenges of a terrestrial habitat?

Test Your Knowledge

FILL IN THE BLANKS

_____ 1. invertebrate chordate group most closely related to craniates

_____ 2. chordate group that includes lancelets

_____ 3. blocks of mesoderm along notochord that develop into muscles

_____ 4. clade of jawed vertebrates

_____ 5. flap over the gills of bony fishes

_____ 6. adaptation that allows reptiles to reproduce on land

_____ 7. egg-laying mammals

_____ 8. structure that helps mammals ventilate their lungs

_____ 9. group of primates that includes monkeys and apes

_____ 10. genus in which the fossil Lucy is placed

MULTIPLE CHOICE: *Choose the one best answer.*

1. Pharyngeal slits appear to have functioned first as
   a. suspension-feeding devices.
   b. gill slits for respiration.
   c. components of the jaw.
   d. portions of the inner ear.
   e. mouth openings.

2. Which of the following is *not* a derived character of craniates?
   a. cranium or skull
   b. neural crest in embryonic development
   c. a mineralized endoskeleton
   d. heart with at least two chambers
   e. cephalization with sensory organs

3. Which of these represents the oldest lineage of vertebrates?
   a. caecilians
   b. sharks and rays
   c. lancelet
   d. hagfishes
   e. lampreys

4. Which of the following is in the lobe-fin clade?
   a. lampreys
   b. sharks and rays
   c. ray-finned fishes
   d. hagfishes
   e. tetrapods

5. Which of the following is *not* in the same lineage as the others?
   a. lizards
   b. birds
   c. dinosaurs
   d. crocodilians
   e. pterosaurs

6. Which of the following is *incorrectly* paired with its gas exchange mechanism?
   a. amphibians—skin and lungs
   b. lungfishes—gills and lungs
   c. reptiles—lungs
   d. bony fishes—swim bladder
   e. mammals—lungs with diaphragm to ventilate

7. Non-bird reptiles have lower caloric needs than do mammals of comparable size because they
   a. are ectotherms.
   b. have waterproof scales.
   c. have a longer digestive tract and obtain more nutrients from their food.
   d. move by bending their vertebral column back and forth.
   e. have a more efficient respiratory system.

8. Which of the following best describes the earliest mammals?
   a. large, herbivorous
   b. large, carnivorous
   c. small, insectivorous
   d. small, herbivorous
   e. small, carnivorous

9. Oviparity is a reproductive strategy that
   a. allows mammals to bear well-developed young.
   b. is used by both reptiles and some sharks.
   c. is necessary for vertebrates to reproduce on land.
   d. is a necessity for all flying vertebrates.
   e. protects the embryo inside the mother and uses the food resources of the egg.

10. In Australia, marsupials fill the niches that eutherians (placental mammals) fill in other parts of the world because
    a. they are better adapted and have outcompeted eutherians.
    b. their offspring complete their development attached to a nipple in a marsupium.
    c. they originated in Australia.
    d. they evolved from monotremes that migrated to Australia about 65 million years ago.
    e. after Pangaea broke up, they diversified in isolation from eutherians.

11. Which of the following is closely associated with language development?
    a. four clusters of *Hox* genes
    b. gene duplication involving *Dlx* genes
    c. reduced sexual dimorphism and postnatal care
    d. *FOXP2* gene
    e. shorter hinged jaws and a hole in the skull behind each eye socket through which jaw muscles pass

12. Which of the following is an *incorrect* statement concerning human evolution?
    a. *Homo sapiens* is the only surviving member of a highly branched evolutionary tree.
    b. Erect posture preceded the enlargement of the brain in human evolution.
    c. The key differences between chimpanzees and humans appear to involve duplications of *Hox* genes.
    d. *H. sapiens* first evolved in Africa and later dispersed to other regions.
    e. The rapid expansion of *H. sapiens* may be tied to the evolution of human cognition, including symbolic thought, language, and complex tool construction.

UNIT 6 — Plant Form and Function

Chapter 35

Plant Structure, Growth, and Development

Key Concepts

35.1 The plant body has a hierarchy of organs, tissues, and cells

35.2 Meristems generate cells for new organs

35.3 Primary growth lengthens roots and shoots

35.4 Secondary growth adds girth to stems and roots in woody plants

35.5 Growth, morphogenesis, and differentiation produce the plant body

Framework

A rather large new vocabulary is needed to name the specialized cells and structures in a study of plant structure and growth. Focus your attention on how the roots, stems, and leaves of a plant are specialized to function in absorption, support, transport, protection, and photosynthesis. The plant body is composed of dermal, vascular, and ground tissue systems. Apical meristems at the tips of roots and shoots create primary growth. The lateral meristems—vascular cambium and cork cambium—create secondary growth that adds girth to stems and roots. New techniques and model systems such as *Arabidopsis* are allowing researchers to explore the molecular bases for plant growth, morphogenesis, and cellular differentiation.

Chapter Review

Plant development and structure are controlled by both genetic and environmental factors. Due to their **developmental plasticity,** plants can respond to their local environments, and, thus, no two plants are alike. Evolutionary adaptation to terrestrial environments has shaped distinctive plant forms or **morphologies.**

35.1 The plant body has a hierarchy of organs, tissues, and cells

A plant **organ** performs a specific function and is composed of several types of **tissues,** groups of cells with a common form and function.

The Three Basic Plant Organs: Roots, Stems, and Leaves As an evolutionary adaptation to the dispersed resources in a terrestrial environment, vascular plants have an underground **root system** for obtaining water and minerals from the soil and an aerial **shoot system** of stems and leaves for absorbing light and carbon dioxide for photosynthesis. The shoot system includes vegetative shoots and reproductive shoots, which, in angiosperms, are flowers.

The functions of **roots** include anchorage, absorption, and storage of food. A *taproot system* is found commonly in eudicots and gymnosperms. The one main deep root often stores nutrients and gives rise to **lateral roots.** Monocots and seedless vascular plants generally have a more shallow *fibrous root system* in which many small roots grow from the stem. Such roots are called *adventitious,* as they grow in an unusual location.

Most absorption of water and minerals occurs through tiny **root hairs,** extensions of epidermal cells that are clustered near root tips. Modified roots may serve various functions, including support, storage, and oxygen absorption.

A **stem** consists of alternating **nodes,** the points at which leaves are attached, and segments between nodes, called **internodes.** An **axillary bud** is found in the angle (axil) between a leaf and the stem. An **apical bud,** or terminal bud, with developing leaves and compacted internodes is found at the tip or apex of a shoot. The apical bud may exhibit **apical dominance,** inhibiting the growth of axillary buds and producing a taller plant. Axillary buds may develop into lateral shoots, or branches, with their own apical buds, leaves, and axillary buds.

Modifications of shoots include stolons, rhizomes, tubers, and bulbs. These structures may function in asexual reproduction and food storage.

Leaves, the main photosynthetic organs of most plants, usually consist of a flattened **blade** and a **petiole,** or stalk. Many monocots lack petioles. Monocot leaves usually have parallel major **veins** (vascular tissue of leaves), whereas eudicot leaves have networks of branched veins. Leaves may be simple or compound. Modified leaves may function in support, storage, and reproduction.

INTERACTIVE QUESTION   35.1

Label the parts in this diagram of a flowering plant.

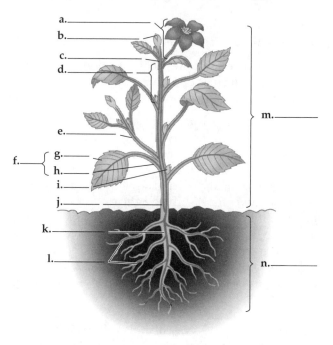

a.
b.
c.
d.
m.
e.
f.
g.
h.
i.
j.
k.
l.
n.

Dermal, Vascular, and Ground Tissues Tissue systems are functional units of tissues that are continuous throughout the plant but have specific characteristics in each plant organ (root, stem, and leaf).

The **dermal tissue system** forms a protective outer layer. The **epidermis** is a single layer of cells that covers nonwoody plants. The epidermis of leaves and most stems is covered with a **cuticle,** a waxy coating that prevents excess water loss. Older stems and roots of woody plants are covered by a protective **periderm.** *Trichomes* are outgrowths of the shoot epidermis that may protect against insects and disease or reduce water loss.

The **vascular tissue system** consists of **xylem** and **phloem** and functions in long-distance transport. In angiosperms, the **stele,** or vascular tissue of a root or stem, is arranged as a solid central *vascular cylinder* in roots, but as *vascular bundles* in stems and leaves.

The **ground tissue system** contains cells that function in photosynthesis, support, and storage. Ground tissue internal to vascular tissue is called **pith;** outside the vascular tissue it is called **cortex.**

Common Types of Plant Cells **Parenchyma cells** carry on most of a plant's metabolic functions, such as photosynthesis and food storage. They usually lack secondary walls and have large central vacuoles. Most retain the ability to divide and differentiate into other types of plant cells.

Collenchyma cells lack secondary walls but have thickened primary walls. Strands or cylinders of these cells function in flexible support for young parts of the plant and elongate along with the plant.

Sclerenchyma cells have thick secondary walls strengthened with lignin. These specialized supporting cells often lose their cell contents at maturity. **Fibers** are long, tapered cells that usually occur in threads. **Sclereids** are shorter and irregular in shape, with very thick, lignified cell walls.

The water-conducting cells of xylem die at functional maturity, leaving behind their secondary walls, which are interrupted by pits. **Tracheids,** found in all vascular plants, are long, thin, tapered cells with lignin-strengthened walls. Water passes through pits from cell to cell. **Vessel elements,** found only in angiosperms, are wider, shorter, and thinner walled, with perforations in their end walls. They align to form long tubes known as **vessels.**

In the phloem of angiosperms, sugars flow through chains of cells called **sieve-tube elements,** or sieve-tube members, which remain alive at functional maturity but lack nuclei, ribosomes, and vacuoles. Fluid flows through pores in the **sieve plates** in the end walls between cells. The nucleus and ribosomes of an adjacent **companion cell,** which is connected to a sieve-tube element by numerous plasmodesmata, serve both cells.

INTERACTIVE QUESTION 35.2

a. Which types of plant cells are dead at functional maturity?

b. Which types of plant cells lack nuclei at functional maturity?

35.2 Meristems generate cells for new organs

Most plants exhibit **indeterminate growth,** continuing to grow as long as they live. Plant organs such as leaves and flowers, as well as animals, have **determinate growth** and stop growing after reaching a certain size. Plants may be **annuals,** which complete their life cycle in a year or less; **biennials,** which have a life cycle spanning two years; or **perennials,** which live many years.

Plants have tissues called **meristems,** which remain embryonic and perpetually divide to form new cells. **Apical meristems,** located at the tips of roots, shoots, and in the axillary buds of shoots, produce **primary growth,** resulting in elongation. Herbaceous (nonwoody) plants usually have only primary growth. **Secondary growth** is an increase in diameter as new cells are produced by **lateral meristems. Vascular cambium** produces secondary xylem and phloem; **cork cambium** produces the periderm. Cells that remain to divide in a meristem are called *initials,* whereas cells that are displaced from the meristem and become specialized in developing tissues are called *derivatives.*

INTERACTIVE QUESTION 35.3

Describe the types of growth that occur in woody plants.

35.3 Primary growth lengthens roots and shoots

The **primary plant body,** produced by primary growth, is the entire herbaceous plant and the youngest parts of a woody plant.

Primary Growth of Roots The apical meristem of the root tip is protected by a **root cap,** which secretes a lubricating polysaccharide slime. The *zone of cell division* includes the apical meristem and its derivatives. In the *zone of elongation,* cells lengthen to many times their original size, which pushes the root tip through the soil. Cells specialize in structure and function as the zone of elongation grades into the *zone of differentiation* (or maturation).

In most roots, xylem cells radiate from the center in spokes, with phloem in between. The vascular tissue of a monocot may have pith, a central core of undifferentiated parenchyma cells, inside a ring of xylem and phloem.

The ground tissue consists of parenchyma cells in the cortex. The innermost layer of cortex is the one-cell-thick **endodermis,** which regulates the passage of materials into the vascular cylinder.

Lateral roots may develop from the **pericycle,** the outer layer of cells in the vascular cylinder. A lateral root pushes through the cortex and retains its connection with the vascular cylinder.

INTERACTIVE QUESTION 35.4

Label the tissues in the cross sections of a eudicot root and its vascular cylinder. Identify the functions of these tissues.

a._____
b._____
c._____

500 μm

d._____
e._____
f._____
g._____

50 μm

Primary Growth of Shoots The dome-shaped mass at the tip of the apical bud is the shoot apical meristem. **Leaf primordia** form on the sides of the apical meristem, and axillary buds develop from clumps of meristematic cells left at the bases of the leaf primordia.

Elongation of the shoot occurs by cell elongation within young internodes. In grasses and some other plants, the base of the leaves and stem internodes have *intercalary meristems* that enable shoots to continue elongation.

The epidermis of the dermal tissue system covers stems. In gymnosperms and most eudicots, vascular bundles may be arranged in a ring, with the ground tissues pith inside and cortex outside the ring. Xylem is located internal to the phloem in the vascular bundles. In most monocot stems, the vascular bundles are scattered throughout the ground tissue. A layer of collenchyma just beneath the epidermis and sclerenchyma fiber cells may strengthen the stem.

Stomata, tiny pores flanked by **guard cells** in the cuticle-covered epidermis, permit both gas exchange and the evaporation of water from the leaf.

Mesophyll consists of parenchyma cells containing chloroplasts. In many eudicot leaves, columnar *palisade mesophyll* is located above *spongy mesophyll*, which has loosely packed cells surrounding many air spaces.

A *leaf trace*, which is a branch of a vascular bundle in the stem, continues into the petiole and divides repeatedly within the blade of the leaf, providing support and vascular tissue to the photosynthetic mesophyll. Each vein is enclosed in a ring of protective cells called a *bundle sheath*.

INTERACTIVE QUESTION   35.5

Name the indicated structures in this diagram of a leaf.

a._____

b._____

c._____

d._____

e._____

f._____

g._____

h._____

i._____

35.4 Secondary growth adds girth to stems and roots in woody plants

The vascular cambium and cork cambium produce the **secondary plant body.** Secondary growth occurs in all gymnosperms and many eudicots but is rare in monocots.

The Vascular Cambium and Secondary Vascular Tissue Vascular cambium forms a continuous cylinder of meristematic cells that produce secondary xylem to the inside and secondary phloem to the outside. Some cambial initials produce radial lines of parenchyma cells called vascular rays, which function in storage and lateral transport of water and nutrients.

Wood is the accumulation of secondary xylem cells with thick, lignified walls. Annual growth rings result from the seasonal cycle of cambium dormancy, early wood production (usually in the spring), and late wood production (in the summer) in temperate regions. Scientists analyzing tree rings (*dendrochronology*) of Mongolian conifers from over the past five centuries found evidence of warming temperatures during the 1900s.

In older trees, a central column of *heartwood* consists of older xylem with resin-filled cell cavities; the *sapwood* consists of secondary xylem that still functions in transport.

Older secondary phloem is sloughed off as bark; only that closest to the vascular cambium functions in sugar transport.

The Cork Cambium and the Production of Periderm
The epidermis splits off during secondary growth and is replaced by new protective tissues produced by the cork cambium, a meristematic cylinder that first forms in the outer cortex in stems and the outer pericycle in roots. The cork cambium produces a thin-layered *phelloderm* to the inside and cork cells to the exterior, which develop suberin-impregnated walls. The protective coat formed by a cork cambium and its tissues is a layer of periderm. As secondary growth continually splits the outer layers of periderm, new cork cambia develop, eventually forming from parenchyma cells in the secondary phloem. **Lenticels** are spaces between cork cells through which gas exchange occurs. **Bark** refers to phloem and periderm.

INTERACTIVE QUESTION 35.6

Starting from the outside, place the letters of the tissues in the order in which they are located in a woody tree trunk.

A. primary phloem	**E.** pith
B. secondary phloem	**F.** cork cambium
C. primary xylem	**G.** vascular cambium
D. secondary xylem	**H.** cork

_____ _____ _____ _____ _____ _____ _____ _____

35.5 Growth, morphogenesis, and differentiation produce the plant body

Growth results from cell division and expansion. The organization of cells to create body form is called **morphogenesis.** The third developmental process, differentiation, creates different types of cells.

Molecular Biology: Revolutionizing the Study of Plants Modern plant biology is using new methods and new research organisms, such as *Arabidopsis thaliana*, to study the genetic control of plant development. The entire genome of *Arabidopsis* has been sequenced, and researchers are working to determine the functions of all of its genes. A major goal of **systems biology** is to track which genes are activated and where during a plant's development.

Growth: Cell Division and Cell Expansion The plane and symmetry of plant cell division are important in determining form. In *asymmetrical cell division*, one daughter cell receives more cytoplasm than the other, often leading to a key developmental event.

The plane of cell division is determined by the **preprophase band,** a ring of cytoskeletal microtubules that forms during late interphase and leaves behind an ordered array of actin microfilaments.

About 90% of plant cell growth is due to the uptake of water into vacuoles. This economical means of cell elongation produces the rapid growth of shoots and roots. When enzymes break the cross-links in the cell wall, the restraint on the turgid cell is reduced and water enters the cell by osmosis. Cells expand in a direction perpendicular to the orientation of the cellulose microfibrils in the inner layers of the cell wall. This orientation of microfibrils was determined by the orientation of microtubules in the outer cytoplasm.

In the squat-bodied *fass* mutants of *Arabidopsis*, preprophase bands are not formed in mitosis and the random distribution of microtubules leads to the random arrangement of cellulose microfibrils. The defects in microtubule organization result in cells that divide in a random arrangement and expand equally in all directions.

INTERACTIVE QUESTION 35.7

Review the role of microtubules in the orientation of plant cell division and expansion.

Morphogenesis and Pattern Formation **Pattern formation** is the characteristic development of structures in specific locations. Pattern formation may depend on **positional information,** signals—such as hormones, proteins, or mRNAs—that indicate a cell's location within a developing organ and direct its growth and differentiation.

Plants typically have an axial **polarity** with a root end and a shoot end. The asymmetric first division of the zygote establishes the shoot–root polarity.

Plants have homeotic genes that regulate key developmental events. In many plants, the *KNOTTED-1* homeotic gene influences leaf morphology. Its overexpression in tomato plants results in "super-compound" leaves.

Gene Expression and Control of Cellular Differentiation Differential gene expression within genetically identical cells leads to their differentiation into the diverse cell types in a plant. In the root epidermis of *Arabidopsis*, the homeotic gene *GLABRA-2* is expressed in epidermal cells that are in contact with only one underlying cortical cell, and these cells do not develop root hairs. The gene is not expressed in those cells in contact with two cortical cells, and they differentiate into root hair cells.

Location and a Cell's Developmental Fate Using *clonal analysis*, researchers are able to identify cells in the apical meristem and follow their lineages to pinpoint when the developmental fate of a cell becomes determined. Apparently a cell's final position in a developing organ determines what type of cell it will become.

Shifts in Development: Phase Changes In a process known as a **phase change,** the apical meristem can switch from one developmental phase to another, such as from producing juvenile vegetative nodes and internodes to laying down mature nodes. A node's developmental phase is fixed when it is produced, and thus both juvenile and mature leaves can coexist along the shoots of a plant.

Genetic Control of Flowering In the transition from a vegetative shoot to a reproductive shoot (flower), **meristem identity genes** are expressed that code for transcription factors that activate the genes necessary for floral meristem development.

The four floral organs develop from floral primordia in four concentric whorls: sepals, petals, stamens, and carpels. Some of the **organ identity genes** that code for transcription factors that regulate this characteristic floral pattern have been identified, and their expression seems to be influenced by positional information. Mutations of organ-identity genes result in the placement of one type of floral organ where another type would normally develop. The **ABC model** of flower formation proposes three classes of organ identity genes, each of which affects two adjacent whorls. This model explains the phenotypes of mutants lacking *A*, *B*, or *C* gene expression, with the additional observation that activity of the *A* or *C* gene inhibits the expression of the other. If either *A* or *C* is nonfunctional, the other gene is expressed in its place.

INTERACTIVE QUESTION   35.8

This diagram of the ABC model shows the active genes in each whorl and the resulting anatomy of a wild-type flower.

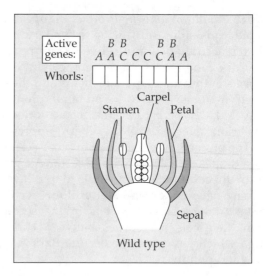

Wild type

a. Fill in the following table to show which organs are produced in the whorls in a normal flower. In a mutant that lacks a functional gene *A*, what gene expression pattern and resulting flower organ arrangement would be produced? (Remember that a lack of *A* activity removes the inhibition of gene *C*.)

Whorl	Genes Active	Organs in Normal	Genes Active in Mutant *A*	Organs in Mutant *A*
1	*A*			
2	*AB*			
3	*BC*			
4	*C*			

b. If you had a double-mutant plant that had no gene activity for *B* or *C*, what would the resulting flower look like?

Word Roots

apic- = the tip; **meristo-** = divided (*apical meristems:* embryonic plant tissue on the tips of roots and in the buds of shoots that supplies cells for the plant to grow)

bienn- = every 2 years (*biennial:* a plant that requires two years to complete its life cycle)

coll- = glue; **-enchyma** = an infusion (*collenchyma cell:* a flexible plant cell type that occurs in strands or cylinders that support young parts of the plant without restraining growth)

endo- = inner; **derm-** = skin (*endodermis:* the innermost layer of the cortex in plant roots)

epi- = over (*epidermis:* the dermal tissue system in plants; the outer covering of animals)

inter- = between (*internode:* the segment of a plant stem between the points where leaves are attached)

meso- = middle; **-phyll** = a leaf (*mesophyll:* the ground tissue of a leaf, sandwiched between the upper and lower epidermis and specialized for photo-synthesis)

morpho- = form; **-genesis** = origin (*morphogenesis:* the development of body shape and organization)

perenni- = through the year (*perennial:* a plant that lives for many years)

peri- = around; **-cycle** = a circle (*pericycle:* a layer of cells just inside the endodermis of a root that may become meristematic and begin dividing again)

phloe- = the bark of a tree (*phloem:* the portion of the vascular system in plants consisting of living cells arranged into elongated tubes that transport sugar and other organic nutrients throughout the plant)

sclero- = hard (*sclereid:* a short, irregular sclerenchyma cell in nutshells and seed coats that is scattered through the parenchyma of some plants)

trachei- = the windpipe (*tracheids:* a water-conducting and supportive element of xylem composed of long, thin cells with tapered ends and walls hardened with lignin)

vascula- = a little vessel (*vascular tissue:* plant tissue consisting of cells joined into tubes that transport water and nutrients throughout the plant body)

xyl- = wood (*xylem:* the tube-shaped, nonliving portion of the vascular system in plants that carries water and minerals from the roots to the rest of the plant)

Structure Your Knowledge

1. How does the indeterminate growth pattern of plants lead to developmental plasticity? What is the adaptive advantage of such plasticity?

2. In the following cross section of a young stem, label the indicated structures. Is this a stem of a monocot or a eudicot? How can you tell?

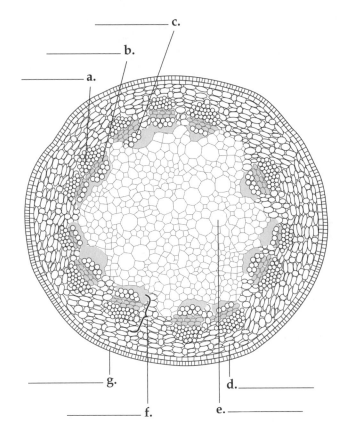

Test Your Knowledge

MATCHING: *Match the plant tissue with its description.*

—— **1.** sclerenchyma

—— **2.** collenchyma

—— **3.** tracheids

—— **4.** fibers

—— **5.** cambial initials

—— **6.** pericycle

—— **7.** mesophyll

—— **8.** periderm

—— **9.** endodermis

—— **10.** pith

A. layer from which lateral roots originate

B. tapered xylem cells with lignin in cell walls

C. parenchyma cells with chloroplasts in leaves

D. protective coat made of cork and cork cambium

E. bundles of long sclerenchyma cells

F. supporting cells with thickened primary walls

G. parenchyma cells inside vascular ring in eudicot stem

H. supporting cells with thick secondary walls

I. cells that produce secondary xylem and phloem

J. cell layer in root regulating movement into vascular cylinder

MULTIPLE CHOICE: *Choose the one best answer.*

1. Which of the following is *incorrect*? Monocots typically have
   a. a taproot rather than a fibrous root system.
   b. leaves with parallel veins rather than branching venation.
   c. no secondary growth.
   d. scattered vascular bundles in the stem rather than bundles in a ring.
   e. pith in the center of the vascular cylinder in the root.

2. Which of the following is *incorrectly* paired with its function?
   a. initials—form radial vascular rays
   b. lenticels—gas exchange in woody stem
   c. root hairs—absorption of water and dissolved minerals
   d. root cap—protects root as it pushes through soil
   e. periderm—meristematic tissue that forms protective layer of cork

3. Axillary buds
   a. may exhibit apical dominance over the apical bud.
   b. form at nodes in the angle where leaves join the stem.
   c. grow out from the pericycle layer.
   d. are formed from intercalary meristems.
   e. only develop into vegetative shoots.

4. A leaf trace is
   a. a petiole.
   b. the outline of the vascular bundles in a leaf.
   c. a vascular bundle that extends into a leaf.
   d. a tiny bulge on the flank of the apical dome that grows into a leaf.
   e. a system of plant identification based on leaf morphology.

5. Which of the following structures of a plant shows determinate growth?
   a. roots
   b. vegetative shoots
   c. adventitious roots
   d. leaves
   e. No parts of a plant show determinate growth.

6. Morphogenesis in plants results from
   a. differences in the plane of cell division and the direction of cell expansion.
   b. migration of cells from the apical meristems to the dermal, vascular, and ground tissue systems.
   c. the differentiation of cells within the apical meristem.
   d. cells moving from the zone of elongation to the zone of differentiation.
   e. the expression of meristem identity genes.

7. Which of the following is *incorrectly* paired with the length of its life cycle?
   a. pine tree—perennial
   b. rose bush—perennial
   c. carrot—biennial
   d. marigold—annual
   e. corn—biennial

8. Secondary xylem and phloem are produced in a root by the
   a. pericycle.
   b. endodermis.
   c. vascular cambium.
   d. apical meristem.
   e. cork cambium.

9. Bark consists of
   a. secondary phloem.
   b. periderm.
   c. cork cells.
   d. cork cambium.
   e. all of the above.

10. Sieve-tube elements
    a. are responsible for lateral transport through a woody stem.
    b. control the activities of phloem cells that have no nuclei or ribosomes.
    c. have spiral thickenings that allow the cell to elongate along with a young shoot.
    d. are transport cells with sieve plates in the end walls between cells.
    e. are tapered water transport cells with pits.

11. Clonal analysis of cells of the shoot apex indicates that
    a. organ-identity gene mutations substitute one body part for another.
    b. a cell's developmental fate is more influenced by position effects than by its meristematic lineage.
    c. cellular differentiation results from regulation of gene expression resulting in production of different proteins in different cells.
    d. all cells were derived from the same parent cell.
    e. the lateral meristem produces the dermal, vascular, and ground tissue systems.

12. In what direction does a plant cell enlarge?
    a. toward the basal end as a result of positional information
    b. parallel to the orientation of the preprophase band of microtubules
    c. perpendicular to the orientation of cellulose microfibrils in the cell wall
    d. in the direction from which water flows into the cell
    e. perpendicular to the internodes

13. What do organ-identity genes code for?
    a. the three tissue systems
    b. pollen and egg cells
    c. transcription factors that control development of floral organs
    d. signals that change a vegetative shoot into a floral meristem
    e. the root and shoot meristems

14. Which of the following is essential to establishing the axial polarity of a plant?
    a. expression of different homeotic genes in the shoot and root meristems
    b. orientation of cortical microtubules perpendicular to the ground
    c. the expression of the *GLABRA-2* gene in the shoot but not the root
    d. the asymmetric first division of the zygote
    e. the proper orientation of microtubules that then orient cellulose microfibrils

Chapter 36

Resource Acquisition and Transport in Vascular Plants

Key Concepts

36.1 Land plants acquire resources from both above and below ground

36.2 Transport occurs by short-distance diffusion or active transport and by long-distance bulk flow

36.3 Water and minerals are transported from roots to shoots

36.4 Stomata help regulate the rate of transpiration

36.5 Sugars are transported from leaves and other sources to sites of use or storage

36.6 The symplasm is highly dynamic

Framework

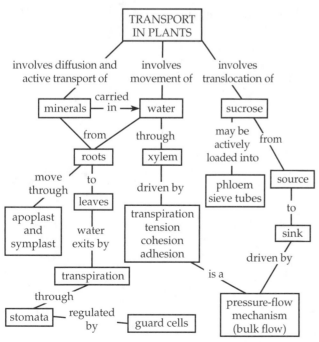

Chapter Review

36.1 Land plants acquire resources from both above and below ground

Adaptations to life on land involved the specialization of roots to absorb water and minerals, shoots to absorb CO_2 and light, and vascular tissue for the transport of materials between roots and shoots.

Shoot Architecture and Light Capture Leaf size is often correlated with water availability. **Phyllotaxy,** the arrangement of leaves on the shoot of a plant, may be one leaf per node (alternate, or spiral), two leaves per node (opposite), or multiple (whorled). Alternate phyllotaxy, with leaves emerging at a light-maximizing 137.5° angle, is most common in angiosperms.

The *leaf area index* is the ratio of the total area of the top surfaces of the leaves to the area of ground covered by the plant or crop. When the leaf area index value is above 7, the shaded, nonproductive lower leaves may undergo programmed cell death and be shed through *self-pruning*. The orientation of leaves affects light absorption. The extent of branching and stem elongation presents a compromise between energy spent growing tall and avoiding shading versus energy spent branching and increasing photosynthetic surface area. Increased stem thickness usually accompanies tall growth.

Root Architecture and Acquisition of Water and Minerals The taproot system of gymnosperms and eudicots better supports tall growth. Evidence indicates that physiological mechanisms prevent competition within the root system of a plant. Symbiotic associations between roots and fungi called **mycorrhizae** greatly increase the surface area for water and mineral absorption.

INTERACTIVE QUESTION 36.1

a. Describe the leaf orientation that would work best in low light levels. In bright sunny regions.

b. Does a high or a low leaf area index result in greater productivity?

36.2 Transport occurs by short-distance diffusion or active transport and by long-distance bulk flow

Transport in plants involves the transport of water and solutes by individual cells, the short-distance movement of substances from cell to cell within tissues, and the long-distance transport of sap in xylem and phloem.

Diffusion and Active Transport of Solutes Solutes may move across the selectively permeable plasma membrane by passive transport when they diffuse down their electrochemical gradient. Active transport requires the cell to expend energy to move a solute against its electrochemical gradient. **Transport proteins** may be either proteins that selectively bind and transport a solute, or selective channels through a membrane. Some channels are gated: They open and close in response to certain stimuli. Transport proteins require energy when they are actively transporting a solute.

A plant **proton pump** uses ATP to pump hydrogen ions (H^+) out of the cell, generating an energy-storing proton gradient and a **membrane potential** resulting from the separation of charges. The membrane potential helps drive positively charged ions (such as K^+) down their electrochemical gradient and into the more negatively charged cell. In **cotransport,** a solute can be pumped against its concentration gradient when a transport protein links its passage to the "downhill" movement of H^+.

Diffusion of Water (Osmosis) The diffusion of water across a membrane is called **osmosis.** The rigid cell wall of plants exerts pressure against a plant protoplast (the living parts of the plant cell) that is expanding by the osmotic uptake of water. **Water potential,** designated by ψ(*psi*), is a useful measurement for predicting the direction that water will move when a plant cell is surrounded by a particular solution. It takes into account both solute concentration and the physical pressure exerted by the plant cell wall. Free water, which is not bound to solutes or surfaces, will flow from a region of higher water potential to one of lower water potential. Water potential is measured in **megapascals (MPa);** 1 MPa is equal to about 10 atmospheres of pressure. The water potential of pure water in an open container under standard conditions is assigned the value of 0 ($\Psi = 0$ MPa).

The combined effect of solute concentration and pressure is shown by the *water potential equation:* $\Psi = \Psi_S + \Psi_P$. **Solute potential** (Ψ_S), also called **osmotic potential,** is proportional to the number of solute molecules. Solutes bind water, reducing the number of free water molecules and the capacity of the water to do work.

Solutes lower the water potential, and a solution's Ψ_S is always negative.

Pressure potential (Ψ_P) measures the pressure on a solution and can be a positive or negative value. Negative pressure is tension. The positive pressure of a cell pushing against the cell wall is called **turgor pressure.**

A **flaccid** plant cell ($\Psi_P = 0$) bathed in a solution more concentrated than the cell will lose water by osmosis because the solution has a lower (more negative) Ψ. The cell's protoplast will shrink and **plasmolyze.** When bathed in pure water, the cell has the lower Ψ. Water will enter the cell until enough turgor pressure builds up so that Ψ_P and Ψ_S are equal and opposite in magnitude, and $\Psi = 0$ both inside and outside the cell. Net movement of water will then stop.

Plant cells are usually **turgid;** they have a greater solute concentration than their extracellular environment and turgor pressure keeps them firm. The loss of turgor can be seen when a plant's leaves and stems **wilt.**

Water-specific transport proteins called **aquaporins** increase the rate of water diffusion across a membrane. Aquaporins may also facilitate absorption of CO_2.

INTERACTIVE QUESTION   36.2

a. A flaccid plant cell has a water potential of -0.6 MPa. Fill in the water potential equation for this cell.

$\Psi_P =$

$+ \ \Psi_S =$

$\overline{\quad\Psi \ = \quad}$

b. The cell is then placed in a beaker of distilled water ($\Psi = 0$). Fill in the equation for the cell after it reaches equilibrium in pure water. Explain what happens to the water potential of this cell.

$\Psi_P =$

$+ \ \Psi_S =$

$\overline{\quad\Psi \ = \quad}$

c. Explain what would happen to the same cell if it is placed in a solution that has a water potential of -0.8 MPa. Fill in the equation for the cell after it is moved to this more concentrated solution and reaches equilibrium.

$\Psi_P =$

$+ \ \Psi_S =$

$\overline{\quad\Psi \ = \quad}$

Three Major Pathways of Transport The three compartments of most mature plant cells are the cell wall, cytosol, and the vacuole. Mineral ions and water move freely through cell walls, intercellular spaces, and interiors of dead xylem, along a continuum called the **apoplast.** The protoplasts of plant cells are connected by plasmodesmata, forming a cytoplasmic continuum called the **symplast.**

Short-distance transport of water and solutes within plant tissues can occur by three routes: *transmembrane*, by crossing plasma membranes and cell walls; *symplastic*, moving through plasmodesmata; and *apoplastic*, the extracellular pathway along cell walls and extracellular spaces. Solutes and water may change routes during transit.

Bulk Flow in Long-Distance Transport Long-distance transport in plants occurs by **bulk flow,** the movement of fluid driven by pressure. Sieve-tube elements lack most cellular organelles, and vessel elements and tracheids are dead (and empty) at maturity, both facilitating more efficient bulk flow.

36.3 Water and minerals are transported from roots to shoots

Absorption of Water and Minerals by Root Cells Much of the absorption of water and mineral ions occurs along young root tips where root hairs are located. The soil solution soaks into the hydrophilic walls of epidermal cells and moves along the apoplast into the root cortex, exposing a large surface area of plasma membrane for the uptake of water and minerals. Active transport of minerals allows cells to accumulate essential minerals.

Transport of Water and Minerals into the Xylem Water and minerals that had entered the symplast through a cortex or epidermal cell pass through plasmodesmata of cells of the **endodermis** and into the vascular cylinder. A ring of suberin around each endodermal cell, called the **Casparian strip,** prevents water and minerals from the apoplast from entering the stele without passing through a selectively permeable plasma membrane.

By a combination of diffusion and active transport, minerals move from endodermal cells to the apoplast and enter the nonliving xylem tracheids and vessel elements along with water.

Bulk Flow Driven by Negative Pressure in the Xylem The water and dissolved minerals of **xylem sap** move from the stele of a root to the branching veins of leaves. Through **transpiration,** the loss of water vapor from leaves, plants lose a tremendous amount of water that must be replaced by water transported up from the roots.

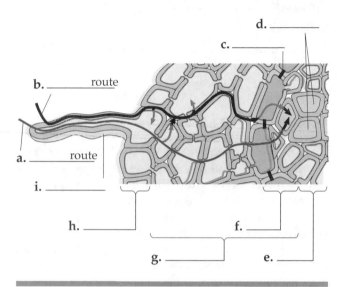
At night, transpiration rate is low, but minerals are still actively pumped into the stele and prevented from leaking out by the endodermis. This lowers the water potential in the stele. Water flowing in from the cortex results in **root pressure,** which pushes xylem sap upward. Root pressure may cause **guttation,** the exudation of water droplets from leaves when more water is forced up the xylem than is transpired by the plant.

By transpiration, water vapor from saturated air spaces within a leaf exits to drier air outside the leaf by way of stomata. The thin layer of water that coats the mesophyll cells lining the air spaces begins to evaporate. The adhesion of the remaining water to the hydrophilic walls causes a curvature in the air-water interface. Because of the high surface tension of water, this curvature produces tension in the water layer. This negative pressure creates the transpiration pull, which draws water from the xylem. Water moves along a gradient toward the most negative water potential, from xylem to neighboring cells to air spaces to the drier air outside the leaf.

The transpirational pull on xylem sap is transmitted from the leaves to the root tips by the cohesiveness of water, which results from hydrogen bonding between molecules. Adhesion of water molecules to the hydrophilic walls of the narrow xylem cells also contributes to overcoming the downward pull of gravity.

The upward transpirational pull on the cohesive sap creates tension within the xylem, lowering the water potential so that water flows passively from the soil, across the cortex, and into the stele.

A break in the chain of water molecules by the formation of a water vapor pocket in a xylem vessel, called cavitation, breaks the transpirational pull, and the vessel cannot function in transport.

Xylem Sap Ascent by Bulk Flow: A Review The transpiration-cohesion-tension mechanism results in the bulk flow of water and solutes from roots to leaves. Water potential differences caused by solute concentration and pressure contribute to the movement of water from cell to cell, but solar-powered tension caused by transpiration is responsible for long-distance transport of water and minerals.

INTERACTIVE QUESTION 36.4

Explain the contribution of each of the following to the long-distance transport of water.

a. Transpiration:

b. Cohesion:

c. Adhesion:

d. Tension:

36.4 Stomata help regulate the rate of transpiration

A plant's tremendous requirement for water is partly a consequence of making food by photosynthesis. To obtain sufficient CO_2 for photosynthesis, leaves must exchange gases through the stomata and provide a large internal surface area for CO_2 uptake, but also from which water may evaporate.

Stomata: Major Pathways for Water Loss A plant loses most of its water through stomata. Guard cells regulate the size of stomatal openings and control the rate of transpiration. Stomatal densities in many plant species are both genetically and environmentally influenced; densities have been shown to relate to CO_2 levels, providing a measure of such levels in past climates.

Mechanisms of Stomatal Opening and Closing When the kidney-shaped guard cells of most angiosperms become turgid and swell, their radially oriented microfibrils cause them to increase in length and bow outward, increasing the size of the gap between them. When the guard cells become flaccid, they sag and close the space.

Guard cells can actively accumulate potassium ions (K^+), which lowers water potential and leads to the osmotic inflow of water and an increase in turgor pressure. The exodus of K^+ (with water following) leads to a loss of turgor. The movement of K^+ into the cell is coupled with the generation of a membrane potential by proton pumps that transport H^+ out of the cell.

Stimuli for Stomatal Opening and Closing The opening of stomata at dawn is related to at least three factors. First, light stimulates guard cells to accumulate K^+, triggered by the illumination of blue-light receptors that activate proton pumps. Second, stomata are stimulated to open when CO_2 within air spaces of the leaf is depleted as photosynthesis begins in the mesophyll. The third factor is a daily rhythm of opening and closing that is endogenous to guard cells. Cycles that have intervals of approximately 24 hours are called **circadian rhythms.**

Environmental stress can cause stomata to close during the day. Guard cells lose turgor when water is in short supply. Also, a hormone called abscisic acid, produced in the roots in response to a lack of water, signals guard cells to close stomata.

Effects of Transpiration on Wilting and Leaf Temperature When transpiration exceeds the water available, leaves wilt as cells lose turgor pressure. Transpiration also produces evaporative cooling, maintaining a cooler temperature for critical enzymes in leaf cells.

Adaptations That Reduce Evaporative Water Loss Many **xerophytes,** plants adapted to arid climates, have leaves that are highly reduced, with photosynthesis carried out in thick, water-storing stems. Leaf cuticles may be thick, and stomata may be sheltered in depressions. Succulent plants of the family Crassulaceae take in CO_2 at night; thus, their stomata can be closed during the day, reducing water loss.

INTERACTIVE QUESTION 36.5

What is meant by saying that plants face a photosynthesis-transpiration compromise? Explain how a sunny, hot, windy, dry day may affect this compromise in a plant.

36.5 Sugars are transported from leaves and other sources to sites of use or storage

Movement from Sugar Sources to Sugar Sinks **Translocation,** the transport of photosynthetic products throughout the plant, occurs in the sieve-tube elements of phloem. **Phloem sap** may have a sucrose concentration as high as 30% and may also contain minerals, amino acids, and hormones.

Phloem sap flows from a **sugar source,** where it is produced by photosynthesis or the breakdown of starch, to a **sugar sink,** an organ that consumes or stores sugar. The direction of transport in any one sieve tube depends on the location of the source and sink connected by that tube, and the direction may change with the season or needs of the plant.

INTERACTIVE QUESTION 36.6

Explain how a root or tuber can serve as both a sugar source and a sugar sink.

In some species, sugar in the leaf moves through the symplast of the mesophyll cells to sieve-tube elements. In other species, sugar first moves through the symplast and then into the apoplast in the vicinity of sieve-tube elements and companion cells, which actively accumulate sugar. In some plants, companion cells are specialized as **transfer cells** with ingrowths of their wall that increase surface area for movement of solutes.

Phloem loading requires active transport in plants that accumulate high sugar concentration in the sieve tubes. Proton pumps and the cotransport of sucrose through membrane proteins along with the returning protons is the mechanism used for active transport. Sugar is moved by various mechanisms out of sieve tubes at the sink end. The concentration gradient favors this movement because sugar is either used or converted into starch within sink cells.

INTERACTIVE QUESTION 36.7

Label the components of this diagram of the mechanism used to actively transport sucrose into companion cells or sieve-tube elements.

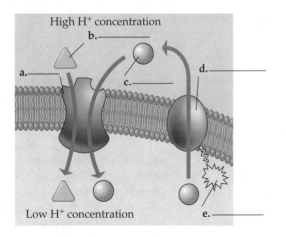

High H$^+$ concentration

b. _____

a. _____

c. _____

d. _____

Low H$^+$ concentration

e. _____

Bulk Flow by Positive Pressure: The Mechanism of Translocation in Angiosperms   The rapid movement of phloem sap from source to sink is due to a pressure flow mechanism. High solute concentration at the source lowers the water potential, and the resulting movement of water into the sieve tube produces positive pressure. At the sink end, the osmotic loss of water following the exodus of sucrose into the surrounding tissue results in a lower pressure. The difference in these pressures causes sap to move by bulk flow from source to sink. Innovative tests of this hypothesis using aphid stylets support it as the explanation for the flow of sap in the phloem of angiosperms.

In a phenomenon known as *self-thinning,* plants abort flowers, seeds, or fruits when there are more sinks than can be supported by sources.

36.6   The symplasm is highly dynamic

Plasmodesmata: Continuously Changing Structures Recent studies have shown that plasmodesmata are dynamic structures that can open and close rapidly, increase in number or size, and be eliminated as developmental needs change. Plant viruses produce *viral movement proteins* that mimic the cell's regulatory machinery that dilates plasmodesmata. Communication during plant development may occur within interconnected groups of cells called *symplastic domains.*

Electrical Signaling in the Phloem   Some studies indicate that electrical signals may be transmitted through phloem that affect gene transcription or other processes in other locations of the plant.

Phloem: An Information Superhighway   The **systemic** transport of macromolecules through phloem, such as flower-inducing signals or defense signals, helps integrate the functions of the whole plant.

Word Roots

apo- = off, away; **-plast** = formed, molded (*apoplast:* in plants, the nonliving continuum formed by the extracellular pathway provided by the continuous matrix of cell walls)

aqua- = water; **-pori** = a pore, small opening (*aquaporin:* a transport protein in the plasma membranes of a plant or animal cell that specifically facilitates the diffusion of water across the membrane)

circa- = a circle (*circadian rhythm:* a physiological cycle of about 24 hours, present in all eukaryotic organisms, that persists even in the absence of external cues)

co- = together; **trans-** = across; **-port** = a gate, door (*cotransport:* the coupling of the "downhill" diffusion of one substance to the "uphill" transport of another against its own concentration gradient)

endo- = within, inner; **-derm** = skin (*endodermis:* the innermost layer of the cortex in plant roots)

gutt- = a drop (*guttation:* the exudation of water droplets caused by root pressure in certain plants)

mega- = large, great (*megapascal:* a unit of pressure equivalent to 10 atmospheres of pressure)

myco- = a fungus; **-rhizo** = a root (*mycorrhizae:* mutualistic associations of plant roots and fungi)

osmo- = pushing (*osmosis:* the diffusion of water across a selectively permeable membrane)

sym- = with, together (*symplast:* in plants, the continuum of cytoplasm connected by plasmodesmata between cells)

turg- = swollen (*turgor pressure:* the force directed against a cell wall after the influx of water and the swelling of a walled cell due to osmosis)

xero- = dry; **-phyto** = a plant (*xerophytes:* plants adapted to arid climates)

Structure your knowledge

1. Describe the ways in which solutes may move across the plasma membrane in plants.

2. Both xylem sap and phloem sap move by bulk flow in an angiosperm. Compare and contrast the mechanisms for their movement.

Test Your Knowledge

MULTIPLE CHOICE: *Choose the one best answer.*

1. Which of the following is *not* a component of the symplast?
   a. sieve-tube elements
   b. xylem tracheids
   c. endodermal cells
   d. cortex cells
   e. companion cells

2. Proton pumps in the plasma membranes of plant cells may
   a. generate a membrane potential that helps drive cations into the cell through their specific channels.
   b. be coupled to the movement of K^+ into guard cells.
   c. drive the accumulation of sucrose in sieve-tube elements.
   d. contribute to the movement of anions through a cotransport mechanism.
   e. be involved in all of the above.

3. The Casparian strip prevents water and minerals from entering the stele through the
   a. plasmodesmata.
   b. endodermal cells.
   c. symplast.
   d. apoplast.
   e. xylem vessels.

4. The water potential of a plant cell
   a. is equal to 0 MPa when the cell is in pure water and is turgid.
   b. is equal to that of air.
   c. is equal to -0.23 MPa.
   d. becomes greater when K^+ ions are actively moved into the cell.
   e. becomes 0 MPa due to loss of turgor pressure in a concentrated sugar solution.

5. Guttation results from
   a. the pressure flow of sap through phloem.
   b. a water vapor break in the column of xylem sap.
   c. root pressure causing water to flow up through xylem faster than it can be lost by transpiration.
   d. a higher water potential of the leaves than of the roots.
   e. specialized structures in transport cells that accumulate sucrose.

6. All of the following increase the surface area available for absorption of water and minerals by a root *except*
   a. mycorrhizae.
   b. numerous branch roots.
   c. root hairs.
   d. cytoplasmic extensions of tracheids.
   e. the large surface area of cortical cells.

7. Adhesion is a result of
   a. hydrogen bonding between water molecules.
   b. the pull on the water column as water evaporates from the surfaces of mesophyll cells.
   c. tension within the xylem caused by a negative pressure.
   d. attraction of water molecules to hydrophilic walls of narrow xylem tubes.
   e. the high surface tension of water.

8. The phyllotaxy of a particular plant is alternate, with leaves emerging at a 30° angle from each previous leaf. What might one surmise about the shape of the leaves of this plant?
   a. The leaves are most likely very broad, indicating that the plant grows where there is sufficient water.
   b. The leaves are most likely very broad, indicating that the plant grows in low light levels.
   c. The leaves are most likely very thin, because emerging at such a close angle they would have to be thin to avoid shading each other.
   d. The leaves are most likely very thin, because the plant grows in a sunny, wet area.
   e. The leaves are most likely intermediate in size, but the leaf area index of the plant must be very high with leaves spaced so close together.

9. Which of these does *not* stimulate the opening of stomata?
   a. daylight in a CAM plant
   b. depletion of CO_2 in the air spaces of the leaf
   c. stimulation of proton pumps that results in the movement of K^+ into the guard cells
   d. circadian rhythm of guard cell opening
   e. an increase in the turgor of guard cells

10. Your favorite houseplant is wilting. What is the most likely cause and remedy for its declining condition?
    a. Water potential is too low; apply sugar water.
    b. The stomata won't open; no remedy available.
    c. Plasmolysis of its cells; water the plant.
    d. Cavitation; perform a xylem vessel bypass.
    e. Circadian rhythm has stomata closed; place it in bright light.

11. A turgid plant cell placed in a solution in an open beaker becomes flaccid.
    a. The water potential of the cell was initially higher than that of the solution.
    b. The water potential of the cell was initially equal to that of the solution.
    c. The pressure (Ψ_P) of the cell was initially lower than that of the solution.
    d. The Ψ_S of the cell was initially more negative than that of the solution.
    e. Turgor pressure disappeared because the cell no longer needed support.

12. What facilitates the movement of K^+ into epidermal cells of the root?
    a. cotransport through a membrane protein
    b. bulk flow of water into the root
    c. passage through selective channels, aided by the membrane potential created by proton pumps
    d. active transport through a potassium pump
    e. simple diffusion across the cell membrane down its concentration gradient

13. What are aquaporins?
    a. cytoplasmic connections between cortical cells
    b. pores through the ends of sieve-tube elements through which phloem sap flows
    c. openings in the lower epidermis of leaves through which water vapor escapes
    d. openings into root hairs through which water enters
    e. water-specific channels in membranes that may speed the rate of osmosis

14. The formation of a curvature in the air-water interface along the cell walls of cells lining the air space of a leaf contributes to water transport by
    a. creating a more positive water potential than in the surrounding mesophyll cells.
    b. creating tension, thus lowering pressure and the water potential of the leaf.
    c. raising the water potential of the surrounding saturated air.
    d. increasing the adhesion of water molecules to the cell walls.
    e. increasing the rate of transpiration from the leaf.

15. Considering an animal cell (first group) and a plant cell (second group) placed in test solutions, which of the following choices gives the *correct* direction for water flow by osmosis?
    a. hypertonic → hypotonic; higher Ψ → lower Ψ
    b. hypertonic → hypotonic; lower Ψ → higher Ψ
    c. hypotonic → hypertonic; higher Ψ → lower Ψ
    d. hypotonic → hypertonic; lower Ψ → higher Ψ
    e. One cannot tell unless told the Ψ_P for the plant cell.

16. Which of these would most likely result in self-pruning or self-thinning?
    a. the transport of viral particles facilitated by viral movement proteins through plasmodesmata
    b. the conduction of electrical signals in phloem
    c. the shading of leaves to the extent that respiration exceeds photosynthesis
    d. a lack of sufficient products of photosynthesis in sources to supply all sinks
    e. both c and d

17. By what method do most mineral anions (negatively charged ions) enter root cells?
    a. apoplastic route
    b. symplastic route
    c. diffusion
    d. cotransport using a proton gradient
    e. bulk flow

18. What mechanism explains the movement of sucrose from source to sink?
    a. evaporation of water and active transport of sucrose from the sink
    b. osmotic movement of water into the sucrose-loaded sieve-tube elements, creating a higher pressure in the source than in the sink
    c. tension created by the differences in pressure in the source and sink
    d. active transport of sucrose through the sieve-tube cells driven by proton pumps
    e. the hydrolysis of starch to sucrose in the mesophyll cells that raises their water potential and drives the bulk flow of sap to the sink

Chapter 37

Soil and Plant Nutrition

Key Concepts

37.1 Soil is a living, finite resource
37.2 Plants require essential elements to complete their life cycles
37.3 Plant nutrition often involves relationships with other organisms

Framework

Soil fertility is influenced by its texture and composition. The nutritional requirements of plants include essential macronutrients and micronutrients. Carbon dioxide enters the plant through the leaves, but water and minerals must be absorbed through the roots. The rhizosphere includes bacteria and fungi that are supported by plant roots and may enhance plant growth. Nitrogen assimilation by plants is made possible by the decomposition of humus by microbes and nitrogen fixation by bacteria. Mycorrhizae are important associations between fungi and plant roots that increase mineral and water absorption.

Chapter Review

Soil mismanagement leads to low productivity. Healthy soil promotes plant **nutrition,** the taking in and use of nourishing substances.

37.1 Soil is a living, finite resource

Soil Texture The texture of soil depends on particle size, which varies from coarse sand to silt to fine clay. The formation of soil begins with the weathering of rock as frozen water in crevices mechanically fractures rock and soil acids break down rocks chemically. Plant roots accelerate these processes. **Topsoil** is a mixture of mineral particles from rock, living organisms, and **humus** (decomposing organic matter). Two other distinct soil layers, or **soil horizons,** are found under the topsoil layer.

Loams, made up of a mixture of sand, silt, and clay, are often the most fertile soils, having enough fine particles to provide a large surface area for retaining water and minerals but enough coarse particles to provide air spaces for respiring roots.

Topsoil Composition Positively charged minerals, such as K^+, Ca^{2+}, and Mg^{2+}, adhere to the negatively charged surfaces of soil particles and are not easily lost by *leaching*. In **cation exchange,** hydrogen ions displace positively charged mineral ions from soil particles, making the ions available for absorption. Negatively charged ions, such as nitrate (NO_3^-), phosphate ($H_2PO_4^-$), and sulfate (SO_4^{2-}), tend to leach away more quickly.

Humus builds a crumbly soil that retains water, provides good aeration of roots, and supplies mineral nutrients.

The activities of the numerous soil inhabitants, such as bacteria, fungi, algae, other protists, insects, worms, nematodes, and plant roots, affect the physical and chemical properties of soil.

INTERACTIVE QUESTION 37.1

Describe the characteristics of a fertile soil.

Soil Conservation and Sustainable Agriculture As ancient farmers learned to fertilize their soil, humans were able to establish permanent residences.

Irrigation can make farming possible in arid regions, but it places a huge drain on surface and underground water resources and leads to soil *salinization*. Excess water removal from *aquifers* results in land subsidence, the sinking of Earth's surface. New methods such as *drip irrigation* may reduce some of these problems.

Agriculture removes mineral nutrients from the soil; **fertilization** replaces them. In industrialized nations, commercially produced fertilizers, usually containing nitrogen, phosphorus, and potassium, are used. Manure, fishmeal, and compost are called "organic" fertilizers because they contain organic material that is in the process of decomposing. These fertilizers decompose into inorganic nutrients, which are taken up by the plant in the same form supplied by commercial fertilizers. Commercial fertilizers may be rapidly leached from the soil, polluting streams and lakes.

The acidity of the soil affects cation exchange and can alter the chemical form of minerals, and thus their ability to be absorbed by the plant. Managing the pH of soil is an important aspect of maintaining fertility.

A large amount of topsoil is lost to water and wind erosion each year. Agricultural use of cover crops, windbreaks, and terracing can minimize erosion, as can **no-till agriculture.** Heavy field equipment leads to soil compaction, which reduces water absorption and drainage, interferes with root gas exchange, and reduces root growth.

The use of plants to extract heavy metals and other pollutants from contaminated soils is an emerging technology known as **phytoremediation.**

INTERACTIVE QUESTION 37.2

a. What is **sustainable agriculture?**

b. What is the N-P-K ratio?

37.2 Plants require essential elements to complete their life cycles

Early scientists speculated on whether soil, water, or air provides the substance for plant growth. Inorganic nutrients absorbed from the soil make only a small contribution to the dry mass of a plant. Water makes up 80–90% of the weight of a fresh plant and supplies most of the hydrogen incorporated into organic compounds. By weight, CO_2 from the air is the source of most of the organic material of a plant. Carbon, oxygen, and hydrogen, as the components of carbohydrates such as cellulose, are the most abundant elements making up the dry weight of a plant.

Macronutrients and Micronutrients **Essential elements** are those required for a plant to complete its life cycle from a seed to an adult that produces more

seeds. **Hydroponic culture** has been used to determine which of the mineral elements found in plants are essential nutrients. Seventeen elements have been identified as essential in all plants.

Nine **macronutrients** are required by plants in relatively large amounts and include the six major elements of organic compounds as well as calcium, potassium, and magnesium.

Eight **micronutrients** have been identified as needed by plants in very small amounts, functioning mainly as cofactors of enzymatic reactions.

INTERACTIVE QUESTION 37.3

a. What is a function of the macronutrient magnesium in plants?

b. What is a function of the macronutrient phosphorus?

c. What is a function of the micronutrient iron?

Symptoms of Mineral Deficiency A mobile nutrient will move to young, growing tissues, so that a deficiency will show up first in older parts of the plant. Symptoms of a mineral deficiency may be distinctive enough for the cause to be diagnosed by a plant physiologist or farmer. Soil and plant analysis can confirm a specific deficiency. Nitrogen, potassium, and phosphorus deficiencies are most common.

INTERACTIVE QUESTION 37.4

a. What is *chlorosis?*

b. Where would you expect a deficiency of a relatively immobile element to be seen first?

Improving Plant Nutrition by Genetic Modification: Some Examples Aluminum resistance has been engineered into tobacco and papaya plants by introducing a citrate synthase gene from a bacterium. Citric acid secreted by roots binds to Al ions, lowering the level of free Al in the soil.

The gene *Submergence 1A-1* (*Sub1A-1*) codes for proteins that regulate genes normally activated under anaerobic flooding conditions, such as those for enzymes that break down toxic products of alcoholic fermentation. Increased expression of *Sub1A-1* in flooding-intolerant varieties of rice conferred tolerance to submergence.

By linking a promoter that responds to a decrease in a nutrient to a reported gene, a grower can tell when a deficiency is imminent, thus only supplying fertilizer in response to need.

37.3   Plant nutrition often involves relationships with other organisms

Soil Bacteria and Plant Nutrition   The **rhizosphere** is the layer of soil bound to plant roots, which is inhabited by diverse microbes that are supported by plant secretions.

Some of the **rhizobacteria** found in the rhizosphere, called *plant-growth-promoting rhizobacteria,* enhance plant growth by producing growth-stimulating hormones or antibiotics, by absorbing toxins, or by making nutrients more available.

The **nitrogen cycle** describes the movement of nitrogen in an ecosystem, which involves microbes, plants, consumers, and the atmosphere. To be absorbed by plants, nitrogen must be converted to nitrate (NO_3^-) or ammonium (NH_4^+) by the action of microbes. Ammonifying bacteria decompose humus and form NH_3; other bacteria convert N_2 to NH_3. In the soil solution, ammonia forms ammonium, which plants can absorb. Nitrifying bacteria oxidize ammonium, producing nitrate, the form of nitrogen most readily absorbed by roots. Most plants incorporate nitrogen into amino acids or other organic compounds in the roots and then transport it through the xylem to shoots. Some nitrate is lost to the atmosphere by the action of denitrifying bacteria.

INTERACTIVE QUESTION   37.5

Fill in the types of bacteria (a–d) that participate in the nitrogen nutrition of plants. Indicate the form (e) in which nitrogen is transported in xylem to the shoot system.

Free-living bacteria in the rhizosphere, called **nitrogen-fixing bacteria,** and bacteria associated with legumes are able to convert atmospheric nitrogen into ammonia through the process of **nitrogen fixation.** Nitrogenase is an enzyme complex that reduces N_2 by adding H^+ and electrons to form ammonia (NH_3). This process is energetically expensive, and nitrogen-fixing bacteria rely on plant roots for carbohydrates.

Plants of the legume family have root swellings, called **nodules,** composed of plant cells with nitrogen-fixing *Rhizobium* bacteria in a form called **bacteroids** contained in vesicles. Nitrogen fixation requires an anaerobic environment. Lignified external layers of nodule cells limit gas exchange. In some nodules, the presence of oxygen-binding leghemoglobin keeps the concentration of free O_2 low and regulates the oxygen supply for the bacterial respiration required to provide ATP for nitrogen fixation.

Mutual signaling between roots and bacteria lead to root hair elongation and formation of an infection thread within a root hair. Bacteria form bacteroids as they bud into dividing cortical cells from the infection thread. Dividing cells of the cortex and pericycle fuse to form the nodule, which develops vascular tissue into the vascular cylinder. The plant provides the bacteria with carbohydrates and other organic molecules. Nodule cells use most of the fixed nitrogen to make amino acids, which are then transported throughout the plant.

Legume seeds are often coated with their specific *Rhizobium* before planting. Rice farmers culture a water fern that has symbiotic cyanobacteria that fix nitrogen, improving the fertility of rice paddies.

INTERACTIVE QUESTION 37.6

a. What is **crop rotation?**

b. What is "green manure"?

Fungi and Plant Nutrition Most plants have symbiotic associations of roots and fungi called **mycorrhizae.** The fungus receives food from the plant, while providing a large surface area for the absorption of water and minerals, especially phosphate, which it provides to the plant. The fungus secretes growth hormones that stimulate root growth and branching and may help protect the plant from some soil pathogens.

Especially common in woody plants, **ectomycorrhizae** form a dense sheath or mantle of mycelium over the root surface. Hyphae extending from the mantle provide a huge absorptive surface area. Hyphae also grow into extracellular spaces in the root cortex, facilitating exchange between plant and fungus.

 Arbuscular mycorrhizae, also called endomycorrhizae, do not form a sheath but extend fine fungal hyphae into the soil. Hyphae also penetrate through root cell walls and form tubes that invaginate the cell membrane of cortex cells. Some of these tubes form dense branched structures called arbuscles that facilitate nutrient exchange. Over 85% of plant species have arbuscular mycorrhizae.

Most plants form mycorrhizae when they grow in their natural habitat. When seeds are planted in foreign soil where their particular species of fungus may not be found, plants show signs of malnutrition. The invasive plant garlic mustard has been shown to interfere with the growth of arbuscular mycorrhizae on native trees.

INTERACTIVE QUESTION 37.7

Why is it thought that mycorrhizae were important to the colonization of land by the first plants?

Epiphytes, Parasitic Plants, and Carnivorous Plants **Epiphytes** are plants that grow on the surface of another plant but do not take nourishment from it. Parasitic plants, such as mistletoe or dodder, produce haustoria that may invade a host plant and siphon xylem or phloem sap from its vascular tissue.

Living in acid bogs or other nutrient-poor soils, carnivorous plants obtain nitrogen and minerals by killing and digesting insects that are caught in traps formed from modified leaves.

Word Roots

ecto- = outside; **-myco-** = a fungus; **-rhizo** = a root (*ectomycorrhizae:* a type of mycorrhizae in which the mycelium forms a dense sheath, or mantle, over the surface of the root; hyphae extend from the mantle into the soil, greatly increasing the surface area for water and mineral absorption)

endo- = inside (*endomycorrhizae:* a type of mycorrhizae that, unlike ectomycorrhizae, does not have a dense mantle ensheathing the root; instead, microscopic fungal hyphae extend from the root into the soil)

macro- = large (*macronutrient:* elements required by plants and animals in relatively large amounts)

micro- = small (*micronutrient:* elements required by plants and animals in very small amounts)

-phyto = a plant (*phytoremediation:* an emerging, nondestructive technology that seeks to cheaply reclaim contaminated areas by taking advantage of the remarkable ability of some plant species to extract heavy metals and other pollutants from the soil and to concentrate them in easily harvested portions of the plant)

Structure Your Knowledge

1. Develop a concept map that organizes your understanding of the basic nutritional requirements of plants.

2. What are the differences between root nodules and mycorrhizae? How is each beneficial to plants?

3. What is a similarity between root nodules and mycorrhizae?

Test Your Knowledge

MULTIPLE CHOICE: *Choose the one best answer.*

1. The inorganic compound that contributes most of the mass to a plant's organic matter is
   a. H_2O.
   b. CO_2.
   c. NO_3^-.
   d. protein.
   e. carbohydrate.

2. The effects of mineral deficiencies involving fairly mobile nutrients will first be observed in
   a. older portions of the plant.
   b. new leaves and shoots.
   c. the root system.
   d. the color of the leaves.
   e. the flowers.

3. Micronutrients may be
   a. cofactors in enzymes.
   b. required in very minute quantities.
   c. components of cytochromes.
   d. identified by hydroponic culture.
   e. all of the above.

4. The most fertile type of soil is usually
   a. sand because its large particles allow room for air spaces.
   b. loam, which has a mixture of fine and coarse particles.
   c. clay, because the fine particles provide much surface area to which minerals and water adhere.
   d. humus, which is decomposing organic material.
   e. moist and alkaline.

5. Chlorosis is
   a. a symptom of a mineral deficiency indicated by yellowing leaves due to decreased chlorophyll production.
   b. the uptake of the micronutrient chlorine by a plant, which is facilitated by symbiotic bacteria.
   c. the production of chlorophyll within the thylakoid membranes of a plant.
   d. a contamination of glassware in hydroponic culture.
   e. a mold of roots caused by wet soil conditions.

6. Negatively charged minerals
   a. are released from clay particles by cation exchange.
   b. are reduced by cation exchange before they can be absorbed.
   c. are converted into amino acids before they are transported through the plant.
   d. are bound when roots release acids into the soil.
   e. are leached away by the action of rainwater more easily than positively charged minerals.

7. Nitrogenase
   a. is an enzyme complex that reduces atmospheric nitrogen to ammonia.
   b. is found in *Rhizobium* and other nitrogen-fixing bacteria.
   c. catalyzes the energy-expensive fixation of nitrogen.
   d. requires an anaerobic environment.
   e. is or does all of the above.

8. Epiphytes
   a. have haustoria for anchoring to their host plants and obtaining xylem or phloem sap.
   b. are symbiotic relationships between leaves and fungi.
   c. live in poor soil and digest insects to obtain nitrogen.
   d. grow on other plants but do not obtain nutrients from their hosts.
   e. are able to fix their own nitrogen.

9. The nitrogen content of some agricultural soils may be improved by
   a. the synthesis of leghemoglobin by ammonifying bacteria.
   b. mycorrhizae on legumes.
   c. water ferns with symbiotic cyanobacteria or legumes with nitrogen-fixing bacteria.
   d. cation exchange.
   e. the action of denitrifying bacteria.

10. An advantage of organic fertilizers over chemical fertilizers is that they
    a. are more natural.
    b. release their nutrients over a longer period of time and are less likely to be lost to runoff.
    c. provide nutrients in the forms most readily absorbed by plants.
    d. are easier to mass produce and transport.
    e. are all of the above.

11. What result would you predict if a gardener sterilizes all of her soil in an effort to reduce plant diseases?
    a. decreased plant growth due to a decrease in cation exchange
    b. increased plant growth due to lack of soil microbes draining away the photosynthetic output of the plants
    c. decreased growth due to lack of water-retaining bacteria
    d. increased growth due to the destruction of disease-causing microbes
    e. overall decreased plant growth due to destruction of mutualistic associations with microbes in the rhizosphere

12. Which of the following takes the form of a mycelial sheath over plant roots with hyphae extending out that increase the surface area for absorption?
    a. root nodules
    b. haustoria
    c. arbuscular mycorrhizae
    d. ectomycorrhizae
    e. infection threads

13. Which of the following is an example of phytoremediation?
    a. dusting legume seeds with spores of *Rhizobium* to increase nodule formation
    b. inoculating seeds with fungal spores to ensure mycorrhizal formation
    c. using plants to remove toxic heavy metals from contaminated soils
    d. genetic engineering of plants to increase protein content
    e. seeding the ocean with iron to create algal blooms that reduce atmospheric CO_2 levels

14. What created the Dust Bowl in the Great Plains in the 1930s?
    a. several years of drought
    b. building dams that removed the normal supply of water to the region
    c. removal of prairie grasses for wheat and cattle farming
    d. lack of crop rotation so that soil fertility was destroyed
    e. both a and c were important factors

Chapter 38

Angiosperm Reproduction and Biotechnology

Key Concepts

38.1 Flowers, double fertilization, and fruits are unique features of the angiosperm life cycle

38.2 Plants reproduce sexually, asexually, or both

38.3 Humans modify crops by breeding and genetic engineering

Framework

This chapter describes the sexual and asexual reproduction of flowering plants. The flower produces spores that grow into the haploid gametophyte stages of the life cycle: Microspores in the anther develop into pollen grains, and a megaspore in the ovule produces an embryo sac. Pollination and the double fertilization of egg and polar nuclei are followed by the development of a seed with an embryo and endosperm, protected in a seed coat and housed within a fruit. Seed dormancy is broken following proper environmental cues and the imbibition of water.

Vegetative propagation allows successful plants to clone themselves. Agriculture makes extensive use of this type of plant reproduction by using cuttings, grafts, and test-tube cloning.

Plant biotechnologists are creating genetically modified (GM or transgenic) plants that have such traits as insect and disease resistance, herbicide tolerance, and improved nutritional value. Opposition to the development of GM organisms focuses on human health concerns and the unknown dangers of introducing transgenic plants into the environment.

Chapter Review

Many flowering plants have mutualistic relationships with their animal pollinators.

38.1 Flowers, double fertilization, and fruits are unique features of the angiosperm life cycle

Plants exhibit an alternation of generations between haploid (*n*) and diploid (2*n*) generations. The diploid plant, the sporophyte, produces haploid spores by meiosis. Spores develop into multicellular haploid male and female gametophytes, which produce gametes by mitosis. **Fertilization** yields diploid zygotes that grow into new sporophyte plants. In angiosperms, the male and female gametophytes have become reduced to only a few cells.

Flower Structure and Function Flowers are determinant, reproductive shoots that usually contain four whorls of modified leaves called floral organs: **sepals, petals, stamens,** and **carpels,** which attach to the stem at the **receptacle.** Sepals enclose and protect the unopened floral bud. Petals are generally more brightly colored and may attract pollinators. Stamens consist of a filament and an **anther,** which contains pollen sacs (microsporangia). A carpel consists of a sticky **stigma** at the top of a slender **style,** which leads to an **ovary.** The ovary encloses one or more **ovules.** A flower may have a single carpel or multiple fused carpels; either may be referred to as a **pistil.**

A **complete flower** has sepals, petals, stamens, and carpels. **Incomplete flowers** lack one or more of these floral organs. Some incomplete flowers may be sterile or *unisexual*. Floral variations include individual flowers or clusters of flowers called **inflorescences,** as well as diverse shapes, colors, and odors adapted to attract different pollinators.

Identify the flower parts in the following diagram.

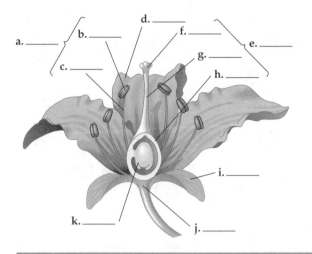

Within the microsporangium (pollen sac), diploid cells called microsporocytes undergo meiosis to form four haploid **microspores.** A microspore divides once by mitosis to produce a *generative cell* and a *tube cell.* The wall surrounding the two cells thickens into the sculptured coat of the **pollen grain.** After the pollen grain lands on a receptive stigma, the tube cell begins to form the pollen tube. The generative cell, which had moved into the tube cell, divides to form two sperm cells. The pollen tube releases the sperm cells near the female gametophyte.

There are many variations in the development of the female gametophyte, also called an **embryo sac.** Two *integuments* surround each megasporangium except at the *micropyle.* The megasporocyte in the megasporangium of each ovule undergoes meiosis to form four haploid **megaspores,** only one of which survives. This megaspore grows and divides by mitosis three times, forming the female gametophyte, which typically consists of eight nuclei contained in seven cells. At one end of the embryo sac, an egg cell is lodged between two cells called synergids; three antipodal cells are at the other end; and two nuclei, called polar nuclei, are in a large central cell.

The transfer of pollen from an anther to a stigma, called **pollination,** may be accomplished by wind, water, or animals, including bees, moths and butterflies, flies, birds, or bats.

a. Describe the male gametophyte.

b. Describe the female gametophyte.

Double Fertilization A pollen grain that lands on a receptive stigma absorbs moisture and germinates. The pollen tube grows through the style, and the generative cell divides to form two sperm. The pollen tube enters the ovule through the micropyle and releases its two sperm in the embryo sac. By **double fertilization,** one sperm fertilizes the egg to form the zygote, and the other combines with the polar nuclei to form a triploid nucleus, which will develop into a food-storing tissue called the **endosperm.**

As shown by *in vitro* studies, gamete fusion is immediately followed by an increase in cytoplasmic Ca^{2+} levels in the egg and the establishment of a block to polyspermy.

a. What role does GABA, a neurotransmitter in animal cells, appear to play in fertilization in *Arabidopsis*?

b. What function does double fertilization serve?

Seed Development, Form, and Function The triploid nucleus divides to form the endosperm, a multicellular mass rich in nutrients that may be stored for later use by the seedling. In many eudicots, the food reserves of the endosperm are transferred to the cotyledons before the seed matures.

In the zygote, the first mitotic division creates a basal cell and a terminal cell. The basal cell divides to produce a thread of cells, called the suspensor, which anchors the embryo and transfers nutrients to it. The terminal cell divides to form a spherical proembryo, on which the cotyledons (two in eudicots, one in monocots) begin to form as bumps. The embryo elongates and apical meristems develop at the apexes of the embryonic shoot and root.

As it matures, the seed dehydrates and the embryo enters **dormancy.** The embryo and its food supply, the endosperm and/or enlarged cotyledons, are enclosed in a **seed coat** formed from the ovule integuments.

In a eudicot seed such as a bean, the embryo is an elongated embryonic axis attached to fleshy cotyledons. The axis below the cotyledonary attachment is called the **hypocotyl;** it terminates in the **radicle,** or embryonic root. The upper axis is the **epicotyl;** it leads to a shoot tip with a pair of leaves. The epicotyl, young leaves, and apical meristem are called the *plumule.*

The monocot seed found in members of the grass family has a single thin cotyledon, called a *scutellum,* which absorbs nutrients from the endosperm during germination. A sheath called a **coleorhiza** covers the root, and a **coleoptile** encloses the young shoot.

INTERACTIVE QUESTION   38.4

Label the parts in these diagrams of a bean and a corn seed.

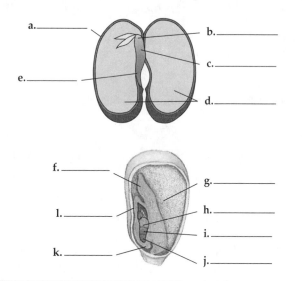

Dormancy increases the chances that the seed will germinate when and where the seedling has a good chance of surviving. The specific cues for breaking dormancy vary with the environment and may include heavy rain, intense heat from fires, cold, light, or chemical breakdown of the seed coat within an animal's digestive tract. The viability of a dormant seed may vary from a few days to decades or longer.

Imbibition, the absorption of water by the dry seed, causes the seed to expand, rupture its coat, and begin a series of metabolic changes. Enzymes digest stored compounds, and nutrients are sent to growing regions.

The radicle emerges from the seed first, followed by the shoot tip. In many eudicots, a hook that forms in the hypocotyl is pushed up through the ground, pulling the delicate shoot and cotyledons behind it. Light stimulates the straightening of the hook, and the first foliage leaves begin photosynthesis.

INTERACTIVE QUESTION   38.5

How does the shoot tip break through the soil in maize and other grasses?

Fruit Form and Function   The ovary of the flower develops into a **fruit,** which both protects and helps to disperse the seeds. Hormonal changes following fertilization cause the ovary to enlarge, its wall becoming the pericarp, or thickened wall of the fruit. Fruit usually does not set if a flower has not been pollinated.

A **simple fruit** is derived from a single carpel or several fused carpels; an **aggregate fruit** results from a flower with more than one separate carpel; a **multiple fruit** develops from an inflorescence, whose many ovaries fuse together to become one fruit. Other floral parts (such as the receptacle) may contribute to what we commonly call a fruit. These fruits, such as apples, are called **accessory fruits.**

Fruits usually ripen as the seeds are completing their development.

INTERACTIVE QUESTION   38.6

a. List some ways in which fruits aid in seed dispersal.

b. What changes usually occur when a fleshy fruit ripens?

38.2   Plants reproduce sexually, asexually, or both

Mechanisms of Asexual Reproduction   Many plant species produce genetically identical copies (clones) of themselves through **asexual reproduction.** Asexual reproduction is an extension of the indeterminate growth of plants in which meristematic tissues can grow indefinitely and parenchyma cells can divide and differentiate into specialized cells. A common type of vegetative reproduction is **fragmentation,** the formation of whole plants from parts of a parent plant. In some species, the root system gives rise to many adventitious shoots that develop into a clone with separate shoot systems. Some plants, such as dandelions, can produce seeds asexually, a process called **apomixes.**

Advantages of asexual or **vegetative reproduction** include the production of clones of plants that are well suited to a certain environment and of progeny that are usually not as frail as seedlings. Sexual reproduction generates variation in a population, an advantage when the environment changes. Seeds, which are almost always produced sexually, provide a means of dispersal to new locations and dormancy during harsh conditions.

INTERACTIVE QUESTION 38.7

a. What is an advantage of apomixis?

b. What is an advantage of "selfing," flowers such as garden peas that self-fertilize?

Mechanisms That Prevent Self-Fertilization Self-fertilization may be prevented by temporal or structural adaptations in flowers. In **dioecious** species, flowers are either staminate or carpellate and cannot self-fertilize because they are on separate plants. In flowers that are **self-incompatible,** a biochemical block prevents the development of pollen that does land on a stigma of the same plant.

The ability of flowers to reject their own pollen, or that of closely related individuals, depends on genes for self-incompatibility called *S*-genes. A plant population may have dozens of alleles of the *S*-gene, and if the pollen's allele matches an allele of the stigma, a pollen tube does not develop. In gametophytic self-incompatibility, RNA-hydrolyzing enzymes produced by the style destroy the RNA of developing pollen tubes that have a matching allele. In sporophytic self-incompatibility, self-recognition activates a signal-transduction pathway in stigma cells that blocks pollen germination.

Further research on the molecular basis of self-incompatibility may allow plant breeders to manipulate crop species to assure hybridization.

Vegetative Propagation and Agriculture New plants may develop from stem cuttings when a **callus,** or mass of dividing cells, forms at the cut end of the shoot and adventitious roots develop from the callus. Adventitious roots can also form from a node in the shoot fragment.

Twigs or buds of one plant can be grafted onto a plant of a different variety or closely related species. The plant that provides the root system is called the **stock,** and the twig is called the **scion.** Grafting can combine the best qualities of different plants.

In test-tube cloning, whole plants can develop from pieces of tissue, called explants, or even from single parenchyma cells. A single plant can be cloned into thousands of plants by subdividing the undifferentiated calluses as they grow in tissue culture. Stimulated by proper hormone balances, calluses sprout shoots and roots and develop into plantlets, which can be transferred to soil to develop.

Foreign DNA may be inserted into individual plant cells, which then grow into **transgenic** or genetically modified (GM) plants using *in vitro* culture.

A technique called **protoplast fusion** is being used to create new plant varieties. Protoplasts are cells whose cell walls have been enzymatically removed. Protoplasts from different species can be fused and cultured to form hybrid plantlets.

38.3 Humans modify crops by breeding and genetic engineering

Almost all our crop species were first domesticated by Neolithic (late Stone Age) humans about 10,000 years ago. Natural hybridization between different species of plants is common, and humans have exploited such genetic variations using selective breeding and artificial selection to develop and improve crops.

Plant Breeding Breeders search for valuable traits in domesticated and wild species and hasten mutations by treating seeds or seedlings with mutagens. Wild plants with desirable traits are repeatedly crossed with the domesticated species until they are agriculturally suitable. Triticale resulted from a cross between wheat and rye, followed by extensive artificial selection to produce this hardy, nutritious grain that grows well on *marginal land.*

Plant Biotechnology and Genetic Engineering Plant biotechnology refers both to the age-old use of plants to make products for human use and to the use of GM organisms in agriculture. Genetic engineering allows plant biotechnologists to transfer genes between species that are not closely related.

The serious malnutrition of millions of people may be due to inequities in distribution of food resources or may signal that the world is already overpopulated. Plant biotechnology can help increase crop yields.

The planting of transgenic crops has increased dramatically over the past several years. Cotton, maize, and potatoes have been engineered to contain genes from *Bacillus thuringiensis* that code for *Bt* toxin, reducing the need for spraying chemical insecticides. Other transgenic crops have been developed that are resistant to a number of herbicides, allowing farmers to "weed" crops without heavy tillage. Some transgenic plants are more resistant to disease or have improved nutritional quality.

Biomass from fast-growing plants, such as switchgrass and poplars, may help meet the world's energy demands. Enzymatic reactions break down the plant cell walls into sugars, which are fermented and distilled to yield **biofuels.**

INTERACTIVE QUESTION   38.8

Biofuels are considered carbon neutral, even though the burning of biofuels releases CO_2, just as the burning of fossil fuels does. Explain.

The Debate over Plant Biotechnology   GM organisms (GMOs) may present an unknown risk to human health or the environment. One concern is the transfer of allergens to a food source. Some GM crops may be safer than regular crops; *Bt* maize contains less of a cancer- and birth defect-causing mycotoxin called fumonisin.

Many ecologists are concerned about the effect of GM crops on nontarget organisms. All GM crops need to be field tested for nontarget effects.

A serious concern is the escape of herbicide- or disease-resistance genes through crop-to-weed hybridization that may create "superweeds." Various techniques are being developed to help reduce the ability of transgenic crops to hybridize, such as developing male sterility so that transgenic pollen is not produced; engineering apomixis into the crop, introducing genes into chloroplast DNA, which is not present in pollen; and developing flowers that self-pollinate and do not open so that pollen cannot escape.

Word Roots

a- = without; **-pomo** = fruit (*apomixis:* the asexual production of seeds)

anth- = a flower (*anther:* the terminal pollen sac of a stamen, inside which pollen grains with male gametes form in the flower of an angiosperm)

carp- = a fruit (*carpel:* The female reproductive organ of a flower, consisting of the stigma, style, and ovary)

coleo- = a sheath; **-rhiza** = a root (*coleorhiza:* the covering of the young root of the embryo of a grass seed)

di- = two (*dioecious:* referring to a plant species that has staminate and carpellate flowers on separate plants)

dorm- = sleep (*dormancy:* a condition typified by extremely low metabolic rate and a suspension of growth and development)

endo- = within (*endosperm:* a nutrient-rich tissue formed by the union of a sperm cell with two polar nuclei during double fertilization, which provides nourishment to the developing embryo in angiosperm seeds)

epi- = on, over (*epicotyl:* the embryonic axis above the point at which the cotyledons are attached)

gamet- = a wife or husband (*gametophyte:* the multicellular haploid form in organisms undergoing alternation of generations, which mitotically produces haploid gametes that unite and grow into the sporophyte generation)

hypo- = under (*hypocotyl:* the embryonic axis below the point at which the cotyledons are attached)

mega- = large (*megaspore:* a large, haploid spore that can continue to grow to eventually produce a female gametophyte)

micro- = small (*microspore:* a small, haploid spore that can give rise to a haploid male gametophyte)

peri- = around; **-carp** = a fruit (*pericarp:* the thickened wall of fruit)

proto- = first; **-plast** = formed, molded (*protoplast:* the contents of a plant cell exclusive of the cell wall)

scutell- = a little shield (*scutellum:* a specialized type of cotyledon found in the grass family)

sporo- = a seed; **-phyto** = a plant (*sporophyte:* the multicellular diploid form in organisms undergoing alternation of generations that results from a union of gametes and that meiotically produces haploid spores that grow into the gametophyte generation)

stam- = standing upright (*stamen:* the pollen-producing male reproductive organ of a flower, consisting of an anther and filament)

uni- = one (*unisexual flower:* a flower missing either stamens or carpels)

Structure Your Knowledge

1. Draw a diagram of the major events in the life cycle of an angiosperm.

2. List the advantages and disadvantages of sexual and asexual reproduction in plants.

3. List some of the potential benefits and dangers of plant biotechnology.

Test Your Knowledge

FILL IN THE BLANKS

_____ 1. structure from which fruit typically develops

_____ 2. generation that produces spores by meiosis

_____ 3. fuel produced from organic matter

_____ 4. female gametophyte of angiosperms

_____ 5. embryonic root

_____ 6. embryonic axis above attachment of cotyledon

_____ 7. protects grass shoot as it breaks through the soil

_____ 8. twig or stem portion of a graft

_____ 9. plant cell from which cell wall is removed

_____ 10. mass of dividing cells at cut end of a shoot

MULTIPLE CHOICE: *Choose the one best answer.*

1. A flower on a dioecious plant would be
   a. complete.
   b. biennial.
   c. incomplete.
   d. staminate or carpellate.
   e. both c and d.

2. Which of the following structures is haploid?
   a. embryo sac
   b. anther
   c. endosperm
   d. microsporocyte
   e. both a and b

3. The terminal cell of an early plant embryo
   a. develops into the shoot apex of the embryo.
   b. forms the suspensor that anchors the embryo and transfers nutrients.
   c. develops into the endosperm when fertilized by a sperm nucleus.
   d. divides to form the proembryo.
   e. develops into the cotyledons.

4. In angiosperms, sperm are formed by
   a. meiosis in the anther.
   b. meiosis in the pollen grain.
   c. mitosis in the anther.
   d. mitosis in the pollen tube.
   e. double fertilization in the embryo sac.

5. The endosperm
   a. may have its nutrients absorbed by the cotyledons in the seeds of eudicots.
   b. is usually a triploid tissue.
   c. is digested by enzymes in monocot seeds following hydration.
   d. develops in concert with the embryo as a result of double fertilization.
   e. is or does all of the above.

6. A seed consists of
   a. an embryo, a seed coat, and a nutrient supply.
   b. an embryo sac.
   c. a gametophyte and a nutrient supply.
   d. an enlarged ovary.
   e. an immature ovule.

7. Which structure protects a bean shoot as it breaks through the soil?
   a. hypocotyl hook
   b. radicle
   c. coleoptile
   d. coleorhiza
   e. seed coat

8. Which of the following is a form of asexual or vegetative reproduction?
   a. apomixis
   b. grafting
   c. test-tube cloning
   d. fragmentation
   e. all of the above

9. Protoplast fusion
   a. is used to study the fertilization of plant egg and sperm.
   b. is the method used to produce test-tube plantlets.
   c. can be used to form new plant species.
   d. occurs within a callus.
   e. occurs following fertilization.

10. In the plant embryo, the suspensor
    a. is produced by vertical mitotic divisions within the proembryo.
    b. connects the early root and shoot apexes.
    c. develops into the endosperm.
    d. is analogous to the umbilical cord in mammals.
    e. is the point of attachment of the cotyledons.

11. Which of the following statements is *false*?
    a. Bird-pollinated flowers are usually light-colored, strong smelling, and produce copious amounts of nectar.
    b. Fly-pollinated flowers may smell like rotten meat.
    c. Insects pollinate approximately 65% of all flowering plants, and bees are the most important of these pollinators.
    d. The flowers of many temperate, wind-pollinated trees appear before the production of leaves, which may interfere with pollen movement.
    e. Moths and bats, which are often active at night, are attracted to light-colored, fragrant flowers.

12. What does self-incompatibility provide for a plant?
    a. a means of transferring pollen to another plant
    b. a means of coordinating the fertilization of an egg with the development of stored nutrients
    c. a means of destroying foreign pollen before it fertilizes the egg cell
    d. a biochemical block to self-fertilization so that cross-fertilization is assured
    e. a means of producing seeds without the need for fertilization

13. Into what does a microspore develop in an angiosperm?
    a. the male gametophyte
    b. a pollen grain
    c. the male sporophyte
    d. the embryo sac
    e. Both a and b are correct.

14. Many plants form clones that develop after shoots emerge from the same root system. What is an advantage of forming such clones?
    a. provide a strong start for new plants
    b. disperse offspring to new habitats
    c. provide a period of dormancy until specific environmental cues signal regrowth
    d. provide a strong root system for a genetically altered shoot system
    e. allow for the production of seeds without fertilization

15. Which of the following is a technique being developed to reduce the threat of introduced genes for herbicide or insect resistance escaping to closely related weed species?
    a. engineering flowers to self-fertilize and remain closed so that pollen is not released
    b. breeding male sterility into transgenic plants so that they have no pollen to be transferred to nearby weeds
    c. engineering the gene of interest into chloroplast DNA, which is inherited from the maternal plant and is not transferred by pollen
    d. engineering crops, such as soybeans, that have no weedy relatives nearby
    e. All of the above would reduce the risk of crop-to-weed transgene escape.

Chapter 39

Plant Responses to Internal and External Signals

Key Concepts

39.1 Signal transduction pathways link signal reception to response

39.2 Plant hormones help coordinate growth, development, and responses to stimuli

39.3 Responses to light are critical for plant success

39.4 Plants respond to a wide variety of stimuli other than light

39.5 Plants respond to attacks by herbivores and pathogens

Framework

Environmental stimuli and internal signals are linked by signal transduction pathways to cellular responses such as changes in gene expression and activation of enzymes. Plant hormones—auxin, cytokinins, gibberellins, brassinosteroids, abscisic acid, and ethylene—control growth, development, flowering, and senescence, as plants respond and adapt to their environments. Plant movements in response to environmental stimuli include phototropism, gravitropism, and thigmotropism. The biological clock of plants controls circadian rhythms, such as stomatal opening and sleep movements. Blue-light photoreceptors are involved in phototropism and stomatal opening. Phytochromes absorb red and far-red light and are involved in the photoperiodic control of flowering. Plants have various physiological responses to environmental stresses and pathogens.

Chapter Review

Plants are able to sense and adaptively respond to their environments, generally by altering their patterns of growth and development.

39.1 Signal transduction pathways link signal reception to response

Plants have cellular receptors that initiate signal transduction pathways, which couple reception of a specific stimulus to a response. The growth pattern of a sprouting potato shoot growing in darkness, called **etiolation**, facilitates the shoot breaking ground. When the shoot reaches sunlight, stem elongation slows, leaves expand, the root system elongates, and chlorophyll production begins—all part of a process known as **de-etiolation** or greening.

Reception Signals are detected by receptors, proteins that change shape in response to a specific stimulus. A *phytochrome* is the photoreceptor involved in de-etiolation, and, unlike receptors built into the plasma membrane, it is located in the cytosol. Researchers have studied the role of phytochrome in de-etiolation using the tomato mutant *aurea,* which has lower than normal levels of phytochrome.

Transduction **Second messengers** are small molecules that amplify the signal from the receptor and transfer it to proteins that produce the specific response. Each light-activated phytochrome may lead to the production of hundreds of second messenger molecules, each of which may activate hundreds of specific enzymes. The second messenger cyclic GMP affects membrane ion channels or activates protein kinases. Phytochrome signal transduction also opens calcium channels. Increases in the second messenger Ca^{2+} also affect ion channels and certain protein kinases.

Response The response to a signal transduction pathway usually involves the activation of specific enzymes, either by transcriptional regulation or by modification of existing enzymes.

Several transcription factors are activated during phytochrome-induced de-etiolation. Changes in gene expression may involve the activation of either positive or negative transcription factors.

Modification of existing proteins usually involves phosphorylations, catalyzed by protein kinases. Cascades of protein kinase activations may lead to the phosphorylation of transcription factors, and thus, ultimately, to changes in gene expression. Protein phosphatases are enzymes that dephosphorylate specific proteins, allowing signal pathways to be turned off when a signal is no longer present.

INTERACTIVE QUESTION 39.1

In the process known as de-etiolation,

a. What is the signal and the receptor?

b. Briefly describe some of the steps in the transduction of this signal.

c. What is the plant's growth response? What types of proteins are involved?

39.2 Plant hormones help coordinate growth, development, and responses to stimuli

Hormones, chemical signals transported through the plant body, trigger responses in target cells and tissues. Because some signal molecules act only locally and others occur at high concentrations, plant biologists may use the term *plant growth regulator* to describe compounds that affect specific physiological processes in plants.

The Discovery of Plant Hormones   A **tropism** is a growth response of plant organs toward or away from stimuli. The growth of a shoot toward light is called positive **phototropism.** A coleoptile, enclosing the shoot of a grass seedling, bends toward the light when illuminated from one side because of the elongation of cells on the darker side.

Darwin and his son observed that a grass seedling would not bend toward light if its tip were removed or covered by an opaque cap. They postulated that a signal must be transmitted from the tip to the elongating region of the coleoptile. P. Boysen-Jensen demonstrated that the signal was a mobile substance, capable of being transmitted through a block of gelatin separating the tip from the rest of the coleoptile. In 1926, F. Went concluded that the chemical produced in the tip, which he

called auxin, promoted growth and that it was in higher concentration on the side away from the light.

Researchers have not found a light-induced asymmetrical distribution of auxin in eudicots, but certain substances that may act as growth inhibitors have been shown to be more concentrated on the lighted sides of such stems.

A Survey of Plant Hormones   Several major classes of plant hormones have been identified. Very low concentrations of hormones, acting through signal transduction pathways, may affect the expression of genes, the activity of enzymes, or the properties of membranes. Depending on the site of action, the developmental stage of the plant, and relative hormone concentrations, the effect of a hormone will vary.

Auxins include any substance that stimulates elongation of coleoptiles. The natural auxin extracted from plants is indoleacetic acid (IAA). Auxin is transported from the shoot tip apical meristem where it is synthesized down the shoot. This polar transport involves auxin transporters located only at the basal ends of cells.

According to the acid growth hypothesis, auxin initiates cell growth in the region of elongation by binding to a plasma membrane receptor and stimulating proton pumps. The proton pumps lower pH in the cell wall, activating enzymes called **expansins** that break cross-links between cellulose microfibrils. The proton pumps also increase the membrane potential, enhancing ion uptake and the resulting osmotic uptake of water. For continued growth after this relatively fast elongation, the cell must produce more cytoplasm and wall material, processes that rely on changes in gene expression that are also stimulated by auxin.

Auxin is involved in root branching and is used commercially to enhance formation of adventitious roots at the cut base of stems. Synthetic auxins, such as 2,4-D, are used as herbicides, killing eudicot (broad-leaf) weeds with a hormonal overdose. Auxin stimulates the vascular cambium in secondary growth. Auxin produced by developing seeds promotes fruit growth; synthetic auxins can induce seedless fruit development.

Cytokinins are modified forms of adenine, named because they stimulate cytokinesis. Zeatin is the most common naturally occurring cytokinin.

Cytokinins are produced in actively growing roots, embryos, and fruits. Acting with auxin, they stimulate cell division and affect differentiation.

According to the direct inhibition hypothesis, the control of apical dominance involves the interaction between auxin, transported down from the terminal bud, which restrains axillary bud development, and cytokinins, transported up from the roots, which stimulate bud growth. Several lines of evidence support this hypothesis. Removal of the apical bud, however, leads to an increase in auxin levels in the axillary buds, exactly opposite the prediction of the direct inhibition hypothesis.

Cytokinins can retard aging of some plant organs, because they stimulate RNA and protein synthesis, mobilize nutrients, and inhibit protein breakdown.

In the 1930s Japanese scientists determined that the fungus *Gibberella* secreted a chemical that caused the hyperelongation of rice stems or "foolish seedling disease." More than 100 different naturally occurring **gibberellins** have now been identified.

Gibberellins, which are produced by roots and young leaves, stimulate growth. Gibberellins may promote cell elongation by stimulating cell wall-loosening enzymes, thus facilitating the penetration of expansins into the cell wall.

Gibberellin applied to dwarf plants may cause them to grow to normal height. Bolting, the growth of an elongated floral stalk, is caused by a surge of gibberellins. In many plants, both auxin and gibberellins contribute to fruit set. Gibberellins are sprayed in the production of Thompson seedless grapes. The release of gibberellins from the embryo signals seeds of many plants to break dormancy.

Brassinosteroids are steroids that have effects very similar to those of auxin: They promote cell elongation and division, retard leaf abscission, and promote xylem differentiation. Identification of a brassinosteroid-deficient mutant of *Arabidopsis* helped to establish these compounds as nonauxin plant hormones.

The hormone **abscisic acid (ABA)** generally slows growth. The high concentration of ABA in maturing seeds inhibits germination and stimulates production of proteins that protect the seeds during dehydration. For dormancy to be broken in some seeds, ABA must be removed or inactivated, or the ratio of gibberellins to ABA must increase.

ABA also reduces drought stress. In a wilting plant, ABA causes stomata in the leaves to close. ABA may be produced in the roots in response to water shortage and transported to the leaves.

Plants produce the gas **ethylene** in response to stress and during fruit ripening and programmed cell death. Ethylene production may be induced by a high concentration of auxin.

The mechanical stress of a seedling pushing against an obstacle as it grows upward through the soil induces the production of ethylene. Ethylene then initiates a growth pattern called the **triple response,** consisting of a slowing of stem elongation, a thickening of the stem, and initiation of horizontal growth. As the effects of ethylene lessen, normal upward growth resumes. Researchers have identified *Arabidopsis* mutants that are ethylene insensitive *(ein),* ethylene overproducing *(eto),* and that undergo the triple response in the absence of ethylene. In these latter

constitutive triple response *(ctr)* mutants, the ethylene signal transduction pathway is permanently turned on, even though ethylene is not present. Their mutant gene codes for a protein kinase, suggesting that the normal kinase product is a negative regulator of ethylene signal transduction. Binding of ethylene to the ethylene receptor may lead to the inactivation of the negative kinase, which allows the synthesis of the proteins involved in the triple response.

Senescence is the programmed death of plant organs or the whole plant. **Apoptosis,** or programmed cell death, requires the synthesis of new enzymes that break down many cellular components, which the plant may salvage. Ethylene is almost always associated with this programmed death.

Deciduous leaf loss protects against winter desiccation. Before leaves abscise in the autumn, many of their compounds are stored in the stem awaiting recycling to new leaves. A change in the balance of auxin and ethylene initiates changes in the abscission layer located near the base of the petiole, including the production of enzymes that hydrolyze polysaccharides in cell walls.

Ethylene initiates the breakdown of cell walls and conversion of starches to sugars associated with fruit ripening. In a rare example of positive feedback, ethylene triggers ripening, and ripening triggers even more ethylene production. Many commercial fruits are ripened in huge containers perfused with ethylene gas.

INTERACTIVE QUESTION 39.2

Fill in the name of the hormone that is responsible for each of the following functions:

a. _____ promotes fruit ripening; initiates triple response; involved in apoptosis

b. _____ stimulate cell division, growth, and germination; anti-aging

c. _____ inhibits growth; maintains dormancy; closes stomata during water stress

d. _____ stimulates stem elongation, root branching, fruit development; apical dominance

e. _____ promote cell elongation and division, xylem differentiation; retard leaf abscission

f. _____ promote stem elongation, seed germination; contributes to fruit set

Systems Biology and Hormone Interactions   Systems biology studies properties that emerge from interactions of many system elements. Biologists can identify all the genes in a plant, and microarray and proteomic techniques can indicate which genes are activated or inactivated in response to environmental stimuli or during development. A systems-based approach may allow biologists to determine how these gene products interact in producing plant responses and to predict the results of genetic manipulation.

39.3   Responses to light are critical for plant success

The effect of light on plant morphology is called **photomorphogenesis**. An **action spectrum** graphs a physiological response across different wavelengths of light. An absorption spectrum shows the wavelengths of light a pigment absorbs. Close correlation between an action spectrum for a plant response and the absorption spectrum of a pigment may indicate that the pigment is the photoreceptor involved in the response. **Blue-light photoreceptors** and **phytochromes,** which absorb mostly red light, are the two major classes of light receptors.

Blue-Light Photoreceptors   Molecular biologists have determined that plants use at least three different types of pigments to detect blue light: *cryptochrome* (for inhibition of hypocotyl elongation when a seedling breaks ground), *phototropin* (for phototropism), and *zeaxanthin* or perhaps phototropin (for stomatal opening).

Phytochromes as Photoreceptors   Studies of lettuce seed germination in the 1930s determined that red light (660 nm wavelength) increased germination the most and far-red light (730 nm) inhibited germination. The effects of red and far-red light are reversible, with the seed's response determined by the last flash of light it receives.

Five different phytochromes have been identified in *Arabidopsis*. A phytochrome is a photoreceptor that consists of a protein component bonded to a nonprotein light-absorbing chromophore. The chromophore alternates between two isomers, one of which absorbs red light and the other, far-red light. These two variations of phytochrome are photoreversible; the P_r to P_{fr} interconversion acts as a switch, controlling various events in the life of the plant.

Phytochrome tells the plant that light is present by the conversion of P_r, which is the form the plant synthesizes, to P_{fr} in the presence of sunlight. P_{fr} triggers many plant responses to light, such as breaking seed dormancy.

INTERACTIVE QUESTION   39.3

Explain how phytochrome can also indicate the quality of light available to a plant. What effect might this have on a shaded tree?

Biological Clocks and Circadian Rhythms   A **circadian rhythm** is a physiological cycle with about a 24-hour frequency. These rhythms persist, even when the organism is sheltered from environmental cues. Research indicates that the circadian clock is endogenous, although it is set (entrained) to a 24-hour period by daily environmental signals.

INTERACTIVE QUESTION   39.4

What are free-running periods?

The molecular mechanism of the biological clock may be cyclical changes in the concentration of a protein that is a transcription factor that inhibits the expression of its own gene.

Researchers have identified clock mutants of *Arabidopsis* by splicing the gene for luciferase to the promoter for genes for photosynthesis-related proteins that follow a circadian rhythm in their production. When the biological clock turned on the promoter for these genes, the plants glowed. Plants that glowed for a nonnormal amount of time were isolated as clock mutants. Some of these mutants had defects in proteins that normally bind photoreceptors, pointing to a light-dependent mechanism that sets the biological clock.

The Effect of Light on the Biological Clock   Both blue-light photoreceptors and phytochrome can entrain the biological clock in plants. In darkness, the phytochrome ratio shifts toward P_r, in part because P_{fr} converts slowly to P_r in some plants, and also because P_{fr} is degraded and new pigment is synthesized as P_r. When the sun rises, P_r is rapidly converted to P_{fr}, resetting the biological clock each day at dawn.

Photoperiodism and Responses to Seasons Seasonal events in the life cycle of plants usually are cued by photoperiod, the relative length of night and day. A physiological response to photoperiod is called **photoperiodism.**

Researchers discovered that a variety of tobacco plant flowered only when the day length was 14 hours or shorter. They termed it a **short-day plant. Long-day plants** flower when days are longer than a certain number of hours. The flowering of **day-neutral plants** is unaffected by photoperiod.

Researchers have found that night length controls flowering and other photoperiod responses. If the dark period is interrupted by even a few minutes of light, a short-day (long-night) plant will not flower. Photoperiodic responses thus depend on a critical night length.

Red light was found to be the most effective in interrupting a plant's perception of night length. A brief exposure to red light breaks a dark period of sufficient length and prevents short-day plants from flowering, whereas a flash of red light during a dark period longer than the critical length will induce flowering in a long-day plant. Due to the photoreversibility of phytochrome, a subsequent flash of far-red light negates the effect of the red light.

Some plants bloom after a single exposure to the required photoperiod. Others respond to photoperiod only after exposure to another environmental stimulus. **Vernalization** is the need for pretreatment with cold to induce flowering.

Leaves detect the photoperiod. The signal for flowering, called **florigen,** travels from leaves to buds. Recent research indicates that CONSTANS, the protein product of the *CONSTANS* gene, increases the expression of *FLOWERING LOCUS T* (*FT*) in the leaf, and the FT protein travels to the shoot meristem and initiates flowering.

In order for flowering to occur, the bud meristem must transition from a vegetative to a flowering state, a change that requires the activation of meristem-identity genes, followed by organ-identity genes. Research is focusing on identifying the signal transduction pathways that link photoperiod and hormonal cues to such changes in gene expression.

INTERACTIVE QUESTION 39.5

Indicate whether a short-day plant (a–e) and a long-day plant (f–j) would flower or not flower under the indicated light conditions.

a. _____ b. _____ c. _____ d. _____ e. _____

Short-day (long-night) plant

f. _____ g. _____ h. _____ i. _____ j. _____

Long-day (short-night) plant

How do these results demonstrate the red/far-red photoreversibility of phytochrome?

39.4 Plants respond to a wide variety of stimuli other than light

Gravity Roots exhibit positive **gravitropism,** whereas shoots show negative gravitropism. According to one hypothesis, the settling of **statoliths,** plastids containing dense starch grains, in cells of the root cap triggers movement of calcium, which causes the lateral transport of auxin. Accumulating on the lower side of the growing root, auxin inhibits cell elongation, causing the root to curve downward. The settling of the protoplast and large organelles may distort the cytoskeleton and also signal gravitational direction.

Mechanical Stimuli The stunting of growth in height and increase in girth of plants that are exposed to wind is an example of **thigmomorphogenesis,** changes in plant form in response to mechanical stimulation.

Most climbing plants have tendrils that coil rapidly around supports, exhibiting **thigmotropism** or directional growth in response to touch.

The sensitive plant *Mimosa* folds its leaves after being touched due to the rapid loss of turgor by cells in specialized motor organs called pulvini, located at the joints of the leaf. The message travels through the plant from the point of stimulation, perhaps as the result of electrical impulses, called **action potentials.** These electrical messages may be used in plants as a form of internal communication.

INTERACTIVE QUESTION   39.6

a. Name the types of tropisms that a stem exhibits.

b. Name and describe the growth mechanism that produces the coiling of a tendril.

Environmental Stresses An **abiotic** environmental stress may be severe enough to threaten a plant's growth, reproduction, and survival. Plants also respond to **biotic** stresses such as pathogens and herbivores.

Mechanisms that reduce transpiration help a plant respond to water deficit. Guard cells lose turgor and stomata close. Abscisic acid, produced by leaves in response to water deficit, acts on guard cell membranes to keep stomata closed. Growth of young leaves is inhibited by abscisic acid and the lack of cell-expanding water, and wilted leaves may roll up, further reducing transpiration. All of these responses, however, reduce photosynthesis. During a drought, root growth in shallow, dry soil decreases while deeper roots in moist soil continue to grow.

Plants adapted to wet habitats may have aerial roots that provide oxygen to their submerged roots. When roots of other plants are in waterlogged soils, oxygen deprivation may stimulate ethylene production, causing some root cortical cells to undergo apoptosis, opening up air tubes within the roots.

Excess salts in the soil may lower the water potential of the soil solution, reducing the water potential gradient and water uptake by roots. The plasma membranes of root cells can reduce the uptake of ions that are toxic to plants in high concentrations. Plants may respond to moderate soil salinity by producing organic solutes that lower the water potential of root cells. Special adaptations for dealing with high soil salinity have evolved in halophytes.

Transpiration creates evaporative cooling for a plant, but this effect may be lost on hot, dry days when stomata close to reduce water loss. Above critical temperatures, plant cells produce **heat-shock proteins** that may provide temporary support to reduce protein denaturation.

Plants respond to cold stress by increasing the proportion of unsaturated fatty acids in membrane lipids in order to maintain the fluidity of cell membranes. Subfreezing temperatures cause ice to form in cell walls, lowering the extracellular water potential and causing cells to dehydrate. Plants adapted to cold winters have adaptations to deal with freezing stress, such as changing the solute composition of the cytosol.

INTERACTIVE QUESTION   39.7

a. A plant is exposed to a spell of very hot and dry weather. How might it survive this stress?

b. How might a plant adapt to an unusually cold and very wet fall?

39.5   Plants respond to attacks by herbivores and pathogens

Defenses Against Herbivores Physical defenses, such as thorns, and chemical defenses, such as distasteful or toxic compounds, help plants cope with herbivory. Some plants produce *canavanine,* which resembles arginine and may be incorporated into an insect's proteins, altering protein shape and causing death.

Some plants recruit predators of their herbivores. A combination of a compound in a caterpillar's saliva and the damage caused by its eating stimulates leaves to release volatile compounds that attract parasitoid wasps. The wasps lay their eggs inside the caterpillar, and the wasp larvae eat their host. These volatile compounds may also signal neighboring plants to activate defense genes. These gene activations are similar in pattern to those produced by exposure to the important plant defense molecule **jasmonic acid.**

Defenses Against Pathogens The physical barrier of the plant's epidermis and/or periderm is the first line of defense against pathogenic viruses, bacteria, and fungi. Should pathogens enter the plant due to injuries or through openings such as stomata, the plant mounts a chemical defense.

Virulent pathogens are those to which the plant has no specific defense. Plants are generally resistant to most pathogens. These **avirulent** pathogens do not extensively harm or kill the plant.

Specific resistance is based on **gene-for-gene recognition** between the protein products of a plant's disease resistance (*R*) genes and a pathogen molecule coded for by an *Avr* gene. Plants have many different *R* genes, corresponding to the many different potential pathogens. The binding of a pathogen molecule called an *elicitor* to an R protein receptor triggers a signal transduction pathway that produces a hypersensitive response and systemic acquired resistance.

In the **hypersensitive response** (HR), plant cells produce antimicrobial *phytoalexins* and *PR proteins* (pathogenesis-related proteins), which may attack the pathogen's cell wall. The HR also stimulates strengthening of the cell walls to slow the spread of the pathogen, after which the plant cells destroy themselves.

A hypersensitive response also involves the release of alarm signals, which stimulate PR protein production throughout the plant, creating a **systemic acquired resistance** (SAR). This generalized defense response helps protect uninfected tissue. One of the signals involved in activating SAR is probably **salicylic acid,** which humans have modified as the active ingredient in aspirin.

Word Roots

aux- = grow, enlarge (*auxins:* a class of plant hormones, including indoleacetic acid, having a variety of effects, such as phototropic response through the stimulation of cell elongation, stimulation of secondary growth, and the development of leaf traces and fruit)

circ- = a circle (*circadian rhythm:* a physiological cycle of about 24 hours, present in all eukaryotic organisms, that persists even in the absence of external cues)

crypto- = hidden; **-chromo** = color (*cryptochrome:* the name given to one of the blue-light photoreceptors)

cyto- = cell; **-kine** = moving (*cytokinins:* a class of related plant hormones that retard aging and act in concert with auxins to stimulate cell division, influence the pathway of differentiation, and control apical dominance)

gibb- = humped (*gibberellins:* a class of related plant hormones that stimulate growth in the stem and leaves, trigger the germination of seeds and breaking of bud dormancy, and stimulate fruit development with auxin)

hyper- = excessive (*hypersensitive response:* a vigorous, localized defense response to a pathogen that is avirulent based on an *R-Avr* match)

photo- = light; **-trop** = turn, change (*phototropism:* growth of a plant shoot toward or away from light)

phyto- = a plant; **-alexi** = to ward off (*phytoalexin:* an antibiotic, produced by plants, that destroys microorganisms or inhibits their growth)

stato- = standing, placed; **-lith** = a stone (*statolith:* specialized plastids that help a plant tell up from down)

thigmo- = a touch; **morpho-** = form; **-genesis** = origin (*thigmomorphogenesis:* a response in plants to chronic mechanical stimulation, resulting from increased ethylene production; an example is thickening stems in response to strong winds)

zea- = a grain; **-xantho** = yellow (*zeaxanthin:* a blue-light photoreceptor probably involved in stomatal opening)

Structure Your Knowledge

1. List some agricultural uses of plant hormones.

2. Develop a concept map to illustrate your understanding of photoperiodism and the control of flowering. Do not forget to include the role of the biological clock.

3. Briefly describe the steps in the hypersensitive response and systemic acquired resistance that result from a plant's encounter with an avirulent pathogen.

Test Your Knowledge

TRUE OR FALSE: *Indicate T or F and then correct the false statements.*

_____ 1. Abscisic acid is necessary for a seed to break dormancy.

_____ 2. Cytokinins, synthesized in the root, counteract apical dominance.

_____ 3. Thigmomorphogenesis is a growth response of a seedling that encounters an obstacle while pushing upward through the soil.

_____ 4. Action potentials are electrical impulses that may be used as internal communication in plants.

_____ 5. A physiological response to day or night length is called a circadian rhythm.

_____ 6. Vernalization is the need for pretreatment with cold before seed germination.

_____ 7. A virulent pathogen is one to which a plant has a specific resistance based on gene-for-gene recognition.

_____ 8. A cryptochrome is the light-absorbing portion of a phytochrome that reverts between two isomeric forms.

MULTIPLE CHOICE: *Choose the one best answer.*

1. The body form within a species of plant may vary more than that within a species of animal because
   a. growth in animals is indeterminate.
   b. plants respond adaptively to their environments by altering their patterns of growth and development.
   c. plant growth and development are governed by many hormones.
   d. plants can respond to environmental stress.
   e. All of the above are reasons why plant form varies more than animal body form.

2. Polar transport of auxin involves
   a. the accumulation of higher concentrations on the side of a shoot away from light.
   b. the movement of auxin from roots to shoots.
   c. movement of auxin through transport proteins located only at the basal end of cells.
   d. the unidirectional active transport of auxin into and out of parenchyma cells.
   e. the settling of statoliths.

3. According to the acid growth hypothesis,
   a. auxin stimulates membrane proton pumps.
   b. a lowered pH outside the cell activates expansins that break cross-links between cellulose microfibrils.
   c. the membrane potential created by the proton pumps enhances ion uptake, increasing the osmotic movement of water into cells.
   d. cells elongate when they take up water by osmosis.
   e. All of the above are involved in cell elongation.

4. Which of the following situations would most likely stimulate the development of axillary buds?
   a. a large quantity of auxin traveling down from the shoot and a small amount of cytokinin produced by the roots
   b. a small amount of auxin traveling down from the shoot and a large amount of cytokinin traveling up from the roots
   c. an equal ratio of gibberellins to auxin
   d. the absence of cytokinins caused by the removal of the terminal bud
   e. an increase in the concentration of brassinosteroids produced by leaves

5. The growth inhibitor in seeds is usually
   a. abscisic acid.
   b. ethylene.
   c. gibberellin.
   d. a small amount of ABA combined with a larger concentration of gibberellins.
   e. a high cytokinin-to-auxin ratio.

6. A circadian rhythm
   a. is controlled by an external oscillator.
   b. is a physiological cycle of approximately a 24-hour frequency.
   c. involves an internal biological clock that is not set by daily environmental signals.
   d. provides the signal for seasonal flowering.
   e. involves all of the above.

7. Which of the following is *not* a component of the signal transduction pathway involved in the de-etiolation process when a shoot breaks ground?
   a. reception of light by a phytochrome located in the cytosol
   b. inhibition of the triple response and initiation of phototropism as the shoot bends toward light
   c. production of second messengers such as cGMP and Ca^{2+}
   d. cascades of protein kinase activations that phosphorylate transcription factors
   e. activation of genes that code for photosynthesis-related enzymes

8. Injecting phytochrome into cells of the tomato mutant *aurea*
   a. causes the plant to undergo the triple response above ground when exposed to ethylene.
   b. causes the plant to undergo the triple response above ground without exposure to ethylene.
   c. helps researchers identify clock mutants.
   d. initiates flowering even when nights are longer than a critical period.
   e. causes the plant to develop a normal de-etiolation response to light.

9. A flash of far-red light during a critical-length dark period will
   a. induce flowering in a long-day plant.
   b. induce flowering in a short-day plant.
   c. not influence flowering.
   d. increase the P_{fr} level suddenly.
   e. be negated by a second flash of far-red light.

10. The conversion of P_r to P_{fr}
    a. occurs slowly at night.
    b. may be the way in which plants sense daybreak and serves to entrain the biological clock.
    c. is the molecular mechanism responsible for the biological clock.
    d. occurs when P_r absorbs far-red light.
    e. occurs in flower buds and induces flowering.

11. A plant may withstand salt stress by
    a. releasing abscisic acid that closes stomata to salt accumulation.
    b. the production of organic solutes that lower the water potential of root cells.
    c. wilting, which reduces water and salt uptake by reducing transpiration.
    d. producing canavanine that reduces the toxic effect of sodium ions.
    e. releasing ethylene that leads to apoptosis of damaged cells.

12. Which of the following hormones would be sprayed on barley seeds to speed germination in the production of malt for making beer?
    a. abscisic acid
    b. auxin
    c. cytokinin
    d. ethylene
    e. gibberellin

13. Many plants will flower in response to a specific
    a. flowering hormone that is produced in the apical bud.
    b. elicitor.
    c. minimal temperature that appears to signal a seasonal change.
    d. photoperiod, which seems to be measured by the length of darkness to which the leaves of the plant are exposed.
    e. combination of a high level of auxin and a low level of cytokinin.

14. Which of the following is *not* a plant defense against herbivory?
    a. production of distasteful compounds.
    b. production of toxic compounds such as canavanine.
    c. physical defenses such as thorns.
    d. initiation of a hypersensitive response with production of phytoalexins and PR proteins.
    e. release of volatile compounds that recruit parasitoid wasps.

15. Which of the following is *not* true of the hypersensitive response?
    a. It relies on an *R-Avr* recognition between a specific plant cell receptor protein and a pathogen molecule.
    b. It increases the production of PR proteins that may have antimicrobial functions.
    c. It enhances the production of ethylene that serves as a signal molecule transported throughout the plant to activate SAR (systemic acquired resistance).
    d. It contains an infection by stimulating cross-linking of cell wall molecules and production of lignin.
    e. The plant cells involved in the defense destroy themselves, leaving lesions that indicate the site of the contained infection.

UNIT 7

Animal Form and Function

Chapter 40

Basic Principles of Animal Form and Function

Key Concepts

40.1 Animal form and function are correlated at all levels of organization

40.2 Feedback control loops maintain the internal environment in many animals

40.3 Homeostatic processes for thermoregulation involve form, function, and behavior

40.4 Energy requirements are related to animal size, activity, and environment

Framework

Most animal bodies have highly folded exchange surfaces and a circulatory system that distributes materials throughout the body. The body structures of an animal are built from four basic tissues: epithelial, connective, muscle, and nervous. Organs function together in organ systems. Structure correlates with function in these hierarchical levels of organization.

The internal environment is carefully regulated by feedback mechanisms of homeostasis. Thermoregulatory mechanisms include insulation, circulatory adaptations, evaporative heat loss, behavioral responses, and regulation of metabolic heat production. Endotherms use metabolic heat to maintain body temperature; ectotherms gain their heat mostly from the environment.

The functions of an animal are powered by chemical energy derived from food. Metabolic rate, the amount of energy used in a unit of time, is higher for endothermic animals and inversely related to body size.

Chapter Review

Anatomy is the study of an organism's structure; **physiology** is the study of biological function.

40.1   Animal form and function are correlated at all levels of organization

Physical Constraints on Animal Size and Shape   The laws of hydrodynamics drove the natural selection for a fusiform body shape in fast-swimming fishes, birds, and mammals, an example of convergent evolution. Physical laws also influence body size, as larger bodies require stronger skeletons and adequate muscle mass for locomotion.

Exchange with the Environment   Every cell must be in an aqueous medium to allow for exchange across its plasma membrane. Animals with two-layered, saclike bodies or thin, flat bodies can maintain sufficient cellular contact with the aqueous environment.

Animals with compact bodies, however, must provide extensively branched or folded internal membranes for exchanging materials with the environment. The circulatory system connects these exchange surfaces with the **interstitial fluid** bathing the body's cells.

INTERACTIVE QUESTION   40.1

List some advantages of a compact, complex body form in which most of the body cells are not in contact with the external environment.

Hierarchical Organization of Body Plans   Groups of similar cells with a common function form **tissues. Organs** often consist of a layered arrangement of tissues. Groups of organs are integrated into **organ systems,** which perform the major functions required for life.

Tissue Structure and Function   **Epithelial tissue** lines the outer and inner surfaces of the body in sheets of tightly packed cells. A *simple epithelium* has one layer; a *stratified epithelium* has multiple layers of cells; and a *pseudostratified epithelium* has a single layer of cells that vary in length. The shape of cells at the *apical* surface facing the lumen or outside of an organ may be *squamous* (flat), *cuboidal* (boxlike), or *columnar* (pillarlike). Cells at the *basal* surface of an epithelium are attached to a *basal lamina,* a dense layer of extracellular matrix.

INTERACTIVE QUESTION   40.2

Name the two types of epithelia illustrated below. One of these epithelia lines the esophagus, while the other lines the intestines. Explain why each would be found in its location.

a. _____   b. _____

Connective tissue binds and supports other tissues and is characterized by relatively few cells suspended in an extracellular matrix of fibers embedded in a liquid, jellylike, or solid foundation.

Connective tissue fibers are of three types. *Collagenous fibers* are made of collagen and have nonelastic strength. *Elastic fibers,* made of the protein elastin, can stretch and provide resilience. Branched and thin *reticular fibers* are composed of collagen and form a tightly woven connection with adjacent tissues.

Loose connective tissue attaches epithelia to underlying tissues and holds organs in place. The most common types of cells enmeshed in such tissues are **fibroblasts,** which secrete the protein of the extracellular fibers, and **macrophages,** which engulf bacteria and cellular debris.

Fibrous connective tissue, with its dense arrangement of parallel collagenous fibers, is found in **tendons,** which attach muscles to bones, and in **ligaments,** which join bones together at joints.

Bone is formed by *osteoblasts,* which deposit a matrix of collagen. Calcium, magnesium, and phosphate ions form a hard mineral within the matrix. *Osteons* consist of concentric layers of matrix around a central canal containing blood vessels and nerves.

Cartilage is composed of collagenous fibers embedded in a rubbery substance; both are secreted by *chondrocytes.* Cartilage is a strong but somewhat flexible support material.

Adipose tissue pads and insulates the body and stores fat. Adipose cells each contain a large fat droplet.

Blood has a liquid extracellular matrix called plasma. Erythrocytes (red blood cells) carry oxygen; leukocytes (white blood cells) function in defense; and cell fragments called platelets are involved in the clotting of blood.

INTERACTIVE QUESTION 40.3

Identify the types of connective tissue and their components in the following three micrographs.

a. _____

b. _____

c. _____

d. _____

e. _____

f. _____

g. _____

h. _____

i. _____

The cells of **muscle tissue** contain contracting filaments of actin and myosin. **Skeletal muscle**—also called *striated muscle*—is responsible for voluntary body movements. **Cardiac muscle,** forming the wall of the heart, is also striated, but its cells are branched and joined by intercalated discs. **Smooth muscle** is composed of spindle-shaped cells lacking striations. It is found in the walls of the digestive tract, arteries, and other internal organs, and is under involuntary control.

Nervous tissue senses stimuli and transmits electrical signals. The **neuron,** or nerve cell, consists of a cell body and two or more processes that conduct impulses toward (dendrites) and away from (axons) the rest of the neuron. **Glial cells (glia)** support and nourish neurons.

INTERACTIVE QUESTION 40.4

Identify these types of vertebrate muscle. What is the dark band at the end of the line in micrograph **a,** and what is its function?

a. _____

b. _____

Coordination and Control The endocrine and nervous systems coordinate the body's activities. The bloodstream carries **hormones** to target tissues; neurons deliver electrical signals directly to target cells.

INTERACTIVE QUESTION 40.5

Compare the speed of signaling and the duration of the response in the endocrine and nervous systems. How do the functions of these two communication systems differ?

40.2 Feedback control loops maintain the internal environment in many animals

Regulating and Conforming For any given environmental variable, an animal may be a **regulator,** using mechanisms to control internal fluctuations, or a **conformer,** allowing internal conditions to vary with certain environmental changes.

Homeostasis The maintenance of a "steady state" or relative internal balance is called **homeostasis.** Homeostatic control mechanisms involve a receptor or **sensor,** which detects a **stimulus,** a fluctuation above or below a particular **set point** for a variable, and triggers a **response** that returns the variable to the set point.

Most homeostatic mechanisms operate by **negative feedback,** in which the response reduces the stimulus, which then reduces the response. Variables may have a **normal range** within which they fluctuate instead of a single set point.

Homeostatic mechanisms allow for *regulated change* in the body's internal environment when necessary. In **acclimatization,** an animal's normal range of homeostasis may change as the animal adjusts to external environmental changes.

INTERACTIVE QUESTION   40.6

What is **positive feedback**? Give an example. Are positive feedback mechanisms usually involved in maintaining homeostasis? Explain.

40.3   Homeostatic processes for thermoregulation involve form, function, and behavior

Changes in temperature affect most biochemical and physiological processes. Animals maintain internal temperatures within an optimal range through the process of **thermoregulation.**

Endothermy and Ectothermy   Animals that use metabolic heat in generating body temperature are **endothermic** (birds, mammals, some fishes, a few other reptiles, and some insects). Animals that gain their heat mostly from the environment are **ectothermic** (most invertebrates, fishes, amphibians, lizards, snakes, and turtles). Ectothermy requires far less energy than endothermy.

Variation in Body Temperature   *Poikilotherm* refers to animals with varying body temperatures; *homeotherm* refers to animals with relatively stable body temperatures.

INTERACTIVE QUESTION   40.7

Explain the following statement: Some ectotherms are homeotherms, and some even have higher body temperatures than those of endotherms.

Balancing Heat Loss and Gain   The flow of heat within an organism and between an organism and the external environment is affected by the physical processes of conduction, the direct transfer of thermal motion between objects in contact; convection, the transfer of heat by the flow of air or water past a surface; radiation, the emission of electromagnetic waves by all objects warmer than absolute zero; and evaporation, the loss of heat due to the conversion of the surface molecules of a liquid to a gas.

The mammalian **integumentary system** consists of skin, hair, and nails. The skin serves both thermoregulatory and protective functions.

Insulation reduces the rate of heat exchange with the environment. Hair, feathers, and fat layers have evolved in mammals and birds. Fur and feathers insulate mainly by trapping still air. A very thick layer of blubber insulates marine mammals.

How can circulatory adaptations contribute to thermoregulation? *Vasodilation* of superficial blood vessels increases blood flow in the skin and the transfer of body heat to the environment. *Vasoconstriction* reduces blood flow and heat transfer. In **countercurrent exchange,** common in marine mammals and birds, the close association of blood vessels servicing the extremities allows heat in arterial blood leaving the body core to be transferred to returning venous blood. Many endothermic insects maintain high temperatures for the flight muscles in the thorax with a countercurrent heat exchanger.

Mammals and birds may use the cooling mechanism of evaporative heat loss. Evaporation across the skin and respiratory surface can prevent body temperatures from rising when environmental temperatures are above body temperature. Panting, sweating, and bathing may enhance evaporative cooling.

Behavioral responses contribute to thermoregulation in both endotherms and ectotherms. Amphibians easily lose heat by evaporation from their moist skin and seek suitable microclimates. Reptiles other than birds often thermoregulate by orienting the body to the sun or moving to cool areas. Many terrestrial invertebrates use similar behavioral mechanisms. The social organization of honeybees enables them to retain heat in cold temperatures by huddling together and to cool their hive during hot weather by bringing in water and fanning their wings.

Mammals and birds are able to regulate the rate of metabolic heat production through activity and shivering. Some mammals generate heat through *nonshivering thermogenesis,* a rise in metabolic rate and the production of heat instead of ATP. Some mammals have *brown fat* stores that are specialized for rapid heat production.

A few large reptiles use shivering to generate heat and become endothermic while incubating eggs. Endothermic flying insects may warm up before taking off by contracting their powerful flight muscles.

Acclimatization in Thermoregulation Birds and mammals often adjust the thickness of their insulating coats to meet seasonal changes. Ectotherms may adjust their cellular physiology (changes in membrane composition or production of "antifreeze" compounds) to acclimatize to changes in environmental and thus body temperature.

Physiological Thermostats and Fever In mammals, a group of neurons in the **hypothalamus** functions as a thermostat that triggers heat loss or gain mechanisms in response to a body temperature outside the normal range. Mammals and birds develop a fever in response to certain infections, apparently due to the resetting of the hypothalamus set point.

INTERACTIVE QUESTION 40.8

a. What cooling mechanisms are activated when the hypothalamus senses a body temperature above the set point?

b. What mechanisms are activated when body temperature falls below the set point?

40.4 Energy requirements are related to animal size, activity, and environment

Energy Allocation and Use An animal's **bioenergetics** involves the overall flow and transformation of energy—the input of energy in the form of food, its use in the body, and its return to the environment as heat or in waste products. Organic fuel molecules are absorbed into body cells and used to generate ATP for cellular work and biosynthesis.

Quantifying Energy Use The total energy an animal uses in a unit of time is its **metabolic rate.** Metabolic rate can be measured by placing an animal in a calorimeter and measuring heat loss. The rate of oxygen consumption or CO_2 production can also be used to determine metabolic rate.

Minimum Metabolic Rate and Thermoregulation The minimal metabolic rate for a nongrowing endotherm at rest, fasting, and nonstressed is called the **basal metabolic rate (BMR).** The **standard metabolic rate (SMR)** is the metabolic rate of a resting, fasting, nonstressed ectotherm determined at a specific temperature.

A human adult male's BMR is about 1,600–1,800 kcal/day; a female's is about 1,300–1,500 kcal/day. The SMR of a comparably sized ectotherm would be much less.

Influences on Metabolic Rate The energy required to maintain each gram of body weight is inversely related to body size. Smaller mammals have higher metabolic rates, breathing rates, volume of blood, and heart rates, and require more food per unit of body mass. With a greater surface-to-volume ratio, small endotherms may have a higher energy cost to maintain a stable body temperature. But the factors that contribute to this inverse relationship among ectotherms as well as endotherms are not fully understood.

Maximal metabolic rates occur during intense activity. The sustainable maximum metabolic rate is inversely related to the duration of the activity. Most terrestrial animals have an average daily rate of energy consumption that is 2 to 4 times BMR or SMR.

Energy Budgets Different species vary in the allocation of their food energy for BMR or SMR, activity, growth, reproduction, and temperature regulation.

Torpor and Energy Conservation Torpor is a physiological state characterized by decreases in metabolism and activity that allow an animal to save energy while avoiding difficult conditions.

INTERACTIVE QUESTION 40.9

Describe the following types of torpor. What challenges does each one help an animal meet?

a. Hibernation

b. *Estivation*

c. Daily torpor

Word Roots

chondro- = cartilage; **-cyte** = cell (*chondrocytes*: cartilage cells)

con- = with; **-vect** = carried (*convection*: the mass movement of warmed air or liquid to or from the surface of a body or object)

counter- = opposite (*countercurrent heat exchanger*: a special arrangement of blood vessels that helps trap heat in the body core and is important in reducing heat loss in many endotherms)

-dilat = expanded (*vasodilation:* an increase in the diameter of superficial blood vessels triggered by nerve signals that relax the muscles of the vessel walls)

ecto- = outside; **-therm** = heat (*ectotherm:* an animal, such as a reptile (other than birds), fish, or amphibian, that must use environmental energy and behavioral adaptations to regulate its body temperature)

endo- = inner (*endotherm:* an animal, such as a bird or mammal, that uses metabolic energy to maintain a constant body temperature)

fibro- = a fiber (*fibroblast:* a type of cell in loose connective tissue that secretes the protein ingredients of the extracellular fibers)

homeo- = same; **-stasis** = standing, posture (*homeostasis:* the steady-state physiological condition of the body)

inter- = between (*interstitial fluid:* the internal environment of vertebrates, consisting of the fluid filling the space between cells)

macro- = large (*macrophage:* an amoeboid cell that moves through tissue fibers, engulfing bacteria and dead cells by phagocytosis)

osteo- = bone; **-blast** = a bud, sprout (*osteoblasts:* bone-forming cells that deposit a matrix of collagen)

Structure Your Knowledge

1. Fill in the table below on the structure and function of the four types of animal tissues.

2. Organs are composed of layers of several different tissues. Which of the four animal tissues do you think would be included in all organs? In what types of organs would you predict the remaining tissues would be included? Give an example of an organ composed of several layers of tissues.

3. Explain this statement: Endotherms can tolerate wider external fluctuations in temperature; ectotherms may be able to tolerate wider internal temperature fluctuations.

4. An animal's energy budget may include basal (or standard) metabolism, reproduction, temperature regulation, growth, and activity. Describe some of the factors that could make the allocation of energy for these functions differ for different animals.

Tissue	Structural Characteristics	General Functions	Specific Examples

Test Your Knowledge

MULTIPLE CHOICE: *Choose the one best answer.*

1. Which of the following is *not* an organ system?
   a. skeletal
   b. connective
   c. digestive
   d. excretory
   e. immune and lymphatic

2. Which of the following is the best description of smooth muscle?
   a. spindle-shaped cells; involuntary control
   b. striated, branching cells; involuntary control
   c. spindle-shaped cells connected by intercalated disks; voluntary control
   d. striated cells containing overlapping filaments; involuntary control
   e. spindle-shaped striated cells; voluntary control

3. Which of the following is *not* true of connective tissue?
   a. It consists of few cells surrounded by fibers in a matrix.
   b. It includes such diverse tissues as bone, cartilage, adipose, and loose connective tissue.
   c. It connects and supports other tissues.
   d. It forms the internal and external lining of many organs.
   e. It can have a matrix that is a liquid, gel, or solid.

4. Which of the following are *incorrectly* paired?
   a. blood—erythrocytes, leukocytes, and platelets in plasma

b. bone—osteoblasts embedded in mineral matrix

c. loose connective tissue—collagenous, elastic, reticular fibers

d. adipose tissue—loose connective tissue with fat-storing cells

e. fibrous connective tissue—chondrocytes embedded in chondroitin sulfate

5. The interstitial fluid of vertebrates
   a. is the internal environment within cells.
   b. surrounds cells and provides for the exchange of nutrients and wastes.
   c. makes up the plasma of blood.
   d. is not necessary in flat, thin vertebrates.
   e. is less abundant in ectotherms than in endotherms.

6. Which of the following would be *least* likely to be true of an animal that is a regulator?
   a. It can live in a variable climate because of its homeostatic mechanisms.
   b. It may have a larger geographic range than a conformer.
   c. It may acclimatize to winter by increasing the thickness of its insulating coat.
   d. Much of its energy budget can be allocated to reproduction.
   e. It has behavioral as well as physiological mechanisms for responding to changing conditions.

7. Tuna, dolphins, penguins, and sharks all have a fusiform body shape. What is the best explanation for this similarity?
   a. This shape is an example of divergent evolution.
   b. They are all vertebrates and evolved from a common ancestor.
   c. The physical laws of hydrodynamics drove the natural selection for this shape.
   d. All aquatic animals have this shape, an example of convergent evolution.
   e. This shape is determined by the thermodynamics of maintaining body temperature in cold water.

8. Negative feedback circuits are
   a. mechanisms that most commonly maintain homeostasis.
   b. activated only when a physiological variable rises above a set point.
   c. analogous to a radiator that heats a room.
   d. involved in maintaining contractions during childbirth.
   e. found in endotherms but not in ectotherms.

9. The basal metabolic rate
   a. is constant for each species.
   b. may vary depending on the sex or size of an organism.
   c. is highest when an animal is actively exercising.
   d. is lower than the standard metabolic rate for ectotherms.
   e. may be measured from the quantity of food an animal eats after exercising.

10. Which of the following would most likely have the highest metabolic rate?
    a. whale
    b. dog
    c. snake
    d. tuna
    e. bat

11. Which of the following would be likely to produce antifreeze compounds in their cells?
    a. polar bear or artic wolf
    b. hibernating mammal
    c. estivating amphibian
    d. ectotherm in subzero environment
    e. chickadee in very cold winter

12. Of the following animals, whose percentage of the energy budget available for growth would be the largest?
    a. deer mouse from temperate forest
    b. python from tropical India
    c. human from temperate climate
    d. penguin from Antarctica
    e. They would all be the same, because most energy is used to maintain their basic metabolic rate.

13. Which of the following is used as a mechanism to dissipate heat?
    a. hibernation
    b. countercurrent heat exchange between vessels that service the extremities
    c. trapping air by raising fur or feathers
    d. vasodilation of surface vessels
    e. vasoconstriction of surface vessels

14. The rate of metabolic heat production can be increased by
    a. nonshivering thermogenesis.
    b. adding to the layer of brown fat.
    c. vasoconstriction.
    d. thick layers of blubber and countercurrent heat exchangers.
    e. all of the above.

15. Which of these animals would be likely to experience a daily torpor?

   a. bear

   b. frog

   c. lizard

   d. Belding's ground squirrel

   e. hummingbird

16. For most terrestrial animals, the average daily rate of energy consumption is 2–4 times BMR or SMR. Which of the following animals has an unusually low average daily metabolic rate?

   a. human in developed country

   b. wild turkey

   c. domesticated turkey

   d. forest-dwelling salamander

   e. desert snake

Chapter 41

Animal Nutrition

Key Concepts

41.1 **An animal's diet must supply chemical energy, organic molecules, and essential nutrients**

41.2 **The main stages of food processing are ingestion, digestion, absorption, and elimination**

41.3 **Organs specialized for sequential stages of food processing form the mammalian digestive system**

41.4 **Evolutionary adaptations of vertebrate digestive systems correlate with diet**

41.5 **Homeostatic mechanisms contribute to an animal's energy balance**

■ **Framework**

■ **Chapter Review**

Nutrition is the process by which organisms take in and use food. **Herbivores** eat plants or algae; **carnivores** eat animals; and **omnivores** consume both autotrophs and animals. Most animals are opportunistic and eat foods from different categories when they are available.

41.1   An animal's diet must supply chemical energy, organic molecules, and essential nutrients

An animal's diet provides fuel for ATP production, organic carbon and nitrogen molecules for biosynthesis, and **essential nutrients,** which are molecules that an animal requires but cannot synthesize itself.

Essential Nutrients   Eight of the 20 amino acids that are needed to make proteins are **essential amino acids** required in the diet of most animals and adult humans. Protein deficiency develops from a lack of one or more essential amino acids in the diet and produces retarded physical and perhaps mental development.

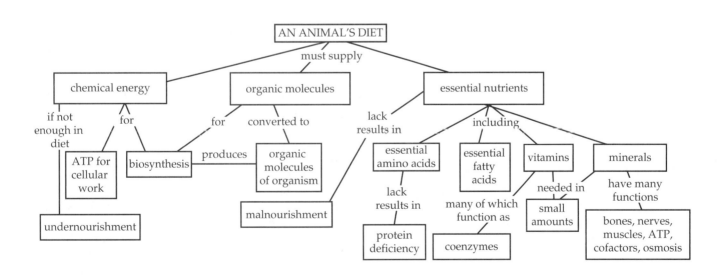

Meat, eggs, and cheese contain "complete" proteins with all essential amino acids in proportions that meet human requirements. Most plant proteins are "incomplete," deficient in one or more essential amino acids.

INTERACTIVE QUESTION   41.1

How can vegetarians avoid protein deficiencies?

Animals are able to make most of the fatty acids they need. Linoleic acid is required in the human diet to make some membrane phospholipids. Deficiencies of **essential fatty acids** are rare.

Vitamins are essential organic molecules required in small amounts in the diet. Thirteen vitamins essential to humans have been identified. Water-soluble vitamins include the B complex, most of which function as coenzymes, and vitamin C, which is required for the production of connective tissue. The fat-soluble vitamins are A, incorporated into visual pigments; D, aiding in calcium absorption and bone formation; E, antioxidant protecting cell membranes; and K, required for blood clotting.

Minerals are inorganic nutrients, usually needed in very small amounts. Calcium is needed for normal nerve and muscle function, and phosphorus is found in ATP and nucleic acids. Iron is a component of the cytochromes and hemoglobin. Other minerals function as cofactors of enzymes. Iodine is needed by vertebrates to make the metabolism-regulating thyroid hormones. Sodium, potassium, and chloride ions are important in nerve function and osmotic balance.

Dietary Deficiencies   **Undernourishment** results from a diet insufficient in calories. With severe deficiency, the body breaks down its own proteins for energy, eventually causing irreversible damage or death. **Malnourishment** can lead to disease, deformity, and death.

INTERACTIVE QUESTION   41.2

a. What is the difference between undernourishment and malnourishment?

b. Which two minerals do vertebrates require in relatively large amounts? Why?

Assessing Nutritional Needs   Research on human dietary requirements involves many challenges. Techniques include the study of genetic defects that affect nutrient uptake or use and *epidemiology*, the study of disease among various populations, which may have different dietary practices.

41.2   The main stages of food processing are ingestion, digestion, absorption, and elimination

Ingestion is the act of consuming food. Many aquatic animals are **suspension feeders,** sifting small food particles from the water. **Substrate feeders** live in or on their food, eating their way through it. **Fluid feeders** suck fluids from a living plant or animal host. Most animals are **bulk feeders,** which eat relatively large pieces of food.

Digestion splits food into small molecules by **enzymatic hydrolysis,** the addition of a water molecule when the bond between components is broken. Mechanical fragmentation often precedes chemical digestion.

In the last two stages of food processing, **absorption** and **elimination,** small molecules are taken into the cells of the animal, and the undigested remainder of the food leaves the digestive system.

INTERACTIVE QUESTION   41.3

Why must macromolecules be digested into monomers or smaller components?

Digestive Compartments   **Intracellular digestion** of food molecules in food vacuoles is typical of sponges. Following phagocytosis or pinocytosis, food vacuoles fuse with enzyme-containing lysosomes, and digestion occurs safely within a membrane-enclosed compartment.

Most animals break down their food, at least initially, by **extracellular digestion** within a separate compartment of the body that connects to the external environment.

A **gastrovascular cavity** functions in both digestion and transport of nutrients throughout the body. In the small cnidarian hydra, food breakdown is initiated by digestive enzymes secreted into the gastrovascular cavity. Gastrodermal cells take in food particles, and hydrolysis of macromolecules occurs within food vacuoles. Undigested materials are expelled through the mouth/anus.

Most animals have **complete digestive tracts** or **alimentary canals,** with two openings, a mouth and an anus, and specialized regions allowing for the sequential processing of nutrients. Food, ingested through the mouth and pharynx, passes through an esophagus that leads to a crop, stomach, or gizzard—organs specialized for storing or grinding food. In the intestine, digestive enzymes hydrolyze macromolecules, and nutrients are absorbed across the tube lining. Undigested material exits through the anus.

INTERACTIVE QUESTION 41.4

Label and list the functions for the five organs of an earthworm's digestive tract. What is the function of the typhlosole?

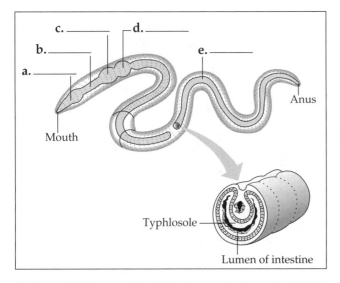

41.3 Organs specialized for sequential stages of food processing form the mammalian digestive system

The mammalian digestive system consists of the alimentary canal and accessory glands. Rhythmic waves of muscular contraction called **peristalsis** push food through the tract. Ringlike valves called **sphincters** regulate the passage of material between some segments. Accessory glands—the salivary glands, pancreas, liver, and gallbladder—secrete digestive juices into the alimentary canal through ducts.

The Oral Cavity, Pharynx, and Esophagus Physical and chemical digestion begins in the mouth, where teeth grind food to expose a greater surface area to enzyme action. The presence of food in the **oral cavity** triggers the release of saliva from **salivary glands.**

Saliva has several components: mucin, a glycoprotein that protects the mouth lining from abrasion and lubricates food for swallowing; buffers to neutralize

acidity; antibacterial agents; and **amylase,** which begins the hydrolysis of starch and glycogen. The tongue tastes and manipulates food, and pushes the food ball, or **bolus,** into the pharynx for swallowing.

The **pharynx** is the intersection leading to both the esophagus and the trachea. During swallowing, the **larynx,** the upper part of the respiratory tract, moves up so that the cartilaginous *epiglottis* covers and blocks the *glottis,* the vocal cords and the opening between them.

Food moves down through the **esophagus** to the stomach, squeezed along by a wave of peristalsis.

Digestion in the Stomach In the expandable **stomach,** food is combined with **gastric juice,** producing a mixture called **chyme.**

Gastric glands in the stomach secrete hydrochloric acid, which breaks down food tissues, kills bacteria, and denatures proteins, and a **protease** called **pepsin,** an enzyme that hydrolyzes specific peptide bonds in proteins. Pepsin is synthesized and secreted by chief cells in an inactive form called **pepsinogen.** It is activated by HCl (secreted by parietal cells) and by pepsin itself—an example of positive feedback.

The **mucus** secreted by the epithelium protects the stomach lining from digestion. Gastric ulcers, mainly caused by the bacterium *Helicobacter pylori,* may worsen when the lining is eroded faster than it can be regenerated.

Smooth muscles churn the contents of the stomach. The opening to the esophagus is closed except when a bolus arrives to prevent backflow of the acid chyme. A sphincter regulates passage of chyme into the intestine.

INTERACTIVE QUESTION 41.5

List the two types of macromolecules that have been partially digested by the time chyme moves into the intestine. Where did this digestion take place, and what enzymes were involved?

a.

b.

Digestion in the Small Intestine Digestive juices are mixed with chyme in the **duodenum,** the first section of the **small intestine.** The **pancreas** produces digestive enzymes and a bicarbonate-rich alkaline solution that offsets the acidity of the chyme. The **liver** produces **bile,** which is stored in the **gallbladder** until needed.

Bile aids in the digestion of fats and contains pigments that are by-products of the breakdown of red blood cells in the liver. The brush border of the duodenum secretes some enzymes; others are bound to the surfaces of these epithelial cells.

Hormones, such as gastrin, secretin, and cholecystokinin (CCK), regulate the release of digestive secretions from the stomach, pancreas, and gallbladder.

Most digestion is completed in the duodenum. The *jejunum* and *ileum* function in nutrient and water absorption.

INTERACTIVE QUESTION   41.6

Describe the enzymatic digestion of the following molecules in the small intestine.

a. polysaccharides

b. polypeptides

c. nucleic acids

d. fats

Absorption in the Small Intestine   Large folds of the small intestine lining are covered with fingerlike projections called **villi,** on which the epithelial cells have microscopic extensions called **microvilli,** creating a huge surface area for absorption.

The core of each villus has a net of capillaries and a lymph vessel called a **lacteal.** Nutrients are absorbed across the epithelium of the villus and then across the single-celled epithelium of the capillaries or lacteal. Transport may be passive by diffusion or active by pumping against a gradient.

Monoglycerides and fatty acids are absorbed by epithelial cells where they recombine to form fats and are mixed with cholesterol and coated with proteins to make tiny globules called **chylomicrons.** These packages move out of the epithelial cells by exocytosis and into a lacteal, and they are then transported by the lymphatic system to veins near the heart.

Absorbed amino acids and sugars enter capillaries. The nutrient-laden blood from the small intestine is carried directly to the liver by the **hepatic portal vein.** The liver regulates the nutrient content of the blood and detoxifies foreign molecules.

Absorption in the Large Intestine   The small intestine leads into the **large intestine** at a T-shaped junction with a sphincter. A pouch, called the **cecum,** with a fingerlike extension in humans, called the **appendix,** attaches at this juncture. The **colon** finishes the reabsorption of the large quantity of water secreted with digestive enzymes into the digestive tract.

Escherichia coli and other mostly harmless bacteria live on unabsorbed organic matter in the large intestine. Some of these bacteria produce B vitamins and vitamin K, which are absorbed by the host. The **feces** contain cellulose, other undigested materials, and a large proportion of intestinal bacteria. Feces are stored in the **rectum,** the final section of the large intestine.

41.4   Evolutionary adaptations of vertebrate digestive systems correlate with diet

Some Dental Adaptations   Dentition, the type and arrangement of teeth, correlates with diet.

Stomach and Intestinal Adaptations   Herbivores have longer alimentary canals because plant material is more difficult to digest than meat.

Symbiotic Adaptations   Many herbivorous mammals have fermentation chambers filled with symbiotic bacteria and protists. These microorganisms, often housed in the cecum, digest cellulose to simple sugars and produce a variety of essential nutrients.

Ruminants have an elaborate system involving several stomach chambers, regurgitation and rechewing of the cud, and digestion of their symbiotic bacteria to maximize the nutrient yield of their grass diet.

INTERACTIVE QUESTION   41.7

Compare and contrast the dentition and alimentary canals of carnivores and herbivores.

41.5   Homeostatic mechanisms contribute to an animal's energy balance

Energy Sources and Stores   Carbohydrates and fats are used preferentially to produce ATP by cellular respiration. When an animal consumes more calories than are needed to meet its energy requirements, the excess can be stored in the liver and muscles as glycogen or in adipose tissue as fat when the glycogen stores are full. The level of glucose in the blood is carefully regulated by the pancreatic hormones insulin, which lowers blood glucose, and glucagon, which raises blood glucose.

Overnourishment and Obesity Overnourishment, consuming more calories than the body needs, can lead to obesity. This excessive accumulation of fat is associated with a number of health problems.

Appetite-regulating hormones act on the brain's "satiety center," which generates impulses that make one feel hungry or satiated. These hormones include ghrelin, which is secreted by the stomach and triggers hunger, and three hormones that suppress appetite: insulin from the pancreas, PYY from the small intestine, and leptin produced by adipose tissue.

Research with mutant mice led to the identification of the *ob* gene that produces leptin and the *db* gene that codes for the leptin receptor.

Obesity and Evolution Sometimes obesity is adaptive, as in the obese chicks of some petrel species. Gorging on the lipid-rich food supplied by parent birds enables them to obtain enough protein for growth and survive times when food is scarce.

INTERACTIVE QUESTION 41.8

How might the human craving for fatty foods, which is helping to fuel the current rise in obesity, have evolved through natural selection?

Word Roots

chylo- = juice; **micro-** = small (*chylomicron:* small globules composed of fats that are mixed with cholesterol and coated with special proteins)

chymo- = juice; **-trypsi** = wearing out (*chymotrypsin:* an enzyme found in the duodenum; it is specific for peptide bonds adjacent to certain amino acids)

di- = two (*dipeptidase:* an enzyme found attached to the intestinal lining; it splits small peptides)

epi- = over; **-glotti** = the tongue (*epiglottis:* a cartilaginous flap that blocks the top of the windpipe, the glottis, during swallowing)

extra- = outside (*extracellular digestion:* the breakdown of food outside cells)

gastro- = stomach; **-vascula** = a little vessel (*gastrovascular cavity:* pouch that serves as the site of extracellular digestion and a passageway to disperse materials throughout most of an animal's body)

herb- = grass; **-vora** = eat (*herbivore:* a heterotrophic animal that eats plants)

hydro- = water; **-lysis** = to loosen (*hydrolysis:* a chemical process that lyses or splits molecules by the addition of water)

intra- = inside (*intracellular digestion:* the joining of food vacuoles and lysosomes to allow chemical digestion to occur within the cytoplasm of a cell)

micro- = small; **-villi** = shaggy hair (*microvilli:* many fine, fingerlike projections of the epithelial cells in the lumen of the small intestine that increase its surface area)

omni- = all (*omnivore:* an animal that consumes both meat and plant material)

peri- = around; **-stalsis** = a constriction (*peristalsis:* rhythmic waves of contraction of smooth muscle that push food along the digestive tract)

Structure Your Knowledge

1. Label the indicated structures in this diagram of the human digestive system. Review the functions of these structures.

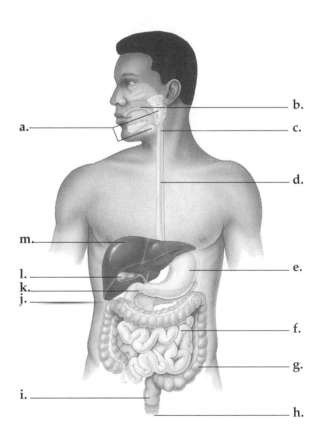

MATCHING: *Match the description with the correct enzyme, hormone, or other substances.*

_____ 1. enzyme that hydrolyzes peptide bonds, works in the stomach

_____ 2. hormone that stimulates secretion of gastric juice

_____ 3. hormone that stimulates release of bile and pancreatic enzymes, also inhibits peristalsis

_____ 4. enzyme that begins digestion of starch in the mouth

_____ 5. enzymes specific for hydrolyzing various disaccharides

_____ 6. substances that emulsify fats

_____ 7. enzyme that hydrolyzes fats

_____ 8. hormone secreted by stomach that triggers hunger

_____ 9. hormone that stimulates secretion of bicarbonate ions from the pancreas

_____ 10. intestinal enzyme specific for peptide bonds adjacent to specific amino acids

A. aminopeptidase
B. amylase
C. bile salts
D. cholecystokinin
E. chymotrypsin
F. disaccharidases
G. gastrin
H. ghrelin
I. insulin
J. lipase
K. maltase
L. nuclease
M. pepsin
N. secretin

MULTIPLE CHOICE: *Choose the one best answer.*

1. Substrate feeders such as earthworms
   a. eat mostly mineral substrates.
   b. filter small organisms from water.
   c. eat autotrophs.
   d. move and eat through their food.
   e. are bulk feeders.

2. In animals, a distinct advantage of extracellular digestion over intracellular digestion is that
   a. polymers are hydrolyzed to monomers by digestive enzymes.
   b. there is a greater surface area for absorption of digested nutrients.
   c. larger pieces of food can be ingested and then digested.
   d. all four types of macromolecules can be digested instead of just carbohydrates.
   e. the products of extracellular digestion can be absorbed into all body cells, without the need for a transport system.

3. Why do many vegetarians combine different protein sources in their diet or eat some animal products such as eggs or milk products?
   a. to make sure they obtain sufficient calories
   b. to provide sufficient vitamins
   c. to make sure they ingest all essential fatty acids
   d. to make their diet more interesting
   e. to provide all essential amino acids

4. Which of the following statements is *false*?
   a. The average human has enough stored fat to supply calories for several weeks.
   b. An increase in leptin levels leads to an increase in appetite and weight gain.
   c. Conversion of glucose and glycogen takes place in the liver.
   d. After glycogen stores are filled, excessive calories are stored as fat, regardless of their original food source.
   e. Carbohydrates and fats are preferentially used as fuel before proteins are used.

5. The purpose of antidiarrhea medicine would most likely be to
   a. speed up peristalsis in the small intestine.
   b. speed up peristalsis in the large intestine.
   c. kill *E. coli* in the intestine.
   d. increase water reabsorption in the large intestine.
   e. increase salt secretion into the feces.

6. Incomplete proteins are
   a. lacking in essential vitamins.
   b. a cause of undernourishment.
   c. found in meat, eggs, and cheese.
   d. deficient in one or more essential amino acids.
   e. a result of overcooking vegetables.

7. Which of the following is a *true* statement about vitamins?
   a. They may be produced by intestinal micro-organisms.
   b. Requirements for them are the same from one species to the next.
   c. They are stored in large quantities in the liver.
   d. They are inorganic nutrients, needed in small amounts, which often function as cofactors.
   e. They are water soluble and must be replaced every day.

8. You and your friends are trying a cereal diet to lose weigh. You each eat 30 g of cereal, six times a day. You choose an organic cereal with nuts, grains, and no added sugar that has 3 g of fat, 20 g of carbohydrates (4 g of which are nondigestible insoluble fiber), and 7 g of protein. Your first friend chooses a high fiber cereal with 1 g of fat, 27 g of carbohydrates (9 g of insoluble fiber), and 2 g of protein. Your second friend chooses a sugary cereal with 0 g of fat, 28 g of carbohydrates (3 g of insoluble fiber), and 2 g of protein. In which order, from the fastest to slowest, would you predict that the three of you will lose weight?
   a. you (because you have the lowest carbs), then friend one (because 9 g of carbohydrates are not digestible), then friend two
   b. friend one (9 g of insoluble fiber), then you, then your sugary friend two
   c. friend one (9 g of insoluble fiber), then friend two, then you (because your fat grams contain approximately two times the energy content of carbohydrate or protein)
   d. friend two (sugar burns the fastest), then friend one (with 9 g insoluble fiber), then you
   e. you (because proteins are not used as fuel, these 7 g do not count), then friend two, then friend one

9. Ruminants
   a. eat antlers or bones in order to obtain phosphorus.
   b. use microorganisms to digest cellulose.
   c. eat their feces to obtain nutrients digested from cellulose by microorganisms.
   d. house symbiotic bacteria and protists in a cecum.
   e. get all of their nutrition from digested plant material.

10. How do molting penguins obtain enough protein to produce new feathers?
    a. They become obese right before molting in order to store protein.
    b. They combine different protein sources to make up for the incomplete proteins in their food.
    c. They become malnourished during the period of new feather production.
    d. They feed on fish, an excellent protein source, while molting.
    e. They stockpile extra muscle proteins prior to molting, which then supply amino acids for protein synthesis.

11. Why does salivary amylase not hydrolyze starch in the duodenum?
    a. Starch is completely hydrolyzed into maltose in the oral cavity.
    b. The acid pH of the stomach denatures salivary amylase, and pepsin begins hydrolyzing it.
    c. Salivary amylase is produced by salivary glands and never leaves the oral cavity.
    d. Pancreatic amylase is a more effective enzyme in the pH of the duodenum.
    e. Salivary amylase can hydrolyze glycogen but not starch.

12. After a meal of greasy french fries, which enzymes would you expect to be most active?
    a. salivary and pancreatic amylase, disaccharidases, lipase
    b. lipase, lactase, maltase
    c. pepsin, trypsin, chymotrypsin, dipeptidases
    d. gastric juice, bile, bicarbonate
    e. sucrase, lipase, bile

13. Which of the following is mismatched with its function?
    a. most B vitamins—coenzymes
    b. vitamin E—antioxidant
    c. vitamin K—blood clotting
    d. iron—component of thyroid hormones
    e. phosphorus—bone formation, nucleotide synthesis

14. Which of the following is *not* a common component of feces?
    a. intestinal bacteria
    b. cellulose
    c. saturated fats
    d. bile pigments
    e. undigested materials

15. The hepatic portal vein
    a. supplies the capillaries of the intestines.
    b. carries absorbed nutrients to the liver for processing.
    c. carries blood from the liver to the heart.
    d. drains the lacteals of the villi.
    e. supplies oxygenated blood to the liver.

16. An organism that exclusively uses extracellular digestion would
    a. be a sponge.
    b. be a hydra with a gastrovascular cavity.
    c. not need a circulatory system.
    d. have to be very thin and elongated.
    e. have a specialized body compartment with digestive enzymes.

17. What moves a bolus of food down the esophagus to the stomach?
    a. the closing of the epiglottis over the glottis
    b. a lower pressure in the abdominal cavity compared to the oral cavity
    c. the contraction of the diaphragm and the opening of the sphincter to the stomach
    d. peristalsis caused by waves of smooth muscle contraction
    e. the action of cilia coated by mucus

18. What are villi, and what do they do?
    a. folds in the stomach that allow the stomach to expand
    b. extensions of the lymphatic system that pick up digested fats for transport to the circulatory system
    c. fingerlike projections of the small intestine lining that increase the surface area for absorption
    d. microscopic extensions of epithelial cells lining the small intestine that provide more surface area for digestion
    e. projections of the intestinal capillaries that join to form the hepatic portal vein

Chapter 42

Circulation and Gas Exchange

42.1 Circulatory systems link exchange surfaces with cells throughout the body

42.2 Coordinated cycles of heart contraction drive double circulation in mammals

42.3 Patterns of blood pressure and flow reflect the structure and arrangement of blood vessels

42.4 Blood components function in exchange, transport, and defense

42.5 Gas exchange occurs across specialized respiratory surfaces

42.6 Breathing ventilates the lungs

42.7 Adaptations for exchange include pigments that bind and transport gases

Framework

This chapter surveys the circulatory and gas exchange systems found in animals, with special attention given to the human systems and the problems of cardiovascular disease. The following concept map organizes some of the chapter's key ideas.

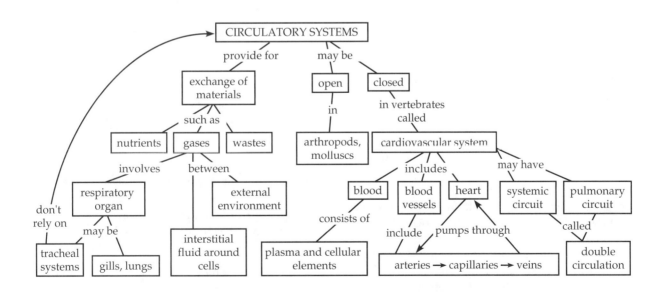

Chapter Review

42.1 Circulatory systems link exchange surface with cells throughout the body

A cell's aqueous environment must provide oxygen and nutrients and permit disposal of carbon dioxide and metabolic wastes. Because diffusion is too slow a process across distances more than a few millimeters, most animals have organs that exchange materials with the environment and an internal transport system to service body cells.

Gastrovascular Cavities   The central gastrovascular cavity inside the two-cell-thick body wall of cnidarians serves for both digestion and transport of materials. The internal fluid exchanges directly with the aqueous environment through the single opening. Flatworms also have gastrovascular cavities that branch throughout their thin, flat body.

Open and Closed Circulatory Systems   The circulatory systems of more complex animals consist of a pumping **heart,** a circulatory fluid, and blood vessels.

In the **open circulatory system** found in arthropods and most molluscs, **hemolymph** (body fluid) in sinuses, or spaces between organs, bathes the internal tissues, providing for chemical exchange.

Annelids, cephalopods, and vertebrates have **closed circulatory systems,** in which the blood remains in vessels and materials are exchanged with the interstitial fluid bathing the cells.

INTERACTIVE QUESTION   42.1

a. How is the hemolymph of an open circulatory system moved through the body?

b. Compare the advantages of open and closed circulatory systems.

Organization of Vertebrate Circulatory Systems   In the **cardiovascular system** of vertebrates, **arteries,** carrying blood away from the heart, branch into tiny **arterioles** within organs, which then divide into microscopic **capillaries,** which infiltrate tissues in networks called **capillary beds.** Capillaries converge to form **venules,** which meet to form the **veins** that return blood to the heart. In a few instances, portal veins connect two capillary beds without the blood returning first to the heart.

The vertebrate heart has one or two **atria,** which receive blood, and one or more **ventricles,** which pump blood out of the heart.

The ventricle of the two-chambered heart of a fish pumps blood to the gills, from which the oxygen-rich blood flows through a vessel to capillary beds in the other organs. Veins return the oxygen-poor blood to the atrium. Passage through two capillary beds slows the flow of blood, but body movements help to maintain circulation.

In the three-chambered heart of amphibians, the single ventricle pumps blood through a forked artery into the **pulmocutaneous circuit,** which leads to the lungs and skin and then back to the left atrium, and the **systemic circuit,** which carries blood to the rest of the body and back to the right atrium. This double circulation repumps blood after it returns from the capillary beds of the lungs or skin, ensuring a strong flow of oxygen-rich blood to the brain, muscles, and body organs.

The three-chambered heart of lizards, snakes, and turtles has a partially divided ventricle that helps to separate blood flow through a **pulmonary circuit** to the gas exchange tissues in the lungs and through a systematic circuit to the body tissues.

Delivery of oxygen for cellular respiration is most efficient in birds and mammals, which, as endotherms, have high oxygen and energy demands. The left side of the large and powerful four-chambered heart handles only oxygen-rich blood, whereas the right side receives and pumps oxygen-poor blood.

INTERACTIVE QUESTION   42.2

a. Explain the difference between a **single circulation** and a **double circulation.**

b. Describe the difference between the circulatory system of amphibians and that of mammals.

42.2 Coordinated cycles of heart contraction drive double circulation in mammals

Mammalian Circulation   Review the circulation of blood through a mammalian cardiovascular system by completing Interactive Question 42.3.

INTERACTIVE QUESTION 42.3

The following diagram will help you review the pathway of blood through a mammalian circulatory system. Label the indicated parts, color the vessels that carry oxygen-rich blood red, and then trace the flow of blood by numbering the circles from 1–11. Start with number 1 in the right ventricle.

p. _____

o. _____

c. _____

n. _____

a. _____

b. _____

d. _____

m. _____

l. _____

k. _____

j. _____

e. _____

f. _____

g. _____

h. _____

i. _____

The Mammalian Heart: **A Closer Look** The human heart, located just beneath the sternum, is composed mostly of cardiac muscle. The atria have relatively thin muscular walls, whereas the ventricles have thicker walls. The **cardiac cycle** consists of **systole,** during which the cardiac muscle contracts and the chambers pump blood, and **diastole,** when the heart chambers are relaxed and filling with blood.

The **cardiac output,** or volume of blood pumped by each ventricle per minute, depends on **heart rate** and **stroke volume,** or quantity of blood pumped by each contraction.

Atrioventricular (AV) valves between each atrium and ventricle are snapped shut when blood is forced against them as the ventricles contract. The **semilunar valves** at the exit of the aorta and pulmonary artery are forced open by ventricular contraction and close when the ventricles relax and blood starts to flow back toward the ventricles.

The heartbeat sounds are caused by the recoil of blood against the closed heart valves. A **heart murmur** is the detectable hissing sound of blood leaking back through a defective valve.

Maintaining the Heart's Rhythmic Beat Certain cells of cardiac muscle have an intrinsic ability to contract. The rhythm of contractions is coordinated by the **sinoatrial (SA) node,** or pacemaker, a region of specialized muscle tissue located in the wall of the right atrium.

The SA node initiates an electrical impulse that spreads via the intercalated disks of the cardiac muscle cells, and the two atria contract. The **atrioventricular (AV) node,** located between the right atrium and ventricle, relays the impulse after a 0.1-second delay through fibers to the ventricles. The electrical currents produced during the heart cycle can be detected by electrodes placed on the skin and recorded in an **electrocardiogram (ECG** or **EKG).**

The SA node is controlled by two sets of nerves with antagonistic signals and is influenced by hormones, temperature, and exercise.

INTERACTIVE QUESTION   42.4

a. Name the valve between an atrium and ventricle. _____

b. Name the valve between a ventricle and the aorta or pulmonary artery. _____

c. Which node initiates atrial systole? _____

d. Which node initiates ventricular systole? _____

42.3   Patterns of blood pressure and flow reflect the structure and arrangement of blood vessels

Blood Vessel Structure and Function   The wall of an artery or vein consists of three layers: an outer connective tissue layer with elastic fibers that allow the vessel to stretch and recoil; a middle layer of smooth muscle and more elastic fibers; and an inner lining of **endothelium,** a single layer of flattened cells. The outer two layers are thicker in arteries. One-way valves in the larger veins assure that blood flows back toward the heart. Capillaries have only the endothelial layer.

Blood Flow Velocity   The enormous number of capillaries creates a large total diameter, and blood flow is very slow, improving opportunity for exchange with interstitial fluid. Blood flow speeds up within venules and veins, due to the decrease in total cross-sectional area.

Blood Pressure   Blood pressure, the force exerted against the wall of a blood vessel, is much greater in arteries than in veins and is greatest during systole **(systolic pressure).** This hydrostatic pressure drives blood from the heart to the capillary beds. Resistance caused by the narrow openings of the arterioles impeding the exit of blood from arteries causes the swelling of the arteries during systole, which can be felt as the **pulse.** The recoiling of the stretched elastic arteries during diastole creates **diastolic pressure** and maintains a continuous blood flow into arterioles and capillaries.

Arterial blood pressure may be measured with a sphygmomanometer. The first number is the pressure during systole (when blood first spurts through the artery that was closed off by the cuff), and the second is the pressure during diastole. **Vasoconstriction,** or contraction of smooth muscles in arteriole walls, increases resistance and thus increases blood pressure, whereas **vasodilation** of arterioles as smooth muscles relax increases blood flow into the arterioles and thus lowers blood pressure. Nervous and hormonal signals control these muscles. Changes in cardiac output also affect blood pressure.

Endothelin is a peptide produced and secreted by endothelial cells that induces vasoconstriction.

Blood pressure is negligible by the time blood exits the capillary beds. Contraction of the smooth muscle walls of venules and veins and the contraction of skeletal muscles between which veins are embedded force blood to flow back to the heart. Pressure changes during breathing also draw blood into the large veins in the thoracic cavity.

INTERACTIVE QUESTION   42.5

Complete the following concept map to help you organize your understanding of blood pressure.

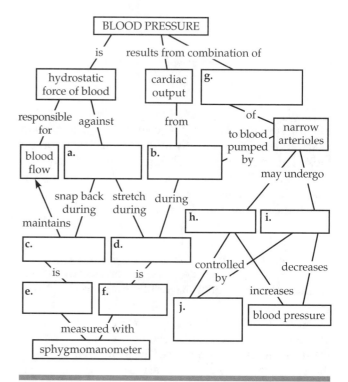

Capillary Function   Only about 5% to 10% of the body's capillaries have blood flowing through them at any one time. Capillaries branch off thoroughfare channels that are direct connections between arterioles and venules. The distribution of blood to capillary beds varies with need and is regulated by nerve signals and hormones that control contraction of smooth muscle in arteriole walls and precapillary sphincters at the entrance to capillary beds.

The exchange of substances between blood and interstitial fluid may involve transport by endocytosis and exocytosis by the endothelial cells, passive diffusion of small molecules, and bulk flow through the openings between cells due to hydrostatic pressure. Water, sugars, salts, oxygen, and urea move through these openings, whereas blood cells and proteins are too large to fit through. Blood pressure at the upstream end of a capillary forces fluid out, whereas osmotic pressure at the downstream end tends to draw about 85% of that fluid back into the capillary.

Fluid Return by the Lymphatic System   The fluid that does not return to the capillary and any proteins that may have leaked out are returned to the blood through the **lymphatic system.** Fluid diffuses into tiny lymph vessels intermingled with blood capillaries. The fluid, called **lymph,** moves through lymph vessels with one-way valves, mainly as a result of the movement of skeletal muscles, and is returned to the circulatory system near the junction of the venae cavae with the right atrium. In **lymph nodes,** the lymph is filtered, and white blood cells attack viruses and bacteria.

42.4   Blood components function in exchange, transport, and defense

Blood Composition and Function   Cells and cell fragments make up about 45% of the volume of blood; the rest is a liquid matrix called **plasma.**

The plasma consists of a large variety of solutes dissolved in water. The collective and individual concentrations of the dissolved ions of inorganic salts are important to osmotic balance, pH levels, and the functioning of muscles and nerves. The kidney is responsible for the homeostatic regulation of these electrolytes.

Plasma proteins function in the osmotic balance of blood and as buffers, antibodies, escorts for lipids, and clotting factors. Blood plasma from which clotting factors have been removed is called serum. Nutrients, metabolic wastes, gases, and hormones are transported in the plasma.

Erythrocytes (red blood cells) transport oxygen. They lack mitochondria and generate ATP by anaerobic metabolism. The red blood cells of mammals lack nuclei. Their small size and biconcave shape create a large surface area of plasma membrane across which oxygen can diffuse. Erythrocytes are packed with **hemoglobin,** an iron-containing protein that binds oxygen.

There are five major types of **leukocytes,** or white blood cells, all of which fight infections. Some white cells are phagocytes that engulf bacteria and cellular debris; lymphocytes give rise to cells that produce the immune response. Leukocytes are also found in interstitial fluid and lymph nodes.

Platelets, pinched-off fragments of large cells in the bone marrow, are involved in blood clotting. The clotting process usually begins when platelets, clumped together along a damaged endothelium, release clotting factors. By a series of steps, the plasma protein fibrinogen is converted to its active form, **fibrin.** The threads of fibrin form a patch. Hemophilia is caused by an inherited defect in any step of the complex clotting process. Anticlotting factors normally prevent clotting of blood in the absence of injury. A **thrombus** is a clot that occurs within a blood vessel and blocks the flow of blood.

Erythrocytes circulate for about 3 to 4 months before they are phagocytosed by cells in the liver and spleen and their components recycled. Red and white blood cells and platelets are produced from multipotent **stem cells** in red bone marrow. The production of red blood cells is controlled by a negative feedback mechanism involving the hormone **erythropoietin (EPO),** which is produced by the kidney in response to low oxygen supply in tissues.

INTERACTIVE QUESTION   42.6

Complete this concept map of the components of vertebrate blood and their functions.

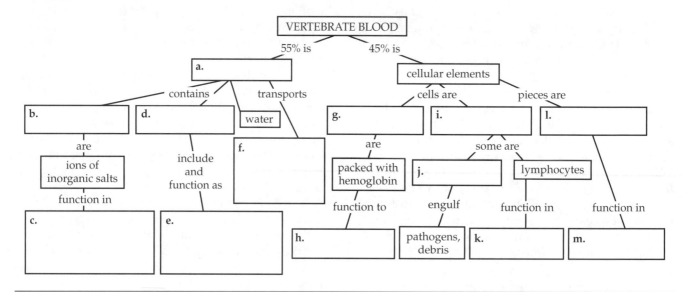

Cardiovascular Disease   Diseases of the heart and blood vessels are called cardiovascular diseases. Both lifestyle and genetics contribute to these diseases.

In the chronic cardiovascular disease known as **atherosclerosis,** plaques narrow the arteries, whose rough lining may trigger an inflammatory response and clot formation.

Blockage of coronary arteries leads to the death of cardiac muscle in a **heart attack;** blockage or rupturing of arteries in the head leads to a **stroke,** the death of nervous tissue in the brain. A heart attack or stroke may result from blockage by a thrombus or by a clot released when a plaque ruptures, which may then become lodged in a vessel in the heart or brain.

Cholesterol is carried in the blood as **low-density lipoproteins (LDLs)** and this type is associated with cholesterol deposits in plaques. **High-density lipoproteins (HDLs)** are cholesterol particles that appear to reduce cholesterol deposition in plaques.

Hypertension, or high blood pressure, is thought to damage the endothelium and initiate plaque formation, promoting atherosclerosis and increasing the risk of heart attack and stroke. This condition can be easily diagnosed and controlled by drugs, diet, and exercise.

INTERACTIVE QUESTION   42.7

What lifestyle choices have been correlated with increased risk of cardiovascular disease?

42.5   Gas exchange occurs across specialized respiratory surfaces

Gas exchange, or respiration, is the uptake of O_2 from the environment and the discharge of CO_2 to the environment.

Partial Pressure Gradients in Gas Exchange   The concentration of a gas in air or dissolved in water is measured as **partial pressure.** At sea level, the partial pressure of O_2, which makes up 21% of the atmosphere, is 160 mm Hg (0.21×760 mm—atmospheric pressure at sea level). The partial pressure of CO_2 is 0.23 mm Hg. A gas will diffuse from a region of higher partial pressure to lower partial pressure.

Respiratory Media   Air is a respiratory medium with plentiful O_2. Water has much less O_2 and greater density and viscosity, making gas exchange more challenging in water than in air.

Respiratory Surfaces The respiratory surface, where gas exchange with the respiratory medium occurs, must be moist, thin, and large enough to supply the whole body.

In sponges, cnidarians, and flatworms, gas exchange takes place across the entire surface of the body. In animals with denser bodies, a localized region of the body surface is usually specialized as a respiratory surface with a thin, moist epithelium associated with a rich blood supply. Most animals have an extensively branched or folded respiratory organ.

Gills in Aquatic Animals Gills are outfoldings of the body surface, ranging from simple bumps on echinoderms to the more complex gills of molluscs, crustaceans, and fishes. Due to the low oxygen concentration in a water environment, gills usually require **ventilation,** or movement of the respiratory medium across the respiratory surface. This is often an energy-intensive process due to the density of water.

In the gills of a fish, blood flows through capillaries in a direction opposite the flow of water. This arrangement sets up a **countercurrent exchange,** in which the diffusion gradient favors the movement of oxygen into the blood along the length of the capillary.

Tracheal Systems in Insects **Tracheal systems** are tiny air tubes, which branch throughout the body and come into contact with nearly every cell. Thus, an insect's open circulatory system does not function in O_2 and CO_2 transport. In large insects, rhythmic muscle contractions ventilate tracheal systems.

Lungs **Lungs** are invaginated respiratory surfaces restricted to one location from which oxygen is transported by the circulatory system.

The lungs of mammals are located in the thoracic cavity. Air, entering through the nostrils, is filtered, warmed, humidified, and smelled in the nasal cavity. Air passes through the pharynx and enters the respiratory tract through the glottis. The **larynx** moves up and tips the epiglottis over the glottis when food is swallowed. The larynx functions as a voicebox in most mammals; exhaled air vibrates a pair of **vocal cords.** The **trachea,** or windpipe, branches into two **bronchi,** which then branch repeatedly into **bronchioles** within each lung. Ciliated, mucus-coated epithelium lines the major branches of the respiratory tree. **Alveoli** are multilobed air sacs encased in a web of capillaries at the tips of the tiniest bronchioles. Gas exchange takes place across their thin, moist epithelium. These tiny sacs are coated with **surfactants,** which decrease surface tension and help keep them from sticking shut.

42.6 Breathing ventilates the lungs

Vertebrate lungs are ventilated by **breathing,** the alternate inhalation and exhalation of air.

How an Amphibian Breathes A frog uses **positive pressure breathing.** It lowers the floor of the oral cavity, expanding it and drawing air into the mouth; raising the floor of the oral cavity then pushes air into the lungs.

How a Mammal Breathes Mammals ventilate their lungs by **negative pressure breathing.** The lungs are enclosed in a double-walled pleural sac. Surface tension holds the two layers together, which also adhere to the lungs and the chest-cavity wall. Contraction of the rib muscles and the **diaphragm** expands the chest cavity and increases the volume of the lungs. Air pressure is reduced within this increased volume, and air is drawn into the lungs. Relaxation of the rib muscles and diaphragm compresses the lungs, increasing the pressure and forcing air out.

Tidal volume is the volume of air inhaled and exhaled by an animal during normal breathing. The maximum volume during forced breathing is called **vital capacity.** The **residual volume** is the air that remains in the alveoli and lungs after forceful exhaling.

How a Bird Breathes Birds have air sacs on either side of their lungs that act as bellows to maintain air flow through the lungs. The air sacs and lungs are ventilated through a circuit that includes a one-way passage through tiny, gas-exchange channels, called parabronchi, in the lungs.

Control of Breathing in Humans Breathing is under automatic regulation by the **breathing control centers** in the medulla oblongata and the pons. Nerve impulses from the medulla instruct the rib muscles and diaphragm to contract. A drop in the pH of the blood and cerebrospinal fluid when CO_2 concentration increases is sensed by the medulla's control center, which increases the depth and rate of breathing. Although breathing centers respond primarily to CO_2 levels, they are alerted by oxygen sensors in the aorta and carotid arteries that react to severe deficiencies of O_2.

INTERACTIVE QUESTION   42.8

For the following animals, indicate the respiratory medium, respiratory surface, and means of ventilation.

Animal	Respiratory Medium	Respiratory Surface	Ventilation
Fish	a.	b.	c.
Grasshopper	d.	e.	f.
Frog	g.	h.	i.
Human	j.	k.	l.

42.7 Adaptations for exchange include pigments that bind and transport gases

Coordination of Circulation and Gas Exchange
Blood entering the capillaries of the lungs has a lower P_{O_2} and a higher P_{CO_2} than does the air in the alveoli, so oxygen diffuses into the capillaries and carbon dioxide diffuses out. In the tissue capillaries, pressure differences favor the diffusion of O_2 out of the blood into the interstitial fluid and CO_2 into the blood.

Respiratory Pigments   In most animals, O_2 is carried by **respiratory pigments** in the blood. Copper is the oxygen-binding component in the blue respiratory protein hemocyanin, common in arthropods and many molluscs.

Hemoglobin, contained in erythrocytes, is the respiratory pigment of almost all vertebrates. Hemoglobin is composed of four subunits, each of which has a cofactor called a heme group with iron at its center. The binding of O_2 to the iron atom of one subunit induces a change in shape of the other subunits, and their affinity for oxygen increases. Likewise, the unloading of the first O_2 lowers the other subunits' affinity for oxygen.

The **dissociation curve** for hemoglobin shows the relative amounts of O_2 bound to hemoglobin under varying O_2 partial pressures. In the steep part of this S-shaped curve, a slight change in partial pressure will cause hemoglobin to load or unload a substantial amount of O_2. Because of the **Bohr shift** in response to the lower pH of active tissues, hemoglobin releases more O_2 in such tissues.

About 70% of CO_2 is transported in blood as bicarbonate ions. Carbon dioxide enters red blood cells, where it first combines with H_2O to form carbonic acid (catalyzed by carbonic anhydrase) and then dissociates into H^+ and HCO_3^-. Bicarbonate moves into the plasma for transport. The H^+ binds to hemoglobin and other proteins, which reduces changes in pH during transport

of CO_2. In the lungs, the diffusion of CO_2 out of the blood shifts the equilibrium in favor of the conversion of HCO_3^- back to CO_2, which is unloaded from the blood.

INTERACTIVE QUESTION   42.9

The following graph shows dissociation curves for hemoglobin at two different pH values. Explain the significance of these two curves.

Elite Animal Athletes   Pronghorns are very fast long-distance runners. Adaptations for speed and endurance include a large exchange surface in the lungs, high cardiac output, large muscle mass, and large numbers of mitochondria, all of which contribute to a very high rate of O_2 consumption.

Special physiological adaptations have enabled some air-breathing animals to make sustained underwater dives. Weddell seals store twice the amount of O_2/kg body weight as do humans, mostly by having a larger volume of blood, a huge spleen that stores blood, and a higher concentration of **myoglobin,** an oxygen-storing muscle protein. Oxygen-conserving adaptations during a dive include decreasing the heart rate, rerouting blood to the brain and essential organs, and restricting blood supply to the muscles, which can use fermentation to produce ATP during long dives.

Word Roots

alveol- = a cavity (*alveoli:* one of the dead-end, multi-lobed air sacs that constitute the gas exchange surface of the lungs)

atrio- = a vestibule; **-ventriculo** = ventricle (*atrioventricular node:* a region of specialized muscle tissue between the right atrium and right ventricle; it generates electrical impulses that primarily cause the ventricles to contract)

cardi- = heart; **-vascula** = a little vessel (*cardiovascular system:* the closed circulatory system characteristic of vertebrates)

counter- = opposite (*countercurrent exchange:* opposite flow of adjacent fluids that maximizes transfer rates)

endo- = inner (*endothelium:* the innermost, simple squamous layer of cells lining the blood vessels; the only constituent structure of capillaries)

erythro- = red; **-poiet** = produce (*erythropoietin:* a hormone produced in the kidney when tissues of the body do not receive enough oxygen. This hormone stimulates the production of erythrocytes.)

fibrino- = a fiber; **-gen** = produce (*fibrinogen:* the inactive form of the plasma protein that is converted to the active form fibrin, which aggregates into threads that form the framework of a blood clot)

hemo- = blood; **-philia** = loving (*hemophilia:* a human genetic disease caused by a sex-linked recessive allele, characterized by excessive bleeding following injury)

leuko- = white; **-cyte** = cell (*leukocyte:* a white blood cell)

multi- = many; **-potent** = powerful (*multipotent stem cell:* a cell within bone marrow that is a progenitor for any kind of blood cell)

myo- = muscle (*myoglobin:* an oxygen-storing, pigmented protein in muscle cells)

para- = beside, near (*parabronchi:* the sites of gas exchange in bird lungs; they allow air to flow past the respiratory surface in just one direction)

pulmo- = a lung; **-cutane** = skin (*pulmocutaneous:* the route of circulation that directs blood to the skin and lungs)

semi- = half; **-luna** = moon (*semilunar valve:* a valve located at the two exits of the heart, where the aorta leaves the left ventricle and the pulmonary artery leaves the right ventricle)

thrombo- = a clot (*thrombus:* a clump of platelets and fibrin that blocks the blood flow through a vessel)

Structure Your Knowledge

1. Identify the labeled structures in this diagram of a human heart. Draw arrows to trace the flow of blood.

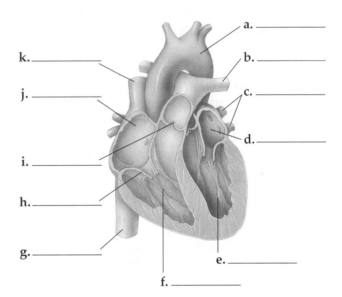

2. Trace the path of a molecule of oxygen being inhaled through the nasal cavity to delivery to the kidney.

Test Your Knowledge

MULTIPLE CHOICE: *Choose the one best answer.*

1. A gastrovascular cavity
   a. is found in cnidarians and annelids.
   b. functions to pump fluid in short vessels from which it spreads throughout the body.
   c. functions in both digestion and distribution of nutrients.
   d. has a single opening for ingestion and elimination, but a separate opening for gas exchange.
   e. involves all of the above.

2. Which of the following is *not* a similarity between open and closed circulatory systems?
   a. Some sort of pumping device helps to move blood through the body.
   b. Some of the circulation of blood is a result of movements of the body.
   c. The blood and interstitial fluid are distinguishable from each other.
   d. All tissues come into close contact with the circulating body fluid so that the exchange of nutrients and wastes can take place.
   e. All of these apply to both open and closed circulatory systems.

3. In a system with double circulation,
   a. blood is usually pumped at two separate locations as it circulates through the body.
   b. there is a countercurrent exchange in the gills.
   c. hemolymph circulates both through a pumping blood vessel and through body sinuses.
   d. blood is pumped to the gas-exchange organ and returning blood is then pumped through systemic vessels.
   e. there are always two ventricles and often two atria.

4. During diastole,
   a. the atria fill with blood.
   b. blood flows passively into the ventricles.
   c. the elastic recoil of the arteries maintains hydrostatic pressure on the blood.
   d. semilunar valves are closed, but atrioventricular valves are open.
   e. all of the above are occurring.

5. An atrioventricular valve prevents the backflow or leakage of blood from
   a. the right ventricle into the right atrium.
   b. the left atrium into the left ventricle.
   c. the aorta into the left ventricle.
   d. the pulmonary vein into the right atrium.
   e. the right atrium into the vena cava.

6. Breathing rate will increase when _____ CO_2 level in your blood causes a _____ in pH.
   a. increased/rise
   b. increased/drop
   c. decreased/rise
   d. decreased/drop
   e. The level of O_2 in the blood normally signals changes in breathing rate.

7. A heartbeat is initiated in humans by the
   a. SA node.
   b. AV node.
   c. right atrium.
   d. superior and inferior venae cavae.
   e. left ventricle.

8. Blood flows more slowly in the arterioles than in the arteries because the arterioles
   a. have thoroughfare channels to venules that are often closed off.
   b. collectively have a larger cross-sectional area than do the arteries.
   c. must provide opportunity for exchange with the interstitial fluid.
   d. have sphincters that restrict flow to capillary beds.
   e. are narrower than arteries.

9. If all the body's capillaries were open at one time,
   a. blood pressure would fall dramatically.
   b. resistance to blood flow would increase.
   c. blood would move too rapidly through the capillary beds.
   d. the amount of blood returning to the heart would increase.
   e. the increased gas exchange would allow for strenuous exercise.

10. Which of the following is *not* a factor in the exchange of substances in capillary beds?
    a. endocytosis and exocytosis
    b. passive diffusion
    c. hydrostatic pressure
    d. bulk flow through spaces between endothelial cells
    e. active transport by white blood cells

11. Edema involves the swelling of body tissues. It can result from
    a. swollen lymph glands.
    b. an abnormal increase in blood volume.
    c. too much lymph being returned to the circulatory system.
    d. too low a concentration of blood proteins.
    e. swollen feet.

12. The functions of the kidney include
    a. production of the hormone erythropoietin to increase production of red blood cells.
    b. maintenance of electrolyte balance.
    c. destruction and recycling of red blood cells.
    d. both a and b.
    e. a, b, and c.

13. Fibrinogen is
    a. a blood protein that escorts lipids through the circulatory system.
    b. a cell fragment involved in the blood-clotting mechanism.
    c. a blood protein that is converted to fibrin to form a blood clot.
    d. a leukocyte involved in trapping viruses and bacteria.
    e. a lymph protein that regulates osmotic balance in the tissues.

14. A thrombus
    a. may form at a region of plaque in an artery.
    b. is a traveling clot that may cause a heart attack or stroke.
    c. may cause hypertension.
    d. narrows the diameter of arteries due to the deposition of cholesterol.
    e. is often associated with high levels of HDLs in the blood.

15. In countercurrent exchange,
    a. the flow of fluids in opposite directions maintains a favorable diffusion gradient along the length of an exchange surface.
    b. oxygen is exchanged for carbon dioxide.
    c. double circulation keeps oxygenated and deoxygenated blood separate.
    d. oxygen moves from a region of high partial pressure to one of low partial pressure, but carbon dioxide moves in the opposite direction.
    e. the capillaries of the lung pick up more oxygen than do tissue capillaries.

16. The tracheae of insects
    a. are stiffened with chitinous rings and lead into the lungs.
    b. are filled by positive pressure breathing.
    c. ramify along capillaries of the circulatory system for gas exchange.
    d. are highly branched, coming into contact with almost every cell for gas exchange.
    e. provide an extensive, fluid-filled system of tubes for gas exchange.

17. Which of the following is *not* involved in speeding up breathing?
    a. nervous and chemical signals
    b. stretch receptors in the lungs
    c. impulses from the breathing centers in the medulla
    d. severe deficiencies of oxygen
    e. a drop in the pH of cerebrospinal fluid

18. Which of the following would tend to reduce the blood supply to capillaries?
    a. release of EPO (erythropoietin)
    b. relaxation of precapillary sphincters
    c. vasodilation
    d. closing of thoroughfare channel
    e. release of endothelin

19. Fluid moves back into the capillaries at the venous end of a capillary bed as a result of
    a. active transport.
    b. a blood pressure that is higher than the osmotic pressure.
    c. an osmotic pressure that is higher than the blood pressure.
    d. a partial pressure that is lower than that at the arteriole end.
    e. the squeezing of surrounding muscles on the interstitial fluid.

20. As a general rule, blood leaving the right ventricle of a mammal's heart will pass through how many capillary beds before it returns to the right ventricle?

    a. one
    b. two
    c. three
    d. one or two, depending on the circuit it takes
    e. at least two, but almost always three

21. Gas exchange between maternal and fetal blood takes place in the placenta. Fetal and adult hemoglobin differ slightly in composition, and their dissociation curves are different so that oxygen can be transferred from maternal blood to fetal blood. If curve *c* in the graph that follows represents the dissociation curve of adult hemoglobin, which curve would represent fetal hemoglobin? (*Hint:* Must fetal hemoglobin have a higher or lower affinity for oxygen than maternal hemoglobin has?)

Use the following to answer questions 22–25.

1. vena cava	4. right atrium
2. left ventricle	5. aorta
3. pulmonary vein	6. pulmonary capillaries

22. Which of the following sequences represents the flow of blood through the human body? (Obviously, many structures are not included.)

    a. 1-5-3-6-2-4   d. 1-2-3-6-4-5
    b. 1-2-6-3-4-5   e. 1-4-6-3-2-5
    c. 1-4-3-6-2-5

23. In which vessel would blood pressure be the highest?

    a. 1
    b. 3
    c. 4
    d. 5
    e. 6

24. In which vessel would velocity of blood flow be the lowest?

    a. 1
    b. 3
    c. 4
    d. 5
    e. 6

25. Of these structures, which would have the thickest muscle layer?

    a. 1
    b. 2
    c. 3
    d. 4
    e. 5

26. The nurse tells you that your blood pressure is 112/70. What does the 70 refer to?

    a. your heart rate
    b. the velocity of blood during diastole
    c. the systolic pressure from ventricular contraction
    d. the diastolic pressure from the recoil of the arteries
    e. the venous pressure caused by the compression of the blood pressure cuff

27. What do the alveoli of mammalian lungs, the gill filaments of fish, and the tracheoles of insects all have in common?

    a. use of a circulatory system to transport absorbed O_2
    b. respiratory surfaces that are invaginated
    c. countercurrent exchange
    d. a large, thin surface area for exchange
    e. all of the above

28. Which of the following would *not* be a possible function of a plasma protein in a mammal?

    a. pH buffer   d. oxygen transport
    b. osmotic balance   e. antibody
    c. clotting agent

29. What is the function of the cilia in the trachea and bronchi?
    a. movement of air into and out of the lungs
    b. increase the surface area for gas exchange
    c. vibrate when air rushes past them to produce sounds
    d. filter the air that rushes through them
    e. sweep mucus with its trapped particles up and out of the respiratory tract

30. Both birds and mammals have four-chambered hearts. Which of the following is the best explanation for this similarity?
    a. They shared a common ancestor that had a four-chambered heart.
    b. They are the only vertebrates with double circulation, and this double pumping requires four chambers.
    c. They are both endotherms, and the evolution of efficient circulatory systems supported the high metabolic rate of endotherms.
    d. This is an example of convergent evolution, because animals that obtain their oxygen from air require both a pulmonary and a systemic circuit.
    e. The more inefficient single atrium of amphibians and most reptiles could not supply the higher oxygen needs of these endotherms.

31. More O_2 is unloaded from hemoglobin in an actively metabolizing tissue than in a resting tissue, even at the same P_{O_2}. Why?
    a. The pH of an active tissue is higher as more CO_2 combines with water to produce H^+ and HCO_3^-.
    b. At the lower pH of an actively metabolizing tissue, hemoglobin has a lower affinity for O_2.
    c. Hemoglobin can bind four O_2 due to a change in conformation caused by the lower pH of the active tissue.
    d. More oxygen is bound to myoglobin at lower levels of oxygen concentration.
    e. More CO_2 is drawn out of tissue cells as it is converted to bicarbonate in red blood cells.

32. Which of the following does *not* contribute to the respiratory adaptations of deep-diving animals?
    a. an unusually large volume of blood
    b. large quantity of myoglobin in the muscles
    c. reduction in blood flow to noncritical body areas
    d. exceptionally large lungs
    e. large spleen that stores blood and releases it during a dive

Chapter 43

The Immune System

Key Concepts

43.1 In innate immunity, recognition and response rely on shared traits of pathogens

43.2 In acquired immunity, lymphocyte receptors provide pathogen-specific recognition

43.3 Acquired immunity defends against infection of body cells and fluids

43.4 Disruptions in immune system function can elicit or exacerbate disease

Framework

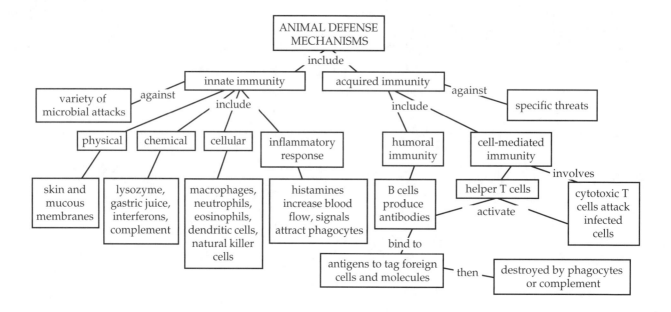

Chapter Review

The **immune system** defends against **pathogens,** infectious agents such as bacteria, viruses, protists, and fungi that cause disease. **Innate immunity** includes external physical barriers and internal defenses of immune cells that have a small group of receptor proteins that recognize a broad range of pathogens. **Acquired immunity,** also called *adaptive immunity,* is a line of defense in vertebrates in which immune cells react specifically to pathogens. A vast array of acquired immune receptors allow recognition and response to specific pathogens.

43.1 In innate immunity, recognition and response rely on shared traits of pathogens

Innate Immunity of Invertebrates Insect defenses begin with their protective exoskeleton. **Lysozyme,** an enzyme that attacks microbial cell walls, a low pH, and the chitin lining of the intestine all protect the digestive system. *Hemocytes* are circulating cells that can engulf and destroy bacteria by **phagocytosis,** trigger production of chemicals that entrap multicellular parasites, and secrete *antimicrobial peptides* that kill fungi and bacteria. Different classes of pathogens bind to distinct Toll receptors that activate pathways for the production of antimicrobial peptides effective against that group of pathogens.

INTERACTIVE QUESTION 43.1

Explain why the data presented in this graph indicate that defensin and not drosomycin is the antimicrobial peptide in fruit flies effective against infection by the bacteria *M. luteus.*

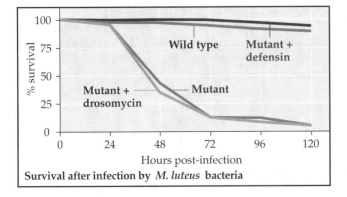

Survival after infection by *M. luteus* bacteria

Innate Immunity of Vertebrates In mammals, skin and the mucous membranes lining the digestive, respiratory, and genitourinary tracts are physical barriers to microbes. The ciliated, *mucus*-coated epithelial lining of the respiratory tract traps and removes microbes. Skin secretions maintain a low pH, which discourages colonization by microbes. The acidity of gastric juice kills most microorganisms that reach the stomach. Lysozyme present in tears, saliva, and mucus kills bacteria.

Cellular innate defenses involve specific **Toll-like receptors (TLRs)** that recognize molecules that are common to a set of pathogens. Recognition by a TLR triggers phagocytosis of microbes by phagocytic white blood cells (leukocytes). Microbes are engulfed and the resulting vacuoles fuse with lysosomes, which contain toxic gases (such as nitric oxide) and digestive enzymes.

Neutrophils are the most numerous phagocytic white blood cells. **Macrophages** may migrate through the body or become permanently attached in various organs of the lymphatic system. These macrophages attack microbes filtered from blood in the spleen and from interstitial fluid that flows as lymph through lymph nodes.

Eosinophils are leukocytes that attack multicellular parasitic invaders with destructive enzymes. The primary role of **dendritic cells,** located in tissues in contact with the environment, is to stimulate acquired immunity.

Various antimicrobial peptides and proteins are released in response to pathogen recognition. Virus-infected cells produce **interferons,** which stimulate neighboring cells to produce substances that inhibit viral reproduction in those cells. Another type of interferon activates macrophages. Interferons produced by recombinant DNA technology are used in treating certain viral infections.

The **complement system** is a group of about 30 proteins in the blood plasma that, when activated by contact with microbes, may lyse cells, trigger inflammation, or assist in acquired defenses.

Chemicals released in response to physical injury or pathogen entry can trigger an **inflammatory response,** characterized by redness, swelling, and heat. Damaged **mast cells** in connective tissue release **histamine,** which triggers dilation and leakiness of blood vessels; activated macrophages and other cells release signals that promote blood flow to the damaged area. Vasodilation as well as signaling proteins result in the congregation of phagocytic cells and the production of antimicrobial compounds. *Pus* is an accumulation of fluid, white blood cells, and dead microbes.

A systemic response to an infection may include an increase in the number of circulating white blood cells and a fever. Fever may be triggered by toxins produced by pathogens or by *pyrogens* released by macrophages. Fevers may stimulate phagocytosis and speed tissue repair. *Septic shock* is a dangerous condition resulting from an overwhelming systemic inflammatory response.

Natural killer (NK) cells recognize an absence of class I MHC molecule on virus-infected or cancer cells, attaching to them and triggering cell death.

Innate Immune System Evasion by Pathogens Certain pathogens may evade the innate immune system with an outer capsule that covers their surface molecules or by resisting breakdown in lysosomes and growing within the host's cells.

INTERACTIVE QUESTION 43.2

Complete the following table that summarizes the functions of the cells and chemicals of the innate defense mechanisms.

Cells or Compounds	Functions
Neutrophils	a.
Macrophages	b.
Eosinophils	c.
Dendritic cells	d.
Natural killer (NK) cells	e.
Mast cells	f.
Histamine	g.
Interferons	h.
Complement system	i.
Lysozyme	j.

43.2   In acquired immunity, lymphocyte receptors provide pathogen-specific recognition

Lymphocytes are the key cells of acquired immunity. **T cells** migrate to the **thymus** to mature, while **B cells** remain and mature in the bone marrow. B cells and T cells are activated by contact with foreign molecules and by **cytokines,** proteins secreted by macrophages and dendritic cells after they engulf microbes. Both B and T cells are involved in *immunological memory,* an enhanced response to a previously encountered foreign molecule.

Acquired Immunity: An Overview   Each B cell and T cell has many surface receptors that can bind to a particular foreign molecule. Such binding activates the cells, stimulating cell division and the production of some cells that will fight future infections by the same microbe. Some T cells help to activate other lymphocytes. Other T cells kill infected host cells. B cells secrete proteins that attach to foreign molecules and cells in body fluids.

Antigen Recognition by Lymphocytes   Most **antigens** are proteins or polysaccharides, often protruding from the surfaces of microbes. B cells and T cells have membrane-bound **antigen receptors** that allow them to recognize a specific antigen. Each B or T lymphocyte carries about 100,000 identical receptors. B cells may give rise to *plasma cells* that secrete **antibodies,** or **immunoglobulins (Ig),** which are soluble antigen receptors. The small region of an antigen to which a lymphocyte or secreted antibody binds is called an **epitope,** or *antigenic determinant.*

Each Y-shaped **B cell receptor** consists of four polypeptide chains: two identical **light chains** and two identical **heavy chains,** linked together by disulfide bridges. Both heavy and light chains have *variable (V) regions* at the ends of the two arms of the Y, which form two identical antigen-binding sites. The *constant (C) regions* of the molecule vary little from cell to cell. Secreted antibodies lack an anchoring transmembrane region and cytoplasmic tail but are otherwise structurally similar to B cell receptors.

A **T cell receptor,** also anchored by a transmembrane region, consists of one α *chain* and one β *chain,* linked by a disulfide bridge. The variable regions at the tip of the molecule form a single antigen-binding site. T cells recognize small fragments of antigens complexed with MHC molecules. A group of genes called the **major histocompatibility complex (MHC)** codes for these cell-surface proteins.

In a process called **antigen presentation,** MHC molecules bind with antigen fragments within the cell and then display them on the cell's surface, where a T cell receptor can recognize the antigen and MHC molecule complex.

Class I MHC molecules are found on almost all nucleated cells. They bind foreign antigens that an infected cell has produced. **Cytotoxic T cells** recognize class I MHC molecules displaying peptide antigens and kill the infected cell.

Class II MHC molecules are made by dendritic cells, macrophages, and B cells. These cells, called **antigen-presenting cells,** engulf and fragment microbes, then display their antigens in class II MHC molecules.

Cytotoxic T cells and **helper T cells,** which assist B cells and cytotoxic T cells, recognize class II MHC molecule-peptide antigen complexes.

INTERACTIVE QUESTION 43.3

Describe the differences between the antigens that B cell receptors and antibodies recognize and the antigens that T cell receptors on cytotoxic T cells and helper T cells recognize.

Lymphocyte Development   Each person may have as many as a million different B cells and 10 million different T cells. The genes coding for this diversity have numerous coding segments that are randomly and permanently rearranged.

For example, *immunoglobulin (Ig)* genes code for secreted antibodies. In the Ig light-chain gene, 40 variable *(V)* gene segments and five joining *(J)* gene segments are separated by a long stretch of DNA. The *J* segments are followed by an intron and a C exon that codes for the constant region. Early in B cell development, one *V* gene segment is randomly linked to one *J* segment by a set of enzymes called recombinase, producing one of 200 possible light chains. After transcription, the pre-mRNA is processed, and the mRNA is translated into a light chain with a variable and a constant region. These light chains combine with heavy chains that were produced the same way.

In the bone marrow and thymus, respectively, the antigen receptors of maturing B and T cells are tested for self-reactivity. Lymphocytes with receptors specific for the body's own molecules are either inactivated or destroyed by apoptosis. This critical *self-tolerance* means that normally there are no mature lymphocytes that react against self components.

When an antigen interacts with receptors on B cells or T cells that are specific for epitopes of that antigen, those particular lymphocytes are activated to divide repeatedly and differentiate into two clones—a large number of short-lived **effector cells,** which combat the antigen, and a clone of long-lived **memory cells,** all of which carry receptors for that antigen. By this **clonal selection,** a small number of cells is selected by their interaction with a specific antigen to produce thousands of cells keyed to that particular antigen.

The body mounts a **primary immune response** upon first exposure to an antigen. About 10 to 17 days are required for selected lymphocytes to proliferate and differentiate to yield the maximum response produced by effector T cells and the antibody-producing effector B cells, called **plasma cells.** Should the body reencounter the same antigen, the **secondary immune response** is more rapid, effective, and prolonged. The long-lived T and B memory cells are responsible for this immunological memory. This secondary immune response provides long-term protection against a previously encountered pathogen.

INTERACTIVE QUESTION 43.4

Answer the following questions concerning the three major stages in the development of lymphocytes.

a. How is the great diversity of B and T cells produced?

b. What prevents B and T cells from reacting against the body's own molecules?

c. Describe clonal selection.

43.3   Acquired immunity defends against infection of body cells and fluids

The **humoral immune response,** also called the antibody-mediated response, involves B cell activation and production of antibodies that circulate in the blood and lymph. The **cell-mediated immune response** involves cytotoxic T cells that destroy target cells. Helper T cells activate both B cells and cytotoxic T cells.

Helper T Cells: A Response to Nearly All Antigens   Helper T cells recognize specific class II MHC molecule-antigen complexes on antigen-presenting cells. A T cell surface protein called **CD4** enhances the binding by attaching to the class II MHC molecule. Signaling between the two cells results in the proliferation and differentiation of a clone of activated helper T cells and memory cells. Activated helper T cells secrete cytokines, which stimulate both the cell-mediated and humoral responses.

INTERACTIVE QUESTION   43.5

Label the components in this diagram that show a helper T cell being activated by interaction with a dendritic cell and the central role of the helper T cell in activating both humoral and cell-mediated immunity.

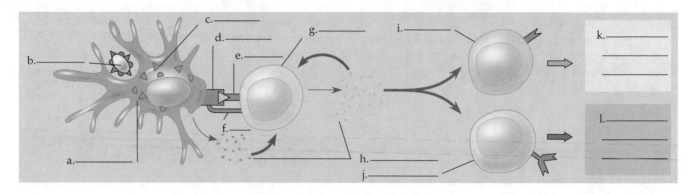

Cytotoxic T Cells: A Response to Infected Cells   Cytotoxic T cells recognize nonself proteins synthesized in infected cells and displayed with class I MHC molecules. **CD8** surface proteins on cytotoxic T cells bind to the side of class I MHC molecules and enhance the interaction between the cells, while the cytotoxic T cells, also stimulated by cytokines from nearby helper T cells, differentiate into active killers. The activated cell secretes proteins that kill the target cell. Pathogens released from the destroyed cell are marked by circulating antibodies for destruction.

INTERACTIVE QUESTION   43.6

a. What surface molecule of a helper T cell facilitates the interaction with a class II MHC molecule of an antigen-presenting cell and the helper T cell?

b. What surface molecule on a cytotoxic T cell assists in the interaction with a class I MHC molecule displayed on infected cells?

c. What does an activated helper T cell release?

d. What does a cytotoxic T cell attached to an infected body cell release?

B Cells: A Response to Extracellular Pathogens
B cells are selectively activated by antigens on the surface of bacteria. The activation is aided by cytokines released from helper T cells also activated by that antigen. Upon first binding antigen, the B cell takes in a few foreign molecules by receptor-mediated endocytosis and presents antigen fragments in its class II MHC molecules to helper T cells. Most protein antigens require the aid of helper T cells to stimulate antibody production. The activated B cell then proliferates into a clone of plasma cells and a clone of memory B cells. When certain antigens bind with multiple receptors on a single cell, a B cell response may not involve cytokines or helper T cells.

The antigen-binding sites on the arms of an antibody allow it to identify a specific antigen. The heavy-chain constant regions in the tail determine the antibody's distribution in the body and its function. There are five major types of constant regions, creating five classes of antibodies: IgM, IgG, IgA, IgD, and IgE. IgG is the most abundant antibody in blood and confers passive immunity on a fetus; IgA is present in tears, saliva, mucus, and breast milk.

Researchers can use the specificity of antigen-antibody binding for biological research, clinical testing, and medical applications. Some antibody tools are *polyclonal* because they were formed by several different B cell clones, each specific for a different epitope. A technique for making **monoclonal antibodies** can supply quantities of identical antibodies.

Antibodies label antigens for disposal by one of several mechanisms. In *neutralization*, antibodies may block the ability of a virus or bacterium to infect a host cell by binding to its surface. In *opsonization*, antibodies coat microbes and enhance phagocytosis by macrophages. Because each antibody molecule has at least two antigen-binding sites, the formation of antigen-antibody complexes produces clumps, which are then engulfed by phagocytes.

Antigen-antibody complexes on microbes may activate the complement system by binding with complement proteins and triggering the generation of a *membrane attack complex (MAC)*, which produces a pore in the membrane and causes the cell to lyse. Complement proteins can be activated as part of the innate or acquired defenses. In addition to lysing microbes, activated complement proteins promote inflammation and phagocytosis.

INTERACTIVE QUESTION 43.7

List three ways in which antibodies mediate the disposal of antigens.

a.

b.

c.

Active and Passive Immunization **Active immunity** can be acquired when the body produces antibodies and develops immunological memory from either exposure to an infectious agent or from **immunization,** also called **vaccination.** A vaccine may be an inactivated toxin, a killed or weakened microbe, a portion of a microbe, or even genes for microbial proteins. In **passive immunity,** temporary immunity is provided by antibodies supplied through the placenta to a fetus, through milk to a nursing infant, or by an antibody injection.

Immune Rejection The immune response to the chemical markers that determine ABO blood groups must be considered in blood transfusions. Antibodies to other blood group antigens (other than the individual's own antigens) arise in response to normal bacterial flora and circulate in the blood plasma, where they will induce a devastating transfusion reaction to transfused blood cells with matching antigens.

INTERACTIVE QUESTION 43.8

Fill in the following table to review your understanding of the antigens and antibodies of the ABO blood groups. Remember to compare the antigens on the donor cells with the antibodies in the recipient's plasma. (Assume these are packed cell transfusions in which the antibodies in the donor's plasma are not a risk factor.)

Blood Type	Antigens on RBCs	Antibodies in Plasma	Can Receive Blood from	Can Donate Blood to

---

Transplanted tissues and organs are rejected because the foreign MHC molecules are antigenic and trigger immune responses. The use of closely related donors, as well as drugs that suppress immune responses, helps to minimize rejection.

In bone marrow transplants, used to treat leukemia and blood cell diseases, the graft itself may be the source of immune rejection. The recipient's bone marrow cells are destroyed by irradiation, eliminating the recipient's immune system. The lymphocytes in the bone marrow transplant, however, may produce a *graft versus host reaction* if the MHC molecules of the donor and recipient are not well matched.

43.4 Disruptions in immune system function can elicit or exacerbate disease

Exaggerated, Self-Directed, and Diminished Immune Responses Allergies are hypersensitivities to certain environmental antigens, or **allergens.** IgE antibodies created in response to initial exposure to an allergen may bind to mast cells in connective tissue. When allergens then bind to these cell surface antibodies, the mast cells, in a process called *degranulation*, release histamines. The resulting inflammatory response may include sneezing, a runny nose, and difficulty in breathing due to smooth muscle contractions. Antihistamines are drugs that combat these symptoms by blocking receptors for histamine. *Anaphylactic shock* is a severe allergic response in which the abrupt dilation of peripheral blood vessels caused by widespread mast cell degranulation leads to a life-threatening drop in blood pressure.

Sometimes the immune system turns against self, leading to **autoimmune diseases,** such as *lupus, rheumatoid arthritis, insulin-dependent diabetes mellitus,* and *multiple sclerosis*. These diseases may be caused by a failure in the regulation of self-reactive lymphocytes.

Exercising to exhaustion and psychological stress have been shown to impair immune system function.

An **immunodeficiency** is the inability of the immune system to protect against pathogens. A genetic or developmental defect in the immune system is called an *inborn immunodeficiency*; a defect that arises later in response to chemical or biological agents is called an *acquired immunodeficiency*. An inborn immunodeficiency may occur in any of the components of the immune system. In the rare congenital disease known as severe combined immunodeficiency (SCID), lymphocytes and the acquired immune response are lacking. Acquired immunodeficiency may be caused by drugs used against autoimmune diseases or to suppress transplant rejection. Certain cancers, such as Hodgkin's disease, and **acquired immunodeficiency syndrome,** or **AIDS,** damage the immune system.

Acquired Immune System Evasion by Pathogens
Some viruses and parasites evade the body's defenses through *antigenic variation*, changing their surface epitopes so that the immune system cannot develop an immunological memory of the pathogen. Examples include the trypanosome that causes sleeping sickness and the influenza virus.

Some viruses enter an inactive state called *latency*, in which viral DNA remains in a host cell's nucleus. Should conditions change, the virus can replicate and spread to neighboring cells or to new hosts.

The infectious agent responsible for AIDS is HIV (human immunodeficiency virus). HIV infects cells with surface CD4 molecules, including helper T cells, macrophages, and brain cells. HIV RNA is reverse-transcribed into DNA, which is integrated into the host cell genome, from where it directs production of new viral particles. HIV escapes the immune system through both antigenic variation, when it mutates rapidly during replication, and latency, when it integrates into host cell DNA. HIV infection kills helper T cells through the effects of virus reproduction or by undergoing apoptosis, triggered by the virus. Individuals with AIDS are highly susceptible to opportunistic infections and cancers that take advantage of a suppressed immune system.

While not able to cure HIV, new drug combinations are slowing the progression to AIDS. The frequent mutational changes during replication generate drug-resistant strains of HIV. And frequent mutational changes in surface antigens have made development of

an effective vaccine difficult. HIV is transmitted by transfer of body fluids containing infected cells, such as blood or semen.

INTERACTIVE QUESTION   43.9

a. Why is AIDS such a deadly disease?

b. Why has it proved so difficult to prevent and cure this disease?

Cancer and Immunity   The incidence of certain cancers increases when the immune system is impaired. A common hypothesis is that a robust immune system destroys cancer cells. An alternative hypothesis is that the infections and inflammatory responses resulting from an impaired immune system contribute to the development of many cancers.

Word Roots

an- = without; **-aphy** = suck (*anaphylactic shock:* an acute, life-threatening, allergic response)

anti- = against; **-gen** = produce (*antigen:* a foreign macromolecule that does not belong to the host organism and that elicits an immune response)

cyto- = cell (*cytokines:* in the vertebrate immune system, protein factors secreted by macrophages and helper T cells as regulators of neighboring cells)

epi- = over; **-topo** = place (*epitope:* a localized region on the surface of an antigen that is chemically recognized by antibodies)

immuno- = safe, free; **-glob** = globe, sphere (*immunoglobulin:* one of the class of proteins comprising the antibodies)

macro- = large; **-phage** = eat (*macrophage:* an amoeboid cell that moves through tissue fibers, engulfing bacteria and dead cells by phagocytosis)

neutro- = neutral; **-phil** = loving (*neutrophil:* the most abundant type of leukocyte; neutrophils tend to self-destruct as they destroy foreign invaders, limiting their life span to but a few days)

perfora- = bore through (*perforin:* a protein that forms pores in a target cell's membrane)

Structure Your Knowledge

This chapter contains a wealth of information that is probably fairly new to you. If you take a little time and pull out the key players of the immune system and organize them first into very basic concept clusters and then develop more interrelated concept maps, you will find that this information is both understandable and fascinating.

1. Create a concept map outlining acquired immunity, showing the cells involved in humoral and cell-mediated responses and their functions.

2. Describe the structure of an antibody molecule and relate this structure to its function.

3. While we presented innate and acquired immunity separately, these defense mechanisms interact in several ways. Describe a few of the chemical and cellular players they share.

Test Your Knowledge

MULTIPLE CHOICE: *Choose the one best answer.*

1. Which of the following is *incorrectly* paired with its effect?
   a. gastric juice—kills bacteria in the stomach
   b. fever—may stimulate phagocytosis
   c. histamine—causes blood vessels to dilate
   d. vaccination—creates passive immunity
   e. lysozyme—attacks cell walls of bacteria

2. Which of the following would release interferon?
   a. a macrophage that has become an antigen-presenting cell
   b. an injured epithelial cell of a blood vessel
   c. a cell infected by a virus
   d. a mast cell that has bound an antigen
   e. a helper T cell bound to an antigen-presenting cell

3. Antibodies are
   a. proteins or polysaccharides usually found on the cell surface of invading bacteria or viruses.
   b. proteins embedded in T cell membranes.
   c. proteins circulating in the blood that tag foreign cells for complement destruction.
   d. proteins that consist of two light and two heavy polypeptide chains.
   e. c and d are both correct.

4. A secondary immune response is more rapid and greater in effect than a primary immune response because
   a. histamines cause rapid vasodilation.
   b. the second response is an active immunity, whereas the primary one was a passive immunity.
   c. helper T cells are available to activate other blood cells.
   d. chemical signals cause the rapid accumulation of phagocytic cells.
   e. memory cells respond to the pathogen and rapidly clone more effector cells.

5. Which of the following is *not* true of Toll-like receptors?
   a. They are found on phagocytic white blood cells.
   b. They recognize molecules specific to individual pathogens.
   c. They resemble the Toll receptor on immune response cells of insects, which initiate signal transduction pathways that result in synthesis of antimicrobial peptides.
   d. TLR proteins may be located on the plasma membrane or inside vesicles of leukocytes.
   e. They trigger innate immune defenses upon binding to molecules common to a set of pathogens.

6. Major histocompatibility complex molecules
   a. are involved in the ability to distinguish self from nonself.
   b. are a collection of cell surface proteins.
   c. may trigger T cell responses after transplant operations.
   d. present antigen fragments on infected cells.
   e. All of the above are correct.

7. In opsonization,
   a. antibodies coat microorganisms and help phagocytes bind to and engulf the foreign cell.
   b. a set of complement proteins lyses a hole in a foreign cell's membrane.
   c. antibodies coat binding sites of viruses.
   d. a flood of histamines is released that may result in anaphylactic shock.
   e. *V* gene segments and *J* gene segments are joined by recombinase.

8. Severe combined immunodeficiency
   a. is an inborn autoimmune disease.
   b. is a form of cancer in which MHC proteins are absent on transformed cells.
   c. is a disease in which both T and B cells are absent or inactive.
   d. is an immune disorder in which the number of helper T cells is greatly reduced.
   e. results from a few types of cancers, such as Hodgkin's disease.

9. A transfusion of type B blood given to a person who has type A blood would result in
   a. the recipient's anti-B antibodies reacting with the donated red blood cells.
   b. the recipient's B antigens reacting with the donated anti-B antibodies.
   c. the recipient forming both anti-A and anti-B antibodies.
   d. no reaction, because B is a universal donor type of blood.
   e. the introduced blood cells being destroyed by innate defense mechanisms.

10. Which of the following are *incorrectly* paired?
    a. variable region—determines antibody specificity for an epitope
    b. immunoglobulins—proteins that form epitopes
    c. constant region—determines class and function of antibody
    d. IgG—most abundant circulating antibodies, confer passive immunity to fetus
    e. IgE—antibody molecules attached to mast cells

11. Which of the following destroys a target cell by phagocytosis?
    a. neutrophil   d. complement proteins
    b. cytotoxic T cell   e. plasma cell
    c. natural killer cell

12. Which of the following best describes an insect's immune system?
    a. They rely on the barrier defense of an exoskeleton.
    b. Lysozyme and a chitin-lined intestine with a low pH protect their digestive system.
    c. Hemocytes can carry out phagocytosis of bacteria and foreign substances.
    d. They produce different antimicrobial peptides in response to binding of molecules from a particular type of pathogen to recognition proteins, which activate the protein Toll and a signal transduction pathway.
    e. All of the above are part of an insect's innate immunity.

13. What do IgE antibodies, T cell receptors, and MHC molecules have in common?
    a. They are found exclusively in cells of the immune system.
    b. They are all part of the complement system.
    c. They are antigen-presenting molecules.
    d. They are or can be membrane-bound proteins.
    e. They are involved in the cell-mediated portion of the immune system.

14. Which of the following is an effective defense against bacteria but does *not* work against viral particles?
    a. secretion of interferon by an infected cell
    b. neutralization by antibodies
    c. the enzyme lysozyme
    d. a secondary immune response
    e. humoral immunity

15. How are antibodies and complement related?
    a. They are both coded for by genes that have hundreds of alleles.
    b. They are both involved in innate defenses.
    c. They are both produced by plasma cells.
    d. Antibodies bound to antigens on a pathogen's membrane may activate complement proteins to form a membrane attack complex.
    e. Complement proteins tag foreign cells for destruction; antibodies destroy cells by opsonization.

16. Which of the following describes the main difference between an innate immune response and an acquired immune response?
    a. An innate immune response is only to free pathogens in a localized area; an acquired immune response is only to pathogens that have entered body cells.
    b. An innate immune response involves only leukocytes, whereas an acquired immune response involves only lymphocytes.
    c. An innate immune response relies on phagocytes to destroy pathogens, whereas an acquired immune response does not involve phagocytes.
    d. An innate immune response involves recognition of molecules common to a set of pathogens, whereas an acquired immune response reacts to specific microbes on the basis of their different antigens.
    e. Complement proteins participate in an acquired immune response but not in an innate immune response.

17. Clonal selection is responsible for the
    a. proliferation of clones of effector and memory cells specific for an encountered antigen.
    b. recognition of class I MHC molecules by cytotoxic T cells.
    c. rearrangement of antibody genes for the light and heavy chains.
    d. formation of cell cultures in the commercial production of monoclonal antibodies.
    e. transformation of a clone of helper T cells into cytotoxic T cells keyed to a specific antigen.

18. What role does a dendritic cell play in an acquired immune response?
    a. activates complement proteins to form a membrane attack complex
    b. binds to the CD8 receptors on cytotoxic T cells to activate their production of perforin
    c. releases cytokines to activate B cells to produce clones of plasma cells
    d. activates both humoral and cell-mediated immunity by releasing interferons after it has ingested a virus
    e. presents peptide antigens of an engulfed pathogen in its class II MHC molecules to helper T cells, and releases cytokines

19. All of the following are involved with innate immunity *except*
    a. the inflammatory response.
    b. the complement system.
    c. antimicrobial proteins such as lysozyme.
    d. chemicals that attract phagocytes.
    e. plasma cells.

20. Place the following steps in the helper T cell activation of cell-mediated and humoral immunity in the correct order:
    1. Helper T cell secretes cytokines.
    2. Macrophage engulfs pathogen and presents antigen in class II MHC.
    3. Plasma cells secrete antibodies and cytotoxic T cells attack cells with class I MHC molecule-antigen complex.
    4. T cell receptor recognizes class II MHC molecule-antigen complex.
    5. Macrophage secretes cytokines.
    6. Activated B cells form plasma and memory cells, activated T cells form cytotoxic T cells and memory cells.
    a. 1, 3, 5, 6, 2, 4
    b. 5, 1, 2, 6, 4, 3
    c. 2, 4, 5, 1, 6, 3
    d. 5, 2, 4, 1, 3, 6
    e. 2, 1, 4, 5, 6, 3

21. Helper T cells play which of the following roles in an acquired immune response?
    a. bind to class I MHC molecules and activate complement proteins to attack and lyse cancer cells.
    b. bind to the antigens presented in the CD8 receptors on cytotoxic T cells and release perforin.
    c. produce interferons and histamines that help initiate a specialized inflammatory response.
    d. present antigens of an engulfed pathogen in its class II MHC molecules to B cells, which are then stimulated to develop into a clone of plasma cells.
    e. activate both the humoral and cell-mediated immunities by releasing cytokines after recognizing class II MHC molecule-antigen complexes on an antigen-presenting cell.

22. Which of the following statements about humoral immunity is *correct*?
    a. It is a form of passive immunity produced by vaccination.
    b. It defends against free pathogens with effector mechanisms such as neutralization, opsonization, or complement activation.
    c. It protects the body against pathogens that have invaded body cells as well as against abnormal body cells.
    d. It is mounted by lymphocytes that have matured in the thymus.
    e. It depends on the recognition of class I MHC molecules that are bound to a specific antigen to activate its effector mechanism.

23. What accounts for the huge diversity of antigens to which B cells can respond?
    a. The antibody genes have millions of alleles.
    b. The rearrangement of the antibody genes during development results in millions of possible combinations of randomly combined light and heavy polypeptide chains.
    c. The antigen-binding sites at the arms of the molecule can assume a huge diversity of shapes in response to the specific antigen encountered.
    d. B cells have thousands of copies of antibodies bound to their plasma membranes.
    e. B cells can be antigen-presenting cells when they take in antigens by endocytosis and display fragments in their class II MHC molecules.

24. What is the function of CD4?

    a. a surface molecule on a cytotoxic T cell that enhances its binding to a class I MHC molecule displaying a foreign antigen

    b. a membrane protein on an antigen-presenting cell that helps a helper T cell recognize the MHC molecule-antigen complex

    c. a receptor that normally functions for cytokines, but which HIV uses as its receptor

    d. a surface molecule on a helper T cell that enhances its binding to a class II MHC molecule displaying a foreign antigen

    e. a portion of the class I MHC molecule found on all nucleated cells that identify cells as "self"

25. In which circumstance would B cells display antigens to T cells?

    a. They take in a few antigen molecules by endocytosis and display them in class II MHC molecules to helper T cells.

    b. They phagocytose bacteria and display bacterial peptide antigens in class II MHC molecules to helper T cells.

    c. After being infected by a virus, they display viral peptides they have synthesized in class I MHC molecules to cytotoxic T cells.

    d. They bind free antigens and display them to helper T cells attached to their B cell receptors.

    e. Both a and c are possible ways that B cells can display antigens.

26. All of the following are considered diseases or malfunctions of the immune system *except*

    a. MHC-induced transplant rejection.

    b. SCID (severe combined immunodeficiency).

    c. lupus, multiple sclerosis, and insulin-dependent diabetes.

    d. AIDS.

    e. allergic anaphylactic shock.

27. Which of the following may induce a graft-versus-host reaction?

    a. organ transplant

    b. blood transfusion

    c. bone marrow transplant

    d. skin transplant

    e. gene therapy

28. From which of the following would an AIDS patient be *least* likely to suffer?

    a. Kaposi's sarcoma or other cancers

    b. tuberculosis

    c. pneumonia

    d. rheumatoid arthritis

    e. yeast infections of mucous membranes

Chapter 44

Osmoregulation and Excretion

Key Concepts

44.1 Osmoregulation balances the uptake and loss of water and solutes

44.2 An animal's nitrogenous wastes reflect its phylogeny and habitat

44.3 Diverse excretory systems are variations on a tubular theme

44.4 Nephron structure is organized for stepwise processing of blood filtrate

44.5 Hormonal circuits link kidney function, water balance, and blood pressure

Framework

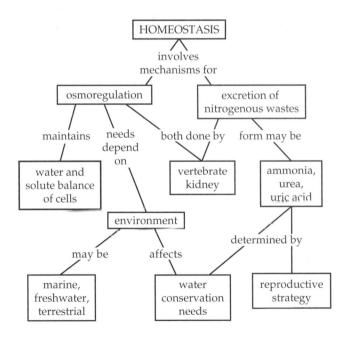

Chapter Review

Animals are able to tolerate the hydration challenges in their external environment by regulating solute and water balance (**osmoregulation**). Animals must also dispose of *nitrogenous* (nitrogen-containing) wastes (**excretion**).

44.1 Osmoregulation balances the uptake and loss of water and solutes

Osmosis   Whatever an animal's habitat or nitrogenous waste product, its water gain must balance water loss.

Osmosis is the diffusion of water across a selectively permeable membrane that separates two solutions differing in **osmolarity** (moles of solute per liter). Osmolarity is expressed in units of milliosmoles per liter (mosm/L). *Isoosmotic* solutions are equal in osmolarity, and there is no net osmosis between them. There is a net flow of water from a *hypoosmotic* (more dilute) to a *hyperosmotic* (more concentrated) solution.

Osmotic Challenges   **Osmoconformers** are isoosmotic with their surroundings. **Osmoregulators** regulate their internal osmolarity. Osmoregulators must get rid of excess water if they live in a hypoosmotic medium or take in water to offset osmotic loss if they inhabit a hyperosmotic environment.

Most animals are **stenohaline,** able to tolerate only small changes in external osmolarity. Animals that are **euryhaline** can survive large differences in the osmotic environment.

Most marine invertebrates are osmoconformers, whereas marine vertebrates are osmoregulators. Marine bony fishes are hypoosmotic to seawater. They gain salt by diffusion, from food, and from the large quantities of seawater they drink to replace the water they lose by osmosis through their skin and gills. Excess salt is pumped out through the gills and other ions are excreted in the scanty urine.

Sharks have an internal salt concentration lower than that of seawater. Salt tends to diffuse into the body and is excreted by the rectal glands and kidneys. A shark's osmolarity is slightly higher than that of seawater because it maintains high concentrations of urea in its tissues, as well as trimethylamine oxide (TMAO), which protects proteins from the damaging effects of urea. Water that enters a shark's body by osmosis or in food is disposed of in urine produced in the kidneys.

Freshwater animals constantly take in water by osmosis and lose salts by diffusion. Freshwater fishes excrete large quantities of dilute urine. Salt supplies are replaced from their food or by active uptake of ions across the gills.

Some animals are capable of **anhydrobiosis,** surviving dehydration, or *dessication*, in a dormant state. Anhydrobiotic roundworms produce large quantities of trehalose, a disaccharide that replaces water around membranes and proteins during dehydration. Many insects survive freezing in the same manner.

Adaptations to prevent dehydration in terrestrial animals include water-impervious coverings and behavioral adaptations such as nocturnal lifestyles. Water is lost in urine and feces and through evaporation, and gained by drinking and eating moist food and by metabolic production during cellular respiration.

Osmoregulation is energetically costly because animals must actively transport solutes in order to maintain osmotic gradients needed to gain or lose water. The energy cost is related to the difference in osmolarity between an animal and its environment, and the body fluid composition of many animals is adapted to the salinity of their habitats.

INTERACTIVE QUESTION   44.1

Indicate whether the following animals are isoosmotic, hyperosmotic, or hypoosmotic to their environment. Then briefly list their mechanisms of osmoregulation.

Animal	Osmotic Relation to Environment	Osmoregulatory Mechanisms
Marine invertebrate	a.	b.
Shark	c.	d.
Marine bony fish	e.	f.
Freshwater fish	g.	h.
Terrestrial animal	i.	j.

Transport Epithelia   The transport of specific solutes across a **transport epithelium** is essential for both osmotic regulation and disposing of metabolic wastes. Transport epithelia are usually arranged in tubular networks that provide large surface areas for exchange. Nasal glands in marine birds utilize a countercurrent exchange in removing salt from the blood and producing a concentrated salt secretion.

44.2   An animal's nitrogenous wastes reflect its phylogeny and habitat

Ammonia, a small and toxic molecule, is produced when proteins and nucleic acids are metabolized. Some animals excrete ammonia; others expend energy to convert it to less toxic wastes.

Forms of Nitrogenous Waste   Aquatic animals can excrete nitrogenous wastes as ammonia because it is very soluble and easily permeates membranes. Many invertebrates lose ammonia across their whole body surface. In fishes, most nitrogenous wastes pass as NH_4^- across the epithelia of the gills.

Mammals, most adult amphibians, sharks, and some marine bony fishes and turtles produce **urea**, a much less toxic compound than ammonia. Urea can be tolerated in a more concentrated form and excreted with less loss of water. Ammonia and carbon dioxide are combined in the liver to produce urea, which is then carried by the circulatory system to the kidneys. The production of urea requires energy.

Insects, land snails, and many reptiles, including birds, produce **uric acid**, a compound of low solubility in water that can be excreted as a semisolid with very little water loss. Its synthesis, however, is energetically expensive.

The Influence of Evolution and Environment on Nitrogenous Wastes The mode of reproduction seems to have determined which form of nitrogenous waste product evolved in various animal groups.

The form of nitrogenous wastes also relates to habitat. Terrestrial turtles excrete mainly uric acid, whereas aquatic turtles excrete both urea and ammonia. Some animals actually shift their nitrogenous waste product depending on environmental conditions.

INTERACTIVE QUESTION 44.2

a. Which would produce more nitrogenous waste for the same size animal:

an endotherm or an ectotherm?

an herbivore or a carnivore?

Explain your answers.

b. Vertebrates that produce shelled eggs excrete_____.

Mammals produce_____.

What adaptive advantage do these types of nitrogenous wastes provide?

44.3 Diverse excretory systems are variations on a tubular theme

Excretory Processes During **filtration**, water and small solutes are forced out of the blood or body fluids across a transport epithelium into an excretory tubule. Two mechanisms transform this **filtrate** to urine: Valuable solutes are returned from the filtrate in selective **reabsorption.** In selective **secretion**, excess salts, toxins, and other solutes are added to the filtrate. The osmotic movement of water into or out of the filtrate follows the pumping of solutes.

Survey of Excretory Systems The **protonephridia** of flatworms are branched systems of closed tubules. Water and solutes from the interstitial fluid pass into the flame bulbs at the ends of the tubules, propelled by cilia. The processed filtrate, or urine, is excreted into the external environment. The flame-bulb system of freshwater flatworms is primarily osmoregulatory. In parasitic flatworms, however, protonephridia function mainly in excretion of nitrogenous wastes.

Metanephridia are found in most annelids. They occur in pairs in each segment of an earthworm. An open ciliated funnel collects coelomic fluid, which then moves through a folded tubule encased in capillaries. The transport epithelium of the tubule reabsorbs most solutes, which then reenter the blood. Hypoosmotic urine, carrying nitrogenous wastes and excess water absorbed by osmosis, exits to the outside.

Malpighian tubules are excretory organs in insects and other terrestrial arthropods. Transport epithelia lining these blind sacs, which open into the digestive tract, secrete solutes from the hemolymph into the tubule. The fluid passes into the rectum, where most of the solutes are pumped back into the hemolymph, and water follows by osmosis. In this water-conserving system, uric acid is eliminated as dry matter along with the feces.

The numerous excretory tubules of most vertebrates are enclosed in a network of capillaries and arranged into compact kidneys. The kidneys, their associated blood vessels, and the structures that carry urine out of the body comprise the vertebrate excretory system.

INTERACTIVE QUESTION 44.3

Indicate whether these tubular systems function in osmoregulation, excretion, or both. If they function in osmoregulation, do they help conserve water or remove excess water that had entered by osmosis?

a. Protonephridia of freshwater planaria

b. Metanephridia of earthworms

c. Malpighian tubules of insects

d. Kidneys of terrestrial mammals

Structure of the Mammalian Excretory System Blood enters each of the pair of bean-shaped kidneys through a **renal artery** and leaves by way of a **renal vein.** Urine exits the kidney through a **ureter** and is temporarily stored in the **urinary bladder.** Urine exits the body through the **urethra.**

The outer **renal cortex** and inner **renal medulla** of the kidney are packed with excretory tubules and blood vessels. A **nephron** consists of a long tubule with a cuplike **Bowman's capsule** at the blind end that encloses a ball of capillaries called the **glomerulus.**

Filtration of the blood occurs as blood pressure forces water, urea, salts, glucose, amino acids, and other small molecules through the porous capillary walls and specialized capsule cells into Bowman's capsule, forming the filtrate. The filtrate passes through the **proximal tubule,** the **loop of Henle** with a descending limb and an ascending limb, and the **distal tubule. Collecting ducts** receive processed filtrate from many nephrons and pass the urine into the **renal pelvis.**

In the human kidney, 85% of the nephrons are **cortical nephrons** located entirely in the renal cortex. **Juxtamedullary nephrons,** found only in mammals and some birds, have long loops of Henle that extend into the renal medulla and allow these animals to form hyperosmotic urine. The transport epithelium lining the nephrons and collecting ducts processes about 180 L of filtrate each day to produce about 1.5 L of urine.

An **afferent arteriole** supplies each nephron, subdividing to form the capillaries of the glomerulus, which then converge to form an **efferent arteriole.** This arteriole divides into a second capillary network—the **peritubular capillaries**—that surrounds the proximal and distal tubules. The **vasa recta** includes descending and ascending capillaries that parallel the loop of Henle. Exchange between the tubules and capillaries is via the interstitial fluid.

INTERACTIVE QUESTION   44.4

a. What is found in the filtrate that moves from Bowman's capsule into the tubule?

b. What remains in the capillaries?

44.4 Nephron structure is organized for stepwise processing of blood filtrate

From Blood Filtrate to Urine: A Closer Look   The numbered steps correspond to the figure in Interactive Question 44.5.

1. Reabsorption and secretion both take place in the proximal tubule. Salt diffuses from the filtrate into the cells of the proximal tubule, which then actively transport Na^+ across the membrane into the interstitial fluid; Cl^- is transported passively out of the cells to balance the positive charge; and water follows by osmosis. Salt and water diffuse from the interstitial fluid into the peritubular capillaries. Glucose, amino acids, potassium ions, and the buffer bicarbonate (HCO_3^-) are actively or passively reabsorbed. The transport epithelium selectively secretes H^+, ammonia, and drugs and poisons (processed by the liver and delivered via peritubular capillaries) into the filtrate.

2. The descending limb of the loop of Henle has numerous **aquaporins** and is thus freely permeable to water. It is not permeable to salt or other small solutes. The interstitial fluid is increasingly hyperosmotic toward the inner medulla, so water exits by osmosis, leaving behind a filtrate with a high solute concentration.

3. The ascending limb of the loop of Henle is not permeable to water but is permeable to salt, which diffuses out of the thin lower segment of the loop, increasing the osmolarity of the medulla. NaCl is actively transported out of the thick upper portion of the ascending limb. As salt leaves, the filtrate becomes less concentrated.

4. The distal tubule regulates K^+ secretion and NaCl reabsorption, and it helps to regulate pH by secreting H^+ and reabsorbing bicarbonate.

5. As the filtrate moves in the collecting duct through the osmotic gradient of the medulla, more and more water exits by osmosis through aquaporin channels. Urea diffuses out of the lower portion of the duct and helps to maintain the osmotic gradient in the medulla. Salt excretion is controlled by the transport epithelium actively reabsorbing NaCl.

INTERACTIVE QUESTION 44.5

In this diagram of a nephron and collecting tubule, label the parts on the indicated lines. Label the arrows to indicate the movement of salt, water, nutrients, K^+, HCO_3^-, H^+, and urea into or out of the tubule.

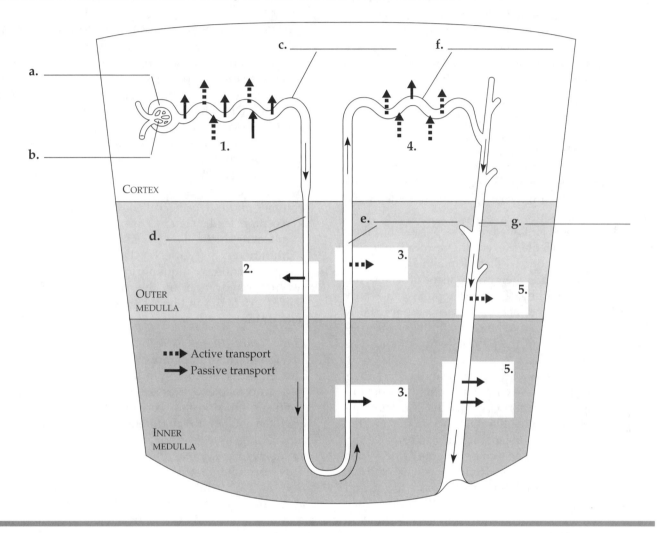

Solute Gradients and Water Conservation   The osmolarity gradient of NaCl and urea in the renal medulla is produced by the loops of Henle and collecting ducts and enables the human kidney to produce urine up to four times as concentrated as blood and normal interstitial fluid.

Filtrate leaving Bowman's capsule has an osmolarity of about 300 mosm/L, the same as blood. Both water and salt are reabsorbed in the proximal tubule; thus, the volume decreases but the osmolarity remains the same. In the trip down the descending limb of the loop of Henle, water exits by osmosis, and the filtrate becomes more concentrated. Salt, which is now in high concentration in the filtrate, diffuses out as the filtrate moves up the salt-permeable but water-impermeable ascending limb—helping to create the osmolarity gradient.

The flow of filtrate in the loop of Henle, along with the energy-consuming active transport of NaCl from the thick segment of the ascending limb, maintain the steep osmotic gradient between the cortex and medulla. This **countercurrent multiplier system** expends energy to create the gradient of the high salt concentration.

The vasa recta also has a countercurrent flow through the loop of Henle. As the blood in these vessels moves down into the inner medulla, water leaves by osmosis and salt enters; as the blood moves back up toward the cortex, water moves back in and salt diffuses out. Thus the capillaries can carry nutrients and other supplies to the medulla without disrupting the osmolarity gradient.

The filtrate is hypoosmotic to body fluids after losing salt in the ascending limb. The filtrate makes one final pass through the medulla, this time in the collecting duct, which is permeable to water and urea but not to salt. Water flows out by osmosis, and as the filtrate becomes more concentrated, urea leaks into the interstitial fluid, adding to the high osmolarity of the inner medulla. The resulting concentrated urine may achieve an osmolarity as high as 1,200 mosm/L, isoosmotic to the interstitial fluid of the inner medulla.

Adaptations of the Vertebrate Kidney to Diverse Environments   Modifications of nephron structure and function correspond to osmoregulatory requirements in various habitats. Mammals in dry habitats, needing to conserve water and excrete hyperosmotic urine, have exceptionally long loops of Henle. Mammals that spend much of their time in fresh water have nephrons with very short loops.

Birds also have juxtamedullary nephrons, but their shorter loops of Henle prevent them from producing urine as hyperosmotic as that of mammals. The production of uric acid is their main means of water conservation. The kidneys of other reptiles have only cortical nephrons, and their urine is isoosmotic to body fluids; water is conserved, however, through reabsorption in the cloaca and the production of uric acid as their nitrogenous waste.

Freshwater fishes are hyperosmotic to their environment and must excrete large quantities of very dilute urine. Salts are conserved by the efficient reabsorption of ions from the filtrate. When in fresh water, frogs and other amphibians actively absorb certain salts across their skin and excrete a dilute urine. When on land, water is reabsorbed from the urinary bladder. The kidneys of marine bony fishes have fewer and smaller nephrons and excrete very little urine. They function mainly to get rid of divalent ions taken in by drinking seawater.

INTERACTIVE QUESTION   44.6

a. What type of nephron allows the production of hyperosmotic urine?

b. Which animal groups have this type of nephron?

44.5   Hormonal circuits link kidney function, water balance, and blood pressure

Volume and osmolarity of urine in mammals can vary depending on the animal's water and salt balance and urea production.

Antidiuretic Hormone   **Antidiuretic hormone (ADH),** also called *vasopressin*, is produced in the hypothalamus and stored in the pituitary gland. When blood osmolarity rises above a set point of 300 mosm/L, osmoreceptor cells in the hypothalamus trigger the release of ADH. This hormone increases water permeability of the distal tubules and collecting ducts, increasing water reabsorption from the urine. As blood osmolarity lowers, negative feedback decreases the release of ADH. When blood osmolarity is low, little ADH is released, resulting in the production of large volumes of dilute urine, called diuresis. ADH binding to receptors leads to an increase in aquaporin proteins in collecting duct cells.

Mutations that affect ADH production, ADH receptor action, or aquaporin formation can cause severe dehydration in the condition, *diabetes insipidus.* Alcohol inhibits ADH release and can cause dehydration.

The Renin-Angiotensin-Aldosterone System   The **juxtaglomerular apparatus (JGA),** located near the afferent arteriole leading to the glomerulus, responds to a drop in blood pressure or volume by releasing renin, an enzyme that converts the plasma protein angiotensinogen to **angiotensin II.** Angiotensin II, functioning as a hormone, constricts arterioles and stimulates the adrenal glands to release **aldosterone.** This hormone stimulates Na^+ and water reabsorption in the distal tubules. The **renin-angiotensin-aldosterone system (RAAS)** is a homeostatic feedback circuit that maintains adequate blood pressure and volume.

Homeostatic Regulation of the Kidney   Both ADH and the RAAS increase water absorption, but in response to different problems. ADH counters dehydration by responding to an increase in blood osmolarity. The RAAS responds to a loss of salt and fluid by sensing a drop in blood volume or pressure.

Atrial natriuretic peptide (ANP) counters the RAAS. ANP, released by the atria of the heart in response to increased blood volume and pressure, inhibits the release of renin from the JGA, inhibits NaCl reabsorption by the collecting ducts, and reduces the release of aldosterone from the adrenals.

INTERACTIVE QUESTION 44.7

This concept map may look overwhelming, but working through it should help organize your understanding of the nervous and hormonal control of water and salt reabsorption in the kidneys.

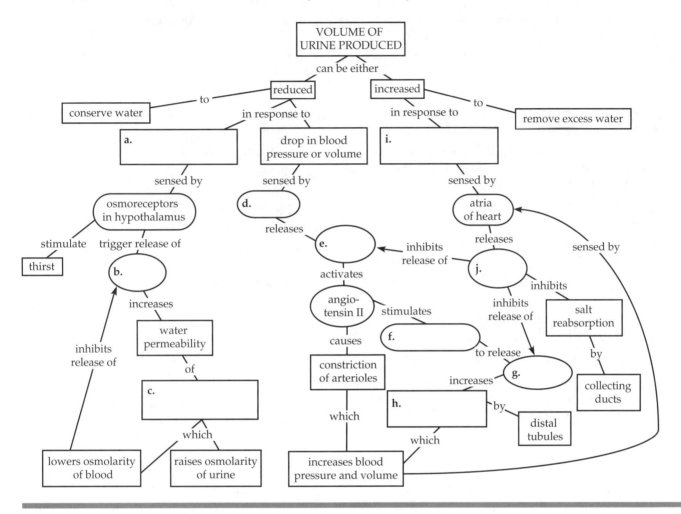

Word Roots

an- = without, **hydro-** = water; **-bios** = life (*anhydrobiosis:* the ability to survive in a dormant state when an organism's habitat dries up)

anti- = against; **-diure** = urinate (*antidiuretic hormone:* a hormone that helps regulate water balance)

eury- = broad, wide; **-halin** = salt (*euryhaline:* organisms that can tolerate substantial changes in external osmolarity)

glomer- = a ball (*glomerulus:* a ball of capillaries surrounded by Bowman's capsule in the nephron and serving as the site of filtration in the vertebrate kidney)

homeo- = like, same; **-stasis** = standing (*homeostasis:* the steady-state physiological condition of the body)

juxta- = near to (*juxtaglomerular apparatus:* a specialized tissue in nephrons that releases the enzyme renin in response to a drop in blood pressure or volume)

meta- = with; **-nephri** = kidney (*metanephridium:* a type of excretory organ in many invertebrates that consists of tubules connecting internal openings that collect body fluids to external openings)

osmo- = pushing; **-regula** = regular (*osmoregulation:* process by which organisms regulate solute concentrations and balance the gain and loss of water)

peri- = around (*peritubular capillaries:* the network of tiny blood vessels that surrounds the proximal and distal tubules in the kidney)

proto- = first (*protonephridium:* an excretory system, such as the flame-cell system of flatworms, consisting of a network of closed tubules lacking internal openings) reni- = a kidney; -angio = a vessel; -tens = stretched (*renin-angiotensin-aldosterone system:* a part of a complex feedback circuit that normally partners with antidiuretic hormone in osmoregulation)

steno- = narrow (*stenohaline:* organisms that cannot tolerate substantial changes in external osmolarity)

vasa- = a vessel; -recta = straight (*vasa recta:* the capillary system that serves the loop of Henle)

Structure Your Knowledge

1. List the substances that are filtered, secreted, and reabsorbed in the production of urine in mammals.
   a. Substances filtered:
   b. Substances secreted:
   c. Substances reabsorbed:

2. Referring back to the diagram in Interactive Question 44.5, describe the two-solute model that allows the production of concentrated urine.

Test Your Knowledge

MULTIPLE CHOICE: *Choose the one best answer.*

1. A unicellular protist may use a contractile vacuole to expel excess water. Contractile vacuoles most likely would be found in protists that are
   a. in a freshwater environment.
   b. in a marine environment.
   c. internal parasites.
   d. hypoosmotic to their environment.
   e. isoosmotic to their environment.

2. Transport epithelia are responsible for
   a. pumping water across a membrane.
   b. forming an impermeable boundary at an interface with the environment.
   c. the movement of solutes for osmoregulation or excretion.
   d. transporting urine in the ureter and urethra.
   e. the passive transport of H^+ and HCO_3^- for the regulation of pH.

3. Which of the following would be most likely to produce the sugar trehalose?
   a. desert mammal
   b. euryhaline fish
   c. shark

d. anhydrobiotic roundworm
e. vampire bat

4. A freshwater fish would be expected to
   a. pump salt out through salt glands in the gills.
   b. produce copious quantities of dilute urine.
   c. diffuse urea across the epithelium of the gills.
   d. have scales that reduce water loss to the environment.
   e. do all of the above.

5. Which of the following is *not* part of the filtrate entering Bowman's capsule?
   a. water, salt, and electrolytes
   b. glucose
   c. urea
   d. plasma proteins
   e. amino acids

6. Which is the correct pathway for the passage of urine in vertebrates?
   a. collecting tubule → ureter → bladder → urethra
   b. renal vein → renal ureter → bladder → urethra
   c. nephron → urethra → bladder → ureter
   d. cortex → medulla → bladder → ureter
   e. renal pelvis → medulla → bladder → urethra

7. Aldosterone
   a. is a hormone that stimulates thirst.
   b. is secreted by the adrenal glands in response to a high osmolarity of the blood.
   c. stimulates the active reabsorption of Na^+ in the distal tubules.
   d. causes diuresis.
   e. is converted from a blood protein by the action of renin.

8. Which of the following statements is *incorrect?*
   a. Long loops of Henle are associated with steep osmotic gradients and the production of hyperosmotic urine.
   b. Ammonia is a toxic nitrogenous waste molecule that passively diffuses out of the bodies of aquatic invertebrates.
   c. Uric acid is the form of nitrogenous waste that requires the least amount of water to excrete.
   d. In the mammalian kidney, urea diffuses out of the collecting duct and contributes to the osmotic gradient within the medulla.
   e. Uric acid is produced by a mammalian fetus and removed through the placenta to the mother's excretory system.

9. The process of reabsorption in the formation of urine ensures that

   a. excess H^+ is removed from the blood.

   b. drugs and other poisons are removed from the blood.

   c. urine is always hyperosmotic to interstitial fluid.

   d. glucose, salts, and water are returned to the blood.

   e. pH is maintained with a balance of hydrogen ions and bicarbonate.

10. The peritubular capillaries

    a. form the ball of capillaries inside Bowman's capsule from which filtrate is forced by blood pressure into the renal tubule.

    b. intertwine with the proximal and distal tubules and exchange solutes with the interstitial fluid.

    c. form a countercurrent flow of blood through the medulla to supply nutrients without interfering with the osmolarity gradient.

    d. surround the collecting ducts and reabsorb water, helping to create a hyperosmotic urine.

    e. rejoin to form the efferent arteriole.

11. ADH and the RAAS both increase water reabsorption, but they respond to different osmoregulatory problems. Which two of the following statements are true?

    1. ADH will be released in response to high alcohol consumption.

    2. ADH is released when osmoregulatory cells in the hypothalamus sense an increase in blood osmolarity.

    3. The RAAS will increase the osmolarity of urine due to the cooperative action of renin, angiotensin II, and aldosterone.

    4. The RAAS is a response to a rise in blood pressure or volume.

    5. The RAAS is most likely to respond following an accident or severe case of diarrhea.

       a. 1 and 4

       b. 1 and 5

       c. 2 and 3

       d. 2 and 4

       e. 2 and 5

12. Which of the following sections of the mammalian nephron is *incorrectly* paired with its function?

    a. Bowman's capsule and glomerulus—filtration of blood

    b. proximal tubule—secretion of ammonia and H^+ into filtrate and transport of glucose and amino acids out of tubule

    c. descending limb of loop of Henle—diffusion of urea out of filtrate

    d. ascending limb of loop of Henle—diffusion and pumping of NaCl out of filtrate

    e. distal tubule—regulation of pH and K^+

13. One of the advantages of the production of uric acid by birds is that uric acid

    a. has low toxicity and can be safely stored in the egg.

    b. is very soluble in water and takes very little energy to produce.

    c. contributes to the production of the egg shell and can thus serve two purposes.

    d. takes less energy to produce than urea and is nontoxic.

    e. Both a and d are correct.

14. What is the mechanism for the filtration of blood within the nephron?

    a. the active transport of Na^+ and glucose, followed by osmosis

    b. both active and passive secretion of ions, toxins, and NH_3 into the tubule

    c. high hydrostatic pressure of blood forcing water and small molecules out of the capillary

    d. the high osmolarity of the medulla that was created by active and passive transport of salt from the tubule and passive diffusion of urea from the collecting duct

    e. a lower osmotic pressure in Bowman's capsule compared to that in the glomerulus

15. What stimulus causes the atria of the heart to release ANP (atrial natriuretic peptide)?

    a. a rise in blood pressure

    b. a drop in blood pressure

    c. a drop in blood pH

    d. a rise in blood osmolarity

    e. a drop in blood osmolarity

16. What stimulus causes the juxtaglomerular apparatus to release renin?

    a. a drop in blood pH

    b. a rise in blood pH

    c. a drop in blood pressure

    d. a rise in blood osmolarity

    e. the release of ANP by the atria of the heart

17. Which of the following would be a good mechanism for a drug to treat hypertension?
    a. increase production of ADH
    b. block production of ANP
    c. vasoconstriction of renal arteries
    d. inhibit enzyme involved in producing angiotensin II
    e. stimulate production of aquaporins in collecting tubules

18. To produce urine that is hyperosmotic to blood and interstitial fluids requires
    a. juxtamedullary nephrons.
    b. an increasing osmotic concentration down through the medulla.
    c. the inhibition of ADH release.
    d. all three of the above.
    e. only a and b above.

Chapter 45

Hormones and the Endocrine System

Key Concepts

45.1 Hormones and other chemical signals bind to target receptors, triggering specific response pathways

45.2 Negative feedback and antagonistic hormone pairs are common features of the endocrine system

45.3 The endocrine and nervous systems act individually and together to regulate an animal's physiology

45.4 Endocrine glands respond to diverse stimuli in regulating metabolism, homeostasis, development, and behavior

Framework

This chapter introduces the intricate system of chemical control and communication within animals. Endocrine cells or neurosecretory cells produce hormones that regulate the activity of other cells and organs. Steroid hormones bind with receptors within their target cells and influence gene expression. The signal transduction pathways of most polypeptide and amine hormones involve binding with cell surface receptors and triggering metabolic reactions in target cells.

The hypothalamus and pituitary gland play coordinating roles by integrating the nervous and endocrine systems and producing many tropic hormones that control the synthesis and secretion of hormones in other endocrine glands and organs.

Chapter Review

An animal **hormone** is a chemical signal released into the extracellular fluid and usually transported through the circulatory system that elicits a specific response from target cells. Coordination and communication among the specialized parts of complex animals are achieved by the **endocrine system,** which produces hormones that regulate many biological processes, and

the **nervous system**, which conveys high-speed messages along neurons that regulate other neurons, muscle cells, and endocrine cells.

45.1 Hormones and other chemical signals bind to target receptors, triggering specific response pathways

Types of Secreted Chemical Signals Chemical signals bind to specific receptor proteins on or in target cells. Endocrine cells secrete endocrine signals, or hormones, that travel through the bloodstream. **Endocrine glands** are ductless secretory organs composed of groups of endocrine cells. Hormones regulate growth, development, and reproduction, and maintain homeostasis.

Local regulators are chemical signals that reach their target cells by diffusion. **Paracrine** signals act on neighboring cells; **autocrine** signals act on the secreting cell itself.

Neurotransmitters are secreted by neurons at synapses with other neurons and muscles. *Neurosecretory cells* are specialized brain neurons that secrete chemical signals called **neurohormones,** which travel through the bloodstream to target cells.

Pheromones are chemical signals released into the environment that communicate between different individuals.

Chemical Nature of Hormones Three groups of hormones include polypeptides (peptides and proteins), amines, and steroid hormones. Polypeptides and many amines are water-soluble; steroids are lipid-soluble.

Hormone Receptor Localization Research in the 1960s showed that the steroid hormones progesterone and estradiol accumulate within the nuclei of cells in female reproductive organs but not in other cells. Receptors for steroid and lipid-soluble hormones have been identified inside cells. Water-soluble hormones bind to cell membranes. Other research showed that microinjection of the water-soluble hormone melanocyte-stimulating hormone into frog skin cells produced no response, while injection into the interstitial space caused a response, indicating that the receptor for MSH is located on the cell surface.

Cellular Response Pathways Water-soluble hormones are released by exocytosis, travel through the bloodstream, and bind to cell-surface receptors, triggering a cytoplasmic response or change in gene expression. Lipid-soluble hormones travel in the bloodstream bound to transport proteins, enter target cells, and bind to a receptor in the cytoplasm or nucleus, triggering changes in gene transcription.

Binding of water-soluble hormones to plasma membrane receptors initiates **signal transduction,** which converts extracellular signals to specific intracellular responses. For instance, the binding of **epinephrine**, released in response to stress by the adrenal gland, to G-protein coupled receptors on liver cells initiates a signal transduction pathway that results in the release of glucose into the bloodstream.

Steroid hormones bind to cytoplasmic receptors, and the hormone-receptor complex then moves into the nucleus, where it stimulates transcription of specific genes. Thyroxine, vitamin D, and other non-steroid lipid-soluble hormones bind with receptors inside the nucleus.

INTERACTIVE QUESTION 45.1

Label the parts of these diagrams, indicating which represents a water-soluble hormone and which a lipid-soluble hormone.

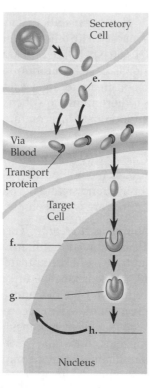

Multiple Effects of Hormones A given signal can have different effects on different target cells and can produce different effects in different species. These differences may result from the types of receptors (for example, the α-type and β-type epinephrine receptors), the specific signal transduction pathways present, or the particular transcription factors that are activated.

Signaling by Local Regulators **Cytokines** are polypeptide regulators involved in immune responses. **Growth factors** are polypeptides that are required in the extracellular environment for many types of cells to divide and develop.

Nitric oxide (NO) is a gas that can serve as a neurotransmitter and as a local regulator that produces vasodilation, increasing blood flow.

Prostaglandins (PGs) are modified fatty acids that have a wide range of effects on nearby target cells. Prostaglandins help to induce labor. Aspirin and ibuprofen inhibit the synthesis of prostaglandins and thus reduce their fever- and inflammation-inducing and pain-intensifying actions.

45.2 Negative feedback and antagonistic hormone pairs are common features of the endocrine system

Simple Hormone Pathways In many hormone pathways, endocrine cells secrete a hormone in response to a stimulus; the hormone travels to target cells, where signal transduction produces a response that reduces the stimulus; and the pathway shuts off. This type of control, called **negative feedback,** is typical of pathways involved in maintaining homeostasis. In many instances, a pair of pathways, each counteracting the other, provides the control system.

Insulin and Glucagon: Control of Blood Glucose Two antagonistic hormones regulate glucose concentration in the blood, and negative feedback controls these pathways. **Insulin** is released in response to a rise in blood glucose above the set point and triggers the uptake of glucose from the blood. **Glucagon,** released in response to a drop in blood glucose concentration, stimulates the release of glucose into the blood. Scattered within the exocrine tissue of the **pancreas** are clusters of endocrine cells known as the **islets of Langerhans.** Within each islet are *alpha cells* that secrete glucagon and *beta cells* that secrete insulin.

Insulin lowers blood glucose levels by promoting the movement of glucose from the blood into body cells, by slowing the breakdown of glycogen in the liver, and by inhibiting the conversion of amino acids and glycerol (from fats) to sugar. Glucagon raises glucose concentrations by stimulating target cells in the liver to increase glycogen hydrolysis, convert amino acids and glycerol to glucose, and release glucose to the blood.

In **diabetes mellitus,** the absence of insulin in the bloodstream or the loss of response to insulin in target tissues reduces glucose uptake by cells. Glucose accumulates in the blood and is excreted in the urine, with an accompanying loss of water. The body must use fats for fuel, and acids from fat breakdown may lower blood pH.

Type 1 diabetes mellitus, also known as insulin-dependent diabetes, is an autoimmune disorder in which pancreatic cells are destroyed. This type of diabetes is treated by regular injections of genetically engineered human insulin. More than 90% of diabetics have *type 2 diabetes,* or non-insulin-dependent diabetes, characterized by reduced responsiveness of target cells to insulin. Exercise and dietary control are often sufficient to manage this disease, although it is becoming a growing public health problem.

INTERACTIVE QUESTION 45.2

Complete this concept map on the regulation of blood glucose levels by hormones of the pancreas.

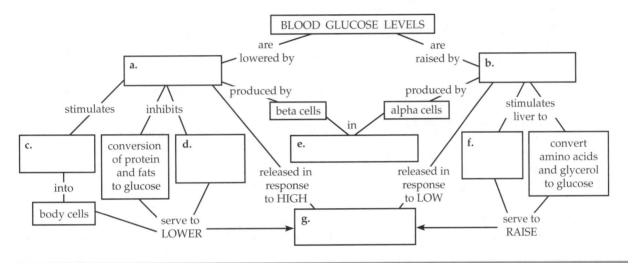

45.3 The endocrine and nervous systems act individually and together to regulate an animal's physiology

Coordination of Endocrine and Nervous Systems in Invertebrates In invertebrates, reproduction and development are usually controlled by an integration of the endocrine and nervous systems.

Hormones control molting and metamorphosis in insects. *Prothoracicotropic hormone* (PTTH), produced by neurosecretory cells in the brain, stimulates the prothoracic glands to secrete ecdysone. **Ecdysone** triggers molts and metamorphosis. **Juvenile hormone** promotes retention of larval characteristics during molting. When its level is high, ecdysone-induced molting produces larger larval stages; only after the juvenile hormone level has decreased does molting result in a pupa.

Coordination of Endocrine and Nervous Systems in Vertebrates The **hypothalamus,** situated in the lower brain, plays a key role in integrating the endocrine and nervous systems. In response to nerve signals it receives from throughout the body, the hypothalamus sends endocrine signals to the pituitary gland located at its base.

The **pituitary gland** has two discrete parts that develop separately and have different functions. The **posterior pituitary,** or *neurohypophysis,* is an extension of the hypothalamus that grows downward during embryonic development. It stores and secretes two hormones that are produced by the hypothalamus. The **anterior pituitary,** or *adenohypophysis,* develops from the roof of the embryonic mouth. The hypothalamus regulates secretion of hormones by the anterior pituitary.

Posterior Pituitary Hormones The posterior pituitary stores and releases two neurohormones that are produced by and delivered from neurosecretory cells of the hypothalamus. **Oxytocin** induces uterine contractions during birth and milk ejection during nursing. Both of these actions are under **positive feedback**. **Antidiuretic hormone (ADH),** or *vasopressin,* functions in osmoregulation. It increases water retention by the kidneys and thus decreases urine volume.

INTERACTIVE QUESTION 45.3

Explain the difference between negative and positive feedback in the control of hormonal pathways.

Anterior Pituitary Hormones Some hypothalamic neurosecretory cells produce *releasing hormones* and *inhibiting hormones* that regulate the hormones of the anterior pituitary. For example, *thyrotropin releasing hormone (TRH)* stimulates the anterior pituitary to secrete thyrotropin, or *thyroid stimulating hormone (TSH)*. These hypothalmic hormones are released into capillaries at the base of the hypothalamus and travel via a short portal vessel to capillary beds in the anterior pituitary.

In a hormone cascade pathway, the hypothalamus secretes a hormone that stimulates or inhibits release of an anterior pituitary hormone, which then acts on a target endocrine tissue, stimulating the release of a hormone that produces a response to the stimulus. For example, TRH delivered to the anterior pituitary stimulates the release of TSH, which stimulates the thyroid gland to release *thyroid hormone.*

Hormones that regulate endocrine glands or cells are called *tropic hormones*. In addition to TSH, the anterior pituitary produces three other tropic hormones. **Follicle-stimulating hormone (FSH)** and **luteinizing hormone (LH),** also called *gonadotropins,* stimulate gonad activity. **Adrenocorticotropic hormone (ACTH)** stimulates the adrenal cortex to produce and secrete its steroid hormones.

The hormone **prolactin (PRL)** is a protein with diverse effects in different vertebrate species, ranging from milk production and secretion in mammals, to delay of metamorphosis in amphibians, to osmoregulation in fishes.

Melanocyte-stimulating hormone (MSH) regulates the activity of pigment-containing cells in the skin of some vertebrates and appears to inhibit hunger in mammals.

Growth hormone (GH) has diverse metabolic nontropic effects. Its main tropic action is signaling the release of *insulin-like growth factors (IGFs)* that are produced by the liver and stimulate bone and cartilage growth. Gigantism and acromegaly are human growth disorders caused by hypersecretion of GH. Pituitary dwarfism can now be treated with genetically engineered GH.

INTERACTIVE QUESTION 45.4

Fill in the following table to review the hormones stored and released by the posterior pituitary (a and b) and secreted by the anterior pituitary (c–h).

Hormone	Main Actions
Oxytocin	**a.**
ADH	**b.**
TSH	**c.**
FSH and LH	**d.**
ACTH	**e.**
Prolactin	**f.**
MSH	**g.**
GH	**h.**

45.4 Endocrine glands respond to diverse stimuli in regulating metabolism, homeostasis, development, and behavior

Thyroid Hormone: Control of Metabolism and Development Thyroid hormone regulates development and contributes to homeostasis in mammals, helping to maintain normal blood pressure, heart rate, muscle tone, digestion, and reproductive functions. The **thyroid gland** produces two hormones, **triiodothyronine (T_3)** and **thyroxine (T_4).** A negative feedback loop controls secretion of thyroid hormones: High levels of T_3 and T_4 inhibit the secretion of TRH and TSH.

Excess thyroid hormone results in hyperthyroidism, with symptoms of weight loss, irritability, and high body temperature and blood pressure. Hypothyroidism can cause weight gain and lethargy in adults.

Thyroid hormones are critical to vertebrate development and maturation. In humans, an inherited deficiency called congenital hypothyroidism results in retarded skeletal and mental development.

INTERACTIVE QUESTION 45.5

Explain how a lack of iodine in the diet may result in goiter, an enlarged thyroid gland.

Parathyroid Hormone and Vitamin D: Control of Blood Calcium The four **parathyroid glands** secrete **parathyroid hormone (PTH)**, which stimulates Ca^{2+} reabsorption in the kidney, and its release from bone to raise blood calcium levels. PTH also activates vitamin D, which then increases Ca^{2+} uptake from food in the intestines. If blood Ca^{2+} rises above the set point, the thyroid gland secretes **calcitonin,** a hormone that lowers calcium levels in the blood. In humans, calcitonin appears to be only required during the bone growth of childhood.

Adrenal Hormones: Response to Stress In mammals, the **adrenal glands** consist of two different glands: the outer *adrenal cortex* composed of endocrine cells and the central *adrenal medulla* derived from neural tissue.

The adrenal medulla produces epinephrine (adrenaline) and **norepinephrine** (noradrenaline)—both of which are **catecholamines,** a class of compounds synthesized from the amino acid tyrosine. Epinephrine and norepinephrine, released in response to positive or negative stress, increase the availability of energy sources by stimulating glycogen hydrolysis in skeletal muscle and the liver, glucose release from the liver, and fatty acid release from fat cells. These hormones increase metabolic rate and the rate and volume of the heartbeat, dilate bronchioles in the lungs, and influence the contraction or relaxation of smooth muscles to increase blood supply to the heart, brain, and skeletal muscles, while reducing the supply to other organs. Norepinephrine's primary role is in sustaining blood pressure. Nerve signals from the hypothalamus stimulate the release of catecholamines from the adrenal medulla.

The adrenal cortex responds to endocrine signals released in response to stress. A releasing hormone from the hypothalamus causes the anterior pituitary to release ACTH. This tropic hormone stimulates the adrenal cortex to secrete **corticosteroids,** a group of steroid hormones.

Glucocorticoids promote synthesis of glucose from noncarbohydrates (such as muscle proteins) and thus increase energy supplies during stress. Cortisol has been used to treat inflammatory diseases such as arthritis, but its side effects can be dangerous.

Mineralocorticoids affect salt and water balance. *Aldosterone,* released in response to low blood volume or pressure, stimulates kidney cells to reabsorb sodium ions and water from the filtrate. ACTH secreted by the anterior pituitary in response to severe stress increases aldosterone secretion. Both glucocorticoids and mineralocorticoids appear to help maintain homeostasis during extended periods of stress.

A third group of corticosteroids produced by the adrenal cortex are sex hormones, in particular androgens, which appear to influence female sex drive.

INTERACTIVE QUESTION 45.6

Describe how the adrenal gland is signaled by the hypothalamus and responds to short-term and long-term stress.

a. Short-term stress:

b. Long-term stress:

Gonadal Sex Hormones The testes of males and ovaries of females produce steroids that affect growth, development, and reproductive cycles and behaviors. The three major categories of gonadal steroids—androgens, estrogens, and progestins—are found in different proportions in males and females.

The testes primarily synthesize **androgens,** such as **testosterone,** which determine the gender of the developing embryo and stimulate development of the male reproductive system and secondary sex characteristics. **Estrogens,** in particular **estradiol,** regulate the maintenance of the female reproductive system and the development of secondary sex characteristics. In mammals, **progesterone** helps prepare and maintain the uterus for the growth of an embryo.

In a hormone cascade pathway, a hypothalamic releasing hormone, GnRH, controls secretion of FSH and LH, gonadotropins from the anterior pituitary gland, which then control the synthesis of estrogens and androgens.

Melatonin and Biorhythms The **pineal gland,** located near the center of the mammalian brain, secretes **melatonin,** which regulates functions related to light and changes in day length. The secretion of melatonin at night may target the suprachiasmatic nuclei (SCN), a biological clock for daily or seasonal activities such as reproduction.

Word Roots

adeno- = gland; **-hypo** = below (*adenohypophysis:* also called the anterior pituitary, a gland positioned at the base of the hypothalamus)

andro- = male; **-gen** = produce (*androgens:* the principal male steroid hormones, such as testosterone, which stimulate the development and maintenance of the male reproductive system and secondary sex characteristics)

anti- = against; **-diure** = urinate (*antidiuretic hormone:* a hormone that helps regulate water balance)

cata- = down; **-chol** = anger (*catecholamines:* a class of compounds, including epinephrine and norepinephrine, synthesized from the amino acid tyrosine)

-cortico = the shell; **-tropic** = to turn or change (*adrenocorticotropic hormone:* a hormone released from the anterior pituitary, it stimulates the production and secretion of steroid hormones by the adrenal cortex)

ecdys- = an escape (*ecdysone:* a steroid hormone that triggers molting in arthropods)

epi- = above, over (*epinephrine:* a hormone produced as a response to stress; also called adrenaline)

gluco- = sweet (*glucagon:* a peptide hormone secreted by pancreatic endocrine cells that raises blood glucose levels; an antagonistic hormone to insulin)

lut- = yellow (*luteinizing hormone:* a gonadotropin secreted by the anterior pituitary)

melan- = black (*melatonin:* a modified amino acid hormone secreted by the pineal gland)

neuro- = nerve (*neurohypophysis:* also called the posterior pituitary, it is an extension of the brain)

oxy- = sharp, acid (*oxytocin:* a hormone that induces contractions of the uterine muscles and causes the mammary glands to eject milk during nursing)

para- = beside, near (*parathyroid glands:* four endocrine glands, embedded in the surface of the thyroid gland, that secrete parathyroid hormone and raise blood calcium levels)

pro- = before; **-lact** = milk (*prolactin:* a hormone produced by the anterior pituitary gland, it stimulates milk synthesis in mammals)

tri- = three; **-iodo** = violet (*triiodothyrodine:* one of two very similar hormones produced by the thyroid gland and derived from the amino acid tyrosine)

Structure Your Knowledge

1. Briefly describe the two general mechanisms by which chemical signals trigger responses in target cells, depending on the location of the receptor.

2. Explain how the action of secretin, which is released in response to a low pH in the intestine and stimulates the release of bicarbonate from the pancreas, illustrates a simple hormone pathway.

3. What is a hormone cascade pathway? Describe how the activation of the thyroid gland illustrates such a pathway.

Test Your Knowledge

MATCHING: *Match the hormone and gland or organ that produces it to the descriptions. Choices may be used more than once, and not all choices are used.*

Hormones	Gland or Organ
A. ACTH	a. adrenal cortex
B. androgens	b. adrenal medulla
C. ADH	c. hypothalamus
D. calcitonin	d. pancreas
E. epinephrine	e. parathyroid
F. glucagon	f. pineal
G. glucocorticoids	g. pituitary
H. insulin	h. testis
I. melatonin	i. thymus
J. oxytocin	j. thyroid
K. PTH	
L. thyroxine	

Hormone	Gland	Hormone Action
_____	_____	1. involved in biological clock and seasonal activities
_____	_____	2. break down muscle protein for conversion to glucose
_____	_____	3. increase blood sugar, glycogen breakdown in liver
_____	_____	4. stimulate development of male reproductive system
_____	_____	5. stimulate adrenal cortex to synthesize corticosteroids
_____	_____	6. increase available energy, heart rate, metabolism
_____	_____	7. regulate metabolism, growth, and development
_____	_____	8. lower blood calcium levels
_____	_____	9. increase reabsorption of water by kidney
_____	_____	10. stimulate contraction of uterus, milk secretion

MULTIPLE CHOICE: *Choose the one best answer.*

1. Which of the following is *not* an accurate statement about hormones?
   a. Not all hormones are secreted by endocrine glands.
   b. Most hormones move through the circulatory system to their destination.
   c. Target cells have specific protein receptors for hormones.
   d. Hormones are essential to homeostasis.
   e. Steroid hormones often function as neurotransmitters.

2. The best description of the difference between pheromones and hormones is that
   a. pheromones are small, volatile molecules, whereas hormones are steroids.
   b. pheromones are involved in reproduction, whereas hormones are not.
   c. pheromones are a form of neural communication; hormones are a form of chemical communication.
   d. pheromones are signals that function between organisms, whereas hormones communicate among the parts within an organism.
   e. pheromones are local regulators, whereas hormones travel greater distances.

3. Which one of the following hormones is *incorrectly* paired with its origin?
   a. releasing hormones—hypothalamus
   b. growth hormone—anterior pituitary
   c. progestins—ovary
   d. TSH—thyroid
   e. mineralocorticoids—adrenal cortex

4. Which of the following is an example of a positive feedback mechanism?
   a. the liver's production of insulin-like growth factors in response to growth hormone, which promote skeletal growth
   b. the ability of the neurotransmitter acetylcholine to cause skeletal muscle to contract, heart muscle to relax, and cells of the adrenal medulla to secrete epinephrine
   c. prostaglandins released from placental cells promoting muscle contraction during childbirth, with muscle contractions stimulating more prostaglandin release
   d. the action of secretin on the pancreas, stimulating the release of bicarbonate
   e. elevated levels of stress resulting in neural stimulation of the adrenal medulla and hormonal stimulation of the adrenal cortex

5. Ecdysone
   a. is a steroid hormone produced in insects that promotes retention of larval characteristics.
   b. is responsible for color changes in amphibians.
   c. is a hormone secreted from specialized neurons that triggers the formation of a pupa.
   d. is secreted by prothoracic glands in insects and triggers molts and development of adult characteristics.
   e. is involved in metamorphosis in amphibians.

6. Which of the following local regulators amplifies the sensation of pain?
   a. prostaglandins
   b. melanocyte-stimulating hormone
   c. growth factors
   d. cytokines
   e. nitric oxide

7. A tropic hormone is a hormone
   a. whose target tissue is another endocrine gland.
   b. that is produced by the hypothalamus but stored and released from the posterior pituitary.
   c. that acts by negative feedback to regulate its own level in the body.
   d. from the hypothalamus that regulates the synthesis and secretion of hormones of the posterior pituitary.
   e. that is released in response to nervous stimulation.

8. The anterior pituitary
   a. stores oxytocin and ADH produced by the hypothalamus.
   b. receives releasing and inhibiting hormones from the hypothalamus through portal vessels connecting capillary beds.
   c. produces several releasing and inhibiting hormones.
   d. is responsible for nervous and hormonal stimulation of the adrenal glands.
   e. produces only tropic hormones.

9. What is the best description of the mechanism of action of steroid hormones?
   a. transported by neurosecretory cells directly to target tissues
   b. form a hormone-receptor complex inside the cell that regulates gene expression
   c. amplified response using second messengers
   d. bind to membrane-bound receptors and initiate a signal transduction pathway
   e. secreted into the interstitial fluid and act as a local regulator

10. Which of the following hormones is *not* involved with increasing the blood glucose concentration?
    a. glucagon
    b. epinephrine
    c. glucocorticoids
    d. ACTH (adrenocorticotropic hormone)
    e. insulin

11. MSH (melanocyte-stimulating hormone) causes frog skin cells to darken when added to the interstitial fluid but has no effect on color when injected into the cells. Which of the following does this evidence support?
    a. MSH is a tropic hormone.
    b. MSH is a neurohormone.
    c. MSH is a steroid hormone.
    d. MSH binds to cell surface receptors.
    e. Both c and d are reasonable conclusions.

12. Which of the following is *not* true of norepinephrine?
    a. It is secreted by the adrenal medulla.
    b. It maintains blood pressure.
    c. Its release is stimulated by ACTH.
    d. It serves as a neurotransmitter.
    e. It is part of the flight-or-fight response to stress.

Choose from the following hormones to answer questions 13–15.
    a. ACTH
    b. parathyroid hormone
    c. epinephrine
    d. estrogen
    e. insulin

13. Which of the above is a tropic hormone?

14. Which hormone is a steroid hormone?

15. Which hormone is released in response to a nervous impulse?

16. A lack of iodine in the diet can lead to formation of a goiter because
    a. excess thyroxine production causes the thyroid gland to enlarge.
    b. with limited iodine available, triiodothyronine (T_3) rather than thyroxine (T_4) is produced.
    c. iodine is a key ingredient of the growth hormones that control growth of the thyroid.
    d. little thyroxine is produced, TRH and TSH production is not inhibited, and thyroid stimulation continues.
    e. the hypothalamus is stimulated to increase its production of TRH.

Chapter 46

Animal Reproduction

Key Concepts

46.1 Both asexual and sexual reproduction occur in the animal kingdom

46.2 Mechanisms for fertilization bring together sperm and eggs of the same species

46.3 Reproductive organs produce and transport gametes

46.4 The timing and pattern of meiosis in mammals differ for males and females

46.5 The interplay of tropic and sex hormones regulates mammalian reproduction

46.6 In placental mammals, an embryo develops fully within the mother's uterus

Framework

This chapter covers the patterns and mechanisms of animal reproduction. In sexual reproduction, fertilization may occur externally or internally. Development of the zygote may take place internally within the female; externally in a moist environment; or in a protective, resistant egg. Placental mammals provide nourishment as well as shelter for the embryo, and they nurse their young.

The human reproductive system is described, including its organs, glands, hormones, gamete formation, and sexual response. The chapter also covers pregnancy and birth, contraception, and recently developed reproductive technologies.

Chapter Review

46.1 Both asexual and sexual reproduction occur in the animal kingdom

In **sexual reproduction,** two meiotically formed haploid gametes fuse to form a diploid **zygote,** which develops into an offspring. Gametes are usually a relatively large, nonmotile **egg** and a small, motile **sperm.** In **asexual reproduction,** a single individual produces offspring, usually using only mitotic cell division.

Mechanisms of Asexual Reproduction Many invertebrates can reproduce asexually by **fission,** in which a parent is separated into two or more individuals of equal size; by **budding,** in which a new individual grows out from the parent's body; or by fragmentation, in which the body is broken into several pieces, each of which develops into a complete animal. **Regeneration,** the regrowth of body parts, is necessary for a fragment to develop into a new organism. In **parthenogenesis,** eggs develop without being fertilized.

Sexual Reproduction: An Evolutionary Enigma Asexual reproduction potentially produces many more offspring, yet most eukaryotic species reproduce sexually. The shuffling of genes in sexual reproduction may allow more rapid adaptation to changing conditions or allow harmful genes to be eliminated more easily.

Reproductive Cycles and Patterns Periodic reproductive cycles may be linked to favorable conditions or energy supplies. A combination of environmental and hormonal cues controls the timing of **ovulation,** the release of eggs.

Animals may reproduce sexually, asexually, or both. The freshwater crustacean *Daphnia,* like aphids and rotifers, produces eggs that develop by parthenogenesis, as well as eggs that are fertilized. In a few parthenogenetic fishes, amphibians, and lizards, doubling of chromosomes after meiosis creates diploid offspring.

In **hermaphroditism,** found in some sessile, burrowing, and parasitic animals, each individual has functioning male and female reproductive systems. Mating results in fertilization of both individuals. In some fishes and oysters, individuals reverse their sex during their lifetime.

INTERACTIVE QUESTION   46.1

a. What adaptive advantages would asexual reproduction provide?

b. What adaptive advantage would sexual reproduction provide?

c. If a hermaphrodite self-fertilizes, is that asexual or sexual reproduction? Explain.

46.2   Mechanisms for fertilization bring together sperm and eggs of the same species

Fertilization is the union of sperm and egg. In **external fertilization,** eggs and sperm are shed, and fertilization occurs in the environment. External fertilization occurs almost exclusively in moist habitats, where the gametes and developing zygote are not in danger of desiccation.

Internal fertilization involves the placement of sperm in or near the female reproductive tract so that the egg and sperm unite internally. Behavioral cooperation, as well as copulatory organs and sperm receptacles, are required for internal fertilization.

Pheromones are small, volatile chemical signals that may function as mate attractants.

INTERACTIVE QUESTION   46.2

List three mechanisms that may help to ensure that gamete release is synchronized when fertilization is external.

a.

b.

c.

Ensuring the Survival of Offspring   Species with external fertilization usually produce huge numbers of gametes; only a small fraction of zygotes survive.

The shells of the amniote eggs of birds, reptiles, and monotremes protect the embryo from water loss or injury. Embryos of marsupial mammals complete development in the mother's pouch, attached to a mammary gland. Embryos of eutherian (placental) mammals develop in the uterus, nourished from the mother's blood through the placenta. Parental care of young is widespread among vertebrates and even among many invertebrates.

Gamete Production and Delivery   Precursor cells for ova or sperm are often determined early in embryogenesis and later *amplified* in number. The simplest reproductive systems do not have **gonads** to produce gametes. In polychaete annelids, eggs or sperm develop from cells lining the coelom.

Most insects have separate sexes and complex reproductive systems. Sperm develop in the testes, are stored in the seminal vesicles, and are ejaculated into the female. Eggs are produced in the ovaries and fertilized in the vagina. Females may have a **spermatheca,** or sperm-storing sac. Parasitic flatworms are hermaphroditic and have complex reproductive systems.

With the exception of mammals, many vertebrates have a common opening, a **cloaca,** for the digestive, excretory, and reproductive systems. Most mammals have a separate opening for the digestive tract. The vagina and excretory opening are separate in most female mammals.

Mechanisms that influence which sperm are successful in sequential matings have evolved in insects.

46.3   Reproductive organs produce and transport gametes

Female Reproductive Anatomy   The external reproductive structures include the clitoris and two sets of labia, which form the **vulva.** The female ovaries contain many **follicles,** which each contain an **oocyte** surrounded by cells that nourish and protect the developing egg during **oogenesis.** During each menstrual cycle, a maturing follicle produces estradiol, the primary female sex hormone. Following ovulation, the follicle forms a solid mass called the **corpus luteum,** which secretes progesterone and estradiol.

The egg cell is expelled into the abdominal cavity and swept by cilia into the **oviduct,** or fallopian tube, through which it is transported to the **uterus.** The **endometrium,** or lining of the uterus, is highly vascularized. The neck of the uterus, the **cervix,** opens into the **vagina,** the elastic-walled birth canal and repository for sperm during copulation.

Two pairs of skin folds, the thick outer **labia majora** and the thin inner **labia minora,** enclose the separate openings of the vagina and urethra. The vaginal opening is initially covered by a membrane called the hymen. The **clitoris,** composed of erectile tissue, consists of a short shaft and a **glans,** or head, covered by the **prepuce.** Bartholin's glands secrete lubricating mucus during sexual arousal.

A **mammary gland,** located within a breast, is composed of fatty tissue and small milk-secreting sacs that drain into ducts that open at the nipple. The low level of estradiol in males prevents the mammary glands from enlarging.

Male Reproductive Anatomy   The external male reproductive organs include the scrotum and penis. Sperm are produced in the highly coiled **seminiferous tubules** of the **testes. Leydig cells** produce testosterone and other androgens. Sperm production requires a cooler temperature than the internal body temperature of most mammals, so the **scrotum** suspends the testes below the abdominal cavity.

Sperm pass from a testis into the coiled **epididymis,** in which they mature and gain motility. During

ejaculation, sperm are propelled through the **vas deferens,** into a short **ejaculatory duct,** and out through the **urethra,** which runs through the penis.

Three sets of accessory glands add secretions to the **semen.** The **seminal vesicles** contribute an alkaline fluid containing mucus, fructose (as an energy source for the sperm), a coagulating enzyme, ascorbic acid, and prostaglandins. The **prostate gland** produces a secretion that contains anticoagulant enzymes and citrate. Benign enlargement of the prostate and prostate cancer are common medical problems in older men. The *bulbourethral glands* produce a mucus that neutralizes urine remaining in the urethra.

The 2–5 mL of ejaculated semen may contain more than 100 to 650 million sperm. The slightly alkaline semen neutralizes the vagina. Semen initially coagulates, then liquifies as the sperm start swimming through the female tract.

INTERACTIVE QUESTION   46.3

Label the indicated structures in these diagrams of the human male and female reproductive systems.

The **penis** is composed of spongy tissue that engorges with blood during sexual arousal, producing an erection that facilitates insertion of the penis into the vagina. Some mammals have a baculum, a bone that helps stiffen the penis. A fold of skin called the prepuce, or foreskin, covers the head, or glans, of the penis.

Human Sexual Response   The same embryonic tissues give rise to the glans of the penis and clitoris, the scrotum and the labia majora, and the skin on the penis and the labia minora.

The human sexual response cycle includes two types of physiological reactions: **vasocongestion,** increased blood flow to a tissue, and **myotonia,** increased muscle tension. The excitement phase involves vasocongestion and vaginal lubrication, preparing the vagina and penis for **coitus,** or sexual intercourse. The plateau phase continues vasocongestion and myotonia, and breathing rate and heart rate increase. In both sexes, **orgasm** is characterized by rhythmic, involuntary contractions of reproductive structures. In males, emission deposits semen in the urethra, and ejaculation occurs when the urethra contracts and semen is expelled. In the resolution phase, vasocongested organs return to normal size, and muscles relax.

46.4   The timing and pattern of meiosis in mammals differ for males and females

Gametogenesis, the production of gametes, differs in females and males. **Spermatogenesis** occurs continuously in the seminiferous tubules of the testes as **spermatogonia,** the stem cells that give rise to sperm, generate spermatocytes that undergo meiosis to produce four spermatids. The haploid nucleus of a sperm is contained in a head, tipped with an **acrosome,** which contains enzymes that help the sperm penetrate the egg. Mitochondria provide ATP for movement of the flagellum, or tail.

Oogenesis begins in the female embryo. **Oogonia,** produced from primordial germ cells, divide and differentiate into **primary oocytes,** which are arrested in prophase I of meiosis before birth. Beginning at puberty, FSH periodically stimulates a primary oocyte to finish meiosis I and develop into a **secondary oocyte,** which is arrested in metaphase II. In humans, meiosis is completed if a sperm penetrates the oocyte. Meiotic cytokinesis is unequal, producing one large ovum and up to three small haploid polar bodies that disintegrate.

INTERACTIVE QUESTION   46.4

List three important ways in which oogenesis differs from spermatogenesis.

a.

b.

c.

46.5   The interplay of tropic and sex hormones regulates mammalian reproduction

Gonadotropin-releasing hormone (GnRH) from the hypothalamus directs the anterior pituitary to produce follicle-stimulating hormone (FSH) and luteinizing hormone (LH), gonadotropins that regulate gametogenesis and sex hormone production in both males and females. The steroid sex hormones, androgens (primarily testosterone) and estrogens (chiefly estradiol and progesterone), also regulate gametogenesis.

Androgens are responsible for the embryonic development of male primary sex characteristics. At puberty, androgens direct formation of secondary sex characteristics, such as voice deepening, pubic and facial hair, and muscle growth. Estrogens stimulate breast and pubic hair development at puberty. Sex hormones influence sexual and other behaviors in both sexes.

Hormonal Control of the Male Reproductive System   FSH acts on Sertoli cells, which nourish developing sperm. LH acts on Leydig cells, which secrete testosterone and other androgens, promoting spermatogenesis. Testosterone inhibits production of GnRH, FSH, and LH; inhibin, produced by Sertoli cells, reduces FSH secretion. These feedback mechanisms maintain proper levels of androgen production.

INTERACTIVE QUESTION 46.5

Fill in the blanks in the following description of the control of male reproductive hormones.

a. _____, produced by the hypothalamus, regulates the release of gonadotropic hormones from the anterior pituitary.

b. _____ stimulates Sertoli cells of the seminiferous tubules, which nourish sperm and promote spermatogenesis.

c. _____ stimulates the production of

d. _____ by Leydig cells.

The production of GnRH, LH, and FSH is regulated by **e.** _____.

The Reproductive Cycles of Females   In females, the **endometrium** thickens to prepare for the implantation of the embryo and then is shed in a process called **menstruation** if fertilization does not occur. The human female reproductive cycle involves the integration of the **uterine cycle** (**menstrual cycle**), which involves these cyclic changes in the uterus, and the **ovarian cycle,** which involves follicle growth and ovulation.

The ovarian cycle begins with the release from the hypothalamus of GnRH, which stimulates the anterior pituitary to secrete small amounts of FSH and LH. In this **follicular phase**, FSH stimulates follicular growth, and the cells of the follicle secrete estradiol. The low level of estradiol inhibits the release of FSH and LH.

When the secretion of estradiol by the growing follicle rises sharply, the hypothalamus is stimulated to increase GnRH output, which results in a rise in LH and FSH release. The increase in LH induces maturation of the follicle, which enlarges and forms a bulge near the ovary surface. Ovulation occurs about a day after the LH surge, with the rupture of the follicle and adjacent ovary wall.

In the **luteal phase,** LH stimulates the transformation of the ruptured follicle to the corpus luteum, which secretes estradiol and progesterone. The rising level of these hormones exerts negative feedback on the hypothalamus and pituitary, inhibiting secretion of LH and FSH. Without LH to maintain it, the corpus luteum disintegrates, thereby dropping the levels of estradiol and progesterone. This drop releases the inhibition of the hypothalamus and pituitary, and FSH and LH secretion begins again, stimulating growth of new follicles and the start of the next ovarian cycle.

The uterine (menstrual) cycle is controlled by the hormones secreted by the ovaries. Estradiol secreted by the growing follicles causes the endometrium to begin to thicken in the **proliferative phase,** which correlates with the follicular phase of the ovarian cycle. After ovulation, estradiol and progesterone stimulate increased vascularization of the endometrium and the development of glands that secrete a nutrient fluid. The luteal phase of the ovarian cycle corresponds with the **secretory phase** of the uterine cycle.

The rapid drop in ovarian hormones caused by the disintegration of the corpus luteum reduces blood supply to the endometrium and begins its disintegration, leading to the **menstrual flow phase** of the uterine cycle. By convention, the uterine and ovarian cycles begin with the first day of menstruation.

Endometriosis is a condition in which some endometrial tissue becomes **ectopic**, located outside the uterus. The cyclic swelling and breakdown of this tissue causes pain and bleeding.

Menopause, the cessation of ovulation and menstruation, occurs in older women as the ovaries become less responsive to FSH and LH.

INTERACTIVE QUESTION   46.6

In the diagram of the human female reproductive cycle, label the lines indicating the levels of the gonadotropic and the ovarian hormones in the blood and the phases of the ovarian and uterine cycles.

(a) **Pituitary gonadotropins in blood**

a. _____

b. _____

(b) **Ovarian cycle**

Growing follicle   Mature follicle   Corpus luteum   Degenerating corpus luteum

c. _____   d. _____   e. _____

(c) **Ovarian hormones in blood**

f. _____   g. _____

(d) **Uterine (menstrual) cycle**

Endometrium

h. _____   i. _____   j. _____

Days 0   5   10   14 15   20   25   28

Menstrual versus Estrous Cycles Humans and some other primates have menstrual cycles. Other mammals have **estrous cycles,** during which the endometrium thickens, but if fertilization does not occur, it is reabsorbed. Estrous cycles are often coordinated with season, and females are receptive to sexual activity only during estrus, or heat, the period surrounding ovulation.

46.6 In placental mammals, an embryo develops fully within the mother's uterus

Conception, Embryonic Development, and Birth Fertilization, or **conception,** occurs in the oviduct. **Cleavage** produces a ball of 16 cells by the time the embryo reaches the uterus in about 3 to 4 days. The **blastocyst,** a sphere of cells around a cavity, implants into the endometrium about 7 days after conception. The embryo secretes **human chorionic gonadotropin (hCG),** which maintains the corpus luteum's secretion of progesterone and estrogens through the first few months.

The development of one or more embryos in the uterus is called **pregnancy** or **gestation.** The gestation period correlates with body size and the degree of development of the young at birth. Human pregnancy averages 266 days (38 weeks).

Human gestation can be divided into three **trimesters.** After implantation, the outer layer of the blastocyst, the **trophoblast,** grows and mingles with the endometrium, helping to form the **placenta,** a disk-shaped organ in which gas and nutrient exchange and waste removal take place between maternal blood and embryonic capillaries. Splitting of the embryo in the first month can produce *monozygotic* or identical twins. Independent fertilization of two eggs can lead to *dizygotic* or fraternal twins.

Development proceeds to **organogenesis,** and by the eighth week, the embryo has all the rudimentary structures of the adult and is called a **fetus.** High progesterone levels stimulate growth of the uterus and maternal part of the placenta, cessation of menstrual cycling, and breast enlargement.

During the second trimester, hCG declines, the corpus luteum degenerates, and the placenta secretes its own progesterone, which maintains the pregnancy. The fetus grows rapidly and is quite active. The third trimester is a period of rapid fetal growth.

Through a positive feedback system, hormones and local regulators induce **labor.** High levels of estradiol stimulate the development of oxytocin receptors on the uterus. Oxytocin produced by the fetus and the mother's posterior pituitary stimulates contractions and placental secretion of prostaglandins, which further enhance uterine contractions.

A series of strong contractions of the uterus results in birth, or *parturition.* During the first stage of labor, the cervix dilates. The second stage consists of the contractions that force the fetus out of the uterus and through the vagina. The placenta is delivered in the final stage of labor.

Lactation is unique to mammals. Prolactin secretion by the anterior pituitary stimulates milk production, and oxytocin controls the release of milk during nursing.

INTERACTIVE QUESTION 46.7

List the functions of the following hormones and the locations that secrete them:

a. human chorionic gonadotropin

b. progesterone

c. oxytocin

The Mother's Immune Tolerance of the Embryo and Fetus Current research is attempting to explain why a mother does not reject an embryo, which has paternal as well as maternal chemical markers. Changes in immune system regulation appear to occur during pregnancy.

Contraception and Abortion **Contraception,** the deliberate prevention of pregnancy, can be accomplished by several methods: preventing release of egg or sperm, preventing fertilization, or preventing implantation. Complete abstinence from sexual intercourse is the most effective means of birth control. The **rhythm method,** or **natural family planning,** is based on refraining from intercourse during the period in which conception is most likely, several days before and after ovulation.

Barrier methods of contraception include **condoms** and **diaphragms,** which, when used in conjunction with spermicidal foam or jelly, present a physical and chemical barrier to fertilization. Condoms are the only form of birth control that offers some protection against sexually transmitted diseases.

IUDs (intrauterine devices) interfere with fertilization and implantation. The release of gametes may be prevented by chemical contraception, as in **birth control pills,** and by sterilization. Most birth control pills are combinations of a synthetic estrogen and progestin, which act by negative feedback to stop the release of GnRH by the hypothalamus and thus of FSH and LH by the pituitary, resulting in a cessation of ovulation and follicle development. High doses of combinational birth control pills can be taken as morning-after pills (MAPS) following unprotected intercourse. A progestin-only minipill alters the cervical mucus so that it blocks sperm entry to the uterus. Longer-lasting progestin treatments include capsules implanted under the skin and injections. Cardiovascular problems are a potential effect of combination birth control pills.

Tubal ligation in women and **vasectomy** in men are sterilization procedures that permanently prevent gamete release.

Abortion is the termination of a pregnancy. Spontaneous abortion, or miscarriage, occurs in as many as one-third of all pregnancies. The drug mifepristone (RU-486) is an analog of progesterone that blocks progesterone receptors in the uterus and can be used to terminate a pregnancy within the first 7 weeks.

INTERACTIVE QUESTION 46.8

List the three general types, along with examples, of birth control methods. Which examples are most likely to prevent pregnancy? Which are least likely to do so?

a.

b.

c.

Modern Reproductive Technology   Some genetic diseases and congenital defects can be detected while the fetus is in the uterus, using such techniques as amniocentesis, chorionic villus sampling, ultrasound imaging, and a new technique that identifies fetal cells in the mother's blood, which can then be tested.

Reproductive technology can help couples unable to conceive. Fertilization procedures called **assisted reproductive technology** include *in vitro* **fertilization (IVF),** which involves the removal of ova from a woman, fertilization within a culture dish, and implantation of the developing embryo in the uterus. In **intracytoplasmic sperm injection (ICSI),** a spermatid or sperm is directly injected into an oocyte.

Word Roots

a- = not, without (*asexual reproduction:* a type of reproduction involving only one parent that produces genetically identical offspring)

acro- = tip; **-soma** = body (*acrosome:* an organelle at the tip of a sperm cell that helps the sperm penetrate the egg)

bacul- = a rod (*baculum:* a bone that is contained in, and helps stiffen, the penis of rodents, raccoons, walruses, and several other mammals)

blasto- = produce; **-cyst** = sac, bladder (*blastocyst:* a hollow ball of cells produced one week after fertilization in humans)

coit- = a coming together (*coitus:* the insertion of a penis into a vagina, also called sexual intercourse)

contra- = against (*contraception:* the prevention of pregnancy)

-ectomy = cut out (*vasectomy:* the cutting of each vas deferens to prevent sperm from entering the urethra)

endo- = inside (*endometrium:* the inner lining of the uterus, which is richly supplied with blood vessels)

epi- = above, over (*epididymis:* a coiled tubule located adjacent to the testes where sperm are stored)

labi- = lip; **major-** = larger (*labia majora:* a pair of thick, fatty ridges that enclose and protect the vulva)

lact- = milk (*lactation:* the production of milk)

menstru- = month (*menstruation:* the shedding of portions of the endometrium during a menstrual cycle)

minor- = smaller (*labia minora:* a pair of slender skin folds that enclose and protect the opening to the urethra and vagina)

myo- = muscle (*myotonia:* increased muscle tension)

oo- = egg; **-genesis** = producing (*oogenesis:* the process in the ovary that results in the production of female gametes)

partheno- = a virgin (*parthenogenesis:* a type of reproduction in which females produce offspring from unfertilized eggs)

partur- = giving birth (*parturition:* the expulsion of a baby from the mother, also called birth)

-theca = a cup, case (*spermatheca:* a sac in the female reproductive system where sperm are stored)

tri- = three (*trimester:* a three-month period)

vasa- = a vessel (*vasocongestion:* the filling of a tissue with blood caused by increased blood flow through the arteries of that tissue)

Structure Your Knowledge

1. Use a simple flow diagram to trace the path of a human sperm from the point of production to the point of fertilization, briefly commenting on the functions of both the structures it passes through and the associated glands.

2. Answer the following questions concerning the human menstrual cycle.
   a. What does GnRH do?
   b. What does FSH stimulate?
   c. What causes the spike in LH level?
   d. What does this LH surge induce?
   e. What does LH maintain during the luteal phase?
   f. What inhibits secretion of LH and FSH?
   g. What allows LH and FSH secretion to begin again?

3. Describe how birth control pills work. How does the drug RU-486 (mifepristone) function?

Test Your Knowledge

FILL IN THE BLANKS

_____ 1. type of asexual reproduction in which a new individual grows while attached to the parent's body

_____ 2. small, volatile chemicals that may act as mate attractants

_____ 3. development of egg without fertilization

_____ 4. individual with functioning male and female reproductive systems

_____ 5. common opening of digestive, excretory, and reproductive systems in non-mammalian vertebrates

_____ 6. type of reproductive cycle in which thickened endometrium is reabsorbed

_____ 7. type of assisted reproductive technology in which ova are fertilized in a culture dish and then implanted in the uterus

_____ 8. common duct for urine and semen in mammalian males

_____ 9. filling of a tissue with blood due to increased blood flow

_____ 10. period when ovulation and menstruation cease in humans

MULTIPLE CHOICE: *Choose the one best answer.*

1. An advantage of sexual reproduction is that
   a. it is easier than asexual reproduction.
   b. offspring have a better start when produced from a union of egg and sperm than when simply budded off a parent.
   c. it produces diploid offspring, whereas asexually produced offspring are haploid.
   d. it produces variation in the offspring, and new genetic combinations may be better adapted to a changing environment.
   e. it requires both male and female members of a species to be present in a population and have behavioral interactions.

2. Which of the following is *least* likely to be hermaphroditic?
   a. earthworm
   b. barnacle (sessile crustacean)
   c. tapeworm
   d. grasshopper
   e. liver fluke

3. Which of the following is *incorrectly* paired with its function?
   a. seminiferous tubules—add fluid containing mucus, fructose, and prostaglandins to semen
   b. scrotum—encases testes and suspends them below abdominal cavity
   c. epididymis—tubules in which sperm gain motility
   d. prostate gland—adds fluid to semen
   e. vas deferens—transports sperm from epididymis to ejaculatory duct

4. The function of the corpus luteum is to
   a. nourish and protect the egg cell.
   b. produce prolactin that stimulates the milk sacs of the mammary gland.
   c. produce progesterone and estradiol.
   d. produce estradiol and disintegrate following ovulation.
   e. maintain pregnancy by production of human chorionic gonadotropin.

5. Which of the following hormones is *incorrectly* paired with its function?
   a. androgens—responsible for primary and secondary male sex characteristics
   b. oxytocin—stimulates uterine contractions during parturition
   c. estradiol—responsible for secondary female sex characteristics
   d. FSH—acts on Sertoli cells that nourish sperm, promoting spermatogenesis
   e. prolactin—stimulates breast development at puberty

6. How does meiosis differ in the production of human sperm and ova?

   a. The meiotic production of eggs occurs while a female is an embryo, but does not begin until puberty in males.

   b. Each meiotic division produces four sperm but only two ova.

   c. Meiosis occurs in the testes of males but in the oviducts of females.

   d. Primary oocytes stop dividing by mitosis before birth, whereas male stem cells continue to divide throughout life.

   e. Meiosis is an uninterrupted process in males, whereas it resumes when a follicle matures and is only completed in human females when a sperm penetrates the egg cell.

7. The secretory phase of the uterine cycle

   a. begins with dropping levels of estradiol and progesterone.

   b. corresponds with the luteal phase of the ovarian cycle.

   c. involves the initial proliferation of the endometrium.

   d. corresponds with the follicular phase of the ovarian cycle.

   e. is when the endometrium begins to degenerate and menstrual flow occurs.

8. Examples of birth control methods that prevent the production or release of gametes are

   a. sterilization and chemical contraception.

   b. birth control pills and IUDs.

   c. condoms and diaphragms.

   d. abstinence and chemical contraception.

   e. the progestin minipill and RU-486.

9. In which location does fertilization usually take place in a human female?

   a. ovary

   b. oviduct

   c. uterus

   d. cervix

   e. vagina

10. Certain maternal diseases, drugs, alcohol, and radiation are most dangerous to embryonic development

    a. during the first 2 weeks when the embryo has not yet implanted and spontaneous abortion may occur.

    b. during the first 2 months when organogenesis is occurring.

    c. during the first and second trimesters when the embryonic liver is not yet filtering toxins.

    d. during the second trimester when the corpus luteum no longer secretes progesterone.

    e. during the third trimester when the most rapid growth is occurring.

Use the following to answer questions 11–15.

   a. estradiol

   b. progesterone

   c. LH (luteinizing hormone)

   d. FSH (follicle-stimulating hormone)

   e. hCG (human chorionic gonadotropin)

11. Which hormone stimulates ovulation and the development of the corpus luteum?

12. Which hormone is produced by the developing follicle and initiates thickening of the endometrium?

13. Which hormone is produced by the embryo during the first trimester and is necessary for maintaining a pregnancy?

14. Which hormone stimulates Leydig cells to make testosterone, which in turn stimulates sperm production?

15. Which hormone is produced by the corpus luteum and later by the placenta and is responsible for maintaining a pregnancy?

16. In the birth process, which of the following is/are involved in triggering and maintaining labor?

    a. hCG produced by the fetus

    b. oxytocin produced by fetus and mother, and prostaglandins produced by the placenta

    c. a drop in progesterone caused by the disintegration of the corpus luteum

    d. prolactin produced by the fetus and mother

    e. a surge in the production of LH

Chapter 47

Animal Development

Key Concepts

47.1 After fertilization, embryonic development proceeds through cleavage, gastrulation, and organogenesis

47.2 Morphogenesis in animals involves specific changes in cell shape, position, and adhesion

47.3 The developmental fate of cells depends on their history and on inductive signals

Framework

Fertilization initiates physical and molecular changes in the egg cell. Early embryonic development includes cleavage to form a blastula, gastrulation, and organogenesis. The amount of yolk in the egg and the phylogeny of the animal determine how these processes occur in any particular animal group.

Development and the differentiation of cells result from the control of gene expression by both cytoplasmic determinants and cell–cell induction. Pattern formation depends on the positional information a cell receives within a developing structure. Developmental biologists are gradually unraveling some of the mechanisms underlying the complex processes of animal development.

Chapter Review

Preformation, or the notion that the egg or sperm contains a miniature embryo—a "homunculus"—was once the favored explanation of animal development. *Epigenesis* explains that the form of an embryo gradually develops from an egg.

The developmental plan of an embryo is predetermined by the zygote genome and by the distribution of **cytoplasmic determinants** (maternal proteins and mRNAs) in the egg. As cell division separates cytoplasmic components and signals arrive from neighboring cells, distinct patterns of gene expression arise in different cells. Differences in gene expression lead to structural and functional specialization, or **cell differentiation**. **Morphogenesis** produces the body shape.

Developmental biologists use molecular genetics, classical studies, and **model organisms**, which are representative of larger groups and easy to study in the lab, to study embryonic development.

47.1 After fertilization, embryonic development proceeds through cleavage, gastrulation, and organogenesis

Fertilization Fertilization, the union of egg and sperm, combines the haploid sets of chromosomes of these specialized cells and activates the egg by initiating metabolic reactions that trigger embryonic development.

In the external fertilization of sea urchins, the egg releases molecules that attract the sperm. When a sperm comes in contact with the jelly coat of an egg, the **acrosome** at the tip of the sperm discharges hydrolytic enzymes. This **acrosomal reaction** allows the *acrosomal process* to elongate through the jelly coat. Fertilization within the same species is assured when proteins on the surface of the acrosomal process attach to specific receptor proteins extending from the plasma membrane through the *vitelline layer* of the egg.

In response to the fusion of sperm and egg plasma membranes, ion channels open and sodium ions flow into the egg. The resulting depolarization of the membrane prevents other sperm from fusing with the egg, providing a **fast block to polyspermy**. The sperm nucleus enters the egg cytoplasm.

Membrane fusion also initiates the **cortical reaction**, which involves release of calcium ions from the endoplasmic reticulum. In response to the Ca^{2+} increase in the cytoplasm, **cortical granules** located in the outer *cortex* release their contents into the *perivitelline space*. The released enzymes and other macromolecules cause the vitelline layer to elevate and harden to form the **fertilization envelope**, which functions as a **slow block to polyspermy**.

The rise in Ca^{2+} concentration increases cellular respiration and protein synthesis, known as *egg activation*. Parthenogenetic development can be initiated by injecting calcium into an egg.

After the sperm nucleus fuses with the egg nucleus, DNA replication begins in preparation for the cell division that begins the development of the embryo.

INTERACTIVE QUESTION   47.1

a. What assures that only sperm of the right species fertilize sea urchin eggs?

b. What assures that only one sperm will fertilize an egg?

Following internal fertilization in mammals, secretions of the female reproductive tract enhance motility of the sperm. A sperm migrates through the follicle cells released with the egg to the **zona pellucida**, the extracellular matrix of the egg. Complementary binding of a molecule on the sperm head to a receptor in the zona pellucida induces an acrosomal reaction. The hydrolytic enzymes released from the acrosome enable the sperm cell to penetrate the zona pellucida.

Binding of a sperm membrane protein with the egg membrane leads to the release of enzymes from cortical granules. This cortical reaction causes alterations of the zona pellucida and a slow block to polyspermy.

The entire sperm enters the egg. The centriole of the basal body of the sperm flagellum generates the mitotic spindle. Nuclear envelopes disperse and the first mitotic division occurs.

Cleavage **Cleavage** is a succession of rapid cell divisions during which the embryo becomes partitioned into many small cells, called **blastomeres**. The first divisions produce a cluster of cells with a fluid-filled cavity called the **blastocoel**, creating the stage called the **blastula**. Cleavage parcels different cytoplasmic determinants into cells and sets the stage for later development.

The polarity of the egg of most animals is established during its development. The **vegetal pole** is where the stored nutrients in **yolk** are most concentrated, and the opposite end—called the **animal pole**—where the polar bodies budded from the egg during meiosis.

In the eggs of many frogs, the animal hemisphere has melanin granules in the cortex, whereas the vegetal hemisphere contains the yellow yolk. The animal-vegetal axis determines the anterior-posterior (head-tail) axis of the embryo. Fertilization results in a rotation of the animal pole cortex toward the point of sperm entry. Rotation of the cortex on the opposite side creates the **gray crescent** and activates vegetal cortex proteins now exposed to inner animal hemisphere molecules. These activated proteins form cytoplasmic determinates that influence the development of dorsal structures. Thus, the *cortical rotation* sets up the dorsal-ventral axis.

The first two cleavage divisions are vertical, or meridional; the third division is equatorial. Yolk impedes cell division, and smaller cells are produced in the animal hemisphere in the frog. In the frog blastula, the blastocoel forms in the animal hemisphere due to unequal cell division.

INTERACTIVE QUESTION   47.2

In the frog embryos diagrammed below, identify the stages and label the indicated regions or structures.

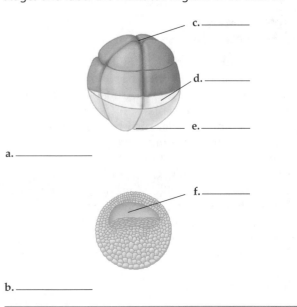

In species in which eggs have little yolk, blastomeres are of equal size, and an animal-vegetal axis results from unequal distribution of other substances. **Holoblastic cleavage** is the complete division of eggs with small or moderate amounts of yolk.

The large amount of yolk in the eggs of birds, other reptiles, many fishes, and insects results in an incomplete division called **meroblastic cleavage**.

Meroblastic cleavage of the fertilized bird egg is restricted to the small disk of cytoplasm on top of the yolk and produces a cap of cells. The blastomeres separate into an upper and lower layer, forming a cavity between them comparable to a blastocoel.

In the yolk-rich eggs of insects, the zygote nucleus undergoes repeated divisions and the nuclei migrate to the egg surface. Plasma membranes eventually form around the nuclei, creating a blastula made of cells surrounding a mass of yolk.

Gastrulation Changes in cell motility, shape, and adhesion are part of the morphogenetic rearrangements in **gastrulation** that result in a three-layered embryo. The **gastrula** consists of the embryonic **germ layers**—the outer **ectoderm**, the middle **mesoderm**, and the inner **endoderm** that lines the embryonic digestive tract. This three-layered body plan is characteristic of most animal phyla.

In a sea urchin, gastrulation occurs as cells from the vegetal pole detach and move into the blastocoel as migratory *mesenchyme cells*. The remaining cells at the vegetal pole flatten into a *vegetal plate* that then buckles inward by a process known as **invagination**. The invagination deepens to form a narrow pouch called the **archenteron**, a cavity that will become the digestive tract. The archenteron opening is the **blastopore**, which develops into the anus. Cytoplasmic extensions (filopodia) from mesenchyme cells at the tip of the archenteron pull the archenteron across the blastocoel, where the endoderm fuses with the ectoderm to form the mouth.

Gastrulation in frog development is more complicated because of the multilayered blastula wall and the large, yolk-filled cells of the vegetal hemisphere. A group of invaginating cells begins gastrulation in the region of the gray crescent, forming a crease. The upper area of the crease is called the **dorsal lip** of the blastopore. In a process called **involution**, surface cells roll over the dorsal lip and move inside, becoming organized into the endoderm and mesoderm. The blastopore eventually forms a circle surrounding a **yolk plug** of large, yolk-filled cells. Gastrulation produces an endoderm-lined archenteron surrounded by mesoderm, with the outer layer of the gastrula composed of ectoderm.

In the chick, a linear thickening, called the **primitive streak**, marks the anterior-posterior axis of the embryo and is the site of gastrulation. Cells of the upper layer of the *blastoderm*, the *epiblast*, move through the primitive streak, forming the mesoderm and the endoderm. Cells remaining in the epiblast become ectoderm. The *hypoblast* cells of the original lower layer form part of the yolk sac and connecting stalk.

INTERACTIVE QUESTION 47.3

Label the parts in these diagrams of gastrulation in a sea urchin and a frog.

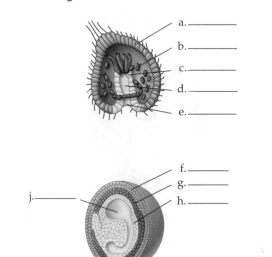

a._____
b._____
c._____
d._____
e._____

f._____
g._____
h._____
j._____
i._____

Organogenesis Organs begin to develop from the embryonic germ layers in a process known as **organogenesis**. In a frog embryo, the **notochord**, the skeletal rod characteristic of chordates, forms from the dense clustering or condensation of dorsal mesoderm just above the archenteron. Ectoderm above the developing notochord forms a *neural plate*, which rolls inward to form a hollow **neural tube**, from which will develop the central nervous system.

In vertebrates, a band of cells called the **neural crest** separates from the side of the neural tube when it forms. These cells migrate to form bones of the skull, teeth, components of the peripheral nervous system, and many other cell types.

Mesoderm along the sides of the notochord condenses and separates into blocks, called **somites**. Mesenchyme (migratory) cells from the somites give rise to vertebrae and muscles associated with the axial skeleton, arranged in segmented fashion. Lateral to the somites, the mesoderm splits to form the lining of the coelom.

Organogenesis in the chick proceeds in a fashion similar to that of the frog embryo. The borders of the blastoderm fold down and join, forming a three-layered tube attached by a stalk to the yolk.

In invertebrates, many of the same mechanisms drive organogenesis: cell migrations, condensations, cell signaling between different tissues, and cell shape changes.

INTERACTIVE QUESTION   47.4

Label the indicated structures in this diagram of organo-genesis in a frog embryo.

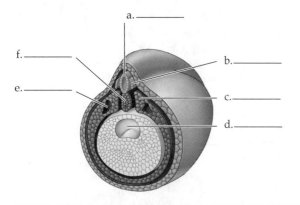

Developmental Adaptations of Amniotes   Reptiles (including birds) and mammals are **amniotes**, meaning that they create the aqueous environment necessary for embryonic development with a fluid-filled sac sur-rounded by a membrane called the amnion. The pri-mary germ layers also give rise to four **extraembryonic membranes**, which are essential to development within the eggs of birds and other reptiles and the uterus of mammals.

INTERACTIVE QUESTION   47.5

List the four extraembryonic membranes found in birds and other reptiles, and briefly describe their functions.

a.

b.

c.

d.

Mammalian Development   Because mammalian eggs contain little yolk, cleavage is holoblastic. Gastrulation and early organogenesis, however, are similar in pat-tern to those of birds and other reptiles.

The **blastocyst** consists of an outer epithelium, the **trophoblast**, surrounding a cavity into which pro-trudes a cluster of cells, called the **inner cell mass**. These inner cells develop into the embryo and some of the extraembryonic membranes. The trophoblast se-cretes enzymes that enable the blastocyst to embed in the uterine lining (endometrium) and extends projec-tions that will develop, along with mesodermal tissue, into the fetal portion of the placenta.

The inner cell mass forms a flat embryonic disk with an upper *epiblast* and lower *hypoblast*, homologous to the two layers in birds. In gastrulation, cells from the epiblast move through a primitive streak to form the mesoderm and endoderm.

The four extraembryonic membranes are homolo-gous to those of birds and other reptiles. The chorion surrounds the embryo and the other membranes and functions in gas exchange. The amnion encloses the embryo in a fluid-filled amniotic cavity. The yolk sac, enclosing a small fluid-filled cavity, is the site of early formation of blood cells. The allantois forms blood ves-sels that connect the embryo with the placenta through the umbilical cord.

INTERACTIVE QUESTION   47.6

Label the indicated cell layers in this diagram of a human blastocyst implanting into the endometrium. List the em-bryonic tissues and membranes that will develop from each cell layer.

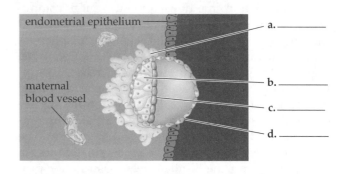

47.2   Morphogenesis in animals involves specific changes in cell shape, position, and adhesion

The Cytoskeleton, Cell Motility, and Convergent Exten-sion   Changes in cell shape usually involve reorganiza-tion of the cytoskeleton. Extension of microtubules to elongate a cell and then contraction of actin filaments at one end create a wedge shape, initiating invaginations and evaginations.

Migrating embryonic cells use their cytoskeleton to extend and retract cellular protrusions. Cells at the leading edge of a migrating tissue may extend filopodia and pull along other cells. By **convergent extension**, cells of a tissue layer crawl between each other, causing the sheet of cells to extend and narrow.

Roles of Cell Adhesion Molecules and the Extracellular Matrix Glycoproteins called **cell adhesion molecules (CAMs)** help regulate movement and tissue building by holding specific cells together. **Cadherins,** which require calcium for proper function, are important adhesion molecules whose genes are differentially expressed in specific locations during development.

Glycoproteins such as fibronectins in the extracellular matrix (ECM) provide anchorage for crawling cells and tracts that direct their movement. Cells may also secrete other ECM substances that inhibit migration. Receptors on migrating cells pick up directional cues that then trigger signaling pathways that direct the orientation of their cytoskeleton and the production of other proteins involved in cell movement.

47.3   The developmental fate of cells depends on their history and on inductive signals

The mechanisms underlying the differentiation of cells during embryonic development are organized into two basic concepts. First, heterogeneous cytoplasm in the egg is partitioned through cleavage into different blastomeres, or, in amniotes, local environmental differences establish early differences between embryonic cells. Second, the interactions between cells, called **induction**, affect a cell's developmental fate. Both of these phenomena influence gene expression, leading to the differentiation of cells.

Fate Mapping In the 1920s, W. Vogt developed a **fate map** for amphibian embryos by labeling regions of the blastula with dye to determine where specific cells showed up in later developmental stages. Cell lineage analysis follows the mitotic descendants of individual blastomeres. The developmental history of every cell has been determined for the transparent nematode, *Caenorhabditis elegans.*

In most animals, early "founder cells" can be identified that will generate specific tissues. Experiments have also shown that a cell's *developmental potential* becomes restricted as development proceeds.

Establishing Cellular Asymmetries In nonamniotic vertebrates, the major axes are established in the egg or at fertilization. In amniotes, polarity does not seem to be established until later. In chicks, gravity may be involved in establishing the anterior-posterior axis while the egg moves down the oviduct, and pH differences may determine the dorsal-ventral axis of the blastoderm. In mammals, polarity becomes obvious after cleavage, but the location of sperm entry may play a role.

The zygote is the only **totipotent** cell in many species. Even though cytoplasmic determinants are asymmetrically distributed in the frog egg, the first cleavage division bisects the gray crescent and equally separates these determinants.

Mammalian blastomeres remain totipotent until the 16-cell stage, although this may be partly due to a separated cell's ability to respond to its environment. The progressive restriction of a cell's developmental potency is characteristic of animal development. Experimental manipulation shows that the tissue-specific fates of cells of a late gastrula are usually fixed.

INTERACTIVE QUESTION   47.7

If an eight-cell stage of a sea urchin embryo is split vertically into two four-cell groups, both develop into normal larvae. If the division is made horizontally, abnormal larvae develop. Explain this experimental result.

Cell Fate Determination and Pattern Formation by Inductive Signals Cells influence the development of other cells through signals that usually switch on sets of genes.

Using transplant experiments with amphibian embryos, H. Mangold and H. Spemann established that the dorsal lip is a primary organizer of the embryo because of its influence in early organogenesis. The molecular basis of induction by *Spemann's organizer* appears to be tied to the inactivation of bone morphogenic protein 4 (BMP-4) on the dorsal side of the embryo. Proteins similar to BMP-4 and its inhibitors appear to play a role in the development of many different organisms.

Development of the vertebrate eye involves a series of inductions from cells in all three germ layers, progressively determining the fate of the ectodermal cells that become lens cells.

Pattern formation is the ordering of cells and tissues into the characteristic structures in the proper locations in the animal. Cells and their offspring differentiate based on **positional information**, molecular cues that indicate their position with respect to body axes.

A limb bud of a chick consists of a core of mesoderm and a layer of ectoderm. Researchers have identified two organizer regions that secrete proteins that provide positional information to cells.

Cells of the **apical ectodermal ridge (AER)** at the tip of the limb bud secrete proteins of the fibroblast growth factor (FGF) family. These growth signals appear to promote limb bud outgrowth and patterning along the proximal-distal axis. An FGF-secreting AER has been identified in the median fin in a shark, suggesting a long evolutionary history of the AER.

The **zone of polarizing activity (ZPA)**, mesodermal tissue located at the posterior attachment of the bud, appears to communicate position along the anterior-posterior axis in a developing limb. Cells of the ZPA secrete a protein growth factor called Sonic hedgehog.

Cells respond to positional information relating to their location in a developing organ according to their developmental histories. Thus, patterns of *Hox* gene expression first determine whether cells will become part of a wing or of a leg; then positional cues indicate what part of the wing or leg the cells will help form. A hierarchy of gene activations leads to the expression of genes that are involved in specifying various regions of the developing limb.

INTERACTIVE QUESTION   47.8

What would be the effect of implanting cells genetically engineered to produce and secrete Sonic hedgehog to the anterior attachment of a limb bud?

Word Roots

acro- = the tip (*acrosomal reaction:* the discharge of a sperm's acrosome when the sperm approaches an egg)

arch- = ancient, beginning (*archenteron:* the endoderm-lined cavity, formed during the gastrulation process, that develops into the digestive tract of an animal)

blast- = bud, sprout; **-pore**= a passage (*blastopore:* the opening of the archenteron in the gastrula that develops into the mouth in protostomes and the anus in deuterostomes)

blasto- = produce; **-mere**= a part (*blastomere:* small cells of an early embryo)

cortex- = shell (*cortical reaction:* a series of changes in the cortex of the egg cytoplasm during fertilization)

ecto- = outside; **-derm** = skin (*ectoderm:* the outermost of the three primary germ layers in animal embryos)

endo- = within (*endoderm:* the innermost of the three primary germ layers in animal embryos)

epi- = above; **-genesis** = origin, birth (*epigenesis:* the progressive development of form in an embryo)

extra- = beyond (*extraembryonic membrane:* four membranes that support the developing embryo in reptiles (including birds) and mammals)

fertil- = fruitful (*fertilization:* the union of haploid gametes to produce a diploid zygote)

gastro- = stomach, belly (*gastrulation:* the formation of a gastrula from a blastula)

holo- = whole (*holoblastic cleavage:* a type of cleavage in which there is complete division of the egg)

in- = into; **vagin-** = a sheath (*invagination:* the infolding of cells)

involut- = wrapped up (*involution:* cells rolling over the edge of a lip into the interior)

mero- = a part (*meroblastic cleavage:* a type of cleavage in which there is incomplete division of yolk-rich egg, characteristic of avian development)

meso- = middle (*mesoderm:* the middle primary germ layer of an early embryo)

noto- = the back; **-chord** = a string (*notochord:* a flexible rod that runs along the dorsal axis of the body in the future position of the vertebral column)

poly- = many (*polyspermy:* fertilization by more than one sperm)

soma- = a body (*somites:* paired blocks of mesoderm lateral to the notochord of a vertebrate embryo)

tropho- = nourish (*trophoblast:* the outer epithelium of the blastocyst, which forms the fetal part of the placenta)

zona = a belt; **pellucid-** = transparent (*zona pellucida:* the extracellular matrix of a mammalian egg)

Structure Your Knowledge

1. Create a flowchart that shows the sequence of key events in the fertilization of a sea urchin egg, and indicate the functions of these events.

2. Fill in this table, briefly describing the early stages of development for sea urchin, frog, bird, and mammalian embryos.

Animal	Cleavage	Blastula	Gastrula
Sea urchin			
Frog			
Bird			
Mammal			

Test Your Knowledge

MULTIPLE CHOICE: *Choose the one best answer.*

1. The fact that an eunucleated sea urchin egg can be activated by the injection of Ca^{2+} is evidence that
   a. development can proceed parthenogenetically.
   b. the fast block to polyspermy may sometimes not work and two sperm may fertilize an egg.
   c. inactive mRNA and other proteins had been stockpiled during development of the egg.
   d. the cortical reaction is not necessary for development to proceed.
   e. development is a cytoplasmic process that does not require genetic regulation.

2. The slow block to polyspermy
   a. is essential to prevent sperm from other species from fertilizing the egg.
   b. is directly produced by the depolarization of the membrane.
   c. is a result of the formation of the fertilization envelope.
   d. is caused by the expulsion of calcium ions.
   e. involves all of the above.

3. The blastocoel
   a. develops into the archenteron or embryonic gut.
   b. is a fluid filled cavity in the blastula
   c. opens to the exterior through a blastopore.
   d. forms a hollow chamber during gastrulation.
   e. is lined with mesoderm.

4. In a frog embryo, gastrulation
   a. is slowed by the large amount of yolk.
   b. proceeds by involution as cells roll over the dorsal lip of the blastopore.
   c. produces a blastocoel displaced into the animal hemisphere.
   d. occurs along the primitive streak.
   e. involves the formation of the notochord and neural tube.

5. Cytoplasmic determinants
   a. are unevenly distributed egg components that influence the developmental fates of cells.
   b. are involved in regulation of gene expression.
   c. usually include maternal mRNA.
   d. are often separated in the first few cleavage divisions.
   e. are all of the above.

6. Which of the following is *incorrectly* paired with its embryonic germ layer?
   a. muscles—mesoderm
   b. central nervous system—ectoderm
   c. liver and pancreas—endoderm
   d. heart—endoderm
   e. notochord—mesoderm

7. The primitive streak of mammalian embryos is analogous to
   a. the dorsal lip of the frog embryo.
   b. the blastoderm of a bird embryo.
   c. the multinucleated stage of a fruit fly embryo.
   d. the archenteron of a sea urchin embryo.
   e. none of the above.

8. A function of the allantois in birds is to
   a. provide for nutrient exchange between the embryo and yolk sac.
   b. store metabolic wastes.
   c. form a respiratory organ in conjunction with the chorion.
   d. provide an aqueous environment.
   e. both b and c.

9. Somites are
   a. blocks of mesoderm circling the archenteron.
   b. condensations of cells from which the notochord arises.
   c. serially arranged mesoderm blocks lateral to the notochord in a vertebrate embryo.
   d. structures arising from neural crest cells.
   e. produced during organogenesis in sea urchins, frogs, birds, and mammals, but not in fruit flies.

10. What forms the fetal portion of the placenta?
    a. the epiblast
    b. the trophoblast and some mesoderm
    c. the allantois and yolk sac
    d. the endometrium
    e. the amnion

11. In her transplant experiments with frog embryos, Mangold found that a part of the dorsal lip moved to another location would result in a second gastrulation and even the development of a doubled, face-to-face tadpole. Which of the following is the best explanation for these results?
    a. morphogenetic movements
    b. separation of cytoplasmic determinants
    c. totipotency of cells
    d. fate map determination
    e. induction by dorsal lip organizer

12. Pattern formation appears to be determined by
    a. positional information a cell receives from gradients of chemical signals called morphogens.
    b. differentiation of cells, which then migrate into developing organs.
    c. the movement of cells along fibronectin fibers.
    d. the induction of cells by mesoderm cells found in the center of organs.
    e. gastrulation and the formation of the three tissue layers.

Use this chick embryo diagram for questions 13–15.

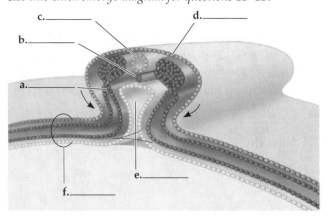

13. Which line points to the notochord?

14. Which line points to the archenteron?

15. To what structure does line f point?
    a. the primitive streak through which gastrulation occurred
    b. the hypoblast
    c. the mesodermal lining of the coelom
    d. the blastocoel enclosed by the epiblast and the hypoblast
    e. the developing extraembryonic membranes

16. The early cleavage divisions of a human zygote most closely resemble
    a. those of a sea urchin because neither egg has a large store of yolk and cleavage is holoblastic.
    b. those of a chick because both birds and mammals share a common reptilian ancestor.
    c. those of a frog because the cells of the animal hemisphere are smaller than those of the vegetal hemisphere in both types of embryos.
    d. those of a *Drosophila* because both pass through an early multinucleated stage.
    e. all of the above animals because as deuterostomes, they have similar embryonic patterns.

17. Gastrulation and organogenesis in mammals most closely resemble
    a. those of a sea urchin because both types of eggs have little yolk and cleavage is holoblastic.
    b. those of a chick because both birds and mammals share a common reptilian ancestor.
    c. those of a frog because gastrulation takes place through the dorsal lip of the blastopore in both.
    d. neither a chick nor a frog because mammals have some embryonic cells that develop into extraembryonic membranes.
    e. those of a *Drosophila* because homeotic genes control pattern development in both organisms.

18. What do cells of the apical ectodermal ridge secrete?
    a. Sonic hedgehog that communicates location along the anterior-posterior axis
    b. cadherins that help to hold migrating cells together as they form tissues of the leg
    c. ZPA, which forms fibronectin fibers that guide migrating cells
    d. fibroblast growth factors that promote limb bud outgrowth
    e. CAMs that cause the limb bud to elongate by convergent extension

Chapter 48

Neurons, Synapses, and Signaling

Key Concepts

48.1 Neuron organization and structure reflect function in information transfer

48.2 Ion pumps and ion channels maintain the resting potential of a neuron

48.3 Action potentials are the signals conducted by axons

48.4 Neurons communicate with other cells at synapses

Framework

This chapter describes the transmission of nervous impulses along and between neurons. Ion channels maintain the membrane potential of a neuron, and the rapid flow of ions through voltage-gated channels creates the rapid depolarization of an action potential. The release of neurotransmitters at synapses converts electrical signals to chemical signals that pass information to receiving cells.

Chapter Review

Neurons transfer long-distance information via electrical signals and usually communicate between cells using short-distance chemical signals. The higher-order processing of nervous signals may involve clusters of neurons called **ganglia** or more structured groups of neurons organized into a **brain**.

48.1 Neuron organization and structure reflect function in information transfer

Information Processing   The three stages of information processing are sensory input, integration, and motor output. Sensors detect external stimuli or internal conditions. **Sensory neurons** transmit this information to the brain or ganglia, where **interneurons** integrate it

and send output through neurons bundled into *nerves* that triggers muscle or gland activity. **Motor neurons** carry impulses to muscle cells.

A **central nervous system (CNS)**, found in many animals, includes the brain and longitudinal nerve cord. The **peripheral nervous system (PNS)** consists of nerves connecting the CNS with the rest of the body.

Neuron Structure and Function   A neuron consists of a **cell body**, which contains the nucleus and organelles, and numerous extensions. The highly branched, short **dendrites** *receive* signals from other neurons; the longer **axon** *transmits* signals to other cells. Axons originate from a region of the cell body called the **axon hillock**.

Terminal branches of axons end in **synaptic terminals**, which usually release **neurotransmitters** that relay signals across the **synapse** to another neuron, muscle, or gland cell. The transmitting cell is called the **presynaptic cell;** the receiving cell is the **postsynaptic cell**.

The very numerous supporting cells called **glia** give structural integrity and physiological support to neurons.

INTERACTIVE QUESTION   48.1

Label the indicated structures on this diagram of a neuron. Indicate the direction of impulse transmission. What happens at part e?

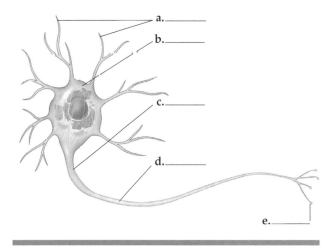

a._____

b._____

c._____

d._____

e._____

48.2   Ion pumps and ion channels maintain the resting potential of a neuron

A voltage, or **membrane potential**, exists across the plasma membrane of all cells. The membrane potential of a typical nontransmitting neuron, called the **resting potential**, is between –60 and –80 mV.

Formation of the Resting Potential   In a mammalian neuron, the concentration of K^+ is 140 mM inside the cell and 5 mM outside. The Na^+ concentration is 150 mM outside and 15 mM inside. *Sodium-potassium pumps* maintain these gradients.

Ion channels that are *selectively permeable* allow specific ions to diffuse across the membrane. The diffusion of K^+ out of a cell results in a buildup of negative charge within the neuron, creating the membrane potential.

Modeling the Resting Potential   Ions diffuse through channels down their concentration gradient until balanced by the electrical gradient across the membrane. The membrane voltage at this equilibrium, or **equilibrium potential (E_{ion})**, is determined for a single ion with a 1+ charge at 37 °C by the Nernst equation: $E_{ion} = 62$ mV $(\log ([ion]_{outside}/[ion]_{inside}))$. The equilibrium potential for K^+ (E_K) in a neuron is –90 mV. Using the concentration gradients for Na^+, E_{Na} is +62 mV (the inside of the membrane is more positive than outside).

Neurons at rest have more K^+ channels open than Na^+ channels, and the resting potential is closer to E_K than E_{Na}.

INTERACTIVE QUESTION   48.2

a. What is the principal cation inside the cell? outside the cell?

b. Which side of the membrane has a negative charge?

c. What change in the permeability of the cell's membrane to K^+ and/or Na^+ could cause the cell's membrane potential to shift from –70 mV to –90 mV?

48.3   Action potentials are the signals conducted by axons

In addition to the ungated potassium and sodium ion channels that create the resting potential, neurons also have **gated ion channels** that open or close in response to stimuli and increase or decrease the membrane potential. Electrophysiologists measure membrane potential by placing microelectrodes connected to a voltage recorder inside and outside a cell.

A stimulus that opens K^+ channels will result in **hyperpolarization**, as K^+ flows out and the membrane potential moves toward E_K (–90 mV). When Na^+ channels open and Na^+ flows in, a **depolarization** occurs as the inside of the cell becomes less negative. With this type of *graded potential*, the magnitude of a voltage change is proportional to the strength of the stimulus: the stronger the stimulus, the more gated ion channels that open.

Production of Action Potentials   **Voltage-gated** channels respond to a change in membrane potential, which may trigger their opening and lead to a massive change in membrane voltage called an **action potential**. Once depolarization of a typical neuron reaches a certain membrane voltage called the **threshold**, an action potential is triggered. The action potential is an *all-or-none* event, always creating the same voltage spike once the threshold is reached.

Time Course of Action Potentials   Action potentials last only about 1–2 milliseconds. The frequency of action potentials increases with the intensity of a stimulus.

Both Na^+ and K^+ voltage-gated channels are involved in an action potential. Sodium channels open rapidly in response to depolarization, but become inactivated when a part of the channel blocks the opening. Potassium channels open slowly in response to depolarization, but remain open throughout the action potential.

As a stimulus depolarizes the membrane to threshold, Na^+ channels open; the influx of Na^+ causes further depolarization, which opens more channels. This positive-feedback cycle during the *rising phase* brings the membrane potential close to E_{Na}. The inactivation of Na^+ channels and opening of voltage-gated K^+ channels rapidly brings the membrane potential back toward E_K during the *falling phase*.

During the *undershoot*, the membrane's permeability to K^+ is higher than at rest, and the continued outflow of K^+ temporarily hyperpolarizes the membrane. During the **refractory period**, which occurs while the Na^+ channels remain inactivated, the neuron cannot respond to another stimulus.

INTERACTIVE QUESTION   48.3

This diagram shows the changes in voltage-gated channels during an action potential. Label the channels and inactivation gate, ions, and the components of the graph. Name and describe the five phases of the action potential. Place numbers on the graph to show where each phase is occurring.

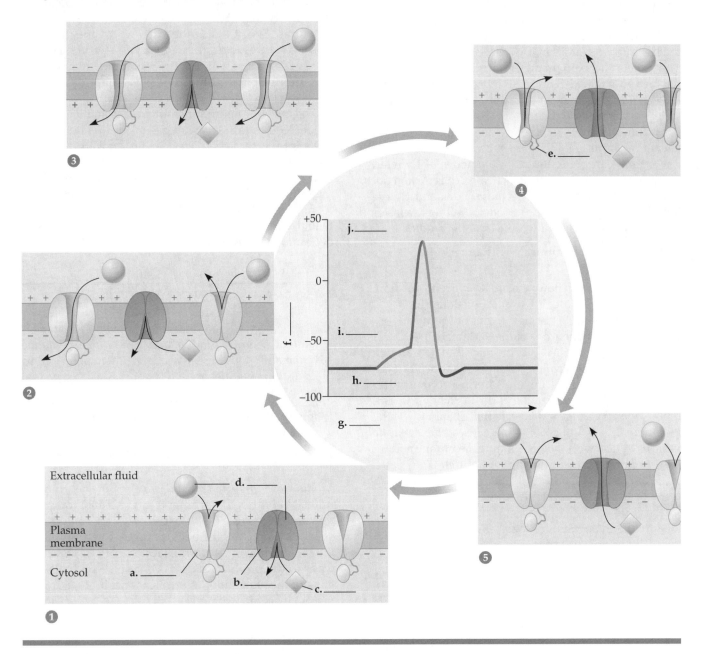

Conduction of Action Potentials Na^+ influx in the rising phase depolarizes adjacent sections of the membrane, bringing them to the threshold. Local depolarizations and action potentials across the membrane result in the propagation of serial action potentials along the length of the neuron. Because of the brief refractory period, the action potential is propagated in only one direction.

Resistance to current flow is inversely proportional to the cross-sectional area of the conducting "wire." The greater the axon diameter, the faster action potentials are propagated. Some invertebrates, such as squid, have giant axons that conduct impulses very rapidly.

In vertebrates, **oligodendrocytes** (in the CNS) and **Schwann cells** (in the PNS) insulate axons in a **myelin sheath** by forming layers of membranes. Voltage-gated ion channels are concentrated in the **nodes of Ranvier**, small gaps between successive Schwann cells. Action potentials can be generated only at these nodes, and a nerve impulse "jumps" from node to node, resulting in a faster mode of transmission known as **saltatory conduction**.

48.4   Neurons communicate with other cells at synapses

Electrical synapses allow electrical current to flow directly from cell to cell via gap junctions. Electrical synapses are found in the giant axons of squid and lobsters and in many neurons of the vertebrate brain.

A *chemical synapse* involves the release of neurotransmitters. A synaptic terminal contains **synaptic vesicles**, in which thousands of molecules of neurotransmitter are stored. The depolarization of the presynaptic membrane opens voltage-gated calcium channels in the membrane. The influx of Ca^{2+} causes the synaptic vesicles to fuse with the presynaptic membrane and release neurotransmitter into the **synaptic cleft**.

INTERACTIVE QUESTION   48.4

Identify the components of this chemical synapse following the depolarization of the synaptic terminal.

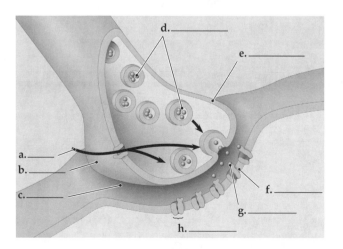

Generation of Postsynaptic Potentials   Binding of neurotransmitter to a receptor on **ligand-gated ion channels** in the postsynaptic cell membrane allows ions to cross the membrane, creating a *postsynaptic potential.* If the channel allows both Na^+ and K^+ to diffuse through, the net inflow of positive charge depolarizes the membrane, creating an **excitatory postsynaptic potential (EPSP)** that brings the membrane potential closer to threshold. If binding of neurotransmitter opens K^+ or Cl^- channels, the membrane hyperpolarizes, producing an **inhibitory postsynaptic potential (IPSP)**.

The effect of neurotransmitters is terminated when the molecules diffuse away, are taken up and repackaged into synaptic vesicles, or are broken down by enzymes.

Postsynaptic potentials are graded potentials. Their magnitude depends on the number of neurotransmitter molecules that bind to receptors. The membrane potential of the axon hillock at any given time is determined by the sum of all EPSPs and IPSPs.

INTERACTIVE QUESTION   48.5

a. _____ **summation** occurs with repeated release of neurotransmitter from one or more synaptic terminals before the postsynaptic potential returns to its resting potential.

b. _____ **summation** occurs when several different presynaptic terminals, usually from different neurons, release neurotransmitter simultaneously.

Modulated Synaptic Transmission   Some neurotransmitters bind to receptors that trigger a signal transduction pathway in the postsynaptic cells. This modulated synaptic transmission begins more slowly but lasts longer. Binding of neurotransmitter to a receptor often leads to the production of cAMP as a second messenger and the phosphorylation of channel proteins, opening or closing ion channels.

Neurotransmitters   Neurotransmitters may have many different receptors and may produce very different effects in postsynaptic cells.

Acetylcholine is a common neurotransmitter in invertebrates and vertebrates. Depending on the type of receptor, it can be inhibitory or excitatory. In neuromuscular junctions, acetylcholine released from a motor axon produces an EPSP in a muscle cell. The enzyme *acetylcholinesterase* hydrolyzes the neurotransmitter.

Biogenic amines, neurotransmitters derived from amino acids, are often involved in modulating synaptic transmission in the CNS. **Epinephrine, norepinephrine,**

and **dopamine** are derived from tyrosine. Norepinephrine acts through a G-protein-coupled receptor to generate EPSPs in the PNS. **Serotonin** (synthesized from tryptophan) and dopamine affect sleep, mood, attention, and learning. Imbalances of these neurotransmitters have been associated with several disorders.

The amino acids **gamma aminobutyric acid (GABA)** and **glutamate** function as neurotransmitters in the CNS. GABA is the most common inhibitory transmitter in the brain. Glutamate is excitatory. Glycine is inhibitory in parts of the CNS outside the brain.

Neuropeptides are short chains of amino acids that activate signal transduction pathways. **Substance P** is an excitatory neurotransmitter that functions in pain perception. **Endorphins** are neuropeptides produced in the brain during physical or emotional stress that have pain-killing and other functions. Opiates bind to endorphin receptors in the brain.

Some neurons use nitric oxide (NO) and carbon monoxide as local regulators. The release of NO triggers relaxation of smooth muscle cells and vessel dilation. CO produced by neurons in the brain affects the release of hypothalamic hormones.

INTERACTIVE QUESTION 48.6

Acetylcholine stimulates skeletal muscle contraction but inhibits or slows cardiac muscle contraction. How can this neurotransmitter have such opposite effects?

Word Roots

bio- = life; **-genic** = producing (*biogenic amines*: neurotransmitters derived from amino acids)

dendro- = tree (*dendrite*: one of usually numerous, short, highly branched processes of a neuron that receive signals from other neurons)

de- = down, out (*depolarization*: an electrical state in an excitable cell whereby the inside of the cell is made less negative relative to the outside)

endo- = within (*endorphin*: a hormone produced in the brain and anterior pituitary that inhibits pain perception)

glia = glue (*glia*: supporting cells that are essential for the structural integrity of the nervous system and for the normal functioning of neurons)

hyper- = over, above, excessive (*hyperpolarization*: an electrical state whereby the inside of the cell is made more negative relative to the outside than at the resting membrane potential)

inter- = between (*interneurons*: an association neuron; a nerve cell within the central nervous system that forms synapses with sensory and motor neurons and integrates sensory input and motor output)

neuro- = nerve; **trans-** = across (*neurotransmitter*: a chemical messenger released from the synaptic terminal of a neuron at a chemical synapse that diffuses across the synaptic cleft and binds to and stimulates the postsynaptic cell)

oligo- = few, small (*oligodendrocytes*: glial cells that form insulating myelin sheaths around the axons of neurons in the central nervous system)

post- = after (*postsynaptic cell*: the target cell at a synapse)

pre- = before (*presynaptic cell*: the transmitting cell at a synapse)

salta- = leap (*saltatory conduction*: rapid transmission of a nerve impulse along an axon resulting from the action potential jumping from one node of Ranvier to another, skipping the myelin-sheathed regions of membrane)

syn- = together (*synapse*: the locus where a neuron communicates with a postsynaptic cell in a neural pathway)

Structure Your Knowledge

1. Develop a flowchart, diagram, or description of the sequence of events in the creation and propagation of an action potential and in the transmission of this potential across a chemical synapse.

Test Your Knowledge

MULTIPLE CHOICE: *Choose the one best answer.*

1. Which of the following is *not* true of the resting potential of a typical neuron?

   a. The inside of the cell is more negative than is the outside.

   b. There are concentration gradients with more sodium outside the cell and a higher potassium concentration inside the cell.

   c. It is about –70 mV and can be measured by using microelectrodes placed inside and outside the cell.

   d. It is formed by the sequential opening of voltage-gated channels.

   e. It results from the combined equilibrium potentials of potassium and sodium.

2. Interneurons
   a. may connect sensory and motor neurons.
   b. are more common in the PNS than the CNS.
   c. are involved in the integration of sensory information.
   d. typically have more axons than dendrites.
   e. Both a and c are correct.

3. Nodes of Ranvier are
   a. gaps where Schwann cells abut and at which action potentials are generated.
   b. neurotransmitter-containing vesicles located in the synaptic terminals.
   c. the parts of neurons where action potentials are initiated.
   d. clusters of receptor proteins located on the post-synaptic membrane.
   e. ganglia adjacent to the spinal cord.

4. After the depolarization of an action potential, the fall in the membrane potential occurs due to the
   a. closing of sodium inactivation gates.
   b. closing of potassium and sodium channels.
   c. refractory period in which the membrane is hyperpolarized.
   d. delay in the action of the sodium-potassium pump.
   e. opening of voltage-gated potassium channels and the closing of sodium inactivation gates.

5. The threshold of a membrane
   a. is exactly midway between E_K and E_{Na}.
   b. opens voltage-gated channels and permits the rapid outflow of sodium ions.
   c. is the depolarization that is needed to generate an action potential.
   d. is a graded potential that is proportional to the strength of a stimulus.
   e. is an all-or-none event.

6. Which of the following is *incorrectly* paired with its function?
   a. axon hillock—region of neuron where action potential originates
   b. Schwann cells—create myelin sheath around axon in PNS
   c. synapse—space between presynaptic and post-synaptic cell into which neurotransmitter is released
   d. synaptic terminal—receptor that is part of an ion channel that is keyed to a specific neuro-transmitter
   e. dendrite—receives signals from other neurons

7. Which of the following is *not* true of chemical synapses?
   a. Synaptic terminals at the ends of branching ax-ons contain synaptic vesicles, which enclose the neurotransmitter.
   b. The influx of sodium when an action potential reaches the presynaptic membrane causes synaptic vesicles to release their neurotransmit-ter into the cleft.
   c. The binding of neurotransmitter to receptors on the postsynaptic membrane usually opens or closes ion channels.
   d. An excitatory postsynaptic potential forms when sodium channels open and the membrane potential moves closer to an action potential threshold.
   e. Neurotransmitter is often rapidly degraded in the synaptic cleft.

8. An inhibitory postsynaptic potential occurs when
   a. sodium flows into the postsynaptic cell.
   b. enzymes do not break down the neurotransmit-ter in the synaptic cleft.
   c. binding of neurotransmitter opens ion gates that result in the membrane becoming hyper-polarized.
   d. acetylcholine is the neurotransmitter.
   e. norepinephrine is the neurotransmitter.

9. Which of the following do (does) *not* function as a neurotransmitter or local regulator released by neurons?
   a. amino acids such as glutamate
   b. neuropeptides such as endorphin
   c. biogenic amines such as dopamine
   d. steroids
   e. NO

10. How is an increase in the strength of a stimulus communicated by a neuron?
    a. The spike of the action potential reaches a higher voltage.
    b. The frequency of action potentials generated along the neuron increases.
    c. The length of an action potential (the duration of the rising phase) increases.
    d. The action potential travels along the neuron faster.
    e. All action potentials are the same; the nervous system cannot discriminate between different strengths of stimuli.

11. Why is signal transmission faster in myelinated axons?

    a. These axons are thinner, and there is less resistance to the voltage flow.

    b. These axons use electrical synapses rather than chemical synapses.

    c. The action potential can jump from node to node along the insulating myelin sheath.

    d. These axons are thicker and provide less resistance to voltage flow.

    e. These axons have higher depolarization values than do unmyelinated axons.

12. What is the main effect of the neurotransmitter GABA in the CNS?

    a. increase pain

    b. create inhibitory postsynaptic potentials

    c. create excitatory postsynaptic potentials

    d. induce sleep

    e. decrease pain and induce euphoria

13. If the binding of a neurotransmitter to its receptor opens Cl⁻ channels, what would be the effect on the postsynaptic cell? (Cl⁻ is in higher concentration outside the cell.)

    a. It would hyperpolarize, receiving an IPSP.

    b. It would reach threshold, and an action potential would fire.

    c. It would depolarize and form an EPSP but probably not fire an action potential.

    d. It would initiate a signal transduction pathway.

    e. Its membrane potential would not change because neither sodium nor potassium channels opened.

14. Movement of an action potential in only one direction along a neuron is a function of

    a. saltatory conduction.

    b. the pathway from dendrite to axon.

    c. the refractory period when sodium inactivation gates are still closed.

    d. the localized depolarization of the surrounding membrane.

    e. the reaching of threshold, which creates an all-or-none firing.

Chapter 49

Nervous Systems

Key Concepts

49.1 Nervous systems consist of circuits of neurons and supporting cells

49.2 The vertebrate brain is regionally specialized

49.3 The cerebral cortex controls voluntary movement and cognitive functions

49.4 Changes in synaptic connections underlie memory and learning

49.5 Nervous system disorders can be understood in molecular terms

Framework

This chapter describes the organization of the vertebrate nervous system and the functional specialization of its components. The branches of the peripheral nervous system include the sensory afferent neurons and the efferent motor system and autonomic nervous system. The central nervous system includes the spinal cord, which mediates reflexes and transports information, and the regionally specialized brain, which regulates homeostatic functions, integrates sensory information, and controls voluntary movement and cognitive functions.

Chapter Review

The estimated 10^{11} neurons of the human brain communicate in complex circuits. Mechanisms to sense and respond to changes in the environment evolved in prokaryotes. Modification of this ability allowed for the development of communication between the cells of multicellular organisms.

49.1 Nervous systems consist of circuits of neurons and supporting cells

A simple diffuse **nerve net** controls the radially symmetrical body and gastrovascular cavity in cnidarians such as hydras and jellies. Sea stars have a central nerve ring with radial **nerves,** bundles of axons of neurons, connected to a nerve net in each arm.

Cephalization, the concentration of neurons in the head, evolved in bilaterally symmetrical animals. Flatworms have a *central nervous system (CNS)* consisting of a simple brain and two longitudinal nerve cords. Annelids and arthropods have a more complicated brain and a ventral nerve cord with segmentally arranged clusters of neurons called ganglia. Vertebrates have a brain and dorsal spinal cord; nerves and ganglia make up the *peripheral nervous system (PNS).*

INTERACTIVE QUESTION 49.1

The organization of an organism's nervous system correlates with how the animal interacts with the environment. How do the different nervous systems of clams and squid illustrate this generalization?

Organization of the Vertebrate Nervous System The complex behavior of vertebrates is integrated by the brain. The spinal cord transports information to and from the brain, controls basic locomotion, and produces **reflexes**, which are automatic responses to stimuli.

The CNS develops from the embryonic dorsal hollow nerve cord. Spaces in the brain called **ventricles** are continuous with the narrow **central canal** of the spinal cord and all are filled with **cerebrospinal fluid.** This fluid, formed by filtration of the blood, carries out circulatory functions and cushions the brain.

Neuron cell bodies, dendrites, and unmyelinated axons make up the **gray matter** of the brain and spinal cord. **White matter** is named for the white color of the myelin sheaths of the bundles of axons.

Glia in the CNS include ciliated ependymal cells, which line the ventricles; microglia, which protect against microorganisms; and oligodendrocytes and Schwann cells, which function in axon myelination. **Astrocytes** are supporting glia with various functions. Some facilitate information transfer at synapses and increase blood flow to active neurons. During development, astrocytes induce the formation of the **blood–brain barrier,** which restricts the passage of most substances into the brain. **Radial glia** guide the embryonic growth of neurons. Radial glia and astrocytes can act as stem cells.

The Peripheral Nervous System Sensory information is carried to the CNS along *afferent* PNS neurons. *Efferent* neurons of the PNS carry instructions to muscles, glands, and endocrine cells. Paired **cranial nerves** and **spinal nerves,** with associated ganglia, make up the vertebrate PNS. Most cranial and all spinal nerves contain both afferent and efferent neurons.

Functionally, the efferent branch of the peripheral nervous system consists of the **motor system,** which carries signals to skeletal muscles, whose movement is largely under conscious control. The **autonomic nervous system** maintains the internal environment by its control over smooth and cardiac muscles and various organs.

The autonomic nervous system is separated into three divisions. The **sympathetic division** accelerates the heart and metabolic rate, generating energy and arousing an organism for action. The **parasympathetic division** carries signals that enhance self-maintenance activities that conserve energy, such as digestion and slowing the heart rate. The **enteric division** includes complex networks of neurons that control the secretions of the digestive tract, pancreas, and gallbladder, and the contractions of smooth muscles that produce peristalsis. The enteric division is normally regulated by the sympathetic and parasympathetic divisions.

49.2 The vertebrate brain is regionally specialized

As a vertebrate embryo develops, the **forebrain, midbrain,** and **hindbrain** become evident as three bilaterally symmetrical, anterior bulges of the neural tube. In humans, five brain regions develop from these initial bulges. The three regions derived from the midbrain and hindbrain give rise to the brainstem. One of these regions also gives rise to the cerebellum. The *telencephalon*

INTERACTIVE QUESTION 49.2

Complete this concept map to help you learn the functional organization of the peripheral nervous system.

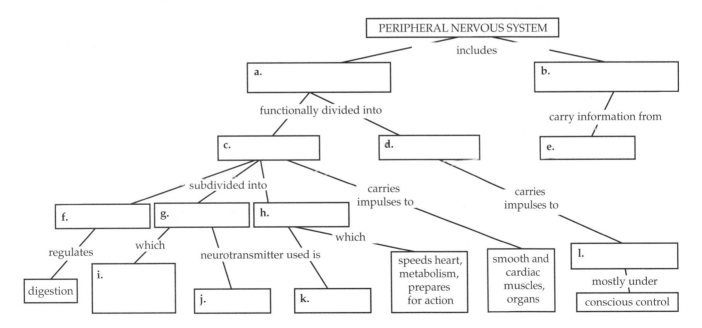

of the forebrain gives rise to the **cerebrum.** During the second and third months of human development, the outer portion of the cerebrum, called the **cerebral cortex,** extends over many of the other brain regions. The thalamus, hypothalamus, and epithalamus develop from the region of the forebrain called the *diencephalon.*

The Brainstem   The **brainstem** is a stalk with caplike swellings at the anterior end of the spinal cord. It consists of the midbrain, the **pons,** and the **medulla oblongata,** or **medulla.** The pons and medulla transfer sensory and motor information between the PNS and the rest of the brain. The axons of neurons from the mid- and forebrain cross in the medulla, so that the right side of the brain controls much of the movement of the left side of the body, and vice versa.

The midbrain receives and integrates sensory information and sends this information to specific regions of the forebrain. Sensory axons involved in hearing pass through or terminate in the midbrain. The midbrain has prominent optic lobes in nonmammalian vertebrates but only coordinates visual reflexes in mammals.

The medulla contains control centers for such homeostatic functions as breathing, swallowing, heart and blood vessel actions, and digestion. The pons functions with the medulla in some of these activities.

Arousal is a state in which an individual is aware of the external world. Several brainstem centers control sleep and arousal. The **reticular formation** is a network of neurons that filters sensory information going to the cerebral cortex, affecting the level of arousal. Sleep-producing centers are located in the pons and medulla, and serotonin may be the neurotransmitter involved in these areas. A center that causes arousal is found in the midbrain.

All birds and mammals have a characteristic sleep–wake cycle. The hormone melatonin, produced by the pineal gland, appears to play an important part in this cycle. Sleep may be important in learning and memory.

The Cerebellum   The **cerebellum** is involved in learning and remembering motor skills, as well as in coordination and error checking during motor and perceptual functions. The cerebellum integrates information from the auditory and visual systems with sensory input from the joints and muscles as well as motor commands from the cerebrum to provide coordination of movements and balance.

The Diencephalon   The diencephalon is the region of the forebrain that evolved earliest in vertebrate history. The *epithalamus* includes the pineal gland and one of several clusters of capillaries producing cerebrospinal fluid. The **thalamus** is the major sorting center for sensory information going to the cerebrum. It also receives input from the cerebrum and other parts of the brain regulating emotion and arousal.

The **hypothalamus** is the major brain region for homeostatic regulation. It produces the posterior pituitary hormones and the releasing hormones that control the anterior pituitary. The hypothalamus contains the regulating centers for many survival functions and also plays a role in sexual and mating behaviors, the alarm response, and pleasure.

Studies of daily behaviors show that an internal timekeeper, called the **biological clock,** is important for maintaining these circadian rhythms. In mammals, the **suprachiasmatic nuclei (SCN),** located in the hypothalamus, function as the biological clock. Sensory neurons provide visual information to the SCN to keep the mammalian clock synchronized with natural cycles of light and dark. Humans kept in free-running conditions have a 24.2-hour cycle.

The Cerebrum   The cerebrum is divided into right and left **cerebral hemispheres,** each with an outer covering of gray matter (the cerebral cortex), inner white matter, and the *basal nuclei.* These groups of neurons, located deep in the white matter, are important in planning and learning movements.

As the largest and most complex part of the mammalian brain, the cerebral cortex is involved in perception, voluntary movement, and learning. The right side of the cortex receives information from and controls the movement of the left side of the body, and vice versa. Communication between the two hemispheres travels through the **corpus callosum,** a thick band of axons.

Evolution of Cognition in Vertebrates   The *neocortex,* the outer six layers of neurons running along the cerebrum, is unique to mammals. Its extensive convolutions, which greatly increase its surface area in primates and cetaceans, were thought to be required for *cognition,* the perception and reasoning of knowledge. The now-documented cognitive ability of birds is linked to the *pallium,* the outer portion of the avian brain, which contains nuclei that receive information from the thalamus.

Identify the structures of the human brain. Match the functions listed below to these structures.

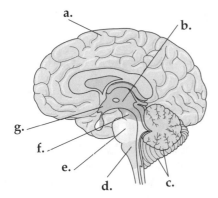

Structure	Function
a. _____	_____
b. _____	_____
c. _____	_____
d. _____	_____
e. _____	_____
f. _____	_____
g. _____	_____

Functions

1. Coordination of balance and movement

2. Aids medulla in some functions, conducts information between brain and spinal cord

3. Sorts and relays information to cerebrum

4. Regulates breathing, heart rate, digestion

5. Integrates sensory and motor information, thinking

6. Produces hormones, homeostatic regulation

7. Sends sensory information to forebrain, involved in hearing and visual/reflexes

49.3   The cerebral cortex controls voluntary movement and cognitive functions

Each side of the cerebral cortex is divided into four lobes—frontal, temporal, occipital, and parietal—with specialized functions localized in each lobe. Functional areas include *primary sensory areas,* which receive and process specific types of sensory information, and *association areas,* which integrate information from parts of the brain.

Information Processing in the Cerebral Cortex   Sensory information (visual, auditory, somatosensory) is directed to primary sensory areas located in different lobes of the cortex. Olfactory information is routed to cortex regions similar in mammals and reptiles, then, via the thalamus, to the interior frontal lobe. Association areas near the primary sensory areas integrate ("associate") the different aspects of sensory inputs. The cerebral cortex generates motor response plans, which the primary motor cortex directs to the brainstem and spinal cord, from which action potentials travel in motor neurons to skeletal muscles.

Language and Speech   Studies of brain injuries and imaging of brain activity using fMRI and PET show that several association areas in different lobes are involved in understanding and generating language. *Broca's area* in the frontal lobe is involved in speech generation; *Wernicke's area* in the posterior temporal lobe is involved in speech comprehension. Other areas are involved in reading, speaking, and attaching meaning to words.

Lateralization of Cortical Function   Lateralization results in the two hemispheres differing in function. The left hemisphere usually becomes specialized for language, math, and logic, whereas the right hemisphere is involved in pattern recognition, spatial perception, and nonverbal thinking. Some of this lateralization relates to handedness.

Emotions   The amygdala, hippocampus, and olfactory bulb, along with sections of the thalamus, form a ring around the brainstem called the *limbic system.* This system interacts with sensory and other areas of the cortex in generating emotions, both basic emotions such as laughing and crying, and emotional feelings associated with the survival-related functions of the brainstem, such as aggression, feeding, and sexuality.

The **amygdala,** located in the temporal lobe, functions in creating emotional memories. The prefrontal cortex is important in the experience of emotions, temperament, and decision-making.

Consciousness   The study of human consciousness, the awareness of self and one's own experiences, has been facilitated by brain-imaging techniques but remains a challenging endeavor. Consciousness appears to be an emergent property of the brain that perhaps involves a repetitive scanning mechanism of multiple brain areas to produce a unified conscious moment.

INTERACTIVE QUESTION   49.4

What are the functions of the primary somatosensory cortex, located along the front of the parietal lobe, and the primary motor cortex, located along the rear of the frontal lobe? Describe the spatial distribution of neurons in these two areas.

INTERACTIVE QUESTION   49.5

Action potentials are initiated more readily in synapses exhibiting long-term potentiation. Explain why it initially takes at least two signaling neurons to establish LTP in a synapse.

49.4   Changes in synaptic connections underlie memory and learning

During embryonic development, neurons that reach the proper target receive growth supporting factors; those that do not undergo programmed cell death. Embryonic neurons form large numbers of synapses; more than half of these synapses are lost, while others are stabilized and retained into adulthood.

Neural Plasticity   The nervous system can be remodeled in response to experience, a capacity called **neural plasticity.** A high level of activity at a synapse may lead to the addition of synaptic terminals with the postsynaptic neuron. A lack of activity may lead to the loss of connections with a neuron. Signals at a synapse can also be strengthened or weakened.

Memory and Learning   Human memory consists of **short-term memory,** information held briefly, and **long-term memory,** the retention of this information for later recall. Transferring information from short-term memory, which involves links in the hippocampus, to long-term memory requires the making of more permanent connections within the cerebral cortex.

Skill memories are formed by repetition of motor activities and do not require conscious recall of specific steps. Learning and remembering skills and procedures seem to involve nerve cells making new connections. Remembering facts and numbers may involve changes in the strength of existing nerve connections.

Long-Term Potentiation   In **long-term potentiation (LTP),** the strength of synaptic transmission is increased. AMPA receptors are inserted in a postsynaptic membrane following a brief burst of action potentials from a presynaptic cell that arrive when the postsynaptic membrane is depolarized by activity at nearby synapses. After LTP is established, glutamate released by the presynaptic neuron binds to and activates both NMDA and AMPA receptors, triggering strong postsynaptic potentials and initiating action potentials. LTP may be involved in learning and memory storage.

49.5   Nervous system disorders can be understood in molecular terms

Chemical or anatomical changes in the brain can lead to nervous system disorders. Research efforts seek to identify genetic and environmental components of these diseases for both prevention and development of better drug treatments.

Schizophrenia   **Schizophrenia** is characterized by psychotic episodes involving hallucinations and delusions. The disease has both environmental and genetic components. Relatives of people with schizophrenia have higher risks of developing the disease. Treatments have focused on drugs that block dopamine receptors, although evidence indicates that glutamate signaling is also involved. Newer drugs can reduce the major symptoms with fewer negative side effects.

Depression   Depressive illnesses include **major depressive disorder,** which involves a persistent low mood with changes in sleep, appetite, and energy; and **bipolar disorder,** which involves swings of mood from very high to low. Both illnesses have a genetic and environmental component. Drug treatments against depression often increase the activity of biogenic amines.

Drug Addiction and the Brain Reward System   Drug addiction is the compulsive consumption of drugs that increase the activity of the brain's reward system. The reward pathways involve neurons of the ventral tegmental area (VTA) near the base of the brain, which release dopamine at synapses in regions of the cerebrum. Addictive drugs such as cocaine, heroin, nicotine, amphetamines, and alcohol affect the reward system in different ways—stimulating dopamine-releasing VTA neurons, blocking the removal of dopamine from synapses, or decreasing the activity of inhibitory neurons in the pathway. Long-lasting changes in the reward circuitry result in increased craving for the drug.

Alzheimer's Disease   **Alzheimer's disease** is dementia characterized by confusion, memory loss, and personality changes. This age-related, progressive disease involves the death of neurons in large areas of the brain. Postmortem diagnosis involves finding neurofibrillary tangles (bundles of processes made up of the protein *tau*) and amyloid plaques (aggregates of β-amyloid). β-amyloid is an insoluble peptide that accumulates outside neurons and appears to trigger death of neighboring neurons.

Parkinson's Disease **Parkinson's disease** is a progressive motor disorder characterized by difficulty in movements, rigidity, and muscle tremors. The death of neurons in the midbrain that normally release dopamine in the basal nuclei leads to the symptoms of this disease. Brain surgery, deep-brain stimulation, and drugs such as L-dopa (a dopamine precursor) are used to manage the symptoms. A potential cure may be to implant dopamine-secreting neurons in the midbrain or basal ganglia.

Neural Stem Cells In 1998, F. Gage and P. Eriksson reported newly formed nerve cells in the hippocampus of cancer victims who had been treated with a chemical marker that labels the DNA of dividing cells. A goal of future research is to find ways to induce neural progenitor cells to differentiate into glia or neurons when needed, and to culture neural stem cells for transplantation into damaged neurological tissue.

Word Roots

astro- = a star; **-cyte** = cell (*astrocytes:* glial cells that provide structural and metabolic support for neurons)

auto- = self (*autonomic nervous system:* the branch of the peripheral nervous system of vertebrates that regulates the internal environment)

cephalo- = head (*cephalization:* the clustering of sensory neurons and other nerve cells to form a brain near the anterior end and mouth of animals with elongated, bilaterally symmetrical bodies)

epi- = above, over (*epithalamus:* a brain region, derived from the diencephalon, that contains several clusters of capillaries that produce cerebrospinal fluid; it is located above the thalamus)

hypo- = below (*hypothalamus:* the ventral part of the vertebrate forebrain that functions in maintaining homeostasis, especially in coordinating the endocrine and nervous systems; it is located below the thalamus)

para- = near (*parasympathetic division:* one of three divisions of the autonomic nervous system)

supra- = above, over (*suprachiasmatic nuclei:* a pair of structures in the hypothalamus of mammals that functions as a biological clock)

Structure Your Knowledge

1. List the location and functions of these important brain nuclei or functional systems.
   a. reticular formation
   b. suprachiasmatic nucleus
   c. basal nuclei
   d. limbic system
   e. amygdala and hippocampus

Test Your Knowledge

MULTIPLE CHOICE: *Choose the one best answer.*

1. Which of the following animals is mismatched with its nervous system?
   a. sea star (echinoderm)—modified nerve net, central nerve ring with radial nerves
   b. hydra (cnidarian)—ring of ganglia leading to multiple nerve cords
   c. annelid worm—brain, ventral nerve cord with segmental ganglia
   d. vertebrate—central nervous system of brain and spinal cord and peripheral nervous system
   e. squid (mollusc)—large complex brain, ventral nerve cord, some giant axons

2. Which of the following is *not* true of the autonomic nervous system?
   a. It is a subdivision of both the central and peripheral nervous system.
   b. It consists of the sympathetic, parasympathetic, and enteric divisions.
   c. Two of its divisions generally have antagonistic effects.
   d. It controls smooth and cardiac muscles.
   e. Control is generally involuntary.

3. If you needed to obtain nuclei from the cell bodies of neurons for an experiment, you would want to make preparations of
   a. the corpus callosum.
   b. cranial nerves.
   c. the neocortex.
   d. sympathetic ganglia near the spinal cord.
   e. either c or d.

4. The amygdala and hippocampus
   a. control biorhythms and are found in the thalamus.
   b. are the part of the reticulum system found in the midbrain.

c. are located in the parietal and frontal lobes, respectively, and are involved in language and speech.

d. are nuclei in the midbrain involved in hearing and vision.

e. are found in the inner cortex and are involved in memory storage.

5. Which of the following structures is *incorrectly* paired with its function?

a. pons—conducts information between spinal cord and brain

b. cerebellum—contains the tracts that cross motor neurons from one side of the brain to the other side of the body

c. thalamus—sorts and relays incoming impulses to the cerebrum

d. corpus callosum—band of axons connecting left and right hemispheres

e. hypothalamus—homeostatic regulation, pleasure centers

6. Which of the following is evidence indicating that schizophrenia, bipolar disorder, and major depression have a genetic as well as an environmental component?

a. All three seem to involve an excess of dopamine.

b. All can be treated with the same drugs.

c. If one identical twin has the illness, there is a 50% chance that the other will, indicating that genes and the environment both contribute to these disorders.

d. The genes have been identified for these illnesses, and they share a common sequence.

e. Identical twins raised separately do not share these illnesses; those raised in the same home are likely to both be affected by the disorder.

7. Long-term potentiation seems to be involved in

a. memory storage and learning.

b. enhanced response of a presynaptic cell resulting from strong depolarization from bursts of action potentials.

c. the attachment of emotion to information to be learned.

d. the release of the excitatory neurotransmitter glutamate that opens voltage-gated channels.

e. all of the above.

8. What makes up the white matter of the spinal cord?

a. the ventricles

b. myelinated sheaths of axons

c. cerebrospinal fluid

d. motor and interneuron cell bodies

e. sympathetic ganglia

9. Lateralization of the cerebrum refers to the fact that

a. visceral functions are controlled by the hindbrain, whereas thinking and language are centered in association areas of the cortex.

b. the right side of the brain controls the left side of the body, and vice versa.

c. the primary somatosensory cortex receives sensory information that is mapped onto one region of the cerebral cortex, whereas the primary motor cortex localized in another region controls motor output to the body.

d. the left hemisphere usually controls verbal and analytic ability, whereas the right hemisphere is specialized for nonverbal thinking and pattern recognition.

e. memory formation involves circuits that include the limbic system, hippocampus, and association areas of the cortex.

10. A function of the reticular formation is to

a. keep you awake during a lecture.

b. coordinate movement and balance.

c. produce emotions.

d. control visceral functions such as breathing and heart rate.

e. create consciousness.

11. Which of these structures is found only in the mammalian brain?

a. telencephalon and diencephalon

b. optic lobes

c. olfactory cortex

d. neocortex

e. hippocampus

12. Neural progenitor cells

a. are responsible for making new synapses during the process of learning.

b. retain the ability to divide and differentiate into both neurons and glia.

c. are similar to embryonic stem cells in that they can differentiate into all kinds of body cells.

d. have been transplanted into individuals with spinal cord injuries and are able to grow and make connections with the proper target cells.

e. are types of glia that form tracks along which newly formed neurons migrate during development.

13. Which of the following does *not* contain efferent neurons?
    a. spinal nerve
    b. motor system
    c. enteric division
    d. sensory neurons
    e. All of the above contain efferent neurons.

14. The pallium of birds is involved in _____ and is homologous to the _____ of mammals.
    a. circadian rhythms ... suprachiasmatic nuclei
    b. flight ... cerebellum
    c. cognition ... neocortex
    d. parenting behavior ... the reward system
    e. vision ... optic lobes

15. Which of the following is descriptive of neural plasticity?
    a. Embryonic neurons compete for growth supporting factors, and those that do not reach target tissues undergo programmed cell death.
    b. Large body movements are controlled by parts of the brainstem, the cerebellum, and the motor cortex.
    c. The neurotransmitter acetylcholine is released by preganglionic neurons in both the parasympathetic and sympathetic division but only by postganglionic neurons in the parasympathetic division.
    d. The number and activity of synapses can change in the process of learning motor skills and developing long-term memories.
    e. Excesses or deficiencies of neurotransmitters such as dopamine or serotonin are implicated in most nervous system disorders.

Chapter 50

Sensory and Motor Mechanisms

Key Concepts

50.1 Sensory receptors transduce stimulus energy and transmit signals to the central nervous system

50.2 The mechanoreceptors responsible for hearing and equilibrium detect moving fluid or settling particles

50.3 The senses of taste and smell rely on similar sets of sensory receptors

50.4 Similar mechanisms underlie vision throughout the animal kingdom

50.5 The physical interaction of protein filaments is required for muscle function

50.6 Skeletal systems transform muscle contraction into locomotion

Framework

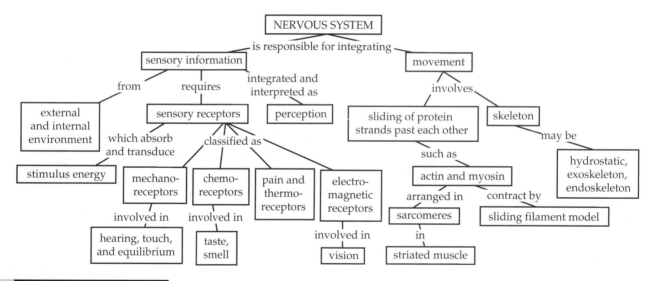

Chapter Review

In a continuous cycle, animals are constantly exploring and sensing their environment and processing that information in the brain, which then directs their next movements.

50.1 Sensory receptors transduce stimulus energy and transmit signals to the central nervous system

Sensory Pathways   **Sensory reception** is the detection of energy from a particular stimulus by sensory cells, which are usually modified neurons or epithelial cells that occur singly or in groups within sensory organs. **Sensory receptors** include organs, cells, or structures within cells that detect specific stimuli from the external or internal environment.

Sensory transduction is the conversion of a physical or chemical stimulus into a **receptor potential,** a graded change in the membrane potential of a receptor cell that is proportional to the strength of the stimulus. Receptor potentials result from the opening or closing of ion channels in response to a stimulus.

Transmission of sensory information to the central nervous system occurs as action potentials. The magnitude of a receptor potential correlates with the frequency of action potentials generated when the sensory receptor is a neuron or the quantity of neurotransmitter released when the receptor synapses with a sensory neuron.

Integration of sensory information begins with the summation of graded potentials from receptors. Sensory information is further integrated within complex receptors and the CNS. Impulses arriving from sensory neurons are routed to different parts of the brain that interpret them, producing **perceptions** of various stimuli.

The strengthening of stimulus energy, either by accessory structures of sense organs or as part of signal transduction pathways, is called **amplification.** Ongoing stimulation may lead to **sensory adaptation** in which continuous stimulation results in a decline in sensitivity of the receptor cell.

INTERACTIVE QUESTION   50.1

List the four functions common to all sensory pathways.

a.

b.

c.

d.

Types of Sensory Receptors   Sensory receptors can be grouped into five categories. **Mechanoreceptors** respond to the mechanical energy of pressure, touch, stretch, motion, and sound. Bending or stretching of external cell structures linked to ion channels increases their permeability to ions, leading to depolarization or hyperpolarization.

Vertebrate stretch receptors monitor the length of skeletal muscles. Dendrites of sensory neurons wound around specialized muscle fibers are stimulated by stretching of the muscles. In humans, dendrites of sensory neurons function as touch receptors; the type of stimulus they respond best to depends on the structure of their connective tissue covering.

Chemoreceptors include both general receptors that monitor the total solute concentration and specific receptors that respond to individual kinds of important molecules. Membrane permeability of the sensory cell changes in response to binding of a stimulus molecule.

Electromagnetic receptors respond to various forms of electromagnetic energy. **Photoreceptors** detect visible light and are often organized into eyes. Some animals have receptors that detect infrared rays, electric currents, and magnetic fields.

Thermoreceptors in the skin and hypothalamus respond to heat or cold and send information to the body's thermostat. Researchers found that receptor proteins for capsaicin, a component of hot peppers, also respond to hot temperatures. Mammals have a number of thermoreceptors that respond to specific temperature ranges.

Pain receptors (nociceptors) in humans are naked dendrites. These receptors respond to excess heat, pressure, or chemicals. Prostaglandins released from injured tissues increase pain by sensitizing receptors.

INTERACTIVE QUESTION   50.2

a. Where are pain receptors located in humans?

b. How do aspirin and ibuprofen reduce pain?

50.2   The mechanoreceptors responsible for hearing and equilibrium detect moving fluid or settling particles

Sensing Gravity and Sound in Invertebrates   Most invertebrates have **statocysts,** which often consist of a layer of ciliated receptor cells lining a chamber containing **statoliths,** dense granules or grains of sand. The stimulation of mechanoreceptors under the statoliths provides positional information to the animal.

Most insects sense sounds with body hairs that vibrate in response to sound waves of specific frequencies. Localized "ears" are found on many insects, consisting of a tympanic membrane with attached receptor cells stretched over an internal air chamber.

Hearing and Equilibrium in Mammals   In mammals and most terrestrial vertebrates, the sensory organs for hearing and balance are in the ear. In **hearing,** pressure waves in the air are converted into nerve impulses that the brain perceives as sound.

The ear consists of three regions: The external pinna and the auditory canal make up the **outer ear.** The **tympanic membrane** (eardrum), separating the outer ear from the **middle ear,** transmits sound pressure waves to three small bones—the **malleus** (hammer), **incus** (anvil), and **stapes** (stirrup)—which conduct the waves to the inner ear by way of a membrane called the **oval window.** The **Eustachian tube,** connecting the pharynx and the middle ear, equalizes pressure within the middle ear. The **inner ear** contains fluid-filled channels. The **semicircular canals** function in equilibrium, and the **cochlea** functions in hearing.

The coiled cochlea has two large, connected fluid-filled canals—an upper vestibular canal and a lower tympanic canal—separated by the smaller cochlear duct. The **organ of Corti,** located on the basilar membrane that forms the floor of the cochlear duct, contains sensory receptors called hair cells. Their hairs extend into the cochlear duct, and some attach to the tectorial membrane, which overhangs the organ of Corti.

Pressure waves in air—transmitted by the tympanic membrane, the three bones of the middle ear, and the oval window—produce pressure waves in the cochlear fluid, which travel from the vestibular canal through the tympanic canal and dissipate when they strike the **round window**. These pressure waves vibrate the basilar membrane, bending the hairs of its receptor cells against the tectorial membrane. The hair cells depolarize, increasing neurotransmitter release and the frequency of action potentials in sensory neurons that travel in the auditory nerve to the brain.

Volume is a result of the amplitude, or height, of the sound wave; a stronger wave bends the hairs more and results in more action potentials. *Pitch* is related to the frequency of sound waves, usually expressed in hertz (Hz). Different regions of the basilar membrane vibrate in response to different frequencies, and neurons associated with the vibrating region transmit action potentials to auditory regions of the cortex where a particular pitch is perceived.

INTERACTIVE QUESTION   50.3

Label the parts in this diagram of the human ear.

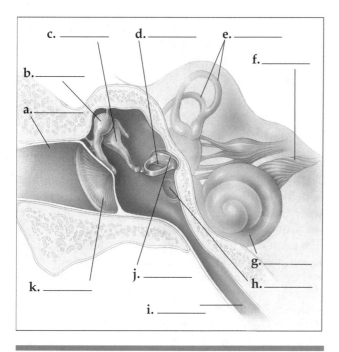

Within the inner ear are two chambers, the **utricle** and **saccule,** and three semicircular canals, responsible for detection of body position and movement. Hair cells in the utricle and saccule project into a gelatinous material containing calcium carbonate particles called otoliths. These heavy particles exert a pull on the hairs in the direction of gravity; different head angles stimulate different hair cells, thereby altering the output of neurotransmitters and the resulting action potentials of sensory neurons of the vestibular nerve. Clusters of hair cells in the swellings at the base of the semicircular canals project into a gelatinous mass called the cupula and are able to sense angular motion in any direction by the pressure of the moving fluid in the semicircular canals against the cupula.

Hearing and Equilibrium in Other Vertebrates   In fish, soundwave vibrations pass through the skeleton of the head to the inner ears, within which hair cells are stimulated by the movement of otoliths. Some fishes have a series of bones that transmits vibrations from the swim bladder to the inner ear.

Clusters of hair cells with sensory hairs embedded in a gelatinous cap or cupula are contained in the **lateral line system** of fishes and aquatic amphibians. Water moving through a tube along both sides of the body bends the cupula and stimulates the hair cells, enabling the fish to perceive its own movement, water currents, and the pressure waves generated by other moving objects.

The inner ear is the organ of hearing and equilibrium in terrestrial vertebrates. In amphibians, sound vibrations are conducted by a tympanic membrane on the body surface and a single middle ear bone to the inner ear. A cochlea has evolved in birds and mammals.

INTERACTIVE QUESTION   50.4

What organs in mammals, fishes, and invertebrates are used to sense movement or equilibrium? How are these organs similar?

a. Mammals:

b. Fishes:

c. Invertebrates:

d. Similarities:

50.3 The senses of taste and smell rely on similar sets of sensory receptors

For terrestrial animals, the sense of **gustation** (taste) detects chemicals called **tastants** present in a solution, whereas **olfaction** (smell) detects airborne **odorants.** Both rely on chemoreceptors.

Sensilla, or sensory hairs containing taste receptors, are found on the feet and mouthparts of insects. Insects also have olfactory hairs, usually located on their antennae.

Taste in Mammals   **Taste buds,** which contain groups of modified epithelial cells, are scattered on the tongue and mouth. The receptor cells recognize five types of tastants, producing the taste perceptions—sweet, sour, salty, bitter, and umami.

A G-protein coupled receptor (GPCR) is involved with sweet, umami, and bitter tastes. Binding of tastant triggers a signal transduction pathway that opens Na^+ channels and depolarizes the membrane. The sour tastant receptor is a transient receptor protein (TRP), similar to the capsaicin and other thermoreceptor proteins. In either type of taste receptor, depolarization leads to activation of a sensory neuron, which transmits action potentials to the brain. Experiments indicate that taste perception depends on which sensory neurons are activated: A sweet taste cell reprogrammed to express a bitter receptor caused mice to be attracted to a bitter chemical.

Smell in Humans   Olfactory receptor cells are neurons that line the upper part of the nasal cavity. Binding of an odorant to a GPCR protein, called an odorant receptor (OR), on an olfactory cilium triggers a signal transduction pathway, which opens ion channels. The membrane depolarizes and sends action potentials to the olfactory bulb of the brain. Approximately 3% of the genes in the human genome are OR genes.

INTERACTIVE QUESTION   50.5

For what purposes do animals use their senses of gustation and olfaction?

50.4 Similar mechanisms underlie vision throughout the animal kingdom

Vision in Invertebrates   The ocelli or eyecups of planarians, which detect light intensity and direction, consist of a layer of cells with a screening pigment that partially encloses photoreceptors. The brain compares impulses from the left and right ocelli to help the animal navigate a direct path away from a light source.

The **compound eye** of insects, crustaceans, and some polychaete worms contains up to thousands of light detectors called **ommatidia,** each with its own lens. The differing intensities of light entering the many ommatidia produce a visual image. The compound eyes of insects are adept at detecting movement and color, and some detect ultraviolet radiation.

Some jellies, polychaetes, spiders, and many molluscs have a **single-lens eye,** in which light is focused through a movable single lens onto a layer of photoreceptors. The **iris** changes the diameter of the **pupil** to adjust the amount of light entering the eye.

The Vertebrate Visual System   The eyeball consists of a tough, outer connective tissue layer called the **sclera** and a thin, pigmented inner layer called the **choroid.** At the front of the eye, the sclera becomes the transparent **cornea,** which functions as a fixed lens, and the choroid forms the colored iris, which regulates the amount of light entering through the pupil. The **retina,** the innermost layer of the eyeball, contains neurons and photoreceptors. The optic nerve attaches to the eye at the optic disk, forming a blind spot on the retina.

The transparent **lens** focuses an image onto the retina. The **ciliary body** produces the **aqueous humor** that fills the anterior eye cavity; jellylike **vitreous humor** fills the posterior cavity. Mammals focus on close objects by accommodation, in which ciliary muscles contract, causing suspensory ligaments to slacken and the elastic lens to become rounder.

Rods and **cones** are the photoreceptors in the retina. The relative proportion of each of these receptors correlates with the activity pattern of the animal: Rods are more light sensitive and enable night vision, whereas cones distinguish colors. In the human eye, rods are most concentrated toward the edge of the retina, whereas the center of the visual field, the **fovea,** is filled with cones.

INTERACTIVE QUESTION   50.6

Label the parts of the vertebrate eye.

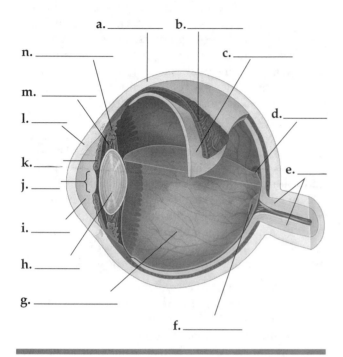

a. _____   b. _____

n. _____

c. _____

m. _____

l. _____

d. _____

k. _____

j. _____

e. _____

i. _____

h. _____

g. _____

f. _____

Both rods and cones have visual pigments embedded in a stack of disks in the outer segment of each cell. **Retinal** is the light-absorbing molecule and is bonded to a membrane protein called an **opsin**. Rods contain the visual pigment **rhodopsin**. When retinal absorbs light, it changes from the *cis* isomer to the *trans* isomer, activating rhodopsin. In the dark, retinal is converted back to its original shape, and it recombines with its opsin.

Light-activated rhodopsin activates a membrane-bound G protein called transducin, which then activates an enzyme that detaches cGMP bound to Na^+ channels in the plasma membrane. Sodium channels close, hyperpolarizing the cell.

The three subclasses of red, green, and blue cones, each with its own type of opsin that binds with retinal to form a **photopsin,** are named for the color of light they optimally absorb.

Visual processing begins in the retina, where rods and cones synapse with neurons called **bipolar cells.** In the dark, rods and cones are depolarized and continually release glutamate, which, depending on the postsynaptic receptor, either depolarizes or hyperpolarizes bipolar cells. When rods and cones hyperpolarize after absorbing light, glutamate release ceases. In response, the bipolar cells reverse their polarization, either becoming hyperpolarized or depolarized, depending on the receptor.

Bipolar cells synapse with **ganglion cells,** whose axons transmit action potentials to the brain. **Horizontal cells** and **amacrine cells** in the retina help to integrate visual information.

Signals from rods and cones may follow a pathway directly from photoreceptors to bipolar cells to ganglion cells. Other pathways involve lateral integration by horizontal cells, which carry signals from one receptor to others and to several bipolar cells, and by amacrine cells, which relay information from one bipolar cell to several ganglion cells. **Lateral inhibition** occurs when a horizontal or amacrine cell inhibits more distant cells and is a form of integration that enhances contrast.

The left and right optic nerves meet at the **optic chiasma** at the base of the cerebral cortex. Information from the left visual field of both eyes travels to the right side of the brain, whereas what is sensed in the right field of view goes to the left side.

Most axons of the ganglion cells go to the **lateral geniculate nuclei**, where neurons lead to the **primary visual cortex** in the cerebrum. Neurons also carry information to other visual processing and integrating centers in the cortex. At least 30% of the cerebral cortex may be devoted to creating the three-dimensional perception of what we see.

Similar developmental genes and pigment molecules indicate that photoreceptors evolved in the earliest bilateral animals.

INTERACTIVE QUESTION   50.7

The *receptive field* of a ganglion cell includes the rods or cones that supply information to it.

a. Would a large or small receptive field produce the sharpest image?

b. Where would the smallest receptive fields be found?

50.5   The physical interaction of protein filaments is required for muscle function

Vertebrate Skeletal Muscle   Vertebrate **skeletal muscle** consists of a bundle of multinucleated muscle cells, called fibers, running the length of the muscle. Each fiber contains a bundle of **myofibrils,** each composed of **thin filaments,** which consist of two strands of actin coiled with one strand of regulatory protein, and **thick filaments,** which are arrays of myosin molecules.

The regular arrangement of filaments produces repeating light and dark bands; thus, skeletal muscle is also called striated muscle. The repeating units are called **sarcomeres** and are delineated by Z lines. Thin filaments attach to the Z line and project toward the center of the sarcomere. Thick filaments lie in the center.

According to the **sliding-filament model** of muscle contraction, the thick and thin filaments do not change length but simply slide past each other, increasing their area of overlap.

The mechanism for the sliding of filaments is the hydrolyzing of ATP by the globular head of a myosin molecule, which then changes to a high-energy form that binds to actin and forms a cross-bridge. The myosin head then bends and pulls the attached thin filament toward the center of the sarcomere. When a new molecule of ATP binds to the myosin head, it breaks its bond to actin, and the cycle repeats with the myosin head attaching to an actin molecule farther along on the thin filament. Energy for muscle contraction comes from creatine phosphate, which regenerates ATP, and from glycogen, which is broken down to glucose. Glucose can generate ATP by glycolysis or aerobic respiration.

INTERACTIVE QUESTION   50.8

Identify the components in this diagram of a section of a skeletal muscle fiber. Explain how and why this diagram will look different when this muscle fiber is fully contracted.

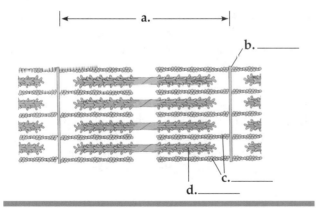

When the muscle is at rest, the myosin-binding sites of the actin molecules are blocked by the regulatory protein **tropomyosin,** associated with another set of regulatory proteins, the **troponin complex**. When troponin binds to Ca^{2+}, the regulatory proteins change shape, and myosin-binding sites on the thin filament are exposed.

Calcium ions are actively transported into the **sarcoplasmic reticulum (SR)** of a muscle cell. When an action potential of a motor neuron causes the release of acetylcholine into the synapse with a muscle fiber, the fiber depolarizes and an action potential spreads into the **transverse (T) tubules,** infoldings of the muscle cell plasma membrane. The action potential opens Ca^{2+} channels in the sarcoplasmic reticulum, releasing Ca^{2+} into the cytosol. Calcium binding with troponin exposes myosin-binding sites, and contraction begins. Contraction ends when motor neuron input stops, the sarcoplasmic reticulum pumps Ca^{2+} back out of the cytosol, and the tropomyosin–troponin complex again blocks the binding sites.

The response of a single muscle fiber to an action potential of a motor neuron is an all-or-none contraction, producing a twitch. Graded contraction of whole muscles may result from involving additional fibers. A single motor neuron and all the muscle fibers it innervates make up a **motor unit**. The strength of a muscle contraction depends on the size and number of motor units involved. Muscle force (tension) can be increased by the activation of additional motor units, called **recruitment**. Prolonged contraction can cause muscle fatigue due to depletion of ATP and loss of ion gradients required for depolarization.

Repeated, rapid-fire action potentials produce summation and a state of smooth and sustained contraction called **tetanus.**

Oxidative fibers, which rely mostly on aerobic respiration, have many mitochondria, a good blood supply, and **myoglobin,** a pigment that extracts oxygen from the blood and stores it. Glycolytic fibers primarily use glycolysis to produce ATP. **Fast-twitch fibers** produce rapid and powerful contractions. **Slow-twitch fibers,** common in muscles that must sustain long contractions, have less sarcoplasmic reticulum and pump calcium more slowly; thus, calcium remains in the cytosol longer and twitches last longer.

INTERACTIVE QUESTION   50.9

a. Identify the three main types of skeletal muscle fibers.

b. Which type of fiber would have few mitochondria; little myoglobin; and produce brief, powerful contractions?

c. Which type of fiber would you expect to be present in the highest proportion in postural muscles?

Other Types of Muscle   Vertebrate **cardiac muscle** fibers are connected by **intercalated disks** through which action potentials spread to all the cells of the heart. Ion channels in the plasma membrane of a cardiac muscle cell produce rhythmic depolarizations that generate action potentials. The action potentials of cardiac muscle cells last quite long.

The scattered arrangement of actin and myosin filaments in **smooth muscle** accounts for its nonstriated appearance. Smooth muscle lacks T tubules, a troponin complex, and a well-developed sarcoplasmic reticulum. Calcium ions enter the cell in response to an action potential and bind to calmodulin, which activates an enzyme that phosphorylates myosin. Contractions are slower and may be triggered by the autonomic nervous system or by the muscle cells themselves.

Invertebrates have muscle cells similar to the skeletal and smooth muscle cells of vertebrates. The flight muscles of insects are capable of independent and rapid contraction.

50.6   Skeletal systems transform muscle contraction into locomotion

Skeletons function in support, protection, and movement. Because muscles can only contract, antagonistic muscle pairs are required to move body parts in opposite directions.

Types of Skeleton Systems   Fluid under pressure in a closed body compartment creates a **hydrostatic skeleton**. As muscles change the shape of the fluid-filled compartment, the animal elongates or moves. Earthworms and other annelids move by **peristalsis,** using rhythmic waves of contractions of circular and longitudinal muscles.

Typical of molluscs and arthropods, **exoskeletons** are hard coverings deposited on the surface of animals.

The cuticle of an arthropod contains fibrils of the polysaccharide **chitin** embedded in a protein matrix. Where protection is most needed, this flexible support is hardened by cross-linking proteins and addition of calcium salts.

The supporting elements of an **endoskeleton** are embedded in the soft tissues of the animal. Sponges have endoskeletons composed of spicules or fibers, and echinoderms have hard plates beneath the skin. The endoskeletons of chordates are composed of cartilage and/or bone. The vertebrate skeleton consists of a skull, vertebral column, rib cage, the pectoral and pelvic girdles, and the limb bones. Joints allow for flexibility in body movement.

INTERACTIVE QUESTION   50.10

List an advantage and a disadvantage for each of the three types of skeletons and indicate in which animal groups each is found.

a. hydrostatic skeleton

b. exoskeleton

c. endoskeleton

Body proportions of small and large animals vary, partly as a result of the physical relationship between the diameter of a support (strength increases with the square of its diameter) and the strain produced by increasing weight (weight increases with the cube of the height). Posture is an important structural factor in supporting body weight. Most of the weight of large land mammals is supported by muscles and tendons, which hold the legs relatively straight and directly under the body.

Types of Locomotion   **Locomotion,** the active movement from place to place, requires energy to overcome the forces of friction and gravity. Animals move to obtain food, escape from danger, and find mates.

Overcoming gravity is less of a problem for swimming animals due to the buoyancy of water. Most fast swimmers have a fusiform shape that reduces friction or drag caused by the density of water. Animals may swim by using their legs as oars, using water to jet-propel themselves, or by undulating their bodies and tails from side to side or up and down.

Walking, running, hopping, or crawling on land requires the animal to support itself and move against gravity; however, air offers little resistance to movement. Powerful leg muscles and strong skeletal support are important for propelling an animal forward and supporting it. A running or hopping animal may store some energy in its tendons to facilitate the next step. Maintaining balance is another requirement for walking, running, or hopping. A crawling animal must overcome the friction produced by its contact with the ground.

A flying animal uses its wings to provide enough lift to overcome the force of gravity. Wings shaped as airfoils attach to lightweight, fusiform bodies.

Energy Costs of Locomotion The cost of locomotion depends on the mode of locomotion and the environment. Adaptations that maximize efficiency of locomotion increase fitness in that more of an animal's energy is available for growth and reproduction.

INTERACTIVE QUESTION 50.11

a. Which mode of locomotion is the most energy efficient?

b. Which mode is the most energetically expensive per distance traveled?

c. Which mode consumes the most energy per unit of time?

d. In animals specialized for the same mode of transport, does a larger animal or a smaller one expend more energy per kilogram of body mass for traveling the same distance?

Word Roots

ama- = together (*amacrine cell:* neurons of the retina that help integrate information before it is sent to the brain)

aqua- = water (*aqueous humor:* the clear, watery solution that fills the anterior cavity of the eye)

bi- = two (*bipolar cell:* neurons that synapse with the axons of rods and cones in the retina of the eye)

chemo- = chemical (*chemoreceptor:* a receptor that transmits information about the total solute concentration in a solution or about individual kinds of molecules)

coch- = a snail (*cochlea:* the complex, coiled organ of hearing that contains the organ of Corti)

electro- = electricity (*electromagnetic receptor:* receptors of electromagnetic energy, such as visible light, electricity, and magnetism)

endo- = within (*endoskeleton:* a hard skeleton buried within the soft tissues of an animal)

exo- = outside (*exoskeleton:* a hard encasement on the surface of an animal that provides protection and points of attachment for muscles)

fovea- = a pit (*fovea:* center of the visual field of the eye)

gusta- = taste (*gustatory receptors:* taste receptors)

hydro- = water (*hydrostatic skeleton:* a skeletal system composed of fluid held under pressure in a closed body compartment; the main skeleton of most cnidarians, flatworms, nematodes, and annelids)

inter- = between; **-cala** = insert (*intercalated disks:* specialized junctions between cardiac muscle cells, which provide direct electrical coupling)

mechano- = an instrument (*mechanoreceptor:* a sensory receptor that detects physical deformations in the body's environment associated with pressure, touch, stretch, motion, and sound)

myo- = muscle; **-fibro** = fiber (*myofibril:* a fibril collectively arranged in longitudinal bundles in muscle cells; composed of thin filaments of actin and a regulatory protein and thick filaments of myosin)

noci- = harm (*nociceptor:* pain receptors in the epidermis of the skin)

olfact- = smell (*olfactory receptor:* smell receptors)

omma- = the eye (*ommatidia:* the facets of the compound eye of arthropods and some polychaete worms)

peri- = around; **-stalsis** = a constriction (*peristalsis:* rhythmic waves of contraction of smooth muscle that push food along the digestive tract)

photo- = light (*photoreceptor:* receptors of light)

rhodo- = red (*rhodopsin:* a visual pigment consisting of retinal and opsin)

sacc- = a sack (*saccule:* a chamber in the vestibule behind the oval window that participates in the sense of balance)

sarco- = flesh; **-mere** = a part (*sarcomere:* the fundamental, repeating unit of striated muscle, delimited by the Z lines)

sclero- = hard (*sclera:* a tough, white outer layer of connective tissue that forms the globe of the vertebrate eye)

semi- = half (*semicircular canals:* a three-part chamber of the inner ear that functions in maintaining equilibrium)

stato- = standing; **-lith** = a stone (*statolith:* sensory organs that contain mechanoreceptors and function in the sense of equilibrium)

tetan- = rigid, tense (*tetanus:* the maximal, sustained contraction of a skeletal muscle caused by a very fast frequency of action potentials elicited by continual stimulation)

thermo- = heat (*thermoreceptor:* a receptor stimulated by either heat or cold)

trans- = across; **-missi** = send (*transmission:* the conduction of impulses to the central nervous system)

tropo- = turn, change (*tropomyosin:* the regulatory protein that blocks the myosin-binding sites on the actin molecules)

tympan- = a drum (*tympanic membrane:* another name for the eardrum)

utric- = a leather bag (*utricle:* a chamber behind the oval window that opens into the three semicircular canals)

vitre- = glass (*vitreous humor:* the jellylike material that fills the posterior cavity of the vertebrate eye)

Structure Your Knowledge

1. Describe the path of a light stimulus from where it enters the vertebrate eye to its transmission as an action potential through the optic nerve.

2. Arrange the following structures in an order that enables you to describe the passage of sound waves from the external environment to the perception of sound in the brain.

round window	basilar membrane
oval window	cerebral cortex
pinna	vestibular and tympanic canals
auditory nerve	tectorial membrane
cochlea	tympanic membrane
auditory canal	malleus, incus, stapes
cochlear duct	organ of Corti

3. Trace the sequence of events in muscle contraction from an action potential in a motor neuron to the relaxation of the muscle.

Test Your Knowledge

MULTIPLE CHOICE: *Choose the one best answer.*

1. What is a sensory perception?
   a. the reception of a stimulus by a sensory cell
   b. the conversion of stimulus energy into a receptor potential by a sensory receptor or sensory neuron
   c. the interpretation of a stimulus according to the region of the brain that receives the nerve impulse
   d. the amplification and transmission of sensory data to the CNS
   e. the emotional response to a stimulus

2. What is the function of the three bones of the middle ear?
   a. respond to different frequencies of sound waves
   b. transmit pressure waves to the oval window
   c. equalize pressure within the inner ear by communicating through the Eustachian tube
   d. sense rotation in three different planes (x, y, and z axes)
   e. support the tympanic membrane (eardrum)

3. The compound eye found in insects and crustaceans
   a. is composed of thousands of prisms that focus light onto a few photoreceptor cells.
   b. cannot sense colors.
   c. contains many ommatidia, each of which contains a lens that focuses light from a tiny portion of the field of view.
   d. may detect infrared and ultraviolet radiation.
   e. does all of the above.

4. Which of the following statements is *not* true?
   a. The iris regulates the amount of light entering the pupil.
   b. The choroid is the thin, pigmented inner layer of the eye that contains the photoreceptor cells.
   c. The ciliary muscle functions in changing the shape of the lens.
   d. Thin aqueous humor fills the anterior eye cavity.
   e. The sclera is the tough outer connective tissue layer of the eye.

5. Which of the following is *incorrectly* paired with its function?
   a. cones—respond to different light wavelengths, producing color vision
   b. horizontal cells—produce lateral inhibition of receptor and bipolar cells
   c. bipolar cells—relay impulses between rods or cones and ganglion cells
   d. ganglion cells—synapse with the optic nerve at the optic disk or blind spot
   e. fovea—center of the visual field, containing only cones in humans

6. The molecule rhodopsin
   a. dissociates when struck by light and is functional only in the dark.
   b. is formed from retinal and opsin by enzymes when struck by light.
   c. is more sensitive to light than are photopsins and is involved in black-and-white vision in dim light.

d. activates a signal transduction pathway that increases a rod cell's permeability to sodium.

e. is or does all of the above.

7. Rotation of the head of a mammal is sensed by

a. the organ of Corti located in the utricle and saccule.

b. changes in the action potentials of hair cells in response to bending against the tectorial membrane.

c. bending of hair cells embedded in a cupula in a semicircular canal in response to the movement of internal fluid.

d. hair cells in response to movement of fluid in the coiled cochlea.

e. statocysts when statoliths stimulate hair cells.

8. Fish can perceive pressure waves from moving objects with their

a. lateral line system containing mechanoreceptor units.

b. semicircular canals in which sensory hairs are embedded in a gelatinous cap.

c. inner ears, within which sensory hairs are stimulated by the movement of otoliths.

d. series of bones that transmits vibrations from the swim bladder to the inner ear.

e. sensilla collected into clusters in the lateral line system.

9. The volume of a sound is determined by

a. the area in the brain that receives the signal.

b. the section of the basilar membrane that vibrates.

c. the number of hair cells that are stimulated.

d. the frequency of the sound wave.

e. the degree to which hair cells are bent and the resulting increase in the number of action potentials produced in the sensory neuron.

10. Hydrostatic skeletons are used for movement by all of the following *except*

a. cnidarians.

b. arthropods.

c. nematodes.

d. annelids.

e. flatworms.

11. When striated muscle fibers contract,

a. the Z lines are pulled closer together.

b. the sarcomere expands.

c. the thin filaments become shorter.

d. the thick filaments become longer.

e. all of the above occur.

12. Tetanus

a. is muscle fatigue resulting from a lack of ATP and loss of ion gradients.

b. is the all-or-none contraction of a single muscle fiber.

c. is the result of stimulation of additional motor neurons.

d. is the result of a volley of action potentials, whose summation produces an increased and sustained contraction.

e. is the rigidity of muscles following death.

13. What is the role of ATP in muscle contraction?

a. to form cross-bridges between thick and thin filaments

b. to release myosin head from actin when it binds to myosin and to provide energy when hydrolyzed to form myosin's high-energy form

c. to remove the tropomyosin–troponin complex from blocking the binding sites on actin

d. to bend the cross-bridge and pull the thick filaments toward the center of the sarcomere

e. to replace the supply of creatine phosphate required for movement of actin past myosin

14. How does calcium affect muscle contraction?

a. It is released from the T tubules in response to an action potential to initiate contraction.

b. The binding of acetylcholine opens calcium channels in the plasma membrane, creating an action potential that travels down the T tubules.

c. It binds to tropomyosin and helps to stabilize cross-bridge formation.

d. Its binding to troponin causes tropomyosin to move away from the myosin-binding sites on the actin filament.

e. Its release from the sarcoplasmic reticulum changes the membrane potential of the muscle cell so that contraction can occur.

15. A motor unit is

a. a sarcomere that extends from one Z line to the next Z line.

b. a motor neuron and the muscle fibers it innervates.

c. the myofibrils of a fiber and the transverse tubules that connect them.

d. the muscle fibers that make up a muscle.

e. the antagonistic set of muscles that flex and extend a body part.

16. Which of the following is *not* a characteristic of cardiac muscle?
    a. scattered arrangement of actin and myosin filaments
    b. intercalated disks that spread action potentials between cells
    c. action potentials that last a long time
    d. ability to generate action potentials without nervous input
    e. striations

17. Smooth muscle contracts relatively slowly because
    a. the only ATP available is supplied by fermentation.
    b. its contraction is stimulated by hormones, not motor neurons.
    c. it does not have a well-developed sarcoplasmic reticulum, and Ca^{2+} enters the cell through the plasma membrane during an action potential.
    d. it is not striated.
    e. it is composed exclusively of slow-twitch muscle fibers.

18. The absorption of light by rhodopsin in a rod cell leads to
    a. an increase in the release of neurotransmitter.
    b. the opening of sodium channels caused by the binding of cGMP.
    c. a hyperpolarized membrane and a decrease in the release of neurotransmitter.
    d. the bleaching of rhodopsin, which increases the sensitivity of rod cells to light.
    e. the closing of sodium channels and the release of neurotransmitter.

19. A receptive field in the retina would involve a *single*
    a. bipolar cell.
    b. photoreceptor (rod or cone) cell.
    c. amacrine cell.
    d. horizontal cell.
    e. ganglion cell.

20. What do the lateral line of fish, invertebrate statocysts, and the cochlea of your ear all have in common?
    a. They are used to sense sound or pressure waves.
    b. They are organs of equilibrium.
    c. They use hair cells as mechanoreceptors.
    d. They use a second-messenger pathway of signal transduction.
    e. They use granules to stimulate their receptor cells.

21. When you pull up your lower arm and "make a muscle" in your biceps, you are
    a. contracting a flexor.
    b. relaxing a flexor.
    c. contracting an extensor.
    d. contracting a tendon.
    e. producing a simple muscle twitch.

22. Which of the following modes of locomotion is the most energy efficient for an animal that is specialized for moving in that fashion?
    a. swimming
    b. running
    c. flying
    d. hopping
    e. crawling

Chapter 51

Animal Behavior

Key Concepts

51.1 Discrete sensory inputs can stimulate both simple and complex behaviors

51.2 Learning establishes specific links between experience and behavior

51.3 Genetic makeup and environment both contribute to the development of behavior

51.4 Selection for individual survival and reproductive success can explain most behaviors

51.5 Inclusive fitness can account for the evolution of altruistic social behavior

Framework

This chapter introduces the complex and fascinating subject of animal behavior.

Behaviors can range from simple fixed-action patterns in response to specific stimuli to problem solving in novel situations. Behaviors result from interactions among environmental stimuli, experience, and individual genetic makeup. The parameters of behavior are controlled by genetics and thus are acted upon by natural selection. Behavioral ecology focuses on the ultimate cause of reproductive fitness, which can be used to interpret foraging behavior, mating patterns, and altruistic behavior. Sociobiology extends evolutionary interpretations to social behavior.

Chapter Review

Behavior is an action performed by muscles or glands under control of the nervous system in response to a stimulus.

51.1 Discrete sensory inputs can stimulate both simple and complex behaviors

The scientific study of how animals behave, particularly in their natural environment, is called **ethology.** Tinbergen's set of questions to guide behavioral studies emphasizes the importance of both **proximate questions,** which explore the immediate cause of a behavior in terms of the stimuli that trigger it, the mechanisms that produce it, or how experience modifies it; and **ultimate questions,** which concern the benefit to survival and reproduction or the evolutionary basis of the behavior. *Behavioral ecology* is the study of the ecological and evolutionary basis of animal behavior.

INTERACTIVE QUESTION 51.1

Many animals breed in the spring or early summer.

a. What is a probable proximate cause of this behavior?

b. What is the probable ultimate cause of this behavior?

Fixed Action Patterns   A **fixed action pattern** is a highly stereotyped sequence of behaviors that, once begun, is usually carried through to completion. It is triggered by a **sign stimulus**—some external stimulus that is often a limited subset of available sensory information.

Oriented Movements   A **kinesis** is a simple change in activity or turning rate in response to a stimulus. Although kinetic movements are randomly directed, they tend to maintain organisms in favorable environments. A **taxis** is an oriented movement toward or away from a stimulus.

Sow bugs are placed in experimental chambers that are either humid or dry and have both light and dark areas. In the humid chamber, the sow bugs move into the dark area and stop moving. In the dry chamber, they move into the dark area and continue to move about in that area. Explain these experimental results.

Migration is a regular, long-distance change in location. Some migratory animals alter their orientation to the sun with the aid of their *circadian*, or 24-hour, clock. Some animals navigate using Earth's magnetic field, perhaps as a result of its effect on magnetite-containing brain structures or on photoreceptors.

Behavioral Rhythms   Many behaviors follow a circadian rhythm mediated by the biological clock. *Circannual* rhythms are linked to the yearly cycle of seasons. The mating behavior of fiddler crabs is linked to the lunar cycle.

Animal Signals and Communication   Communication between animals involves the transmission of, reception of, and response to special behaviors called **signals.** Courtship often involves a *stimulus response chain,* in which each response is the stimulus for the next behavior. Communication may be **visual, auditory, chemical,** or **tactile,** depending on the lifestyle and sensory specializations of a species.

Pheromones are chemical signals commonly used by mammals and insects in reproductive behavior to attract mates and to trigger specific courtship behaviors. Pheromones can also function as alarm signals.

Why is most communication among mammals olfactory and auditory, whereas communication among birds is visual and auditory?

51.2 Learning establishes specific links between experience and behavior

Behavior that is performed virtually the same by all individuals in a population, regardless of developmental or environmental differences, is *developmentally fixed* and called **innate behavior. Learning** is the modification of behavior as a result of experience.

Habituation   A simple type of learning called **habituation** is the loss of sensitivity to unimportant stimuli or to stimuli not associated with appropriate feedback. Habituation may increase fitness by allowing an animal's nervous system to focus on important stimuli.

Imprinting   **Imprinting** is characterized by a limited **sensitive period** during which learning may occur and a behavioral response to an individual or object that is usually long lasting. The ability to respond is innate; the environment provides the *imprinting stimulus.*

Spatial Learning   The establishment of a memory of the spatial structure of the environment is called **spatial learning.** Animals may learn and use a particular set of **landmarks,** or location indicators, to find their way within their area.

Cognitive Maps   More complicated than a set of learned landmarks, **cognitive maps** are neural representations of the spatial relationships of objects in an animal's surroundings. Evidence for cognitive maps comes from research with nutcrackers that are able to retrieve stored food from thousands of caches.

Associative Learning   In **associative learning,** animals learn to associate one stimulus with another. In **classical conditioning,** an arbitrary stimulus is associated with a particular outcome. **Operant conditioning** refers to trial-and-error learning through which an animal associates a behavior with a reward or punishment. An animal's nervous system enables association of behaviors with stimuli most likely to be present in an animal's natural habitat.

Cognition and Problem Solving   **Cognition** refers to an animal's ability to learn that involves awareness, reasoning, recollection, and judgment. Research with bees indicates that they are capable of categorizing environmental objects as "same" or "different."

Problem-solving behavior, the ability to achieve an end in the face of obstacles, is most often observed in mammals, especially in primates and dolphins, although such behavior has also been documented in some bird species. Many animals use the behavior of others as information in problem solving.

Development of Learned Behaviors   Studies show that the learning of bird songs in some species occurs in distinct stages. White-crowned sparrows appear to have a 50-day sensitive period in which they memorize the song of their species. If raised in isolation during this sensitive period, a bird does not develop a typical adult song. In a second learning phase, a juvenile bird sings a subsong, which gradually improves, apparently as the bird compares its own singing with the memorized song. The adult sparrow then sings this "crystallized" song for the rest of its life.

Canaries do not crystallize an adult song. They learn a new, more elaborate song each breeding season.

INTERACTIVE QUESTION   51.4

Indicate the type of learning illustrated by the following examples:

a. Ewes will adopt and nurse a lamb shortly after they give birth to their own lamb but will butt and reject a lamb introduced a day or two later.

b. A dog, whose early "accidents" were cleaned up with paper towels accompanied with harsh discipline, hides any time a paper towel is used in the household.

c. Ducklings eventually ignore a cardboard silhouette of a hawk that is repeatedly flown over them.

d. Honeybees appear to use landmarks to locate their hive.

e. In Pavlov's experiments, the ringing of a bell caused a dog to salivate.

51.3   Genetic makeup and environment both contribute to the development of behavior

Experience and Behavior   In a **cross-fostering study,** young of one species are raised by adults of another species. Such studies examine changes in behavior that are influenced by the social and physical surroundings. California male mice provide extensive parental care and are always highly aggressive toward other mice. When cross-fostered in nests of white-footed mice, California mice showed reduced aggression and reduced parental behavior. Early experiences that alter parental behavior extend this environmental influence to the next generation.

Human **twin studies** have shown that susceptibility to various behavioral disorders in twins ranged from 20%–80%, indicating the contribution of both environment and genetics.

Regulatory Genes and Behavior   The gene called *fru* (for *fruitless*) in fruit flies is a master regulatory gene that controls the sex-specific development of the nervous system and thus the male courtship ritual. Males lacking a functional *fru* gene fail to court, and females expressing a male *fru* gene court other females.

Studies have shown that the courtship song of the green lacewing is genetically controlled by several genes. Hybrid offspring produce songs that have components of the songs of both parent species.

Genetically Based Behavioral Variation in Natural Populations   Closely related species often have behavioral differences. Variations in behavior between populations within a species may correlate with variations in the environment.

Field observations and laboratory studies have documented changes in migratory behavior in European blackcaps over the past few decades. Instead of migrating southwest and then south to Africa for the winter, some German blackcaps are migrating west to Britain. Offspring of these new migrants had similar migratory orientations as documented in funnel cages.

Most laboratory-born garter snakes from coastal areas where banana slugs are an abundant source of prey ate slugs when they were offered, whereas few laboratory-born snakes from inland populations ate the slugs. The researcher proposed that snakes in coastal areas with the ability to recognize slugs by chemoreception had higher fitness, leading to the evolution of this difference in prey selection behavior.

Influence of Single-Locus Variation   Male prairie voles form *pair bonds,* help care for the young, and are aggressive to intruders. Male meadow voles do not show these behaviors. The gene for receptors for the neurotransmitter vasopressin is highly expressed in the brains of prairie voles but not in meadow voles. Male meadow voles with inserted prairie vole receptor genes developed more vasopressin receptors in their brains and showed mating behaviors similar to male prairie voles.

INTERACTIVE QUESTION   51.5

Why do experiments looking at behavioral variations among natural populations often raise and test animals in the laboratory?

51.4   Selection for individual survival and reproductive success can explain most behaviors

Foraging Behavior   Foraging is behavior involved with searching for, recognizing, catching, and consuming food. Natural selection is expected to refine behaviors that improve foraging. Laboratory studies of *Drosophila* have documented an evolutionary change in foraging path length in populations of high or low density. The frequency of the *for*^{R} (for rover) allele increased in high-density populations, whereas the frequency of the *for*^{S} (for sitter) allele increased in low-density populations.

According to the **optimal foraging model,** feeding behaviors will maximize energy intake over expenditure and risk of being eaten while foraging. Mule deer appear to forage more in open areas, where food may be slightly less available but the deer are less likely to fall prey to mountain lions.

INTERACTIVE QUESTION 51.6

Explain how the study of whelk-eating crows supports the optimal foraging model.

Mating Behavior and Mate Choice   Many species have **promiscuous** mating, with no strong pair bonds forming. Longer-lasting relationships may be **monogamous** or **polygamous.** Polygamous relationships are most often **polygynous** (one male and many females), although a few are **polyandrous.** Monogamous species are less likely to be sexually dimorphic.

The needs of offspring are an important factor in the evolution of reproductive patterns. If young require more food than one parent can supply, a male may increase his reproductive fitness by helping to care for offspring rather than going off in search of more mates. With mammals, the female often provides all the food, and males are often polygynous.

Certainty of paternity also influences mating systems and parental care. With internal fertilization, the acts of mating and egg laying or birth are separated, and paternity is less certain than when eggs are fertilized externally.

INTERACTIVE QUESTION 51.7

Exclusive male parental care is observed much more frequently in species with external fertilization, where the male's genetic contribution to the offspring is more certain. Explain how such behavior could evolve.

Sexual selection may be *intrasexual*, involving competition among members of one sex for mates, or *intersexual*, in which mates are chosen by one sex on the basis of particular characteristics.

Female choice in stalk-eyed fruit flies has been a selection factor in the evolution of long eyestalks, which correlate with male quality. Experiments involving feather ornaments added to zebra finch parents suggest that females imprint on their fathers, and that mate choice may play a role in the evolution of ornamentation in male zebra finches.

As with female choice, male competition for females can also reduce variation among males. **Agonistic behavior** involves a contest to determine which competitor gains access to a resource, such as food or a mate. The encounter may include a test of strength or, more commonly, symbolic behavior or ritual.

Behavioral ecologists apply **game theory,** which evaluates strategies in situations where outcome depends on both an individual's strategy and the strategies of others, to mating behavior. Researchers studying the coexistence of three male phenotypes of the side-blotched lizard found that each one's reproductive success depended on the frequency of the other types.

INTERACTIVE QUESTION 51.8

In populations of the side-blotched lizard, aggressive orange throat males defend large territories with many females; blue throat males defend smaller territories and fewer females; and yellow throats mimic females and use "sneaky" tactics to obtain matings. Starting with a high abundance of orange throats in the population, which type of male will tend to increase its mating success and increase in frequency next, and which type of male will then replace this second type after it increases in numbers? Explain your answer.

51.5 Inclusive fitness can account for the evolution of altruistic social behavior

Many social behaviors are selfish, benefiting one individual's reproductive success at the expense of others.

Altruism   Selflessness, or **altruism,** is behavior that reduces an individual's fitness while increasing the fitness of other individuals.

Inclusive Fitness   Natural selection favors traits that increase reproductive success, thus propagating the genes for those traits. W. Hamilton explained altruistic behavior in terms of **inclusive fitness,** the ability of an individual to pass on its genes either by producing its own offspring or by helping close relatives produce their offspring.

A quantitative measure, called **Hamilton's rule,** predicts that natural selection would favor altruistic acts among related individuals if $rB > C$. C and B are the benefit to the recipient and the cost to the altruist, measured by the change in the average number of offspring produced as a result of the altruistic act. The term r refers to the **coefficient of relatedness,** the fraction of genes that are shared by two individuals. **Kin selection** is the term for the natural selection of altruistic behavior that enhances the reproductive success of related individuals.

Studies show that most cases of altruistic behavior involve close relatives, such as females in Belding's squirrel populations and worker bees in a hive, and thus improve the individual's inclusive fitness.

When altruistic behavior involves nonrelated animals, the explanation offered is **reciprocal altruism;** there is no immediate benefit for the altruistic individual, but some future benefit may occur when the helped animal may "return the favor." Cheaters obtain a large benefit, but behavioral ecologists, using game theory, propose that a *tit for tat* behavioral strategy, in which cheating is immediately retaliated against, may enable reciprocal altruism to evolve and persist in a population. Reciprocal altruism often is used to explain altruism in humans.

INTERACTIVE QUESTION 51.9

a. According to kin selection, would an individual be more likely to exhibit altruistic behavior toward a parent, a sibling, or a first (full) cousin?

b. Explain your answer in terms of the coefficient of relatedness and Hamilton's rule.

Social Learning   **Social learning,** learning through the observation of others, forms the basis of **culture**—a system of information transfer that involves teaching and/or social learning and influences behavior in a population.

In **mate choice copying,** individuals may copy the mate choice of others in a population. Experiments with guppies have shown that females will preferentially mate with males who they have observed engaged in courtship with other females. Such mate choice copying was shown to mask a genetically controlled preference for a particular male coloration.

Vervet monkeys have an innate ability to give alarm calls in response to threatening objects. They learn to discriminate in their calls by observing other members of the group and receiving social confirmation of their calls.

Evolution and Human Culture   **Sociobiology** relates evolutionary theory to social behavior and to human culture. In his 1975 book, *Sociobiology,* E. O. Wilson speculated on the evolutionary basis of certain social behaviors of humans.

The parameters of human social behavior may be set by genetics, but the environment undoubtedly shapes behavioral traits just as it influences the expression of physical traits. Due to our capacity for learning, human behavior appears to be quite plastic. Our structured societies, with their codification of acceptable behaviors that exclude some behaviors that might otherwise enhance an individual's fitness, may be the one unique characteristic separating humans and other animals.

Word Roots

agon- = a contest (*agonistic behavior:* a type of behavior involving a contest of some kind that determines which competitor gains access to some resource, such as food or mates)

andro- = a man (*polyandry:* a polygamous mating system involving one female and many males)

etho- = custom, habit (*ethology:* the study of animal behavior in natural conditions)

gyno- = a woman (*polygyny:* a polygamous mating system involving one male and many females)

kine- = move (*kinesis:* a change in activity rate in response to a stimulus)

mono- = one; **-gamy** = reproduction (*monogamous:* a type of relationship in which one male mates with just one female)

poly- = many (*polygamous:* a type of relationship in which an individual of one sex mates with several of the other sex)

socio- = a companion (*sociobiology:* the study of social behavior based on evolutionary theory)

Structure Your Knowledge

1. How does the nature-versus-nurture controversy apply to the study of behavior?

2. How does the concept of fitness in an evolutionary sense apply to all aspects of behavior?

Test Your Knowledge

MULTIPLE CHOICE: *Choose the one best answer.*

1. Behavioral ecology is the
   a. study of the behavior of animals, focusing on stimulus and response.
   b. application of human emotions and thoughts to other animals.
   c. study of animal cognition.
   d. study of animal behavior from an evolutionary perspective of fitness.
   e. behavioral study of ecology.

2. Proximate causes
   a. explain the evolutionary significance of a behavior.
   b. are immediate causes of behavior such as environmental stimuli.
   c. are environmental, whereas ultimate causes are genetic.
   d. are endogenous, although they may be set by exogenous cues.
   e. show that nature is more important than nurture.

3. What is the behavior called that maximizes an animal's energy intake-to-expenditure ratio?
   a. optimal foraging
   b. Hamilton's rule
   c. a fixed-action pattern
   d. cognition
   e. learning

4. Which of the following is an example of a fixed-action pattern?
   a. a crane in a captive-breeding program imprinting on its human caregiver
   b. a goose rolling an egg back toward its nest with its head and beak
   c. a blackcap migrating to its winter territory
   d. a songbird learning its song after listening to a taped song of its species
   e. a digger wasp returning to its nest using landmarks

5. A kinesis
   a. is a randomly directed movement that is not caused by external stimuli.
   b. is a movement that is directed toward or away from a stimulus.
   c. is a change in activity in response to a stimulus.
   d. is illustrated by trout swimming upstream.
   e. may involve landmarks but not cognitive maps.

6. A sensitive period
   a. is the time right after birth.
   b. usually follows the receiving of a sign stimulus.
   c. is a limited time in which imprinting can occur.
   d. is the period during which birds can learn to fly.
   e. is the time during which mate selection occurs.

7. In operant conditioning,
   a. an animal improves its performance of a fixed-action pattern.
   b. an animal learns as a result of trial and error.
   c. sensitivity to unimportant or repetitive stimuli decreases.
   d. a bird can learn the song of a related species if it hears only that song.
   e. an irrelevant stimulus can elicit a response because of its association with a normal stimulus.

8. Which type of intraspecies communication signal would be best suited to a nocturnal species such as an owl?
   a. auditory
   b. visual
   c. chemical
   d. tactile
   e. electrical

9. A female bird would most likely increase her fitness by
   a. mating with as many males as possible.
   b. being polygynous.
   c. reproducing only once in her lifetime.
   d. choosing a mate based on evidence that he has "good genes."
   e. always foraging in a large flock.

10. Which of the following examples of behavior provide evidence of animal cognition?
    a. chimpanzee stacking up boxes to reach a banana
    b. ravens pulling up string to obtain attached food item
    c. trained honeybees that can match colors or patterns, indicating their concept of "same" and "different"
    d. chimpanzee learning to crack palm nuts with stones by watching others
    e. All of the above show evidence of information processing and animal cognition.

11. In a species in which females provide all the needed food and protection for the young,
    a. males are likely to be promiscuous.
    b. mating systems are likely to be monogamous.

c. mating systems are likely to be polyandrous.

d. females will bond in social groups.

e. females will have a higher fitness than males.

12. A crow that aids its parents in raising siblings is increasing its

a. reproductive success.

b. altruistic behavior.

c. inclusive fitness.

d. coefficient of relatedness.

e. certainty of paternity.

13. Sociobiology

a. explores the evolutionary basis of behavioral characteristics within animal societies.

b. applies evolutionary explanations to human social behaviors.

c. studies the roles of culture and genetics in human social behavior.

d. considers communication, mating systems, and altruism from the viewpoint of fitness.

e. does all of the above.

14. According to the concept of kin selection,

a. an animal would be more likely to aid a stranger if the "kindness" could be reciprocated.

b. an animal would aid its parent before it would help its sibling.

c. animals are more likely to choose close relatives as mates.

d. examples of altruism usually involve close relatives and increase an animal's inclusive fitness.

e. evolution is the ultimate cause of animal behavior.

15. According to Hamilton's rule, natural selection would favor altruistic acts when

a. the probability that the altruist will lose its life is less than 0.5 and the coefficient of relatedness is greater than 0.25 ($rC = B$).

b. the cost to the altruist times the coefficient of relatedness is less than the benefit to the receiver ($rC < B$).

c. the benefit to the receiver times the coefficient of relatedness is greater than the cost to the altruist ($rB > C$).

d. the cost to the receiver times the coefficient of relatedness is greater than the cost to the altruist ($rC > B$).

e. the benefit to the altruist times the coefficient of relatedness is less than the cost to the receiver ($rB < C$).

16. The cross-fostering of California mice in white-footed mice nests provides evidence for

a. the genetic control of aggression and parenting behavior.

b. the relationship between the distribution of vasopressin receptors and parenting behavior.

c. an imprinting period during which behaviors related to aggression and parenting are set.

d. the influence of the early social environment on the expression of aggressive and parental behaviors.

e. the cognitive ability of mice to change their behavior in different environments.

17. Which of the following provides a way of analyzing situations in which the fitness of one behavioral phenotype is influenced by other phenotypes in the populations?

a. zero sum game

b. inclusive fitness

c. social learning

d. reciprocal altruism

e. game theory

18. Which of the following is *not* an example of social learning?

a. garter snakes from coastal areas eating slugs

b. mate choice copying in guppies

c. alarm calls of vervet monkeys

d. human culture

e. chimpanzees using stones to crack nuts

Chapter 52

An Introduction to Ecology and the Biosphere

Key Concepts

52.1 Ecology integrates all areas of biological research and informs environmental decision making

52.2 Interactions between organisms and the environment limit the distribution of species

52.3 Aquatic biomes are diverse and dynamic systems that cover most of Earth

52.4 The structure and distribution of terrestrial biomes are controlled by climate and disturbance

Framework

This chapter describes the scope of ecology, the abiotic factors that help determine the distribution of species, and the key characteristics of the major aquatic and terrestrial biomes.

Chapter Review

Ecology studies the distribution and abundance of organisms and their interactions with the environment.

Organismal ecology, which may include the disciplines of behavioral, physiological, and evolutionary ecology, considers the responses and adaptations of an organism to its environment. **Population ecology** is concerned with the factors that affect the size of **populations,** which are groups of individuals of a species occupying a particular area. The **community** includes all the populations of organisms in an area; **community ecology** looks at interactions such as predation and competition and how they affect community structure.

An **ecosystem** includes abiotic factors as well as the biological community, and **ecosystem ecology** addresses such topics as the flow of energy and chemical cycling. A **landscape** or seascape is the arrangement of several ecosystems in a geographic region; **landscape ecology** looks at the flow of energy, materials, and organisms among different ecosystems. The **biosphere** includes all of Earth's ecosystems. **Global**

ecology studies the effect of regional energy and material exchanges on the distribution and functioning of organisms across the biosphere.

52.1 Ecology integrates all areas of biological research and informs environmental decision making

Ecology had its foundation in and continues to make use of the descriptive science of natural history. Rigorous experimental designs, however, are commonly used to investigate complex ecological questions.

Linking Ecology and Evolutionary Biology   Interactions between organisms and their environments occur within ecological time. The effects of these interactions have an evolutionary effect by altering the gene pool of populations.

Ecology and Environmental Issues   The science of ecology is not synonymous with advocacy for protecting nature or *environmentalism*, but it helps provide the scientific knowledge needed to understand complicated environmental problems and their possible solutions.

INTERACTIVE QUESTION   52.1

List the six areas of ecological study and describe the focus of research at each level.

a.

b.

c.

d.

e.

f.

52.2 Interactions between organisms and the environment limit the distribution of species

Biogeography studies the past and present distribution of species. Ecologists often work through a series of logical questions to determine what limits geographic distributions of species. They consider both **biotic** factors, such as the organisms in the environment, and **abiotic** factors, such as the chemical and physical environment.

Dispersal and Distribution   The cattle egret is an example of a species that has naturally extended its range through **dispersal.** Transplants of a species can indicate whether dispersal limits its distribution. A successful transplant shows that the *potential* range of a species is larger than its *actual* range. Introduced species, which are introduced to new areas either purposely or accidentally, often disrupt their new ecosystem.

Behavior and Habitat Selection   Sometimes the behavior of organisms in habitat selection keeps them from occupying all their potential range. Habitat selection by ovipositing insects, which often choose only certain host plants, may limit their distribution.

Biotic Factors   The inability of transplanted organisms to survive and reproduce may be due to predation, disease, parasitism, competition, or lack of mutual symbiosis. "Removal and addition" experiments test whether predators limit the distribution of prey species. Sea urchins were shown to limit the abundance and distribution of seaweeds.

Abiotic Factors   Global patterns of geographic distributions are influenced by abiotic factors such as regional differences in temperature, rainfall, and light. An environment may have both *spatial* and *temporal heterogeneity* (differ in both space and time) for abiotic factors.

Temperature is an important environmental factor because of its effects on metabolism. Most organisms function best within a narrow range of environmental temperatures.

The availability of water in different habitats can vary greatly. The distribution of organisms reflects their ability to obtain sufficient water and avoid desiccation in terrestrial habitats.

The ability of organisms to osmoregulate determines their distribution in freshwater, saltwater, or other high-salinity habitats.

Light energy drives almost all ecosystems. The intensity and quality of light are limiting factors in aquatic environments and on forest floors. Too high a light intensity at high elevations and in deserts also limits distribution.

Soils, which vary in their physical structure, pH, and mineral composition, affect the distribution of plants and, in turn, the distribution of animals. Substrate composition in aquatic environments influences water chemistry and the types of organisms that can inhabit those areas.

INTERACTIVE QUESTION 52.2

List and give examples of the four factors that ecologists examine to explain the geographic distribution of a species.

a.

b.

c.

d.

Climate The **climate,** or prevailing weather conditions of a locality, is determined by temperature, water, sunlight, and wind. **Macroclimate** is the climatic pattern on a local to global level; **microclimate** is the fine variations within a particular habitat.

The absorption of solar radiation heats the atmosphere, land, and water, setting patterns for temperature variations, air circulation, and water evaporation that cause latitudinal variations in climate. The shape of Earth and the tilting of its axis create seasonal variations in day length and temperature that increase with latitude. The **tropics** receive the greatest amount of and least variation in solar radiation.

The global circulation of air begins as intense solar radiation near the equator causes warm, moist air to rise, producing the characteristic wet tropical climate, the arid conditions around 30° north and south as dry air descends, the fairly wet though cool climate about 60° latitude as air rises again, and the cold and rainless climates of the arctic and antarctic regions. Air flowing close to Earth's surface produces predictable global wind patterns, such as the cooling trade winds in the tropics and the prevailing westerlies in temperate zones.

Regional climate is influenced by proximity to water and topographical features. Coastal areas are generally more moist, and large bodies of water moderate the climate. A *Mediterranean climate,* however, is created when cool, dry ocean breezes warm as they cross land, absorbing moisture and creating a hot, rainless climate. Water warmed at the equator flows along the "great ocean conveyor belt" toward the North Atlantic and warms the coast of western Europe as the Gulf Stream. As the water cools, it sinks and may take 1,000 years to recirculate to the surface.

INTERACTIVE QUESTION 52.3

Mountains affect local climate. Describe their influence in the following three areas:

a. solar radiation

b. temperature

c. rainfall

Seasonal changes affect local climate. Seasonal changes in wind patterns can produce wet and dry seasons and affect ocean currents, sometimes causing upwellings of cold, nutrient-rich water.

Climate also varies on a very small scale. Micro-climates within an area have differences in abiotic features that affect the local distributions of organisms.

Global warming will have a great effect on the distributions of plants and animals. Fossil pollen deposits have documented the rates of northward expansion of various tree species following the last continental glacier. As their geographic climatic limits change with global warming, seed dispersal may not be rapid enough for some species of plants to migrate into new ranges.

52.3   Aquatic biomes are diverse and dynamic systems that cover most of Earth

Biomes are major types of ecological groupings that are found in broad geographic regions of land or water.

One of the chemical differences in aquatic biomes is salt concentration—less than 0.1% for freshwater biomes versus an average of 3% for marine biomes. Three-fourths of Earth is covered by oceans, which influence global rainfall, climate, and wind patterns. Marine algae and photosynthetic bacteria produce a large portion of the world's oxygen and consume enormous amounts of carbon dioxide. Freshwater biomes are closely linked with and shaped by the surrounding terrestrial biomes.

Stratification of Aquatic Biomes   Many aquatic biomes are stratified. The **photic zone** receives sufficient light for photosynthesis, whereas little light penetrates into the lower **aphotic zone.** The bottom substrate, called the **benthic zone,** is home to organisms collectively called **benthos.** Settling **detritus** (dead organic material) provides food for the benthos. The very deep ocean benthic zone is called the **abyssal zone.** In both lakes and marine environments, open water is called the pelagic zone.

In the ocean and in many lakes, a narrow **thermocline** separates warmer surface waters from the cold bottom layer. In temperate lakes and oceans, seasonal temperature changes produce the **turnover** of waters that brings oxygenated water to the bottom and nutrient-rich water to the surface.

Oligotrophic lakes are often deep, nutrient poor, fairly nonproductive, and generally oxygen rich. The shallower, nutrient-rich waters of **eutrophic lakes** support large, productive phytoplankton communities. Rooted and floating plants are found in the **littoral zone** close to shore; phytoplankton and cyanobacteria inhabit the **limnetic zone.** Runoff carrying nutrients and sediment may gradually convert oligotrophic lakes into eutrophic lakes. Dumping of municipal wastes and runoff from fertilized lands can cause algal blooms, oxygen depletion, and fish kills.

Defined as areas covered with water often enough to support plants adapted to water-saturated soil, **wetlands** are among the most productive biomes.

Topography creates basin, riverine, and fringe wetlands. Wetlands filter nutrients and chemical pollutants and reduce flooding. Much of these richly diverse biomes has been lost to draining and filling.

Streams and rivers are flowing habitats, whose physical and chemical characteristics vary from the headwaters to the mouth. Overhanging vegetation contributes to nutrient content. Oxygen levels are high in turbulently flowing water and low in murky, warm waters. Human impact on streams and rivers includes pollution and damming.

Where a freshwater river meets the ocean, an **estuary** is formed. Salinity varies both spatially and daily with the rise and fall of tides. Saltmarsh grasses and algae are the major producers. Estuaries serve as feeding and breeding areas for marine invertebrates, fish, and waterfowl. Pollution, filling, and dredging have extensively disrupted these highly productive areas.

INTERACTIVE QUESTION   52.4

Indicate with a + or − whether the following are relatively high or low in oxygen level, nutrient content, and productivity.

Biome	Oxygen Level	Nutrient Content	Productivity
Oligotrophic lake	**a.**		
Eutrophic lake	**b.**		
Headwater of stream	**c.**		
Turbid river	**d.**		
Wetlands	**e.**		
Estuary	**f.**		

In **intertidal zones** the daily cycle of tides exposes the shoreline to variations in water, nutrients, and temperature, and to the mechanical force of wave action. Rocky intertidal communities are vertically stratified, with organisms adapted to attach firmly to the hard substrate. Sandy intertidal zones are home to burrowing worms, clams, and crustaceans. Oil pollutants have damaged many intertidal areas.

The water of the **oceanic pelagic biome** is typically nutrient poor but oxygen rich. Seasonal mixing of temperate oceans stimulates phytoplankton growth. Phytoplankton flourish in the photic region and are grazed on by numerous types of zooplankton and larvae of invertebrates and fish. Overfishing and waste dumping have damaged the Earth's oceans.

Found in the photic zone of clear tropical waters, **coral reefs** are highly diverse and productive biomes. The structure of the reef is produced by the calcium carbonate skeletons of the coral (various cnidarians) and serves as a substrate for red and green algae. The coral animals are nourished by symbiotic unicellular algae. Overfishing, coral collecting, pollution, and global warming are destroying coral reefs.

Nutrients reach the **marine benthic zone** as detritus falling from the waters above. Shallow benthic communities of the **neritic zone** receive sunlight and are very diverse and productive. Inhabitants of the abyssal zone are adapted to cold and high water pressure. Chemoautotrophic prokaryotes form the basis of a collection of organisms adapted to the hot, low-oxygen environment surrounding **deep-sea hydrothermal vents.** Overfishing has eliminated many benthic fish populations.

INTERACTIVE QUESTION 52.5

Different marine environments can be classified on the basis of light penetration, distance from shore, and open water or bottom. Match the following zones to their corresponding numbers on the diagram:

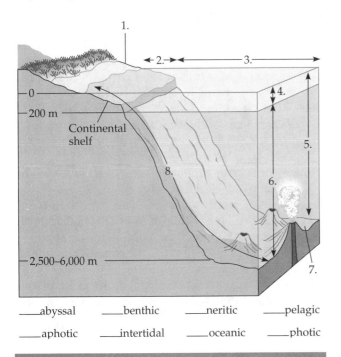

___abyssal   ___benthic   ___neritic   ___pelagic

___aphotic   ___intertidal   ___oceanic   ___photic

52.4 The structure and distribution of terrestrial biomes are controlled by climate and disturbance

The geographic distribution of the world's major terrestrial biomes is related to abiotic factors—in particular, the latitudinal patterns of climate. A **disturbance** is an event such as a fire, storm, or human activity that usually removes organisms and alters resource availability.

Climate and Terrestrial Biomes A **climograph** plots annual mean temperature and rainfall for a region; generally these values correlate with the distribution of various biomes. Overlaps of biomes on a climograph indicate the importance of seasonal patterns of variation in rainfall and temperatures.

General Features of Terrestrial Biomes and the Role of Disturbance Biomes are usually named for their predominant vegetation and major climatic features. Each biome also has characteristic microorganisms, fungi, and animals. Terrestrial biomes may have vertical stratification, such as the layers in a forest from **canopy,** low-tree layer, shrub understory, herbaceous plant ground layer, and forest floor (litter) to root layer. Vertical stratification of vegetation provides diverse habitats for animals. The area where biomes grade into each other is called an **ecotone.**

Species composition of any one biome varies locally. Often, convergent evolution has produced a superficial resemblance of unrelated species.

The extensive patchiness typical of most biomes results from natural or human disturbances. Grasslands, savannas, chaparral, and many coniferous forests are maintained by the periodic disturbance of fire. Urban and agricultural biomes now cover a large portion of Earth's land mass.

Tropical forests occur in equatorial and subequatorial regions. Variations in rainfall result in tropical dry forests, where rainfall is seasonal, and **tropical rain forests,** where rainfall is more constant and abundant. Temperature is uniformly warm. The tropical rain forest has pronounced vertical stratification, and animal diversity is higher than in any other terrestrial biome. Agriculture and development are destroying some tropical forests.

Savannas are equatorial and subequatorial grasslands with scattered trees and rainy and dry seasons. Fires are common, and the dominant grasses and forbs (small broad-leaved plants) are fire-adapted, drought-tolerant, and able to withstand grazing. Large grazing mammals and their predators are common, although insects are the dominant herbivores. Cattle ranching and overhunting have reduced large mammal populations.

Characterized by low and unpredictable precipitation, **deserts** may be hot or cold, depending on location. Plants may use C_4 and CAM photosynthesis, and have reduced leaf surface area, water storage adaptations, and protective spines and toxins. Desert animals have physiological and behavioral adaptations to dry conditions. Irrigated agriculture and urbanization are now common in deserts, reducing natural biodiversity.

Chaparral, common along coastlines in midlatitudes, has cool, rainy winters and hot, dry summers. The dominant vegetation—evergreen shrubs and small trees—is maintained by and adapted to periodic fires. Browsing and small mammals are common. Urbanization and agriculture have reduced areas of chaparral.

Temperate grasslands are maintained by fire, seasonal drought, and grazing by large mammals. Winters are generally cold and dry; summers are hot and wet. Soils are deep and fertile, and most North American grasslands have been converted to farmland.

Characterized by broad-leaved deciduous trees, **temperate broadleaf forests** grow in midlatitude regions that have adequate moisture to support the growth of large trees. These forests have distinct vertical layers. Winters are cold and summers hot and humid. Many mammals hibernate and birds migrate during the cold winters. Humans have heavily logged broadleaf forests, clearing land for agriculture and development.

The largest terrestrial biome, the **northern coniferous forest,** or *taiga,* is characterized by harsh winters with heavy snowfall and periodic droughts. Summers may be hot. Birds and large mammals are common animals. Old-growth stands of trees in coniferous forests are rapidly being logged.

Tundra, covering large areas of the Arctic, is characterized by long, cold winters; short, cool summers; and dwarfed or matlike vegetation. A layer of frozen soil called **permafrost** restricts root growth. Migratory large mammals and birds are common. The *alpine tundra,* found at all latitudes on very high mountains, has similar flora and fauna. Mineral and oil extraction are becoming common in areas of arctic tundra.

INTERACTIVE QUESTION   52.6

Temperature and precipitation are two of the key factors that influence the vegetation found in a biome. On the climograph at the top of the next page, label the North American biomes (arctic and alpine tundra, northern coniferous forest, desert, temperate grassland, temperate broadleaf forest, and tropical forest) represented by each plotted area of temperature and precipitation.

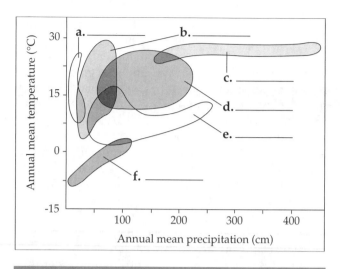

Word Roots

a- = without; **bio-** = life (*abiotic components:* nonliving chemical and physical factors in the environment)

abyss- = deep, bottomless (*abyssal zone:* the very deep benthic communities near the bottom of the ocean; this region is characterized by continuous cold, extremely high water pressure, low nutrients, and near or total absence of light)

bentho- = the depths of the sea (*benthic zone:* the bottom surfaces of aquatic environments)

estuar- = the sea (*estuary:* the area where a freshwater stream or river merges with the ocean)

eu- = good, well; **troph-** = food, nourishment (*eutrophic:* shallow lakes with high nutrient content in the water)

geo- = the Earth (*biogeography:* the study of the past and present distribution of species)

hydro- = water; **therm-** = heat (*deep-sea hydrothermal vents:* a dark, hot, oxygen-deficient environment associated with volcanic activity; the food producers are chemoautotrophic prokaryotes)

inter- = between (*intertidal zone:* the shallow zone of the ocean where land meets water)

limn- = a lake (*limnetic zone:* the well-lit, open surface waters of a lake farther from shore)

littor- = the seashore (*littoral zone:* the shallow, well-lit waters of a lake close to shore)

oligo- = small, scant (*oligotrophic lake:* a nutrient-poor, clear, deep lake with minimum phytoplankton)

micro- = small (*microclimate:* very fine scale variations of climate, such as the specific climatic conditions underneath a log)

pelag- = the sea (*oceanic pelagic biome:* most of the ocean's waters far from shore, constantly mixed by ocean currents)

perman- = remaining (*permafrost:* a permanently frozen stratum below the arctic tundra)

-photo = light (*aphotic zone:* the part of the ocean beneath the photic zone, where light does not penetrate sufficiently for photosynthesis to occur)

thermo- = heat; **-clin** = slope (*thermocline:* a narrow stratum of rapid temperature change in the ocean and in many temperate-zone lakes)

Structure Your Knowledge

1. **a.** Define ecology
   b. How does ecology relate to evolutionary biology?

2. **a.** What are biomes?
   b. What accounts for the similarities in life forms found in the same type of biome in geographically separated areas?

Test Your Knowledge

MULTIPLE CHOICE: *Choose the one best answer.*

1. Which level of ecology considers the effects of predation, parasitism, and competition on species distribution?
   a. landscape
   b. community
   c. ecosystem
   d. organismal
   e. population

2. Ecologists often use mathematical models and computer simulations because
   a. ecological experiments are always too broad in scope to be performed.
   b. most of them are mathematicians.
   c. ecology is becoming a more descriptive science.
   d. these approaches allow them to study the interactions of multiple variables and simulate large-scale experiments.
   e. variables can be manipulated with computers but cannot be manipulated in field experiments.

3. In which of the following biomes would you expect to find organisms with adaptations for tolerating changes in salinity?
   a. oligotrophic lake
   b. wetland
   c. deep-sea hydrothermal vent
   d. estuary
   e. intertidal zone

4. Which of the following would affect the distribution of a species?
   a. dispersal ability
   b. interactions with mutualistic symbionts
   c. climate and physical factors of the environment
   d. behavior, such as habitat selection
   e. All of the above influence where species are found.

5. Why do the tropics and the windward side of mountains receive more rainfall than areas around 30° latitude or the leeward side of mountains?
   a. Rising air expands, cools, and drops its moisture.
   b. Descending air condenses and drops its moisture.
   c. The tropics and the windward side of mountains are closer to the ocean.
   d. There is more solar radiation in the tropics and on the windward side of mountains.
   e. The rotation of Earth determines global wind patterns.

6. Which of the following is a concern about the effects of global warming on tree species?
   a. The increased ozone may damage leaf cells, reducing photosynthetic rates.
   b. Trees may not be able to disperse fast enough to reach new habitats that meet their climatic requirements.
   c. Warmer temperatures may speed tree growth, producing trees that are too tall and spindly.
   d. The additional CO_2 in the atmosphere may actually increase photosynthetic rates and prove beneficial to tree growth.
   e. All of the above are correct.

7. In which of the following biomes is light most likely to be a limiting factor?
   a. desert
   b. estuary
   c. coral reef
   d. grassland
   e. ocean pelagic zone

8. Which of the following is *incorrectly* paired with its description?
   a. neritic zone—shallow region of ocean over continental shelf
   b. abyssal zone—very deep benthic region of ocean
   c. limnetic zone—shallow waters close to shore in a lake
   d. intertidal zone—shallow area where land meets water
   e. pelagic zone—area of open water

9. Phytoplankton are the basis of the food chain in
   a. the headwater of streams.
   b. wetlands.
   c. the oceanic photic zone.
   d. rocky intertidal zones.
   e. deep-sea hydrothermal vents.

10. The ample rainfall of the tropics and the arid areas around 30° north and south latitudes are caused by
    a. ocean currents that flow clockwise in the northern hemisphere and counterclockwise in the southern hemisphere.
    b. the global circulation of air initiated by intense solar radiation near the equator producing wet and warm air.
    c. the tilting of Earth on its axis and the resulting seasonal changes in climate.
    d. the heavier rain on the windward side of mountain ranges and the drier climate on the leeward side.
    e. the location of tropical rain forests and deserts.

11. The permafrost of the arctic tundra
    a. prevents plants from getting established and growing.
    b. protects small animals during the long winters.
    c. anchors plant roots in the frozen soil, helping them withstand the area's high winds.
    d. prevents plant roots from penetrating deep in the soil.
    e. Both b and c are correct.

12. Many plant species have adaptations for dealing with the periodic fires typical of a
    a. savanna.
    b. chaparral.
    c. temperate grassland.
    d. temperate broadleaf forest.
    e. a, b, and c.

13. Two communities have the same annual mean temperature and rainfall but very different biota and characteristics. The best explanation for this phenomenon is that the two
    a. are found at different altitudes.
    b. are composed of species that have very low dispersal rates.
    c. are found on different continents.
    d. receive different amounts of sunlight.
    e. have different seasonal temperatures and patterns of rainfall throughout the year.

14. Upwellings in oceans
    a. support reef communities.
    b. occur over deep-sea hydrothermal vents.
    c. are responsible for ocean currents.
    d. bring nutrient-rich water to the surface.
    e. are most common in tropical waters, where they bring oxygen-rich water to the surface.

MATCHING: *Match the biotic description with its biome.*

Biome	Biotic Description
_____ 1. chaparral	A. broad-leaved deciduous trees
_____ 2. desert	B. lush growth, vertical layers
_____ 3. savanna	C. evergreen shrubs, fire-adapted vegetation
_____ 4. coniferous forest	D. scattered thorny trees, grasses, and forbs
_____ 5. temperate broadleaf forest	E. tall stands of cone-bearing trees
_____ 6. temperate grassland	F. low shrubby or grassland matlike vegetation
_____ 7. tropical rain forest	G. grasses adapted to fire and drought
_____ 8. tundra	H. widely scattered shrubs, cacti, succulents

Chapter 53

Population Ecology

Key Concepts

53.1 Dynamic biological processes influence population density, dispersion, and demographics

53.2 Life history traits are products of natural selection

53.3 The exponential model describes population growth in an idealized, unlimited environment

53.4 The logistic model describes how a population grows more slowly as it nears its carrying capacity

53.5 Many factors that regulate population growth are density dependent

53.6 The human population is no longer growing exponentially but is still increasing rapidly

Framework

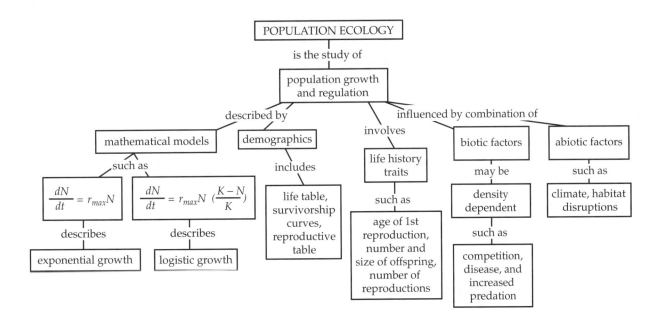

Chapter Review

Population ecology is the study of the influence of the abiotic and biotic environment on population size, density, distribution, and age structure.

53.1 Dynamic biological processes influence population density, dispersion, and demographics

A **population** is a group of individuals of the same species that occupy the same general area, use the same resources, and have a high probability of interacting and breeding with each other.

Density and Dispersion   Every population has geographic boundaries; ecologists define boundaries based upon the type of organism and the research question being asked. The number of individuals per unit area or volume is a population's **density;** the pattern of spacing of those individuals is referred to as **dispersion.**

Population density is often measured by using one of a variety of sampling techniques to count and estimate population size. Indirect indicators, such as burrows or nests, also may be used. In the **mark-recapture method,** animals are trapped and marked (or identified), released, and then captured or sampled again. The proportion of the total second trapping that is marked indicates the proportion of the population that was marked; dividing the initial number marked by this proportion estimates population size.

INTERACTIVE QUESTION   53.1

In a mark-recapture study, an ecologist traps, marks, and releases 25 voles in a small wooded area. A week later she resets her traps and captures 30 voles, 10 of which are marked. What is her estimate of the vole population in that area?

Changes in population density reflect additions of members through birth (including all forms of reproduction) and **immigration,** and removal through death and **emigration.**

Individuals may be dispersed in the population's geographic range in several patterns. *Clumping* may indicate a heterogeneous environment or social interactions between individuals.

Uniform distribution may be related to competition for resources and result from interactions between individuals. **Territoriality,** the defense of a physical space, can lead to uniform dispersion. *Random* spacing, indicating the absence of strong interactions between individuals or a fairly consistent habitat, is not that common.

Demographics   The study of the vital statistics of a population, such as birth and death rates, is called **demography.**

A **life table** presents age-specific survival data for a population. It can be constructed by following a **cohort** of organisms from birth to death, calculating the proportion of the cohort surviving in each age group.

A **survivorship curve** shows the number or proportion of members of a cohort still alive at each age. Survivorship curves are often based on a beginning cohort of 1,000 individuals, with the y axis logarithmic and the x axis on a relative scale, so species with different life spans can be compared on the same graph. There are three general types of survivorship curves. Type I has low mortality during early and middle age and a rapid increase with old age. In a Type II curve, death rate is relatively constant throughout the life span. A Type III curve has high initial mortality. The few offspring that survive are likely to reach adulthood. Many species show intermediate or more complex survivorship patterns.

INTERACTIVE QUESTION   53.2

Identify the types of survivorship curves shown below. Give examples and describe the characteristics of species that exhibit each curve.

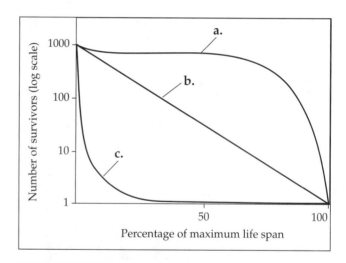

In sexually reproducing species, demographers usually follow only female reproduction, often only of female offspring. A **reproductive table** gives the age-specific reproductive rates in a population.

53.2 Life history traits are products of natural selection

The **life history** of an organism from birth through reproduction to death reflects evolutionary tradeoffs between survival and reproduction. Life history traits include the age at first reproduction, how often an organism reproduces, and the number of offspring produced in each reproductive episode.

Evolution and Life History Diversity   Some species put all their reproductive resources into a single reproductive effort, often called **big-bang reproduction,** or **semelparity.** Other species follow the strategy of **iteroparity,** making repeated reproductive efforts over a span of time. Two key factors in the selection for and evolution of big-bang versus **repeated reproduction** is the survival rate of young offspring and the probability that the adult will survive to reproduce again.

"Tradeoffs" and Life Histories Because organisms have a finite energy budget, they cannot maximize all life history traits simultaneously. Reproductive costs often include a reduction in survival. The production of large numbers of offspring is related to the selective pressures of high mortality rates of offspring in uncertain environments or from intense predation. Parental investments in the size of offspring, incubation or gestation, and parental care increase survival chances of offspring.

INTERACTIVE QUESTION 53.3

a. In what way might high competition for limited resources in a predictable environment influence the evolution of life history traits?

b. Explain why the life history of an organism can't be to reproduce early and often, to have large numbers of offspring at a time, and to live long.

53.3 The exponential model describes population growth in an idealized, unlimited environment

Per Capita Rate of Increase Growth of a small population in a very favorable environment will be restricted only by the biological limitations of its life history traits. Ignoring immigration and emigration, the change in population size during a specific time period is equal to the number of births minus deaths.

Births and deaths can be expressed in terms of the average number per individual during a time period, or a *per capita birth rate (b)* and *per capita death rate (d)*. These rates can be calculated from estimates of population size (N) and data in life tables and reproductive tables. The population growth equation using per capita birth and death rates becomes $\Delta N / \Delta t = bN - dN$. The difference between per capita birth rate and death rate is the *per capita rate of increase*, symbolized by $r (r = b - d)$. **Zero population growth (ZPG)** occurs when $r = 0$. The formula describing change in the population at any one instant uses differential calculus and is written as $dN/dt = r_{inst}N$.

Exponential Growth Under ideal conditions, a population may exhibit **exponential population growth,** or geometric population growth. The fastest per capita rate of increase possible for a species is called r_{max}. This exponential population growth, expressed as $dN/dt = r_{max}N$, produces a J-shaped growth curve when graphed. The larger the population (N) becomes, the faster the population grows. Periods of exponential growth may occur in some populations that exploit an unfilled environment or rebound from a catastrophic event.

53.4 The logistic model describes how a population grows more slowly as it nears its carrying capacity

A population may grow exponentially for only a short time before its increased density limits the resources available for its members. The **carrying capacity** (K) is the maximum population size that a particular environment can support at a particular time. Crowding and resource limitation may lead to decreased per capita birth rates and increased per capita death rates.

The Logistic Growth Model The per capita rate of increase decreases from its maximum at low population size to zero as carrying capacity is reached. The mathematical model of **logistic population growth** ($dN/dt = r_{max}N(K - N)/K$) includes the expression $(K - N)/K$ to reflect the impact of increasing N on the per capita rate of increase as the population approaches the carrying capacity.

When N is small, $(K - N)/K$ is close to 1, and growth is approximately exponential ($r_{max}N$). As population size approaches the carrying capacity, the $(K - N)/K$ term becomes a small fraction, and per capita rate of increase is small. When N reaches K, the term $(K - N)/K$ is 0, and the population stops growing. The logistic model produces an S-shaped growth curve, and maximum increase in population numbers occurs when N is intermediate.

INTERACTIVE QUESTION   53.4

Label the exponential and logistic growth curves, and show the equation associated with each curve. What is *K* for the population shown with curve *b*?

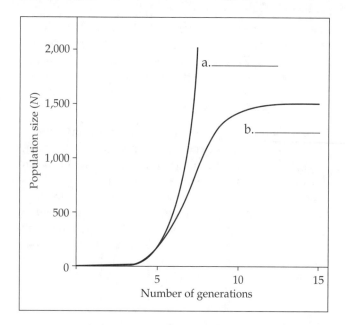

The Logistic Model and Real Populations   Some laboratory populations of small animals and microorganisms show logistic growth. Natural populations may fluctuate around carrying capacity.

The logistic model makes the assumption that any increase in population numbers will have a negative effect on population growth. The *Allee effect* is seen in some populations, however, when individuals have a harder time reproducing if the population size is too small.

The Logistic Model and Life Histories   Natural selection will favor different life history traits depending on population densities and environmental conditions. Populations at high density, close to their carrying capacity, may experience **K-selection** or density-dependent selection for traits such as competitive ability and efficient resource use. In environments in which population density is low, **r-selection** or density-independent selection would favor traits that maximize population growth, such as production of numerous, small offspring.

INTERACTIVE QUESTION   53.5

Indicate whether the following would be considered to be *r*-selected or *K*-selected life history traits.

a. early age at first reproduction; many small offspring produced

b. few, relatively large offspring produced every year

53.5   Many factors that regulate population growth are density dependent

The ecological questions of what eventually stops population growth and what determines population fluctuations or stability have practical applications for conservation and agriculture.

Population Change and Population Density   When the birth rate or death rate does not change as population density changes, it is said to be **density independent.** Rates are said to be **density dependent** if death rate rises and birth rate falls with increasing population density. An equilibrium density may be reached in a population as long as birth rate or death rate or both are density dependent.

Density-Dependent Population Regulation   Density-dependent decreases in birth rate and increases in death rate may regulate populations through negative feedback.

Limited food supply often limits reproductive output. The availability of territorial space may be the limiting resource for some animals. Increased population densities may affect health and the transmission rate of disease in both plant and animal populations. Predation may be a density-dependent factor when a predator feeds preferentially on a prey population that has reached a high density. The accumulation of toxic metabolic wastes may also be a limiting factor.

Intrinsic factors may also regulate population size. Studies of mice have shown that even when food or shelter is not limiting, population size stabilizes when high densities induce a stress syndrome of hormonal changes that inhibit reproduction and increase mortality.

Population Dynamics   All populations show some fluctuations in numbers. **Population dynamics** studies these variations in population size and the factors that cause them. Fluctuations in populations of some mammals such as Soay sheep may be linked to the severity of winter and the availability of food.

The 10-year cycles in density of snowshoe hares and lynx in northern forests have been studied to determine whether winter food shortages, predator overexploitation, or sunspot activity causes the cyclic collapses in hare populations. The decrease in ozone and increase in UV radiation associated with low sunspot activity results in plants producing more sunscreen chemicals and fewer herbivore-discouraging chemicals. Experimentally increasing the food supply raised the carrying capacity for hares, but the density cycles continued to occur in both experimental and control areas. Field ecologists determined that 90% of hare deaths are due to predation. Experiments that excluded predators from one area, and both excluded predators and added food to another area, support the hypothesis that excessive predation mainly drives hare cycles, but available winter food supply is a contributing influence. Low sunspot activity was followed by peaks in the hare population, suggesting that sunspot activity also regulates hare populations.

Predator cycles most likely follow the population cycles of their prey, and these may be accentuated when predators turn on one another as prey become scarce.

A group of populations may form a **metapopulation** in which immigration and emigration may influence individual population sizes and serve to repopulate habitats in which local populations become extinct.

INTERACTIVE QUESTION 53.6

a. List some density-dependent factors that may limit population growth.

b. List some abiotic factors that may cause population fluctuations.

53.6 The human population is no longer growing exponentially but is still increasing rapidly

The Global Human Population From 1650, it took 200 years for the global population to double to 1 billion. In the next 80 years, it doubled to 2 billion; it doubled again in the next 45 years; and is now more than 6.6 billion. The rate of growth has begun to slow, partly due to diseases such as AIDS and voluntary population control, although population ecologists project a population of 7.8–10.8 billion by 2050.

Population stability can be reached in one of two ways: Zero population growth = high birth rates − high death rates; or zero population growth = low birth rates − low death rates. The movement from the first configuration to the second is called the **demographic transition.** Death rates declined rapidly in most developing countries after 1950, but birth rate decline has been more uneven—from rapid in China to slower in Africa and India. The world's annual population growth rate is regional; it is near equilibrium (0.1%) in industrialized nations and 1.15% in developing countries, where 80% of the world's population lives.

Human population growth is unique in that it can be consciously controlled by voluntary contraception and family-planning programs. The key to the demographic transition is reduced family size. In many cultures, women are delaying marriage and reproduction, thus slowing population growth.

The **age structure** of a population affects present and future growth. A large proportion of individuals of reproductive age or younger results in more rapid growth now or in the near future. Age structure also predicts future social conditions and needs.

Infant mortality and *life expectancy at birth* vary among human populations, with mortality being much higher and life expectancy lower in less industrialized countries.

Global Carrying Capacity Estimates of Earth's carrying capacity have varied greatly and average around 10–15 billion. These estimates may use different assumptions, such as the logistic equation, the amount of habitable land, or food as the limiting factor.

The concept of an **ecological footprint** takes into account multiple human needs and the land and water area required to meet those needs and absorb wastes. The estimated ecologically productive land area per person on Earth is calculated to be about 2 hectares (ha). Land for parks and conservation reduces this estimate to 1.7 ha per person. In the United States, a typical person, with an ecological footprint of about 10 ha, is using an unsustainable share of Earth's resources.

Ecological footprints may also be based on the amount of photosynthetic products consumed. The greatest usage is in places of high population density (India and China) or high per capita consumption (parts of the United States and Europe).

The ultimate carrying capacity of Earth may be determined by food supplies, space, nonrenewable resources, degradation of the environment, or several interacting factors. When and how we reach zero population growth is an issue of great social and ecological consequence.

Word Roots

co- = together (*cohort:* a group of individuals of the same age, from birth until all are dead)

demo- = people; **-graphy** = writing (*demography:* the study of statistics relating to births and deaths in populations)

itero- = to repeat (*iteroparity:* a life history in which adults produce large numbers of offspring over many years; also known as repeated reproduction)

semel- = once; **-parity** = to beget (*semelparity:* a life history in which adults have but a single reproductive opportunity to produce large numbers of offspring, such as the life history of the Pacific salmon; also known as "big-bang reproduction")

Structure Your Knowledge

1. Create a concept map to organize your understanding of the exponential and logistic equations—the mathematical models of population growth.

2. What is the best collection of life history traits that would maximize reproductive success?

Test Your Knowledge

MULTIPLE CHOICE: *Choose the one best answer.*

1. In an area with a heterogeneous distribution of suitable habitats, the dispersion pattern of a population probably would be
   a. clumped.
   b. uniform.
   c. random.
   d. unpredictable.
   e. dense.

2. Which of the following is *not* true of life tables?
   a. They were first used by life insurance companies to estimate survival patterns.
   b. They show the age-specific death rate for a population.
   c. They are used to predict logistic growth.
   d. They can be used to construct survivorship curves.
   e. They are often constructed by following a cohort from birth to death.

3. In a population in which offspring survival is quite low and the environment is inconsistent, one might expect
   a. the production of a small number of large offspring.
   b. the production of a large number of large offspring.
   c. iteroparity or repeated reproduction with a small number of offspring.
   d. semelparity or big-bang reproduction.
   e. more *K*-selected traits.

4. A Type I survivorship curve is level at first, with a rapid increase in mortality in old age. This type of curve is
   a. typical of many invertebrates that produce large numbers of offspring.
   b. typical of humans and other large mammals.
   c. found most often in *r*-selected populations.
   d. almost never found in nature.
   e. typical of most species of birds.

5. The middle of the S-shaped growth curve in the logistic growth model
   a. shows that at middle densities, individuals of a population do not affect each other.
   b. is best described by the term rN.
   c. is the point where population growth begins to slow due to the Allee effect.
   d. is the period when competition for resources is highest.
   e. is the period when the population growth rate is the highest.

6. In which of the following would immigration and emigration likely play a role in population dynamics?
   a. metapopulations
   b. exponential growth
   c. demographic transition
   d. big-bang reproduction
   e. territoriality

7. The term $(K - N)/K$
   a. is the carrying capacity for a population.
   b. is greatest when K is very large.
   c. is zero when population size equals carrying capacity.
   d. increases in value as N approaches K.
   e. accounts for the overshoot of carrying capacity.

8. Which of the following would *not* be a density-dependent factor limiting a population's growth?
   a. increased specialization by a predator
   b. a limited number of available nesting sites
   c. a stress syndrome that alters hormone levels
   d. a very early fall frost
   e. intraspecific competition

9. The carrying capacity for a population is estimated at 500; the population size is currently 400; and r_{max} is 0.1. What is dN/dt?

   a. 0.01

   b. 0.8

   c. 8

   d. 40

   e. 50

10. In order to maintain the largest sustainable fish harvest, fishing efforts should

    a. take only postreproductive fish.

    b. maintain the population close to its carrying capacity.

    c. reduce the population to a very low number to take advantage of exponential growth.

    d. maintain the population density close to $\frac{1}{2} K$.

    e. be prohibited.

11. The human population is growing at such a fast rate because

    a. technology has increased our carrying capacity.

    b. the death rate has greatly decreased since the Industrial Revolution.

    c. the age structure of many countries is highly skewed toward younger ages.

    d. fertility rates in many developing countries are above the 2.1 children per female replacement level.

    e. all of the above are true.

12. A few members of a population have reached a favorable habitat with few predators and unlimited resources, but their population growth rate is slower than that of the parent population. What is a possible explanation for this situation?

    a. The genetic makeup of these founders may be less favorable than that of the parent population.

    b. The parent population may still be in the exponential part of its growth curve and not yet limited by density-dependent factors.

    c. The Allee effect may be operating; there are not enough population members present for successful survival and reproduction.

    d. A density-independent factor such as a severe storm has limited their growth.

    e. All of the above may apply.

In questions 13–15, use the following choices to indicate how these factors would be affected by the described changes.

   a. increase

   b. decrease

   c. stay the same

   d. no relationship or unable to predict

13. For a population regulated by density-dependent factors, how might clutch or seed crop size change with increased population density?

14. In the study of European kestrels described in the textbook (Figure 53.8), how did reducing the brood size by transferring chicks to other nests affect the survivorship of those parents in the following winter?

15. In a population showing exponential growth, how would dN/dt be expected to change with an increase in N?

16. Experimental studies of the population cycles of the snowshoe hare and the lynx have shown that

    a. the hare population is regulated by its food resources because adding food increased the carrying capacity of experimental areas.

    b. hares are as likely to die of starvation as of predation.

    c. lynx are the only predators of hares and increases in the lynx population cause the cycles in the hare populations.

    d. the stress of overcrowding causes the population cycles in both hare and lynx.

    e. the hare population is regulated by a combination of predators (not just the lynx) and the effect of sunspot activity on food quality; the lynx population appears to cycle in response to its prey availability.

17. The demographic transition is the gradual shift from

    a. a Type II survivorship curve to a Type I curve.

    b. iteroparity to semelparity.

    c. an age structure skewed toward the younger ages to an even age distribution.

    d. high birth rates and high death rates to low birth rates and low death rates.

    e. exponential growth to logistic growth.

18. An ecological footprint is an estimate of

    a. the carrying capacity of a nation.

    b. the number of offspring an adult produces and the resulting demand on resources.

    c. the amount of land needed per person to meet the current demand on resources.

    d. the size of a population in relationship to the resources it uses.

    e. how much photosynthetic product is used to produce food for a vegetarian versus a meat eater.

Chapter 54

Community Ecology

Key Concepts

54.1 Community interactions are classified by whether they help, harm, or have no effect on the species involved

54.2 Dominant and keystone species exert strong controls on community structure

54.3 Disturbance influences species diversity and composition

54.4 Biogeographic factors affect community biodiversity

54.5 Community ecology is useful for understanding pathogen life cycles and controlling human disease

Framework

Communities are composed of populations of various species that may interact through competition, predation, herbivory, parasitism, or mutualism. The structure of a community—its species composition and relative abundance—is determined by these interactions and the trophic structure of the community. Disturbances keep most communities in a state of nonequilibrium. Species diversity relates to a community's size and geographic location.

Chapter Review

The collection of populations of different species living close enough to interact is called a biological **community**. Community ecology studies the factors involved in determining a community's structure—its composition, relative abundance, and diversity of species.

54.1 Community interactions are classified by whether they help, harm, or have no effect on the species involved

Interspecific interactions occur between the different species living in a community. The effect of these interactions on the survival and reproduction of a population can be signified by $+$ and $-$ signs. For example, in a $+/-$ interaction such as predation, the interaction is beneficial to the predator species and detrimental to the prey population.

Competition If populations of two species use the same limited resource, **interspecific competition** may affect the survival and reproduction of one or both populations.

G. F. Gause's laboratory experiments showed that two species of *Paramecium* that rely on the same limited resource could not coexist in the same community. The **competitive exclusion** principle predicts that the less efficient competitor will be locally eliminated.

An organism's **ecological niche** is described as its role in an ecosystem—its use of biotic and abiotic resources. The competitive exclusion principle holds that two species with identical niches cannot coexist in a community.

Resource partitioning, slight variations in niche that allow ecologically similar species to coexist, provides circumstantial evidence that competition was a selection factor in evolution. As a result of competition, a species' *realized niche* might be smaller than its *fundamental niche*. **Character displacement** of some morphological trait or resource use allows closely related sympatric species to avoid competition. When these species are allopatric (geographically separate), their differences may be much less.

INTERACTIVE QUESTION 54.1

Two species of *Anolis* lizards are often found perched and feeding in the same trees, with species I in the upper and outer branches, and species II occupying the shady inner branches. After removing one or the other species in test trees, an ecologist observes the following results: Species I is found throughout the branches of trees in which it is now the sole occupant. Species II is still found only in the shady interior when it is the sole occupant. What do these results indicate about the niches of these two species?

Predation   **Predation** involves a predator killing and eating prey. Adaptations to increase success in predation may include acute senses, speed and agility, camouflage coloration, and physical structures such as claws, fangs, teeth, and stingers.

Animals can defend against predation by hiding, fleeing, or forming herds or schools. Potential prey may use camouflage in the form of **cryptic coloration** to blend in with the background. Mechanical and chemical defenses discourage predation. Some animals passively accumulate compounds from the food they eat that are toxic to their predators; others may synthesize their own toxins. Bright, conspicuous, **aposematic coloration** warns predators not to eat animals with chemical defenses.

Mimicry may be used by prey to exploit the warning coloration of other species. Predators may use mimicry to "bait" their prey.

INTERACTIVE QUESTION 54.2

Name the following two types of mimicry:

a. harmless species resembling a poisonous or distasteful species

b. mutual imitation by two or more distasteful species

Herbivory   In **herbivory,** an herbivore eats parts of a plant or alga. Most herbivores are small invertebrates such as insects, which may have chemical sensors that recognize their food plants. Herbivores may have teeth or digestive systems adapted for processing vegetation.

Plants may defend themselves with mechanical devices, such as thorns, or chemical compounds. Distasteful or toxic chemicals include such well-known compounds as strychnine, nicotine, tannins, and various spices.

Symbiosis   **Symbiosis** may be defined as the relationship between organisms of two species that live in direct contact.

In **parasitism,** a **parasite** obtains its nourishment from its **host.** Parasites that live within a host are called **endoparasites;** those that feed on the surface of a host are called **ectoparasites.** Parasitoid insects lay eggs on or in hosts, on which their larvae then feed.

Parasites may have complex life cycles with a number of hosts. Parasites can have a substantial influence on their host population.

In **mutualism,** interactions between species benefit both participants. In *obligate mutualism*, at least one species is unable to survive on its own; in *facultative mutualism*, both species can survive alone. Mutualistic interactions may involve the evolution of related adaptations in both species.

In **commensalism,** only one member appears to benefit from the interaction. Examples include "hitchhiking" species and species that feed on food incidentally exposed by another.

INTERACTIVE QUESTION 54.3

Name and give examples of the interspecific interactions symbolized in the table.

	Interaction	Examples
+ / +	**a.**	
+ / 0	**b.**	
− / −	**c.**	
+ / −	**d.**	
+ / −	**e.**	
+ / −	**f.**	

54.2 Dominant and keystone species exert strong controls on community structure

Species Diversity   The **species diversity** of a community is determined both by **species richness,** the number of different species present, and by **relative abundance,** the relative numbers of individuals in each species. Estimating these two aspects of a community requires various sampling techniques and may be difficult due to the rarity of most species in a community. The diversity and richness of bacterial communities may be determined using molecular tools such as RFLP analysis.

A widely used quantitative diversity index is Shannon diversity (H), which is based on species richness and relative abundance:

$$H = - [(p_A \ln p_A) + (p_B \ln p_B) + (p_C \ln p_C) + \cdots]$$

In this equation, p is the proportion of each species in the community and ln is the natural log. The negative sum of ($p \ln p$) for all species equals Shannon diversity (H).

INTERACTIVE QUESTION 54.4

Compute the Shannon diversity of the sea urchin communities in two tide pools.

Pool 1 has three species with the following numbers: A = 8, B = 6, C = 6.

Pool 2 has four species of sea urchins with the following numbers: A = 14, B = 2, C = 2, D = 2.

Which pool has the greater species richness?

Which pool has the higher Shannon diversity?

Trophic Structure   The feeding relationships in a community determine its **trophic structure**. A **food chain** shows the transfer of food energy from one trophic level to the next: from producers to herbivores (primary consumers) to carnivores (secondary, tertiary, or quaternary consumers) and eventually to decomposers.

A **food web** diagrams the complex trophic relationships within a community. The complicated connections of a food web arise because many consumers feed at various trophic levels. Food webs can be simplified by grouping species into functional groups such as primary consumers, or by isolating partial food webs that interact little with the more complex web.

Within a food web, each food chain usually consists of five or fewer links. According to the **energetic hypothesis,** food chains are limited by the inefficiency of energy transfer from one trophic level to the next. Only about 10% of the **biomass** (total mass of all individuals) of one trophic level is converted to the biomass of the next level. The **dynamic stability hypothesis** suggests that short food chains are more stable than long ones. An environmental disruption

that reduces production at lower levels will be magnified at higher trophic levels as food supply is reduced all the way up the chain. The increasing size of animals at successive trophic levels may also limit food chain length, due to the quantity of food required to support large animals.

INTERACTIVE QUESTION 54.5

Experimental data from tree hole communities showed that food chains were longest when food supply (leaf litter) was greatest. Which hypothesis about what limits food chain length do these results support?

Species with a Large Impact   A **dominant species** in a community has the highest abundance or largest biomass and is a major influence on the presence and distribution of other species. A species may become dominant due to its competitive use of resources or success at avoiding predation or disease. **Invasive species** may reach high biomass due to the lack of natural predators and pathogens. The removal of a dominant species from a community may adversely affect any species that relied exclusively on that species, but its role may quickly be filled by other species.

A **keystone species** has a large impact on community structure as a result of its ecological role. Paine's study of a predatory sea star demonstrated its role in maintaining species richness in an intertidal community by reducing the density of mussels, a highly competitive prey species.

Ecosystem "engineers" or foundation species influence community structure by changing the physical environment. **Facilitators** are foundation species that positively affect other species by modifying the environment.

Bottom-Up and Top-Down Controls   Arrows can be used to indicate the effect of an increase in the biomass of one trophic level on another trophic level. $V \rightarrow H$ indicates that an increase in vegetation (V) would increase the number of herbivores (H); $V \leftarrow H$ indicates that an increase in herbivores would decrease vegetation biomass. $V \leftrightarrow H$ means that interactions are reciprocal with each trophic level affected by changes in the other. According to the **bottom-up model** of community organization, $N \rightarrow V \rightarrow H \rightarrow P$, an increase in mineral nutrients (N) yields an increase in biomass at each succeeding trophic level: vegetation, herbivores, and predators (P). The **top-down model,** $N \leftarrow V \leftarrow H \leftarrow P$, assumes that predation controls community organization, with a series of +/− effects cascading down the trophic levels. According to this model, also called the *trophic cascade model*, increasing predators will decrease herbivores, which results in increased vegetation and then lowered levels of nutrients.

INTERACTIVE QUESTION 54.6

Many freshwater lake communities appear to be organized along the top-down model. What actions might ecologists take if they wanted to use **biomanipulation** to control excessive algal blooms in a lake with four trophic levels (algae, zooplankton, primary predator fish, and top predator fish)?

54.3 Disturbance influences species diversity and composition

Traditionally, biological communities were viewed as existing in a state of equilibrium, held there by interspecific interactions. The ability of a community to reach and maintain this relatively constant species composition is known as *stability.* The **nonequilibrium model,** in contrast, emphasizes that communities are constantly changing as a result of **disturbances.**

Characterizing Disturbance Disturbances such as fire, drought, storms, overgrazing, or human activities change resource availability, reduce or eliminate some populations, and may create opportunities for new species. According to the **intermediate disturbance hypothesis,** moderate intensity or frequency of disturbance may result in greater species diversity than low or high levels of disturbance. Small-scale disturbances may enhance environmental patchiness, helping to maintain species diversity. Human prevention of some natural disturbances, such as small fires, may lead to large-scale disturbances, such as large, destructive fires to which species may not be adapted.

Ecological Succession The sequential transitions in species composition in a community, usually following some disturbance, are known as **ecological succession.** If no soil was originally present, as on a new volcanic island or on the moraine left by a retreating glacier, the process is called **primary succession.** A series of colonizers usually begins with autotrophic prokaryotes and moves through lichens, mosses, grasses, shrubs, and trees until the community reaches its prevalent form of vegetation.

Secondary succession occurs when an existing community is disrupted by fire, logging, or farming, but the soil remains intact. Herbaceous species may colonize first, followed by woody shrubs and eventually forest trees.

Early colonizers may *facilitate* the arrival of other species by improving the environment. Or the actions of early species may *inhibit* the establishment of later species. Species may be independent in their colonization and *tolerate* the conditions created by early species.

Ecologists have studied moraine succession over the 250-year retreat of glaciers at Glacier Bay in Alaska. The first pioneering plant species include mosses and fireweed; *Dryas,* a mat-forming shrub, dominates after about 30 years. Alder is the dominant plant a few decades later, followed by Sitka spruce. The community becomes a spruce-hemlock forest by the third century after deglaciation, except in flat, poorly drained areas, where sphagnum mosses invade. These mosses make the soil waterlogged and acidic, killing the trees and creating sphagnum bogs.

INTERACTIVE QUESTION 54.7

a. During the succession following glacial retreat, describe the effects of the alder stage on soil fertility.

b. What is the effect of the spruce forest on soil pH?

Human Disturbance Human activities have altered the structure of communities all over the world through such activities as conversion of land for agriculture, logging, clearing for urban development, and ocean trawling. A common result of human disturbance is a reduction in species diversity.

54.4 Biogeographic factors affect community biodiversity

Latitudinal Gradients Surveys of plant and animal species have documented much greater numbers of species in tropical habitats than in temperate and polar regions. Tropical communities are older, partly because of their longer growing season and partly because they have not had to "start over" after glaciation, as has been the case several times for many polar and temperate communities.

Solar energy input and water availability are important climatic explanations for the latitudinal gradient in biodiversity. **Evapotranspiration** is the amount of water evaporated from soil and transpired by plants, and is determined by solar energy, temperature, and water availability. *Potential evapotranspiration* is determined by just solar radiation and temperature. Evapotranspiration rates have been shown to correlate with species richness of trees and vertebrates in North America.

INTERACTIVE QUESTION   54.8

Why would the fact that tropical communities are "older" than temperate or polar communities contribute to greater species diversity?

Area Effects   A **species-area curve** illustrates the correlation between the size of a community and the number of species found there. In general, the larger the area, the greater the diversity of habitats and the greater the species richness. Use of such curves in conservation biology can allow predictions on how a loss of habitat may affect biodiversity.

Island Equilibrium Model   Any habitat surrounded by a significantly different habitat is considered an island and allows ecologists to study factors that affect species diversity. In the 1960s, R. MacArthur and E. O. Wilson developed a general hypothesis of island biogeography, stating that the size of the island and its closeness to the mainland (or source of dispersing species) are important variables directly correlated with species diversity. Larger islands closer to the mainland will have a higher species diversity than smaller or more distant islands. The rates of immigration and extinction change as the number of species on the island increases. When the rates become equal, species diversity reaches an equilibrium, although species composition may continue to change.

While the island biogeography hypothesis may apply over relatively short time periods in cases where colonization determines species composition, over long periods, adaptive evolutionary changes and abiotic disturbances are probably more important in determining community structure and composition.

INTERACTIVE QUESTION   54.9

Many biogeographic studies have found that large islands have greater species richness than small islands. Label the lines on the following graph that show how immigration rate and extinction rate vary with the number of species on large and small islands. Indicate the location of the equilibrium number on the x axis for a small and a large island.

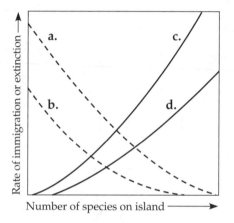

54.5 Community ecology is useful for understanding pathogen life cycles and controlling human disease

Pathogens and Community Structure   **Pathogens** are disease-causing microorganisms, viruses, viroids, or prions. Pathogens have altered terrestrial ecosystems when dominant tree species are killed and are changing the community structure in coral reef communities. The global economy contributes to the spread of pathogens.

Community Ecology and Zoonotic Diseases   Many human diseases are caused by **zoonotic pathogens,** which are transferred from other animals to humans, often by means of a **vector** such as ticks or mosquitoes. Understanding species interactions and the life cycles of vectors help to combat and control the spread of such diseases.

INTERACTIVE QUESTION   54.10

Why are ecologists trapping and testing migrating birds in Alaska?

Word Roots

crypto- = hidden, concealed (*cryptic coloration:* a type of camouflage that makes potential prey difficult to spot against its background)

ecto- = outer (*ectoparasites:* parasites that feed on the external surface of a host)

endo- = inner (*endoparasites:* parasites that live within a host)

herb- = grass; **-vora** = eat (*herbivory:* the consumption of plant material by an herbivore)

inter- = between (*interspecific competition:* competition for resources between plants, between animals, or between decomposers when resources are in short supply)

mutu- = reciprocal (*mutualism:* a symbiotic relationship in which both the host and the symbiont benefit)

Structure Your Knowledge

1. Complete this concept map to organize your understanding of the important factors that structure a community.

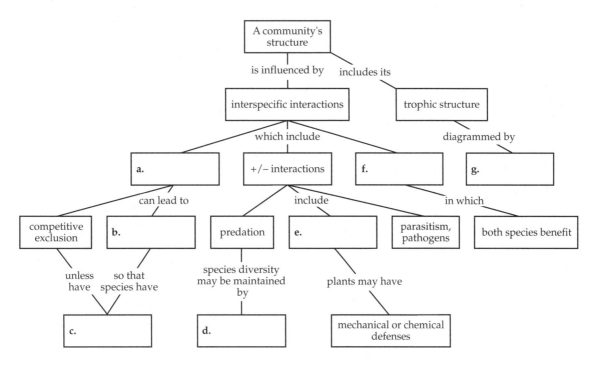

2. Community ecologists develop models or hypotheses to describe community structure and the factors that contribute to such structure. Briefly explain the following models that were described in this chapter.
   a. Competitive exclusion principle
   b. Energetic hypothesis
   c. Dynamic stability hypothesis
   d. Bottom-up model
   e. Top-down (trophic cascade) model
   f. Nonequilibrium model
   g. Island equilibrium model

Test Your Knowledge

MULTIPLE CHOICE: *Choose the one best answer.*

1. Two allopatric species of Galápagos finches have beaks of similar size. There is a significant difference in beak size when the two species occur on the same island. What is this an example of?
   a. competitive exclusion
   b. species diversity
   c. commensalism
   d. character displacement
   e. trophic cascade

2. Two species, A and B, occupy adjoining environmental patches that differ in several abiotic factors. When species A is experimentally removed from a portion of its patch, species B colonizes the vacated area and thrives. When species B is experimentally removed from a portion of its patch, species A does not successfully colonize the area. What might you conclude from these results?
   a. Both species A and species B are limited to their range by abiotic factors.
   b. Species A is limited to its range by competition, and species B is limited by abiotic factors.
   c. Both species are limited to their range by competition.
   d. Species A is limited to its range by abiotic factors, and species B is limited to its range because it cannot compete with species A.
   e. Species A is a predator of species B.

3. The species richness of a community refers to
   a. the relative numbers of individuals in each species.
   b. the number of different species found in a community.
   c. the feeding relationships or trophic structure within the community.
   d. the species diversity of that community.
   e. its stability or ability to persist through disturbances.

4. Aposematic coloring is most commonly found in
   a. prey whose body morphology is cryptic.
   b. predators who are able to sequester toxic plant compounds in their bodies.
   c. prey species that have chemical defenses.
   d. good-tasting prey that evolve to look like each other.
   e. plants that have toxic secondary compounds.

5. Through resource partitioning,
   a. two species can compete for the same prey item.
   b. slight variations in niche allow closely related species to coexist in the same habitat.
   c. two species can share identical niches in a habitat.
   d. competitive exclusion results in the success of the superior species.
   e. two species with identical niches do not share the same habitat and thus avoid competition.

6. Which of the following organisms is mismatched with its *accompanying* trophic level?
   a. algae—producer
   b. phytoplankton—primary consumer
   c. fungi—decomposer
   d. carnivorous fish larvae—secondary consumer
   e. eagle—tertiary or quaternary consumer

7. Ecologists survey the tree species in two forest plots of different ages. Plot 1 has six different species and 95% of all trees belong to just one species. Plot 2 has five different species, each of which is represented by approximately 20% of the trees. How would you describe plot 2 as compared with plot 1?
   a. higher species richness
   b. greater species diversity
   c. lower relative abundance
   d. lower species richness
   e. Both b and d are correct.

8. A palatable (good-tasting) prey species may defend against predation by
   a. Müllerian mimicry.
   b. Batesian mimicry.
   c. secondary compounds.
   d. aposematic coloration.
   e. either a or b.

9. When one species was removed from a tidepool, the species richness became significantly reduced. The removed species was probably
   a. a strong competitor.
   b. a potent parasite.
   c. a resource partitioner.
   d. a keystone species.
   e. the species with the highest relative abundance.

10. Invasive species often reach a large biomass because
    a. they are better competitors than native species.
    b. they are usually producers and not top predators.
    c. they often lack natural predators or pathogens.
    d. their superior ability to disperse enables them to spread to new niches.
    e. they are often protected by the humans who have introduced them.

11. Why do most food chains consist of only three to five links?
    a. There are only five trophic levels: producers; primary, secondary, and tertiary consumers; and decomposers.
    b. Most communities are controlled bottom-up by mineral nutrient supply.
    c. The dominant species in most communities consumes the majority of prey; thus, not enough food is left to support higher predators.

d. According to the energetic hypothesis, the inefficiency of energy transfer from one trophic level to the next limits the number of links that can exist.

e. According to the trophic cascade model, increasing the biomass of top trophic levels causes a decrease in the biomass of lower levels, so that the top levels can no longer be supported.

12. During succession, inhibition by early species
    a. may prevent the achievement of a stable community.
    b. may slow down both the rate of colonization and the rate of extinction, depending on the size of the area and distance from the source of dispersing species.
    c. results from the frequent disturbances that often eliminate early colonizers.
    d. may slow down the successful colonization by other species.
    e. may involve changes in soil pH or accelerated accumulation of humus.

13. According to the nonequilibrium model,
    a. chance events such as disturbances play major roles in the structure and composition of communities.
    b. species composition in a community is always in flux as a result of human interventions.
    c. food chains are limited to a few links because long chains are more unstable in the face of environmental disturbances.
    d. the communities with the most diversity have the least stability or resistance to change.
    e. early colonizers inhibit other species, whereas later colonizers facilitate the arrival of new species.

14. An island that is small and far from the mainland, in contrast to a large island close to the mainland, would be expected to
    a. have lower species diversity.
    b. be in an earlier successional stage.
    c. have higher species diversity but a much lower abundance of organisms.
    c. have a higher rate of colonization but a higher rate of extinction.
    d. have a lower rate of colonization and a lower rate of extinction.

15. According to the top-down (trophic cascade) model of community control, which trophic level would you *decrease* if you wanted to *increase* the vegetation level in a community?
    a. nutrients
    b. vegetation
    c. secondary consumers (carnivores)
    d. tertiary consumers
    e. omnivores

16. A major explanation for the decline in species richness along an equatorial-polar gradient is the correlation of high levels of solar radiation and water availability with biodiversity. Which of the following is also suggested as a factor in the high species richness of tropical communities?
    a. the inverse relationship between biodiversity and evapotranspiration
    b. the greater age of these communities (longer growing season and fewer climatic setbacks), providing more time for speciation events
    c. the larger area of the tropics and corresponding richness predicted by the species-area curve
    d. the lack of disturbances in tropical areas
    e. the greater immigration rate and lower extinction rate found on large tropical islands

17. Which of the following organisms is mismatched with its community role?
    a. beaver—community "engineer"
    b. black rush *Juncus* in salt marsh—facilitator
    c. sea otter in North Pacific—keystone predator
    d. trees in spruce-hemlock forest—dominant species
    e. alder and *Dryas* (a mat-forming shrub)—inhibitor

18. Which of the following best describes a zoonotic pathogen?
    a. a pathogen that affects insects
    b. a pathogen that requires a vector to spread from animal to animal
    c. a disease-causing agent that is transmitted to humans from other animals
    d. a pathogen that is found in zoos due to the unnatural habitat provided for animals
    e. an ectoparasite that is transferred from animals to humans

Chapter 55

Ecosystems

Key Concepts

55.1 Physical laws govern energy flow and chemical cycling in ecosystems

55.2 Energy and other limiting factors control primary production in ecosystems

55.3 Energy transfer between trophic levels is typically only 10% efficient

55.4 Biological and geological processes cycle nutrients between organic and inorganic parts of an ecosystem

55.5 Human activities now dominate most chemical cycles on Earth

Framework

This chapter describes energy flow and chemical cycling through ecosystems. Producers convert light energy into chemical energy, which passes, with a loss of energy at each trophic level, through the ecosystem and dissipates as heat. Chemical elements are cycled within the ecosystem from abiotic reservoirs through producers, consumers, and detritivores, and back to the reservoirs.

An ecosystem's primary production may be limited by nutrients, temperature, or moisture. The low trophic efficiency in the transfer of energy from one level to the next is reflected in pyramids of production and biomass.

Human activities are altering chemical cycles, increasing greenhouse gases, depleting atmospheric ozone, and altering ecosystems.

Chapter Review

An **ecosystem** consists of a community and the abiotic factors with which the organisms interact. Most ecosystems are powered by energy from sunlight, which is transformed to chemical energy by autotrophs, passed to heterotrophs in the organic compounds of food, and continually dissipated in the form of heat. Chemical elements are cycled among the abiotic and biotic components of the ecosystem as autotrophs incorporate them into organic compounds and the processes of metabolism and decomposition return them to the soil, air, and water.

55.1 Physical laws govern energy flow and chemical cycling in ecosystems

Ecosystem ecologists group species into trophic levels to follow energy transformations and chemical movements in a community.

Conservation of Energy According to the first law of thermodynamics, energy cannot be created or destroyed, only transformed. Energy flows through ecosystems from its input as solar radiation to its conversion into chemical energy to its dissipation as heat. The second law of thermodynamics states that every energy exchange increases entropy. Thus, in each energy conversion, some energy is lost as heat.

Conservation of Mass The **law of conservation of mass** states that matter cannot be created or destroyed. Ecosystem ecology measures chemical cycling within ecosystems as well as the input and output of elements.

INTERACTIVE QUESTION 55.1

What may happen if an element's output from an ecosystem is greater than its input?

Energy, Mass, and Trophic Levels Most **primary producers,** or autotrophs, use light energy to synthesize organic compounds for use as fuel and building materials. Heterotrophs depend on autotrophs for their organic compounds. **Primary consumers** are herbivores; **secondary consumers** are carnivores. **Tertiary consumers** eat other carnivores.

Detritivores, or **decomposers,** consume **detritus,** which is organic wastes, fallen leaves, and dead organisms. Detritivores often form a link between producers and consumers. Fungi and prokaryotes are important detritivores in most ecosystems, converting organic materials from all trophic levels to inorganic compounds that can be recycled by autotrophs.

INTERACTIVE QUESTION 55.2

Compare the movement of energy and chemical nutrients in ecosystems.

55.2 Energy and other limiting factors control primary production in ecosystems

Primary production is the amount of light energy converted to chemical energy during a period of time—the photosynthetic output of an ecosystem's autotrophs. Because photosynthesizing vegetation absorbs more wavelengths of visible light, scientists can estimate primary production using data from satellites that compare the wavelengths of light reflected from Earth's surface.

Ecosystem Energy Budgets The intensity of solar energy striking Earth varies by latitude. Only a small portion of incoming solar radiation strikes photosynthetic organisms, and, of that, only about 1% is converted to chemical energy. Nevertheless, worldwide photosynthetic production is about 150 billion metric tons of organic material per year.

Net primary production (NPP) is an ecosystem's **gross primary production (GPP)** minus the energy used by producers in their own cellular respiration (R).

$$NPP = GPP - R$$

Net primary production can be expressed as energy ($J/m^2/yr$) or as new biomass produced measured in terms of dry weight of vegetation ($g/m^2/yr$). *Standing crop* is the total biomass of photosynthetic organisms in an ecosystem. Primary production and the contribution to Earth's total production vary by ecosystem.

Primary Production in Aquatic Ecosystems The depth to which light penetrates and the availability of nutrients affects primary production in oceans and lakes. Nitrogen or phosphorus is often the **limiting nutrient** in the photic zone of the ocean.

In freshwater lakes, nutrient limitation also affects production. **Eutrophication,** the rapid growth or "bloom" of cyanobacteria and algae, has been linked to phosphorus pollution from sewage and fertilizer runoff.

Primary Production in Terrestrial Ecosystems Production in terrestrial ecosystems is related primarily to moisture and temperature. The variation in precipitation and temperature is reflected in a measure called **actual evapotranspiration**—the annual amount of water evaporated from a landscape and transpired by plants. Ecosystems usually show a positive correlation between actual evapotranspiration and primary production.

Nutrients may limit production on a local scale. Nitrogen or phosphorus is the limiting nutrient in many terrestrial and wetland ecosystems.

INTERACTIVE QUESTION 55.3

a. List some ecosystems with high primary production.

b. List some ecosystems with low primary production.

c. The oceans have low primary production yet contribute as much global net primary production as terrestrial areas. Explain.

d. Antarctic seas are often more productive than tropical seas, even though they are colder and receive lower light intensity. Explain.

55.3 Energy transfer between trophic levels is typically only 10% efficient

The rate at which consumers in an ecosystem produce new biomass from their food is called **secondary production.**

Production Efficiency   Herbivores consume only a fraction of the plant material produced; they cannot digest all they eat; and much of the energy they do absorb is used for cellular respiration. Only the chemical energy stored as growth or in offspring is available as food to secondary consumers. The proportion of assimilated food energy that is used for net secondary production (growth and reproduction) is a measure of the efficiency of energy transformation: production efficiency = net secondary production × 100%/assimilation of primary production. Production efficiencies vary from 1–3% for endothermic birds and mammals, to 10% for fishes, to 40% for insects and microorganisms.

Trophic efficiency is the percentage of the production of one trophic level that makes it to the next level, usually ranging from 5–20%. A *pyramid of net production* shows this multiplicative loss of energy.

A *biomass pyramid* illustrates the standing crop biomass of organisms at each trophic level. This pyramid usually narrows rapidly from producers to the top trophic level. Some aquatic ecosystems have inverted biomass pyramids in which zooplankton (consumers) outlive and outweigh the highly productive, but heavily consumed, phytoplankton. Phytoplankton have a short **turnover time,** determined by dividing standing crop biomass by production. The production pyramid for such an aquatic ecosystem, however, is normal in shape.

INTERACTIVE QUESTION   55.4

a. Why is production efficiency higher for fishes than for birds and mammals?

b. Assuming a 10% trophic efficiency (transfer of energy to the next trophic level), approximately what proportion of the chemical energy produced in photosynthesis makes it to a tertiary consumer?

The Green World Hypothesis   Most terrestrial ecosystems are green. The **green world hypothesis** proposes several factors that keep herbivore populations from stripping Earth's vegetation: plant defenses; limited essential nutrients that restrict herbivore growth and reproduction; abiotic fluctuations; intraspecific competition; and interspecific interactions such as predation, parasitism, and disease.

55.4   Biological and geological processes cycle nutrients between organic and inorganic parts of an ecosystem

Chemical elements are passed between abiotic and biotic components of ecosystems through **biogeochemical cycles.**

Biogeochemical Cycles   Gaseous forms of carbon, oxygen, sulfur, and nitrogen have global cycles. Heavier elements, such as phosphorus, potassium, and calcium, have a more local cycle, especially in terrestrial ecosystems.

Most nutrients are found in four types of reservoirs: organic material in living organisms or detritus, available to other organisms; unavailable organic material in "fossilized" deposits; available inorganic elements and compounds in water, soil, or air; and unavailable elements in rocks. Nutrients may leave the unavailable reservoirs through weathering of rock, erosion, or burning of fossil fuels.

Ecologists study the movement of elements by adding radioactive isotopes or by following naturally occurring nonradioactive isotopes through ecosystems.

Exploring Nutrient Cycles   The water cycle involves evaporation from oceans by solar energy, evapotranspiration from land, condensation into clouds, and precipitation. Runoff and groundwater flow return water to the oceans.

In the carbon cycle, producers take CO_2 from the atmosphere for photosynthesis and organisms release it in cellular respiration. Fossil fuel combustion is increasing atmospheric CO_2. Reservoirs include fossil fuels; dissolved carbon in oceans; plant and animal biomass; CO_2 in the atmosphere; and sedimentary rocks such as limestone.

Plants require nitrogen in the form of ammonium (NH_4^+) or nitrate (NO_3^-). Animals obtain nitrogen in organic form from plants or other animals. In a process called *ammonification*, decomposers break down organic compounds and return NH_4^+ to the soil. The major reservoir is the atmosphere, and most nitrogen enters ecosystems through *nitrogen fixation:* Soil bacteria convert N_2 into NH_4^+. In *nitrification*, nitrifying bacteria convert NH_4^+ to NO_3^-. In *denitrification*, the anaerobic metabolism of other bacteria returns N_2 to the atmosphere. Fertilizers add a significant amount of nitrogen to ecosystems.

INTERACTIVE QUESTION 55.5

Label the organisms and compounds that are illustrated in this terrestrial nitrogen cycle.

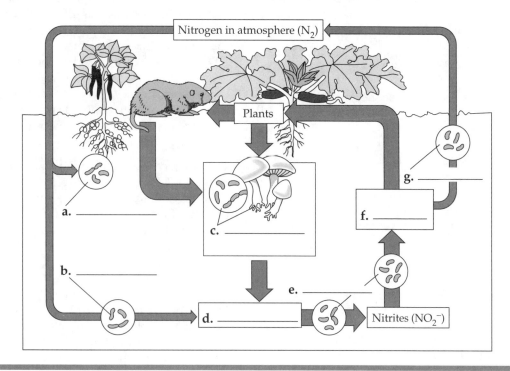

Weathering of rock adds phosphorus to the soil in the form of PO_4^{3-}, which is absorbed by plants. Organic phosphate is transferred from plants to consumers and returned to the soil through the action of decomposers or by animal excretion. Humus and soil particles usually bind phosphate, keeping it available locally for recycling. Sedimentary rocks of marine origin are the largest reservoirs.

INTERACTIVE QUESTION 55.6

What is the biological importance of water, carbon, nitrogen, and phosphorus?

Decomposition and Nutrient Cycling Rates Temperature and the availability of water influence decomposition rates; thus, nutrient cycling times vary in different ecosystems. Nutrients cycle rapidly in a tropical

rain forest; the soil contains only about 10% of the ecosystem's nutrients. In temperate forests where decomposition is slower, 50% of organic nutrients are stored in the detritus and soil. Decomposition rates are slow where oxygen is limited, such as in peat lands and aquatic sediments.

Case Study: Vegetation and Nutrient Cycling in the Hubbard Brook Experimental Forest A team of scientists has looked at nutrient cycling in the Hubbard Brook forest ecosystem since 1963, an example of long-term ecological research (LTER). The mineral budget for each of six valleys was determined by measuring the input of key nutrients in rainfall and their outflow through the creek that drained each watershed. Most minerals were recycled within the forest ecosystem.

The effect of deforestation on nutrient cycling was measured for three years in a valley that was completely logged and sprayed with herbicides. Compared with a control, water runoff from the deforested valley increased 30–40%; net loss of minerals such as Ca^{2+} and K^+ was large. Nitrate increased in concentration in the creek 60-fold, removing this critical soil nutrient and contaminating the water.

Long-term data from Hubbard Brook indicate that acid precipitation has removed most of the Ca^{2+} from the soils. The addition of Ca^{2+} to experimental watersheds has resulted in healthier growth and seedling establishment.

INTERACTIVE QUESTION   55.7

a. In which natural ecosystem do nutrients cycle the fastest? Why?

b. In which natural ecosystems do nutrients cycle slowly? Why?

c. What is the effect of loss of vegetation on nutrient cycling?

55.5   Human activities now dominate most chemical cycles on Earth

Nutrient Enrichment   The harvesting of crops removes nutrients that would otherwise recycle in the soil. After depleting the organic and inorganic reserves, crops require the addition of synthetic fertilizers. The addition of nitrogen fertilizers, increased legume cultivation, and burning of fossil fuels have doubled Earth's supply of fixed nitrogen.

Nitrogen that exceeds the **critical load,** the amount of added nutrient that can be absorbed by plants without damaging the ecosystem, can contaminate groundwater, degrade lakes and rivers, and drain into the ocean. Elevated nitrogen levels create phytoplankton blooms. The decomposition of this phytoplankton lowers oxygen levels, creating "dead zones" in coastal waters. Eutrophication of lakes due to nutrient runoff can kill fish and harm other organisms.

Acid Precipitation   The burning of coal and other fossil fuels as well as wood releases oxides of sulfur and nitrogen, which form sulfuric and nitric acid in the atmosphere. These acids create acid precipitation, defined as rain, snow, sleet, or fog with a pH less than 5.2. Emissions from the tall exhaust stacks of ore smelters and electrical plants drift downwind and create acid precipitation over vast areas. Nutrients leach from soils as acid precipitation changes soil chemistry, and forests have been damaged. Fish populations in lakes across North America and Europe have declined and community compositions have changed.

Toxins in the Environment   Humans release a huge variety of toxic chemicals into the environment. Organisms absorb these toxins from food and water and may retain them within their tissues. In a process known as **biological magnification,** the concentration of such compounds increases in each successive link of the food chain. Chlorinated hydrocarbons, such as DDT, and polychlorinated biphenyls, or PCBs, have been implicated in endocrine system problems in many animal species. Many toxic chemicals dumped into ecosystems are nonbiodegradable; others, such as mercury, may become more harmful as they react with other environmental factors.

Greenhouse Gases and Global Warming   The concentration of CO_2 in the atmosphere has been increasing since the Industrial Revolution as a result of the combustion of fossil fuel and deforestation. If C_3 plants become able to outcompete C_4 plants with the increase in CO_2, species composition in natural and agricultural communities may be altered.

In the Forest–Atmosphere Carbon Transfer and Storage (FACTS-I) experiment, scientists are monitoring the effects of elevated CO_2 levels on sample plots in a forest ecosystem over many years. Trees in experimental plots have produced 15% more wood each year, although this increase is less than had been predicted.

Through a phenomenon known as the **greenhouse effect,** CO_2, other greenhouse gases, and water vapor in the atmosphere absorb infrared radiation reflected from Earth and re-reflect it back to Earth, thus warming Earth.

Scientists use various models to try to estimate the extent and consequences of increasing CO_2 levels. A number of studies predict a doubling of CO_2 levels and a temperature rise of 3°C by the end of the 21st century. Ecologists study the effects on vegetation of previous global warming trends to try to predict the impact of increasing temperatures. Controlling the level of CO_2 emissions in increasingly industrialized societies is a huge international challenge.

INTERACTIVE QUESTION   55.8

List some of the potential consequences of global warming.

Depletion of Atmospheric Ozone   A layer of ozone molecules (O_3) in the lower stratosphere absorbs damaging ultraviolet radiation. This layer has been gradually thinning since 1975, largely as a result of the accumulation of breakdown products of chlorofluoro-carbons in the atmosphere. The dangers of ozone depletion may include increased incidence of skin cancer and cataracts and unpredictable effects on phytoplankton, crops, and natural ecosystems.

Word Roots

auto- = self; **troph-** = food, nourishment (*autotroph:* an organism that obtains organic food molecules without eating other organisms)

bio- = life; **geo-** = Earth (*biogeochemical cycles:* the various nutrient circuits that involve both biotic and abiotic components of ecosystems)

de- = from, down, out (*denitrification:* the process of converting nitrate back to nitrogen)

detrit- = wear off; **-vora** = eat (*detritivore:* a consumer that derives its energy from nonliving organic material)

hetero- = other, different (*heterotroph:* an organism that obtains organic food molecules by eating other organisms or their by-products)

Structure Your Knowledge

1. Two processes that emerge at the ecosystem level of organization are energy flow and chemical cycling. Develop a concept map that explains, compares, and contrasts these two processes.

2. Describe four or five human intrusions in ecosystem dynamics that have detrimental effects.

Test Your Knowledge

MULTIPLE CHOICE: *Choose the one best answer.*

1. Which of the following is absolutely essential to the functioning of an ecosystem?
   a. producers and primary consumers
   b. producers and secondary consumers
   c. primary, secondary, and tertiary consumers
   d. primary consumers and detritivores
   e. producers and detritivores

2. Primary production
   a. is equal to the standing crop of an ecosystem.
   b. is limited by light, nutrients, and moisture in all ecosystems.
   c. is the amount of light energy converted to chemical energy per unit time in an ecosystem.
   d. is inverted in some aquatic ecosystems.
   e. is all of the above.

3. Which of the following is an accurate statement about ecosystems?
   a. Energy is recycled through the trophic structure.
   b. Energy is usually converted from sunlight by primary producers, passed to secondary producers in the form of organic compounds, and lost to detritivores in the form of heat.
   c. Chemicals are recycled between the biotic and abiotic sectors, whereas energy makes a one-way trip through the food web and is eventually dissipated as heat.
   d. There is a continuous process by which energy is lost as heat, and chemical elements leave the ecosystem through runoff.
   e. A food web shows that all trophic levels may feed off each other.

4. In the experiment in which iron was added to areas of the Pacific Ocean, the growth of eukaryotic phytoplankton was stimulated because
   a. iron was the limiting nutrient for eukaryotic phytoplankton growth.
   b. the iron interacted with bottom sediments, releasing nitrogen and phosphorus into the water.
   c. iron interacted with phosphorus, making that nutrient available to the phytoplankton.
   d. the iron reached the critical load necessary to promote photosynthesis.
   e. iron stimulated the growth of nitrogen-fixing cyanobacteria, which then made nitrogen available for phytoplankton growth.

5. The open ocean and tropical rain forest are the two largest contributors to Earth's net primary production because
   a. both have high rates of net primary production.
   b. both cover huge surface areas of Earth.
   c. nutrients cycle rapidly in these two ecosystems.
   d. the ocean covers a huge surface area and the tropical rain forest has a high rate of production.
   e. both a and b are correct.

6. Production in terrestrial ecosystems is affected by
   a. temperature.
   b. light intensity.
   c. availability of nutrients.
   d. availability of water.
   e. all of the above.

7. Secondary production
   a. is measured by the standing crop.
   b. is the rate of biomass production in consumers.
   c. is greater than primary production.
   d. is 10% less than primary production.
   e. is the gross primary production minus the energy used for respiration.

8. Which of the following is *not* true of a pyramid of net production?
   a. Only about 10% of the energy in one trophic level passes into the next level.
   b. Because of the loss of energy at each trophic level, most food chains are limited to three to five links.
   c. The pyramid of production of some aquatic ecosystems is inverted because of the large zooplankton primary consumer level.
   d. Eating grain-fed beef is an inefficient means of obtaining the energy trapped by photosynthesis.
   e. A biomass pyramid is usually the same shape as a pyramid of production.

9. In which of the following would you expect production efficiency to be the greatest?
   a. amphibians
   b. mammals
   c. fishes
   d. insects and microorganisms
   e. birds

10. Biogeochemical cycles are global for elements
    a. that are found in the atmosphere.
    b. that are found mainly in the soil.
    c. such as carbon, nitrogen, and phosphorus.
    d. that are dissolved in water.
    e. Both a and c are correct.

11. Which of these processes is *incorrectly* paired with its description?
    a. nitrification—oxidation of ammonium in the soil to nitrite and nitrate
    b. nitrogen fixation—reduction of atmospheric nitrogen into ammonia
    c. denitrification—return of N_2 to air, occurs when denitrifying bacteria metabolize nitrate
    d. ammonification—decomposition of organic compounds into ammonium
    e. industrial fixation—nitrogen added to soil in rain or dust particles

12. Clear-cutting tropical forests yields agricultural land with limited productivity because
    a. it is too hot in the tropics for most food crops.
    b. the tropical forest regrows rapidly and chokes out agricultural crops.
    c. few of the ecosystem's nutrients are stored in the soil; most are in the forest trees.
    d. phosphorus, not nitrogen, is the limiting nutrient in those soils.
    e. decomposition rates are high but primary production is low in the tropics.

13. Which of the following was *not* shown by the Hubbard Brook Experimental Forest study?
    a. Most minerals recycle within a forest ecosystem.
    b. Deforestation results in a large increase in water runoff.
    c. Mineral losses from a valley were great following deforestation.
    d. Nitrate was the mineral that showed the greatest loss.
    e. Acid rain increased as a result of deforestation.

14. According to the green world hypothesis, herbivores eat only a small portion of an ecosystem's vegetation because
    a. primary production is limited by light, nutrients, or moisture, not by primary consumption.
    b. plants have a very short turnover time.
    c. trophic efficiency is generally only 10%.
    d. predators, parasites, competition, and other factors keep herbivore populations in check.
    e. the production efficiency of herbivores is very low.

15. The finding of harmful levels of DDT in grebes (fish-eating birds) following years of trying to eliminate bothersome gnat populations in a lakeshore town is an example of
    a. eutrophication.
    b. biological magnification.
    c. the biomass pyramid.
    d. chemical cycling.
    e. increasing resistance to pesticides.

16. Which of the following is a direct effect of the thinning of the ozone layer?
    a. a reduction in species diversity
    b. global warming
    c. acid precipitation
    d. an increase in harmful UV radiation reaching Earth
    e. dead zones in the Antarctic Ocean

Chapter 56

Conservation Biology and Restoration Ecology

Key Concepts

56.1 Human activities threaten Earth's biodiversity

56.2 Population conservation focuses on population size, genetic diversity, and critical habitat

56.3 Landscape and regional conservation aim to sustain entire biotas

56.4 Restoration ecology attempts to restore degraded ecosystems to a more natural state

56.5 Sustainable development seeks to improve the human condition while conserving biodiversity

Framework

Biodiversity at the genetic, species, and ecosystem levels is crucial to human welfare. This chapter explores the threats to biodiversity and several of the approaches to preserving the diversity of species on Earth. Conservation biologists focus on determining the habitat needs of endangered species, establishing and managing nature reserves that are often in human-dominated landscapes, and restoring degraded areas. Ecological research and our biophilia may help to achieve the goal of sustainable development—the long-term perpetuation of human societies *and* the ecosystems that support them.

Chapter Review

Conservation biology integrates all areas of biology in the effort to sustain ecosystem processes and biodiversity.

Restoration ecology uses ecological principles to return degraded ecosystems to a more natural state.

56.1 Human activities threaten Earth's biodiversity

Three Levels of Biodiversity Loss of the genetic diversity within and between populations lessens a species' adaptive potential. As populations and species go extinct, we lose genetic resources for agricultural and other human needs.

A second level of biodiversity is species diversity. An **endangered species,** according to the U.S. Endangered Species Act (ESA), is one that is "in danger of extinction throughout all or a significant portion of its range." A **threatened species** is defined as one that is likely to become endangered. There are many well-documented examples of recent extinctions and endangered species in most taxonomic groups.

The third level of biodiversity is ecosystem diversity. A loss of species can negatively affect a community and ecosystem. Human activities have altered many ecosystems.

Biodiversity and Human Welfare There are both moral and practical reasons for preserving biodiversity. A loss of biodiversity is a loss of the genetic potential held in the genomes of species. Biodiversity is a natural resource that can provide medicines, fibers, industrial chemicals, and food.

Humans depend on Earth's ecosystems. **Ecosystem services** include such things as purification of air and water, detoxification and decomposition of wastes, nutrient cycling, flood control, pollination of crops, and soil creation and preservation. Ecosystem services represent a huge monetary value.

Three Threats to Biodiversity The greatest threat to biodiversity is habitat destruction, caused by agriculture, urban development, forestry, mining, pollution, and global warming. Habitat destruction is cited for 73% of the species listed as extinct, endangered, vulnerable, or rare by the International Union for Conservation of Nature and Natural Resources (IUCN). Fragmentation of natural habitats is a common occurrence and almost always leads to species loss. Both terrestrial ecosystems and aquatic habitats have been damaged.

Introduced species, sometimes called nonnative or exotic species, compete with or prey upon native species and have probably contributed to many extinctions. Humans have transplanted thousands of species, intentionally and unintentionally, with huge economic costs in terms of damage and control efforts.

Overexploitation involves harvesting wild plants or animals at rates higher than the populations' abilities to reproduce. Species of large animals with low intrinsic reproductive rates and species on small islands are particularly vulnerable to extinction. Overfishing, particularly using new harvesting techniques, has drastically reduced populations of many commercially important fish species. The tools of molecular genetics allow biologists to track the origin of tissues that may have come from threatened or endangered species.

from millions in the 19th century to 50 in 1993. A comparison with DNA from museum specimens indicated decreased genetic variation in the threatened population. This decline was associated with a decrease in fertility. After transplanting birds from larger populations in other states, researchers noted an increase in egg viability, and the Illinois population rebounded.

Computer models that integrate many factors are used to estimate **minimum viable population (MVP),** the minimum population size necessary to sustain a population.

The **effective population size** (N_e) is based on a population's breeding potential and is determined by a formula that includes data on the number of individuals that breed and the sex ratio of the population: $N_e = (4 N_f N_m)/(N_f + N_m)$. Alternative formulas take into account life history or genetic factors. Conservation efforts should be based on maintaining the minimum number of *reproductively active* individuals needed to prevent extinction.

INTERACTIVE QUESTION   56.1

Give an example of how each of the following threats to biodiversity has reduced population numbers or caused extinctions.

a. habitat destruction

b. introduced species

c. overexploitation

56.2 Population conservation focuses on population size, genetic diversity, and critical habitat

Small-Population Approach   Very small populations are considered endangered. According to the small-population approach, the inbreeding and genetic drift characteristic of a small population may draw it into an **extinction vortex,** in which the loss of genetic variation leads, by positive feedback loops, to smaller and smaller numbers until the population becomes extinct.

For example, as agriculture fragmented their habitat, the number of prairie chickens in Illinois declined

---

INTERACTIVE QUESTION   56.2

Is the effective population size usually larger or smaller than the actual number of individuals in the population? Explain.

Population viability analysis uses MVP to predict long-term viability of a population—the probability of survival over a particular time. M. Shaffer performed a probability viability analysis as part of a long-term study of grizzly bears in Yellowstone National Park. He estimated the minimum viable population size for threatened grizzly bear populations, given a suitable habitat, was 100 bears. Current estimates of the grizzly population in the greater Yellowstone ecosystem indicate a population of about 400. The effective population size, N_e, is only 25% of the total population size, or 100 bears. Genetic analyses indicate that the Yellowstone grizzly population has less genetic variability than other populations in North America. Migration between isolated populations could increase both the effective size and genetic variation of the Yellowstone grizzly bear population.

INTERACTIVE QUESTION 56.3

Explain the basic premise of the small population approach. What conservation strategy is recommended for preserving small populations?

Declining-Population Approach The emphasis of the declining-population approach is to identify populations that may be declining, identify the environmental factors that caused that decline, and then recommend corrective measures.

The following logical steps are part of the declining-population approach: assess population trends and distribution to establish that a species is in decline; determine its environmental requirements; list all possible causes of the decline and the predictions that arise from each of these hypotheses; test the most likely hypothesis to see if the population rebounds if this suspected factor is altered; and apply the results to the management of the threatened species.

Logging and agriculture have fragmented the mature pine forest habitats of the red-cockaded woodpecker, driving this species into decline. Historically, periodic fires kept the understory around the pines low, which is another habitat requirement. Recognition of the social organization of this species and the factors that slow its dispersal to new territories has aided in its recovery. Management strategies now include protection of some longleaf pine forests, controlled fires, and the excavation of breeding cavities in unoccupied habitats to encourage establishment of new breeding groups.

INTERACTIVE QUESTION 56.4

Describe the declining-population approach to the conservation of endangered species.

Weighing Conflicting Demands Preserving habitat for endangered species often conflicts with human economic and recreational desires. Keystone species exert more influence on community structure and ecosystem processes, and prioritizing the species to be saved on the basis of their ecological role may be key to the survival of whole communities.

56.3 Landscape and regional conservation aim to sustain entire biotas

Conservation efforts increasingly are directed at sustaining the biodiversity of whole communities and ecosystems. A goal of landscape ecology is to make biodiversity conservation a part of ecosystem management and landscape use.

Landscape Structure and Biodiversity Landscape dynamics are important to conservation efforts because species often use more than one ecosystem or live on borders between ecosystems. Landscapes include ecosystems separated by boundaries or *edges*, which have their own sets of physical conditions and communities of organisms. Edge communities may be important sites of speciation, but their proliferation due to human fragmentation of habitats may serve to reduce biodiversity as edge-adapted species become predominant. The long-term Biological Dynamics of Forest Fragments Project has shown that species adapted to forest interiors decline the most in small fragments.

Movement corridors are narrow strips or clumps of habitat that connect isolated patches. Artificial corridors are sometimes constructed when habitat patches have been separated by major human disruptions.

INTERACTIVE QUESTION 56.5

What are some potential benefits of corridors? How may they be harmful?

Establishing Protected Areas Currently, about 7% of Earth's land has been set aside as reserves. Conservation biologists apply the study of community, ecosystem, and landscape dynamics to the designation and management of these protected areas. **Biodiversity hot spots,** small areas with very high concentrations of endemic, threatened, and endangered species, are good choices for nature reserves.

Even though nature reserves provide islands of protected habitat, the concept of nonequilibrium ecology with natural disturbances applies to them. Patch dynamics, edges, and corridor effects must be considered in the design and management of preserves.

New information on the requirements for minimum viable population sizes indicates that most national parks and reserves are much too small—the biotic boundary needed to sustain a population is usually much larger than the legal boundary set in a reserve.

INTERACTIVE QUESTION   56.6

What factors would favor the creation of larger, extensive preserves? What factors favor smaller, unconnected preserves?

Zoned reserves have protected core areas surrounded with buffer zones in which the human social and economic climate is stable and activities are regulated to promote the long-term viability of the protected zones. Costa Rica has established eight zoned reserves, but deforestation has continued in some buffer zones. The Florida Keys National Marine Sanctuary, a zoned marine reserve established in 1990, has increased marine life and provided an economic value from recreational divers.

56.4   Restoration ecology attempts to restore degraded ecosystems to a more natural state

Areas degraded by farming, mining, or environmental pollution are often abandoned. Restoration ecologists attempt to identify and manipulate the factors that most limit recovery time in order to speed the successional processes involved in a community's recovery from human disturbances. In some cases, the physical structure of a site must first be reconstructed.

Exploring Restoration   Restoration ecologists often apply adaptive management, in which they experiment with promising management approaches and learn as they work in each unique and complex disturbed ecosystem.

Bioremediation   **Bioremediation** uses prokaryotes, fungi, or plants to detoxify polluted ecosystems. Some plants are able to extract metals from contaminated soils.

Biological Augmentation   **Biological augmentation** uses organisms to add essential substances to degraded ecosystems.

INTERACTIVE QUESTION   56.7

Give an example of bioremediation and of biological augmentation.

56.5   Sustainable development seeks to improve the human condition while conserving biodiversity

Sustainable Biosphere Initiative   **Sustainable development** emphasizes meeting the needs of people today without limiting the long-term ability of future generations to meet their needs. The Ecological Society of America endorses a research agenda, the Sustainable Biosphere Initiative, to encourage studies of global change, biodiversity, and maintenance of the productivity of natural and artificial ecosystems. An important goal is developing the ecological knowledge necessary to make intelligent and responsible decisions concerning Earth's resources.

Case Study: Sustainable Development in Costa Rica   Partnerships between the government, nongovernment organizations (NGOs), and citizens have contributed to the success of conservation in Costa Rica. Living conditions in the country have improved, as evidenced by a decrease in infant mortality, increase in life expectancy, and high literacy rate. A projected increase in population from 4 to 6 million in the next 50 years, however, will present challenges to the goal of sustainable development.

The Future of the Biosphere   E. O. Wilson calls our attraction to Earth's diversity of life and our affinity for natural environments *biophilia*. Perhaps this connection is innate and will provide the ethical resolve to protect species from extinction and ecosystems from destruction. By coming to know and understand nature through the study of biology, we may be more able to appreciate and preserve the processes and diversity of the biosphere.

Word Roots

bio- = life (*biodiversity hot spot:* a relatively small area with an exceptional concentration of species)

Structure Your Knowledge

1. What are the major threats to biodiversity, listed in order of importance?

2. How does the loss of biodiversity threaten human welfare?

3. What do edges and movement corridors have to do with habitat fragmentation?

Test Your Knowledge

MULTIPLE CHOICE: *Choose the one best answer.*

1. According to the Endangered Species Act, what is the definition of a threatened species?
   a. an exotic species that cannot successfully compete with native organisms
   b. an endemic species that is found nowhere else in the world
   c. a species that is found in disturbed habitats
   d. a species that is in danger of extinction in all or a large part of its range
   e. a species that is likely to become endangered

2. Ecosystem services include all of the following *except*
   a. pollination of crops.
   b. production of antibiotics and drugs.
   c. purification of air and water.
   d. decomposition of wastes.
   e. reduced impact of weather extremes.

3. Which of the following is the most serious threat to biodiversity?
   a. competition from introduced species
   b. commercial harvesting
   c. habitat destruction
   d. overexploitation
   e. disruptions of community dynamics

4. Some grassland and conifer forest preserves have effective fire prevention programs. What is the most likely result of such programs?
   a. an increase in species diversity because fires are prevented
   b. a change in community composition because fires are natural disturbances that maintain the community structure
   c. the preservation of endangered species in the area

   d. no change in the species composition of the preserved community
   e. succession to a deciduous forest

5. According to the small population approach, what is the most important remedy for preserving an endangered species?
   a. establish a large nature reserve around its habitat
   b. control the populations of its natural predators
   c. determine the reason for its decline
   d. encourage dispersal and increase in genetic variability
   e. set up artificial breeding programs

6. Which of the following is typical of biodiversity hot spots?
   a. a large number of endemic species
   b. a high rate of habitat degradation
   c. little species diversity
   d. a large land or aquatic area
   e. large populations of migratory birds

7. Which of the following may occur when a population drops below its minimum viable population size?
   a. genetic drift
   b. a further reduction in population size
   c. inbreeding
   d. a loss of genetic variability
   e. All of the above are characteristics of an extinction vortex that the population may enter.

8. What are movement corridors?
   a. strips or clumps of habitat that connect isolated habitats
   b. the routes taken by migratory animals
   c. connections within a landscape that includes several different ecosystems
   d. the areas forming the boundary or edge between two ecosystems
   e. buffer zones for human traffic, which promote the long-term viability of protected areas

9. What does it mean if a population's effective population size (N_e) is the same as its actual population size?
   a. The population is not in danger of becoming extinct.
   b. The population has high genetic variability.
   c. All the members of the population breed.
   d. The population's minimum viable population will not sustain the population.
   e. The population is being drawn into an extinction vortex.

10. The focus of the declining-population approach to conservation is to

 a. predict a species' minimum viable population size.

 b. transplant members from other populations to increase genetic variation.

 c. perform a population viability analysis to predict the long-term viability of a population in a particular habitat.

 d. determine the cause of a species' decline and take remedial action.

 e. establish zoned reserves that ensure that human landscapes surrounding reserves support the protected habitats.

11. With limited resources, conservation biologists need to prioritize their efforts. Of the following choices, which should receive the greatest conservation attention in order to preserve biodiversity?

 a. the northern spotted owl

 b. declining keystone species in a community

 c. a commercially important species

 d. endangered and threatened vertebrate species

 e. all declining species

12. Restoration ecology

 a. uses the zoned reserve system to restore nature reserves.

 b. identifies biodiversity hot spots for protection.

 c. may use bioremediation and biological augmentation to return degraded areas to their natural state.

 d. uses the research agenda of the Sustainable Biosphere Initiative to study biodiversity and preserve Earth's ecosystems.

 e. uses adaptive management to restore and maintain the productivity of agricultural ecosystems.

Answer Section

CHAPTER 1: INTRODUCTION: THEMES IN THE STUDY OF LIFE

INTERACTIVE QUESTIONS

1.1 a. The biosphere includes most regions of land and water and the atmosphere up to several kilometers: all of Earth that is inhabited by life.
b. An ecosystem includes all the living organisms in an area, along with the nonliving parts of the environment with which they interact.
c. A community consists of all the organisms inhabiting a particular area.
d. A population includes the individuals of a single species in an area.
e. An organism is an individual living entity.
f. Organs and organ systems, found in more complex organisms, perform life's functions.
g. Tissues are collections of similar cells. Several tissues make up an organ.
h. Cells are the fundamental unit of life.
i. Organelles are the functional components that make up cells.
j. Molecules are composed of two or more atoms. They are the chemical units of life.

1.2 In predicting how a particular drug or treatment may affect various aspects of the body; in predicting effects on various parts of the ecosystem as CO_2 concentration climbs.

1.3 The order of nucleotides in a DNA molecule that makes up a gene "spells" the instructions, rewritten in RNA, for making a protein with a specific function.

1.4 They are characterized to a large extent by their mode of nutrition. Plants are photosynthetic, fungi absorb their nutrients from decomposing organic material, and animals ingest other organisms.

1.5 Most species tend to produce more offspring than can survive. Organisms with heritable traits best suited to the environment will tend to leave more offspring. Over time, favorable adaptations will accumulate in a population. New species may arise as small populations are exposed to different environments and natural selection favors different traits.

1.6 a. Since most encounters with coral snakes are fatal, predators didn't "learn" to avoid them. But predators with genes that somehow made them instinctively avoid coral snakes would have been more likely to pass on those genes to offspring, gradually adapting the predator population to the presence of poisonous snakes in their area.
b. This experiment could not control for the number of predators in each area and thus the total number of attacks on snakes. By presenting data as the percent of total attacks, the effect of this variable was eliminated.

1.7 a. A hypothesis is less broad in scope than a theory; it is a tentative explanation for a smaller set of observations. A theory can generate many testable hypotheses and is supported by a large body of evidence. Both hypotheses and theories are revised when new data do not support their predictions.
b. Science seeks to understand natural phenomenon; technology is the practical use of scientific knowledge.

SUGGESTED ANSWERS TO STRUCTURE YOUR KNOWLEDGE

1. a. Evolution explains the unity and diversity of life. All organisms are modified descendants of common ancestors. Darwin's theory of natural selection leading to differential reproductive success accounts for the adaptation of populations to Earth's varying environments.

b. With every increase in biological level, the organization and interactions of component parts lead to emergent, new properties of the dynamic system.

c. Organisms exchange materials and energy with the living and nonliving components of their environment. All organisms require energy. Energy flows through ecosystems from sunlight to chemical energy in producers and consumers to escape as heat.

d. At each level of biological organization, structure and function are correlated.

e. Cells are the basic units of life. They come in two distinct forms: prokaryotic and eukaryotic.

f. DNA is the molecule of inheritance, coding information for proteins and the functions of a cell, and passed on from one generation to the next.

g. Organisms regulate their internal environment, usually through negative feedback mechanisms that slow down body processes, matching supply to demand.

ANSWERS TO TEST YOUR KNOWLEDGE

Multiple Choice:

1. d	**3.** b	**5.** a	**7.** e	**9.** e
2. d	**4.** b	**6.** d	**8.** c	

CHAPTER 2: THE CHEMICAL CONTEXT OF LIFE

INTERACTIVE QUESTIONS

2.1 calcium, phosphorus, potassium, sulfur, sodium, chlorine, magnesium

2.2 neutrons; 15, 15, 16, 31

2.3 absorb; released

2.4 **a.** Carbon, $_6$C   **c.** Oxygen, $_8$O

b. Nitrogen, $_7$N   **d.** Magnesium, $_{12}$Mg

2.5 Although each orbital can hold a pair of electrons, each electron fills a separate orbital until no empty orbitals remain.

2.6 a. protons
  b. atomic number
  c. element
  d. neutrons
  e. mass number or atomic mass
  f. isotopes
  g. electrons
  h. electron shells
  i. valence shell

2.7 a. H = 1, O = 2, N = 3, C = 4

b. H:H Ö::Ö: N:::N: H:C:H with H above and below C

2.8 a. Nonpolar; even though N has a high electronegativity, the three pairs of electrons are shared equally between the two N atoms because each atom has an equally strong attraction for the electrons.

b. Polar; N is more electronegative than H and pulls the shared electrons in each covalent bond closer to itself.

c. Nonpolar; C and H have similar electronegativities and share electrons fairly equally between them.

d. The C=O bond is polar because O is more electronegative than C; the C—H bonds are relatively nonpolar.

2.9 a. $CaCl_2$ **b.** Ca^{2+} is the cation.

2.10

2.11 Water's two covalent bonds with H are spread apart at a 104.5° angle due to the hybridization of the *s* and 3 *p* orbitals.

2.12 6, 6, 6

SUGGESTED ANSWERS TO STRUCTURE YOUR KNOWLEDGE

1.

Particle	Charge	Mass	Location
Proton	+1	1 dalton	nucleus
Neutron	0	1 dalton	nucleus
Electron	−1	negligible	orbitals in electron shells

2. a. The atoms of each element have a characteristic number of protons in their nuclei, referred to as the *atomic number*. In a neutral atom, the atomic number also indicates the number of electrons. The *mass number* is an indication of the approximate mass of an atom and is equal to the number of protons and neutrons in the nucleus.

The *atomic mass* is equal to the mass number and is measured in the atomic mass unit of daltons. Protons and neutrons both have a mass of approximately 1 dalton.

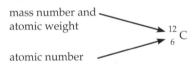

mass number and atomic weight

atomic number

$^{12}_{6}C$

b. The *valence*, an indication of bonding capacity, is the number of unpaired electrons that an atom has in its valence shell.

c. The valence of an atom is most related to the chemical behavior of an atom because it is an indication of the number of bonds the atom will make, or the number of electrons the atom must share in order to reach a filled valence shell.

3. Ionic and nonpolar covalent bonds represent the two ends of a continuum of electron sharing between atoms in a molecule. In ionic bonds the electrons are completely pulled away from one atom by the other, creating negatively and positively charged ions (anions and cations). In nonpolar covalent bonds the electrons are equally shared between two atoms. Polar covalent bonds form when a more electronegative atom pulls the shared electrons closer to it, producing a partial negative charge associated with that portion of the molecule and a partial positive charge associated with the atom from which the electrons are pulled.

ANSWERS TO TEST YOUR KNOWLEDGE

Multiple Choice:

1. b	**6.** e	**11.** b	**16.** c	**21.** b
2. d	**7.** d	**12.** c	**17.** d	**22.** c
3. a	**8.** b	**13.** c	**18.** b	**23.** d
4. e	**9.** e	**14.** b	**19.** a	**24.** e
5. a	**10.** c	**15.** e	**20.** c	**25.** e

CHAPTER 3: WATER AND THE FITNESS OF THE ENVIRONMENT

INTERACTIVE QUESTIONS

3.1

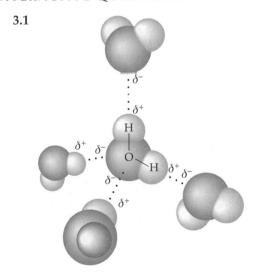

3.2 a. polar water molecules
b. absorbed
c. released
d. specific heat
e. heat of vaporization
f. evaporative cooling
g. solar heat
h. rain
i. ice forms

3.3 a. olive oil: hydrophobic mostly nonpolar
b. sugar: hydrophilic polar
c. salt: hydrophilic ionic
d. candle wax: hydrophobic nonpolar

3.4 a. The molecular mass of $C_3H_6O_3$ is 90 d, the combined atomic masses of its atoms. A mole of lactic acid = 90 g. A 0.5 *M* solution would require ½ mol or 45 g.

b. 3.01×10^{23}

3.5

[H$^+$]	[OH$^-$]	pH	Acidic, Basic, or Neutral?
10^{-3}	10^{-11}	3	acidic
10^{-8}	10^{-6}	8	basic
10^{-7}	10^{-7}	7	neutral
10^{-1}	10^{-13}	1	acidic

3.6 carbonic acid   bicarbonate   hydrogen ion

$$H_2CO_3 \rightleftharpoons HCO_3^- + H^+$$

H$^+$donor   H$^+$ acceptor

a. Bicarbonate acts as a base to accept excess H$^+$ ions when the pH starts to fall; the reaction moves to the left.

b. When the pH rises, H$^+$ ions are donated by carbonic acid, and the reaction shifts to the right.

3.7 a. $CO_2 + H_2O \rightleftharpoons H_2CO_3 \rightleftharpoons HCO_3^- + H^+$
Increasing [CO$_2$] will drive these reactions to the right, increasing [H$^+$].
b. $HCO_3^- \rightleftharpoons CO_3^{2-} + H^+$
Increasing [H$^+$] will drive this reaction to the left, thus decreasing [CO$_3^{2-}$].
c. With less CO_3^{2-} available to react with Ca^{2+}, calcification rates would be expected to decrease, as was shown in the study by Chris Langdon and colleagues.

SUGGESTED ANSWERS TO STRUCTURE YOUR KNOWLEDGE

1.

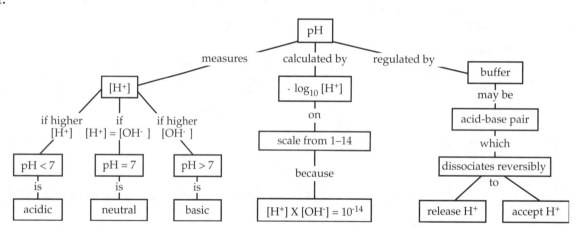

2. a. Cohesion, adhesion
b. A water column is pulled up through plant vessels.
c. Heat is absorbed or released when hydrogen bonds break or form. Water absorbs or releases a large quantity of heat for each degree of temperature change.
d. High heat of vaporization
e. Solar heat is dissipated from tropical seas.
f. Evaporative cooling
g. Evaporation of water cools surfaces of plants and animals.
h. Hydrogen bonds in ice space water molecules apart, making ice less dense.

i. Floating ice insulates bodies of water so they don't freeze solid.
j. Versatile solvent
k. Polar water molecules surround and dissolve ionic and polar solutes.

ANSWERS TO TEST YOUR KNOWLEDGE

Multiple Choice:

1. c	**6.** d	**11.** b	**16.** a	**21.** b
2. a	**7.** e	**12.** d	**17.** b	**22.** e
3. e	**8.** b	**13.** c	**18.** d	
4. e	**9.** e	**14.** c	**19.** d	
5. b	**10.** d	**15.** e	**20.** e	

CHAPTER 4: CARBON AND THE MOLECULAR DIVERSITY OF LIFE

INTERACTIVE QUESTIONS

4.1 A variety of organic compounds were produced in Miller's apparatus, which attempted to simulate conditions on early Earth. The abiotic synthesis of organic compounds may have been a first step in the origin of life.

4.2 Ethanol and dimethyl ether, *structural isomers*, have the same number and kinds of atoms but a different bonding sequence and very different properties. Maleic acid and fumaric acid are *geometric isomers* whose double bonds fix the spatial arrangement of the molecule. Maleic acid is the *cis* isomer; both groups are on the same side of the double bond. Although the *enantiomers* L- and D-lactic acid look similar in a flat representation of their structures, they are not superimposable.

SUGGESTED ANSWERS TO STRUCTURE YOUR KNOWLEDGE

1.

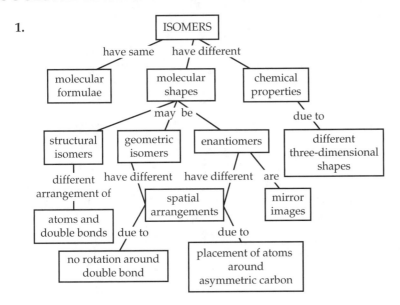

2.

Functional Group	Molecular Formula	Names and Characteristics of Organic Compounds Containing Chemical Group
Hydroxyl	—OH	Alcohols; polar group
Carbonyl	$>C=O$	Aldehyde or ketone; polar group
Carboxyl	—COOH	Carboxylic acid; release H^+
Amino	$—NH_2$	Amines; basic, accept H^+
Sulfhydryl	—SH	Thiols; cross-links stabilize protein structure
Phosphate	$—OPO_3^{2-}$	Organic phosphates; involved in energy transfers, adds negative charge
Methyl	$—CH_3$	Methylated compounds; addition may alter function

ANSWERS TO TEST YOUR KNOWLEDGE

Multiple Choice:

1. c	**4.** a	**7.** d
2. b	**5.** d	**8.** d
3. e	**6.** a	**9.** c

Matching:

1. a, c	**4.** e	**7.** b, f	**10.** a, d
2. b, f	**5.** e	**8.** e	**11.** e
3. a, d, e	**6.** a, c, d, e	**9.** a, c	**12.** c

CHAPTER 5: THE STRUCTURE AND FUNCTION OF MACROMOLECULES

INTERACTIVE QUESTIONS

5.1 a. hydroxyl
  b. carbonyl
  c. aldose
  d. ketose
  e. rings

5.2

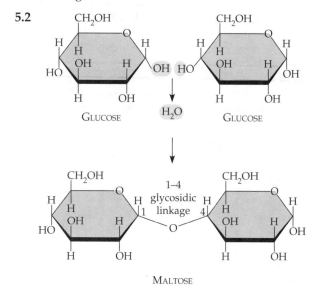

5.3 a. monosaccharides
  b. $(CH_2O)_n$
  c. energy compounds
  d. carbon skeletons, monomers
  e. glycosidic linkages
  f. disaccharides
  g. polysaccharides
  h. glycogen
  i. animals
  j. starch
  k. cellulose
  l. chitin

5.4

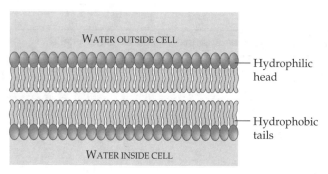

5.5 a. fats, triacylglycerides
  b. phospholipids
  c. glycerol
  d. fatty acids
  e. unsaturated: has some C=C bonds
  f. saturated: no C=C, all possible C—H bonds
  g. phosphate group
  h. cell membranes
  i. steroids
  j. animal cell membrane component (cholesterol), hormones

5.6 a.

  b. serine's R group is polar; alanine's R group is nonpolar
  c. a polypeptide backbone

5.7 a. hydrogen bond
  b. hydrophobic and van der Waals interactions
  c. disulfide bridge
  d. ionic bond
  These interactions between R groups produce tertiary structure.

5.8 a. A change in pH alters the availability of H^+, OH^-, or other ions, thereby disrupting the hydrogen bonding and ionic bonds that maintain protein shape.
  b. A protein in an organic solvent would turn inside out as the hydrophilic regions became clustered on the inside of the molecule and the hydrophobic regions interacted with the nonpolar solvent.
  c. The return to its functional shape indicates that a protein's three-dimensional structure is intrinsically determined by its primary structure—the sequence of its amino acids.

5.9

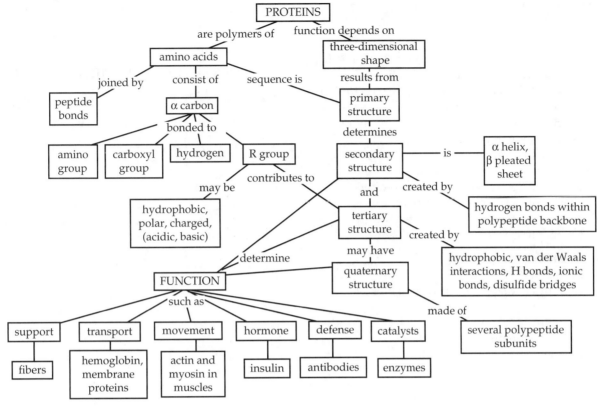

5.10 DNA → RNA → Protein

5.11 a.

b. pyrimidine; base has single ring

c. DNA; 2nd carbon in sugar is lacking O

5.12

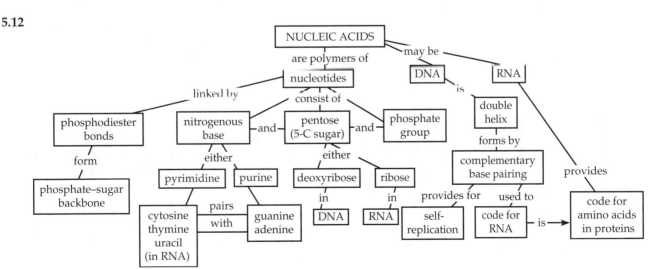

SUGGESTED ANSWERS TO STRUCTURE YOUR KNOWLEDGE

1. b, d	**3.** c, e, f	**5.** c	**7.** e, f
2. a	**4.** f, g	**6.** a	**8.** b

1. The primary structure of a protein is the specific, genetically coded sequence of amino acids in a polypeptide chain. The secondary structure involves the coiling (α helix) or folding (β pleated sheet) of the protein, stabilized by hydrogen bonds along the polypeptide backbone. The tertiary structure involves interactions between the side chains (R groups) of amino acids and produces a characteristic three-dimensional shape for a protein. Quaternary structure occurs in proteins composed of more than one polypeptide chain.

2. **a.** amino acid (glycine)
   b. fatty acid
   c. nitrogenous base, purine (adenine)
   d. glycerol
   e. phosphate group
   f. sugar (pentose, ribose)
   g. sugar (triose)

ANSWERS TO TEST YOUR KNOWLEDGE

Matching:

1. A	**4.** C	**7.** B	**10.** A
2. B	**5.** C	**8.** C	
3. D	**6.** D	**9.** A	

Multiple Choice:

1. e	**6.** b	**11.** c	**16.** a	**21.** d
2. c	**7.** c	**12.** d	**17.** c	**22.** a
3. a	**8.** a	**13.** b	**18.** b	**23.** d
4. e	**9.** d	**14.** c	**19.** c	**24.** e
5. c	**10.** c	**15.** d	**20.** b	

CHAPTER 6: A TOUR OF THE CELL

INTERACTIVE QUESTIONS

6.1 a. the study of cell structure
b. the internal ultrastructure of cells
c. the three-dimensional surface topography of a specimen
d. Light microscopy enables the study of living cells and may introduce fewer artifacts than do TEM and SEM.

6.2 a. a phospholipid bilayer with the hydrophobic tails clustered in the interior and the phosphate heads facing the hydrophilic outside and inside of the cell; proteins are embedded in and attached to the membrane
b. 10^2, or 100 times the surface area
c. 10^3, or 1,000 times the volume

6.3 The genetic instructions for specific proteins are transcribed from DNA into messenger RNA (mRNA), which then passes into the cytoplasm to complex with ribosomes where it is translated into the primary structure of proteins.

6.4 a. smooth ER—in different cells may house enzymes that synthesize lipids; metabolize carbohydrates; detoxify drugs and alcohol; store and release calcium ions in muscle cells
b. nuclear envelope—double membrane that encloses nucleus; pores regulate passage of materials
c. rough ER—attached ribosomes produce proteins that enter cisternae; produces secretory proteins and membranes
d. transport vesicle—carries products of ER and Golgi apparatus to various locations
e. Golgi apparatus—processes products of ER; makes polysaccharides, packages products in vesicles targeted to specific locations
f. plasma membrane—selective barrier that regulates passage of materials into and out of the cell
g. lysosome—houses hydrolytic enzymes to digest macromolecules

6.5

Mitochondrion

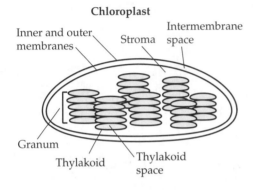

Inner membrane

Outer membrane

Cristae

Intermembrane space

Matrix

Chloroplast

Inner and outer membranes

Stroma

Intermembrane space

Granum

Thylakoid

Thylakoid space

6.6 Peroxisomes do not bud from the endomembrane system, but grow by incorporating proteins and lipids from the cytosol; they increase in number by dividing.

6.7 a. hollow tube, formed from columns of tubulin dimers; 25-nm diameter
b. cell shape and support (compression resistant), tracks for moving organelles, chromosome movement, beating of cilia and flagella
c. two twisted chains of actin molecules; 7-nm diameter
d. muscle contraction, maintain (tension bearing) and change cell shape, pseudopod movement, cytoplasmic streaming, sol-gel transformations
e. supercoiled fibrous proteins of keratin family; 8–12 nm
f. reinforce cell shape, anchor nucleus; nuclear lamina

6.8

Two adjacent plant cells

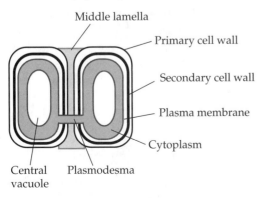

Middle lamella

Primary cell wall

Secondary cell wall

Plasma membrane

Cytoplasm

Central vacuole

Plasmodesma

SUGGESTED ANSWERS TO STRUCTURE YOUR KNOWLEDGE

1. **a.** nucleus, chromosomes, centrioles, microtubules (spindle), microfilaments (actin-myosin aggregates pinch apart cell)
   b. nucleus, chromosomes, DNA → mRNA → ribosomes → enzymes and other proteins
   c. mitochondria
   d. ribosomes, rough and smooth ER, Golgi apparatus, transport vesicles
   e. smooth ER (peroxisomes also detoxify substances)
   f. lysosomes, food vacuoles
   g. peroxisomes
   h. cytoskeleton: microtubules, microfilaments, intermediate filaments; extracellular matrix
   i. cilia and flagella (microtubules), microfilaments (actin) in muscles and pseudopodia
   j. plasma membrane, transport vesicles
   k. desmosomes, tight and gap junctions, ECM

2. **a.** structural support, middle lamella glues cells together
   b. storage, waste disposal, protection, growth
   c. photosynthesis, production of sugars
   d. starch storage
   e. cytoplasmic connections between cells

3. **a.** rough endoplasmic reticulum
   b. smooth endoplasmic reticulum
   c. chromatin
   d. nucleolus
   e. nuclear envelope
   f. nucleus
   g. ribosomes
   h. Golgi apparatus
   i. plasma membrane
   j. mitochondrion
   k. lysosome
   l. cytoskeleton
   m. microtubules
   n. intermediate filaments
   o. microfilaments
   p. microvilli
   q. peroxisome
   r. centrosome (contains a pair of centrioles)
   s. flagellum

4.

ANSWERS TO TEST YOUR KNOWLEDGE

Multiple Choice:

1. c	**3.** d	**5.** a	**7.** c	**9.** b	**11.** e	**13.** b	**15.** e	**17.** c	**19.** b
2. b	**4.** e	**6.** d	**8.** c	**10.** d	**12.** b	**14.** c	**16.** d	**18.** e	**20.** a

CHAPTER 7: MEMBRANE STRUCTURE AND FUNCTION

INTERACTIVE QUESTIONS

7.1 a. phosphate head—hydrophilic
  b. phospholipid bilayer
  c. hydrocarbon tail—hydrophobic
  d. hydrophobic region of protein
  e. hydrophilic region of protein

7.2 In hybrid human/mouse cells, membrane proteins rapidly intermingle (Frye and Edidin experiment).

7.3 transport, enzymatic activity, signal transduction, intercellular joining, cell–cell recognition, attachment to cytoskeleton and ECM

7.4 Ions and larger polar molecules, such as glucose, are impeded by the hydrophobic center of the plasma membrane's lipid bilayer. Passage through the center of a lipid bilayer is not fast even for small, polar water molecules.

7.5 Side A initially has fewer free water molecules; side B initially has more. More water molecules are clustered around the glucose in the 1 *M* solution than around the fructose and sucrose whose combined concentration is 0.9 *M*. Water will move by osmosis from side B to A.

7.6 a. The protists will gain water from their hypotonic environment.
  b. They may have membranes that are less permeable to water and contractile vacuoles that expel excess water.

7.7 a. isotonic   **b.** hypotonic

7.8 Although it may speed diffusion, facilitated diffusion is still passive transport because the solute is moving down its concentration gradient; the process is driven by the concentration gradient and not energy expended by the cell.

7.9 Three sodium ions are pumped out of the cell for every two potassium ions pumped in, resulting in a net movement of positive charge from the cytoplasm to the extracellular fluid.

7.10 a. Human cells use receptor-mediated endocytosis to take in cholesterol.
  b. LDL receptor proteins in the plasma membrane are defective, and low-density lipoproteins cannot bind and be transported from the blood into the cell.

SUGGESTED ANSWERS TO STRUCTURE YOUR KNOWLEDGE

1.

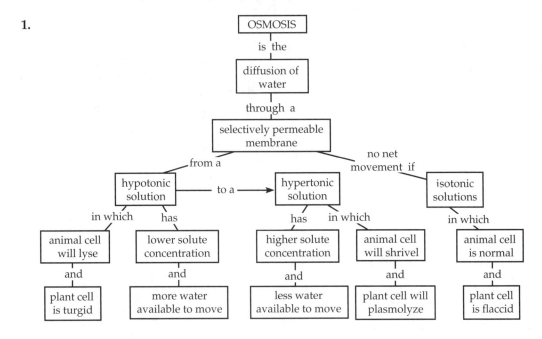

2. a. II represents facilitated diffusion. The solute is moving through a transport protein and down its concentration gradient. The cell does not expend energy in this transport. Polar molecules and ions may move by facilitated diffusion.

b. III represents active transport because the solute is clearly moving against its concentration gradient and the cell is expending ATP to drive this transport against the gradient.

c. I illustrates diffusion through the lipid bilayer. The solute molecules must be nonpolar or very small polar molecules.

d. I and II. Both diffusion and facilitated diffusion are considered passive transport because the solute moves down its concentration gradient and the cell does not expend energy in the transport.

ANSWERS TO TEST YOUR KNOWLEDGE

Multiple Choice:

1. c	**3.** a	**5.** d	**7.** e	**9.** b	**11.** e	**13.** b	**15.** b	**17.** c	**19.** a*
2. b	**4.** e	**6.** e	**8.** e	**10.** a	**12.** a	**14.** c	**16.** b	**18.** e	**20.** d**

*Explanation for answer to question 19: This problem involves both osmosis and diffusion. Although the solutions are initially equal in molarity, glucose will diffuse down its concentration gradient until it reaches dynamic equilibrium with a 1.5 *M* concentration on both sides. The increasing solute concentration on side A will cause water to move into this side, and the water level will rise.

**Explanation for answer to question 20: As the solute in the solution crosses the cell membrane, it increases the concentration of solutes within the cell, reducing the hypertonicity of the solution to the cell. As the solute reaches an equal concentration inside and outside the cell, it no longer causes osmotic changes in the cell.

CHAPTER 8: AN INTRODUCTION TO METABOLISM

INTERACTIVE QUESTIONS

8.1 **a.** capacity to cause change
 b. kinetic
 c. motion
 d. potential
 e. position
 f. conserved
 g. created nor destroyed
 h. first
 i. transformed or transferred
 j. entropy
 k. second

8.2

	System with High Free Energy	System with Low Free Energy
Stability	low	high
Spontaneous	change will be	change will not be
Equilibrium	moves toward	is at
Work capacity	high	low

8.3

8.4 **a.** adenine
 b. ribose
 c. three phosphate groups
 d. A hydrolysis reaction breaks the terminal phosphate bond and releases a molecule of inorganic phosphate: ATP + $H_2O \rightarrow$ ADP + P_i

e. The negatively charged phosphate groups are crowded together, and their mutual repulsion makes this area instable. The chemical change to a more stable state of lower free energy accounts for the relatively high release of energy.

8.5 **a.** free energy
 b. transition state
 c. E_A (free energy of activation) without enzyme
 d. E_A with enzyme
 e. ΔG of reaction

8.6

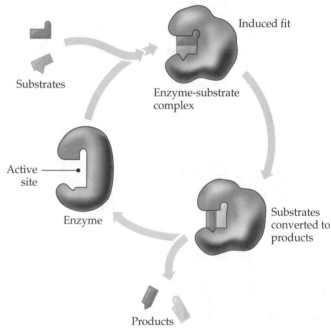

8.7 A competitive inhibitor would mimic the shape of the substrates and compete with them for the active site. A noncompetitive inhibitor would be a shape that could bind to another site on the enzyme molecule and would change the shape of the active site such that the substrates could no longer fit.

8.8 ATP would act as an inhibitor to catabolic pathways, slowing the breakdown of fuel molecules if sufficient energy is available in the cell. ATP may act as an activator of anabolic pathways that store resources in more complex molecules.

SUGGESTED ANSWERS TO STRUCTURE YOUR KNOWLEDGE

1. Metabolism is the totality of chemical reactions that take place in living organisms. To create and maintain the structural order required for life requires an input of free energy—from sunlight for photosynthetic organisms and from energy-rich food molecules for other organisms. A cell couples catabolic, exergonic reactions ($-\Delta G$) with anabolic, endergonic reactions ($+\Delta G$), using ATP as the primary energy shuttle between the two.

2. Enzymes are essential for metabolism because they lower the activation energy of the specific reactions they catalyze and allow those reactions to occur extremely rapidly at a temperature conducive to life. By regulating the enzymes it produces, a cell can regulate which of the myriad of possible chemical reactions take place at any given time. Metabolic control also occurs through allosteric regulation and feedback inhibition. The compartmental organization of a cell facilitates a cell's metabolism.

ANSWERS TO TEST YOUR KNOWLEDGE

Multiple Choice:

1. c	6. c	11. b	16. d	21. e
2. b	7. a	12. c	17. b	22. a
3. a	8. e	13. b	18. d	
4. e	9. e	14. b	19. c	
5. c	10. c	15. e	20. b	

Fill in the Blanks:

1. metabolism
2. anabolic
3. kinetic
4. allosteric
5. entropy
6. free energy of activation
7. competitive inhibitors
8. coenzymes
9. feedback inhibition
10. phosphorylated compound

CHAPTER 9: CELLULAR RESPIRATION: HARVESTING CHEMICAL ENERGY

INTERACTIVE QUESTIONS

9.1 $C_6H_{12}O_6$; $6 CO_2$; energy (ATP + heat)

9.2
a. oxidized
b. reduced
c. donates (loses)
d. oxidizing agent
e. accepts (gains)

9.3
a. oxygen
b. glucose
c. Some is stored in ATP and some is released as heat.

9.4
a. electron carrier (or acceptor) or oxidizing agent
b. NADH

9.5
a. glycolysis: glucose → pyruvate (not technically considered part of cellular respiration)
b. citric acid cycle
c. oxidative phosphorylation: electron transport and chemiosmosis
d. substrate-level phosphorylation
e. substrate-level phosphorylation
f. oxidative phosphorylation

The top two arrows show electrons carried by NADH (and $FADH_2$, another electron carrier) to the electron transport chain.

9.6
a. 2 ATP
b. 2 three-carbon sugars (glyceraldehyde-3-phosphate)
c. $2 NAD^+$
d. $2 NADH + 2H^+$
e. 4 ATP
f. 2 pyruvate

9.7
a. pyruvate
b. CO_2
c. $NADH + H^+$
d. coenzyme A
e. acetyl CoA
f. oxaloacetate
g. citrate
h. $NADH + H^+$
i. CO_2
j. CO_2
k. $NADH + H^+$
l. ATP
m. $FADH_2$
n. $NADH + H^+$

9.8
a. intermembrane space
b. inner mitochondrial membrane
c. mitochondrial matrix
d. electron transport chain
e. $NADH + H^+$
f. NAD^+
g. $FADH_2$
h. chemiosmosis
i. $2 H^+ + 1/2 O_2$
j. H_2O
k. ATP synthase
l. $ADP + Ⓟi$
m. ATP

9.9 a. −2
  b. 4
  c. citric acid cycle
  d. 32 or 34
  e. 38
  f. 2
  g. 6
  h. 2
  i. 2
  j. 2

9.10 Respiration yields up to 19 times more ATP than does fermentation. By oxidizing pyruvate to CO_2 and passing electrons from NADH (and $FADH_2$) through the electron transport chain, respiration can produce a maximum of 38 ATP compared to the 2 net ATP that are produced by fermentation.

SUGGESTED ANSWERS TO STRUCTURE YOUR KNOWLEDGE

1. See Interactive Questions 9.5, 9.6, 9.7, and 9.8.

Process	Main Function	Inputs	Output
Glycolysis	Oxidation of glucose to 2 pyruvate, 2 ATP net	glucose 2 ATP 2 NAD^+ 4 ADP + Ⓟ$_i$	2 pyruvate 4 ATP (2 net) 2 NADH + $2H^+$ 2 H_2O
Pyruvate to acetyl CoA	Oxidation of pyruvate to acetyl CoA, which then enters citric acid cycle	2 pyruvate 2 CoA 2 NAD^+	2 acetyl CoA 2 CO_2 2 NADH + $2H^+$
Citric acid cycle	Acetyl CoA is combined with oxaloacetate to produce citrate, which is cycled back to oxaloacetate as redox reactions produce NADH and $FADH_2$, ATP is formed by substrate-level phosphorylation, and CO_2 is released.	2 acetyl CoA 2 ADP + Ⓟ$_i$ 6 NAD^+ 2 FAD	2 CoA 4 CO_2 2 ATP 6 NADH + $6H^+$ 2 $FADH_2$
Oxidative phosphorylation (Electron transport and chemiosmosis)	NADH (from glycolysis and citric acid) and $FADH_2$ transfer electrons to electron transport chain. In a series of redox reactions, H^+ is pumped into intermembrane space, and electrons are delivered to 1/2 O_2. Proton-motive force drives H^+ through ATP synthase to make ATP.	10 NADH + H^+ 2 $FADH_2$ H^+ + O_2 34 ADP + Ⓟ$_i$	10 NAD^+ 2 FAD H_2O 34 ATP
Fermentation	Anaerobic catabolism: glycolysis followed by oxidation of NADH to NAD^+ so glycolysis can continue. Pyruvate is either reduced to ethyl alcohol and CO_2 or to lactate.	See glycolysis above 2 pyruvate 2 NADH	2 ATP 2 NAD^+ 2 ethanol and 2 CO_2 or 2 lactate

2.

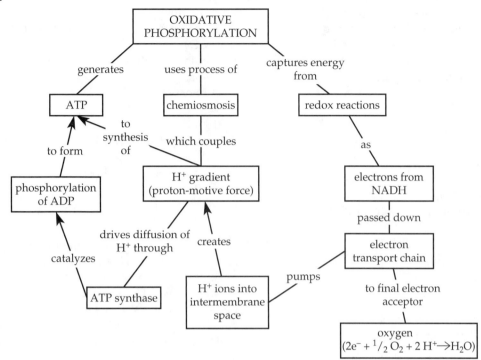

ANSWERS TO TEST YOUR KNOWLEDGE

Multiple Choice:

1. a	**4.** d	**7.** e	**10.** b	**13.** b	**16.** c	**19.** c	**22.** a
2. a	**5.** e	**8.** d	**11.** a	**14.** b	**17.** b	**20.** e	**23.** d
3. c	**6.** d	**9.** b	**12.** e	**15.** c	**18.** d	**21.** c	**24.** e

CHAPTER 10: PHOTOSYNTHESIS

INTERACTIVE QUESTIONS

10.1
  a. outer membrane
  b. granum
  c. inner membrane
  d. thylakoid space
  e. thylakoid
  f. stroma

10.2
  a. light
  b. H_2O
  c. light reactions in thylakoid membranes
  d. O_2
  e. ATP
  f. NADPH
  g. CO_2
  h. Calvin cycle in stroma
  i. CH_2O (sugar)

10.3 The solid line is the absorption spectrum; the dotted line is the action spectrum. Some wavelengths of light, particularly in the blue and the yellow-orange range, result in a higher rate of photosynthesis than would be indicated by the absorption of those wavelengths by chlorophyll *a*. These differences are partially accounted for by accessory pigments, such as chlorophyll *b* and the carotenoids, which absorb light energy from different wavelengths and make that energy available to drive photosynthesis.

10.4 A photosystem contains light-harvesting complexes of pigment molecules (chlorophyll *a*, chlorophyll *b*, and carotenoids) bound to particular proteins and a reaction center, which includes two chlorophyll *a* molecules (P700 or P680) and a primary electron acceptor.

10.5
  a. photosystem II
  b. photosystem I
  c. water (H_2O)
  d. oxygen ($\frac{1}{2} O_2$)
  e. P680, reaction-center chlorophyll *a*
  f. primary electron acceptor
  g. electron transport chain
  h. photophosphorylation by chemiosmosis
  i. ATP
  j. P700, reaction-center chlorophyll *a*
  k. primary electron acceptor
  l. $NADP^+$ reductase
  m. NADPH

  ATP and NADPH provide the chemical energy and reducing power for the Calvin cycle.

10.6
  a. Ferredoxin (Fd) passes the electrons to the cytochrome complex in the electron transport chain, from which they return to $P700^+$.
  b. Electrons from P680 are not passed to P700. Without the oxidizing agent $P680^+$, water is not split. Fd does not pass electrons to $NADP^+$ reductase to form NADPH.
  c. Electrons do pass down the electron transport chain, and the energy released by their "fall" drives photophosphorylation.

10.7
  a. in the thylakoid space (pH of about 5)
  b. (1) transport of protons into the thylakoid space as Pq transfers electrons to the cytochrome complex; (2) protons from the splitting of water remain in the thylakoid space; (3) removal of H^+ in the stroma during the reduction of $NADP^+$.

10.8
  a. carbon fixation
  b. reduction
  c. regeneration of CO_2 acceptor (RuBP)
  d. $3 CO_2$
  e. ribulose bisphosphate (RuBP)
  f. rubisco
  g. 3-phosphoglycerate
  h. $6 ATP \rightarrow 6 ADP$
  i. 1,3-bisphosphoglycerate
  j. $6 NADPH \rightarrow 6 NADP^+$
  k. $6 \circledP_i$
  l. glyceraldehyde-3-phosphate (G3P)
  m. G3P
  n. glucose and other organic compounds
  o. $3 ATP \rightarrow 3 ADP$

10.9 Photorespiration may be an evolutionary relic from the time when there was little O_2 in the atmosphere and the ability of rubisco to distinguish between O_2 and CO_2 was not critical. Photorespiration appears to protect plants from damaging products of the light reactions that build up when the Calvin cycle slows due to a lack of CO_2.

10.10
  a. in the bundle-sheath cells
  b. Carbon is initially fixed into a four-carbon compound in the mesophyll cells by PEP carboxylase. When this compound is broken down in the bundle-sheath cells, CO_2 is maintained at a high enough concentration that rubisco does not accept O_2 and cause photorespiration.
  c. ATP is used to convert pyruvate, returning from the bundle-sheath cells, to PEP in the mesophyll cells.

SUGGESTED ANSWERS TO STRUCTURE YOUR KNOWLEDGE

1.

2.

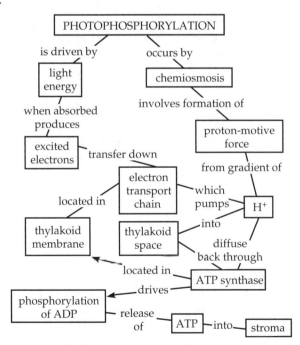

ANSWERS TO TEST YOUR KNOWLEDGE

Multiple Choice:

1. d	**7.** e	**13.** d	**19.** e	**25.** a
2. a	**8.** c	**14.** d	**20.** d	**26.** d
3. d	**9.** a	**15.** d	**21.** d	**27.** a
4. a	**10.** e	**16.** b	**22.** c	
5. b	**11.** a	**17.** c	**23.** a	
6. b	**12.** b	**18.** d	**24.** b	

CHAPTER 11: CELL COMMUNICATION

INTERACTIVE QUESTIONS

11.1 Plant hormones (often called *growth regulators*) may reach their target cells by traveling through vessels, through cells via plasmodesmata, or even through the air as a gas.

11.2 The G protein is also a GTPase enzyme that hydrolyzes its bound GTP to GDP and inactivates itself. It then dissociates from the enzyme it had activated, and that enzyme returns to its original state.

11.3 **a.** signaling molecules
b. α helix in the membrane
c. phosphates
d. ATP → ADP
e. tyrosines
f. activated relay proteins
g. cellular responses

11.4 **a.** A protein kinase transfers a phosphate group from ATP to a protein; adding a charged phosphate group causes a shape change that usually activates the protein.
b. A protein phosphatase removes a phosphate group from a protein, usually inactivating the protein.
c. A phosphorylation cascade is a series of protein kinase relay molecules that are sequentially phosphorylated.

11.5 **a.** signaling molecule (first messenger)
b. G-protein-linked receptor
c. activated G protein (GTP bound)
d. adenylyl cyclase
e. ATP
f. cAMP (second messenger)
g. protein kinase A
h. phosphorylation cascade to cellular response

11.6 **a.** signaling molecule
b. G protein
c. membrane phospholipid
d. IP_3
e. Ca^{2+}
f. endoplasmic reticulum

11.7 **a.** Signaling molecules reversibly bind to receptors, and when they leave a receptor, the receptor reverts to its inactive form. The concentration of signaling molecules influences how many are bound at any time.
b. Activated G proteins are inactivated when the GTPase portion of the protein converts GTP to GDP.
c. Phosphodiesterase converts cAMP to AMP, thus damping out this second messenger.
d. Protein phosphatases remove phosphate groups from activated proteins. The balance of active protein kinases and active phosphatases regulates the activity of many proteins.

11.8 **a.** The basic mechanism for programmed cell death evolved early in animal evolution.
b. Programmed cell death in humans is important in the normal development of the nervous system, fingers, and toes; for normal functioning of the immune system; and to prevent the development of cancerous cells (by the internal triggering of apoptosis in cells with DNA damage).

SUGGESTED ANSWERS TO STRUCTURE YOUR KNOWLEDGE

1. Cell signaling is essential for communication between cells. Unicellular organisms use signals to relay environmental or reproductive information. Communication in multicellular organisms allows for the development and coordination of specialized cells. Extracellular signals control the crucial activities of cells, such as cell division, differentiation, metabolism, and gene expression.

2. Cell signaling occurs through signal transduction pathways that include reception, transduction, and response. First a signaling molecule binds to a specific receptor. The message is transduced as the activated receptor activates a protein that may relay the message through a sequence of activations, finally leading to the activation of proteins that produce the specific cellular response.

3. The sequence described in the previous answer results in the activation of cellular proteins. When the signal is transduced to activate a transcription factor, however, the cellular response

is a change in gene expression and the production of new proteins.

4. In an enzyme cascade, each step in the pathway activates multiple substrates of the next step, thus amplifying the original message to produce potentially millions of activated proteins and thus a large cellular response to a few signals.

ANSWERS TO TEST YOUR KNOWLEDGE

Multiple Choice:

1. c	**4.** b	**7.** c	**10.** a	**13.** b
2. d	**5.** e	**8.** e	**11.** c	**14.** b
3. a	**6.** d	**9.** c	**12.** e	

CHAPTER 12: THE CELL CYCLE

INTERACTIVE QUESTIONS

12.1 a. 46
   b. 23
   c. 92

12.2 a. Growth—most organelles and cell components are produced continuously throughout these subphases.
   b. DNA synthesis

12.3 Refer to main text Figure 12.6 for chromosome diagrams (although you were asked to draw only four chromosomes, and the text shows six).
   a. G_2 of interphase
   b. prophase
   c. prometaphase
   d. metaphase
   e. anaphase
   f. telophase and cytokinesis
   g. centrosomes (with centrioles)
   h. chromatin (duplicated)
   i. nuclear envelope
   j. nucleolus
   k. early mitotic spindle
   l. aster
   m. nonkinetochore microtubules
   n. kinetochore microtubules
   o. metaphase plate
   p. spindle
   q. cleavage furrow
   r. nuclear envelope forming

12.4 a. MPF is a complex of cyclin and Cdk that initiates mitosis by phosphorylating proteins and other kinases.
   b. MPF concentration is high as it triggers the onset of mitosis but is reduced at the end of mitosis because it depends on the concentration of cyclin in the cell. The Cdk level is constant through the cell cycle, but the level of cyclin varies because active MPF starts a process that

degrades cyclin. Thus, MPF regulates its own level and can only become active when sufficient cyclin accumulates after being synthesized in the S and G_2 phases of the next interphase.

SUGGESTED ANSWERS TO STRUCTURE YOUR KNOWLEDGE

1. Interphase: 90% of cell cycle; growth and DNA replication.
   • G_1 phase: The chromosome consists of a long, thin chromatin fiber made of DNA and associated proteins. RNA molecules are being transcribed from genes that are switched on.
   • S phase—synthesis of DNA: The chromosome is duplicated; two exact copies, called sister chromatids, are produced and held together by cohesins along their length. Growth continues.
   • G_2 phase: Growth continues.

   Mitosis phase: cell division
   • Prophase: The sister chromatids, held together by sister chromatid cohesion and at the centromere, become tightly coiled and folded.
   • Prometaphase: Kinetochore fibers from opposite ends of the mitotic spindle attach to the kinetochores of the sister chromatids; the chromosome moves toward midline.
   • Metaphase: The centromere of the chromosome is aligned at the metaphase plate along with the centromeres of the other chromosomes.
   • Anaphase: Cohesins are cleaved and the sister chromatids separate (now considered to be individual chromosomes) and move to opposite poles.
   • Telophase: Chromatin fiber of chromosome uncoils and is surrounded by reforming nuclear membrane.

2.

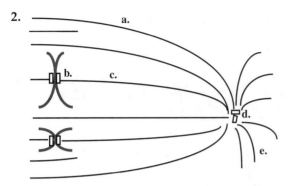

a. nonkinetochore microtubule: push poles apart by elongating and "walking" past microtubules from the opposite pole
b. kinetochore: protein and DNA structure in region of centromere where microtubules attach
c. kinetochore microtubule: move chromosomes to metaphase plate and separate chromosomes as motor proteins of kinetochores "walk" toward the pole and the microtubules disassemble

d. centrosome and centrioles: region of mitotic spindle formation; organizes microtubules
e. aster: radiating spindle fibers (in animal cells)

3. a. anaphase
  b. interphase
  c. late telophase
  d. metaphase

ANSWERS TO TEST YOUR KNOWLEDGE

Fill in the Blanks:

1. G_0	**5.** S phase	**9.** prometaphase
2. anaphase	**6.** metaphase	**10.** G_1 phase
3. prophase	**7.** telophase	
4. cytokinesis	**8.** prophase	

Multiple Choice:

1. d	**5.** b	**9.** b	**13.** e
2. c	**6.** c	**10.** a	**14.** d
3. d	**7.** c	**11.** c	
4. e	**8.** c	**12.** a	

CHAPTER 13: MEIOSIS AND SEXUAL LIFE CYCLES

INTERACTIVE QUESTIONS

13.1 a. 14; 7
  b. 28; 1
  c. 56. Sister chromatids are produced when a chromosome replicates. They are joined at the centromere and attached along their lengths. Nonsister chromatids are found on different chromosomes of a homologous pair (homologs).

13.2 a. meiosis
  b. fertilization
  c. zygote
  d. mitosis
  e. gametes
  f. fertilization
  g. zygote
  h. meiosis
  i. spores
  j. gametes
  k. fertilization
  l. $2n$
  m. zygote
  n. meiosis

13.3 a. metaphase II
  b. prophase I

  c. anaphase I
  d. interphase
  e. metaphase I
  f. anaphase II
  Proper sequence: d. b. e. c. a. f.

13.4 a. 2^{23}, approximately 8.4 million
  b. 70 trillion ($2^{23} \times 2^{23}$)
  c. This number of combinations does not take into account the additional variation of recombinant chromosomes produced by crossing over.

SUGGESTED ANSWERS TO STRUCTURE YOUR KNOWLEDGE

1. a. sister chromatids
  b. centromere
  c. pair of homologous chromosomes
  d. nonsister chromatids

2. a. Chromosome replication, sister chromatids attached at centromere and by sister chromatid cohesion
  b. Synapsis of homologous pairs (synaptonemal complex), crossing over evident at chiasmata

c. Homologous pairs line up independently at metaphase plate

d. Homologous pairs of chromosomes separate and homologs move toward opposite poles, sister chromatids remain attached at centromere

e. Haploid set of chromosomes, each consisting of two sister chromatids, aligns at metaphase

plate; sister chromatids not identical due to crossing over

f. Sister chromatids separate and move to opposite poles as chromosomes

3.

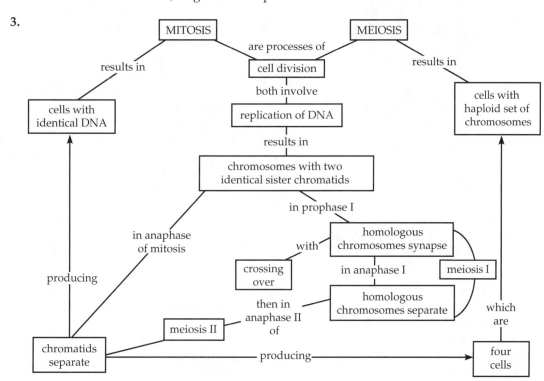

ANSWERS TO TEST YOUR KNOWLEDGE

Multiple Choice:

1. b	**4.** c	**7.** a	**10.** b	**13.** c	**16.** b	**19.** e
2 d	**5.** e	**8.** c	**11.** a	**14.** d	**17.** d	**20.** b
3. e	**6.** b	**9.** b	**12.** b	**15.** e	**18.** b	

CHAPTER 14: MENDEL AND THE GENE IDEA

INTERACTIVE QUESTIONS

14.1 a. *R*
  b. *r*
  c. F$_1$ Generation
  d. *Rr*
  e. F$_2$ Generation
  f. *R*
  g. *r*
  h. *Rr*
  i. *Rr*
  j. *rr*
  k. 3 round:1 wrinkled
  l. 1 *RR*:2*Rr*:1*rr*

14.2 a. all tall (*Tt*) plants
  b. 1:1 tall (*Tt*) to dwarf (*tt*)

You can use a Punnett square to determine the expected outcome of a test cross, but a shortcut is to use only 1 column for the recessive individual's gametes, since they produce only one type of gamete.

14.3 a. all tall purple plants
  b. *TtPp*
  c. *TP, Tp, tP, tp*

d.

sperm

	$\frac{1}{4}$ TP	$\frac{1}{4}$ Tp	$\frac{1}{4}$ tP	$\frac{1}{4}$ tp
$\frac{1}{4}$ TP	TTPP	TTPp	TtPP	TtPp
$\frac{1}{4}$ Tp	TTPp	TTpp	TtPp	Ttpp
$\frac{1}{4}$ tP	TtPP	TtPp	ttPP	ttPp
$\frac{1}{4}$ tp	TtPp	Ttpp	ttPp	ttpp

(eggs)

Phenotype	Genotype	Ratio
Black	M_B_	$\frac{3}{4} \times \frac{3}{4} = \frac{9}{16}$
Brown	M_bb	$\frac{3}{4} \times \frac{1}{4} = \frac{3}{16}$
White	mm_	$\frac{1}{4} \times 1 = \frac{1}{4}$ or $\frac{4}{16}$

e. 9 tall purple:3 tall white:3 dwarf purple:1 dwarf white

f. 12:4 or 3:1 tall to dwarf; 12:4 or 3:1 purple to white

14.4 Dihybrids produce four types of gametes. The probability of getting both recessive alleles in a gamete is $\frac{1}{4}$, and for two such gametes to join is $\frac{1}{4} \times \frac{1}{4}$, or $\frac{1}{16}$.

14.5 a. Consider the outcome for each gene as a monohybrid cross. The probability that a cross of $Aa \times Aa$ will produce an $A\_$ offspring is $\frac{3}{4}$. The probability that a cross of $Bb \times bb$ will produce a $B\_$ offspring is $\frac{1}{2}$. The probability that a cross of $cc \times CC$ will produce a $C\_$ offspring is 1. To have all of these events occur simultaneously, multiply their probabilities: $\frac{3}{4} \times \frac{1}{2} \times 1 = \frac{3}{8}$.

b. Offspring could be $A\_bbC\_$, $aaB\_C\_$, or $A\_B\_C\_$. The genotype A_B_cc is not possible. Can you see why?

Probability of $A\_bbC\_ = \frac{3}{4} \times \frac{1}{2} \times 1 = \frac{3}{8}$

Probability of $aaB\_C\_ = \frac{1}{4} \times \frac{1}{2} \times 1 = \frac{1}{8}$

Probability of $A\_B\_C\_ = \frac{3}{4} \times \frac{1}{2} \times 1 = \frac{3}{8}$

Probability of offspring showing at least two dominant traits is the sum of these independent probabilities, or $\frac{7}{8}$.

c. There is only one type of offspring that can show only one dominant trait ($aabbCc$). Can you see why? Its probability is $\frac{1}{4} \times \frac{1}{2} \times 1 = \frac{1}{8}$.

14.6 a. A: $I^A I^A$ and $I^A i$

b. B: $I^B I^B$ and $I^B i$

c. AB: $I^A I^B$

d. O: ii

14.7 The ratio of offspring from this $MmBb \times MmBb$ cross would be 9:3:4, a common ratio when one gene is epistatic to another. All epistatic ratios are modified versions of 9:3:3:1.

14.8 a. The parental cross produced 25-cm tall F_1 plants, all $AaBbCc$ plants with 3 units of 5 cm added to the base height of 10 cm.

b. As a general rule in the polygenic inheritance of a quantitative character, the number of phenotypic classes resulting from a cross of heterozygotes equals the number of alleles involved plus one. In this case, 6 alleles ($AaBbCc$) + 1 = 7. So, there will be 7 different phenotypic classes in the F_2 among the 64 possible combinations of the 8 types of F_1 gametes. (See why you wouldn't want to go through a Punnett square to figure that out!) These 7 classes will go from 6 dominant alleles (40 cm), 5 dominant (35 cm), 4 dominant (30 cm), and so on, to all 6 recessive alleles (10 cm).

14.9 a. This trait is recessive. If it were dominant, then albinism would be present in every generation, and it would be impossible to have albino children with two nonalbino (homozygous recessive) parents.

b. father Aa; mother Aa, because neither parent is albino and they have albino offspring (aa)

c. mate 1 AA (probably); mate 2 Aa; grandson 4 Aa

d. The genotype of son 3 could be AA or Aa. If his wife is AA, then he could be Aa (both his parents are carriers) and the recessive allele never would be expressed in his offspring. Even if he and his wife were both carriers (heterozygotes), there would be a $\frac{243}{1024}$ ($\frac{3}{4} \times \frac{3}{4} \times \frac{3}{4} \times \frac{3}{4} \times \frac{3}{4}$) or 24% chance that all five children would be normally pigmented.

14.10 a. $\frac{1}{4}$

b. $\frac{2}{3}$. Of offspring with a normal phenotype, $\frac{2}{3}$ would be predicted to be heterozygotes and, thus, carriers of the recessive allele.

14.11 Both sets of prospective grandparents must have been carriers. The prospective parents do not have the disorder so they are not homozygous recessive. Thus, each has a $\frac{2}{3}$ chance of being a heterozygote carrier. The probability that both parents are carriers is $\frac{2}{3} \times \frac{2}{3} = \frac{4}{9}$; the chance that two heterozygotes will have a recessive homozygous child is $\frac{1}{4}$. The overall chance that a child will inherit the disease is

$\frac{4}{9} \times \frac{1}{4} = \frac{1}{9}$. Should this couple have a baby that has the disease, this would establish that they are both carriers, and the chance that a subsequent child would have the disease is $\frac{1}{4}$.

SUGGESTED ANSWERS TO STRUCTURE YOUR KNOWLEDGE

1. Mendel's law of segregation occurs in anaphase I, when alleles segregate as homologs move to opposite poles of the cell. The two cells formed from this division have one-half the number of chromosomes and one copy of each gene (although sister chromatids still have to separate in anaphase II). Mendel's law of independent assortment relates to the lining up of attached homologous chromosomes at the equatorial plate in a random fashion during metaphase I. Genes on different chromosomes will assort independently into gametes.

2. *Aa* = 2 different gametes   *AaBb* = 4 gametes
   AaBbCc = 8 gametes   *AABbCc* = 4 gametes
   A general formula is 2^n, where *n* is the number of gene loci that are heterozygous.

3. Always look to the F_1 heterozygote. If alleles show complete dominance/recessiveness, then the heterozygote will have a phenotype identical to one parent. In incomplete dominance, the F_1 phenotype will be intermediate between that of the parents. If codominant, both alleles will be expressed in the heterozygote.

ANSWERS TO GENETICS PROBLEMS

1. White alleles are dominant to yellow alleles. If yellow were dominant, then you should be able to get white squash from a cross of two yellow heterozygotes.

2. **a.** $\frac{1}{4}$ [$\frac{1}{2}$ (to get AA) $\times \frac{1}{2}$ (bb)]
   b. $\frac{1}{8}$ [$\frac{1}{4}$ (aa) $\times \frac{1}{2}$ (BB)]
   c. $\frac{1}{2}$ [1 (Aa) $\times \frac{1}{2}$ (Bb) $\times$ 1 (Cc)]
   d. $\frac{1}{32}$ [$\frac{1}{4}$ (aa) $\times \frac{1}{4}$ (bb) $\times \frac{1}{2}$ (cc)]

3. Since flower color shows incomplete dominance, use symbols such as C^R and C^W for those alleles. There will be six phenotypic classes in the F_2 instead of the normal four classes found in a 9:3:3:1 ratio. You could find the answer with a Punnett square, but multiplying the probabilities of the monohybrid crosses is more efficient.

Tall red	$T\_C^RC^R$	$\frac{3}{4} \times \frac{1}{4} = \frac{3}{16}$
Tall pink	$T\_C^RC^W$	$\frac{3}{4} \times \frac{1}{2} = \frac{3}{8}$ or $\frac{6}{16}$
Tall white	$T\_C^WC^W$	$\frac{3}{4} \times \frac{1}{4} = \frac{3}{16}$
Dwarf red	$tt\, C^RC^R$	$\frac{1}{4} \times \frac{1}{4} = \frac{1}{16}$
Dwarf pink	$tt\, C^RC^W$	$\frac{1}{4} \times \frac{1}{2} = \frac{1}{8}$ or $\frac{2}{16}$
Dwarf white	$tt\, C^WC^W$	$\frac{1}{4} \times \frac{1}{4} = \frac{1}{16}$

4. Determine the possible genotypes of the mother and child. Then find the blood groups for the father that could not have resulted in a child with the indicated blood group.
   a. no groups exonerated   **d.** AB only
   b. A or O   **e.** B or O
   c. A or O

5. The parents are *CcBb* and *Ccbb*. Right away you know that all *cc* offspring will die and no *BB* black offspring are possible because one parent is *bb*. Only four phenotypic classes are possible. Determine the proportion of each type by applying the law of multiplication.

Lethal (*cc_ _*)	1/4 of embryos do not develop
Normal brown (*CCBb*)	$\frac{1}{4} \times \frac{1}{2} = \frac{1}{8}$
Normal white (*CCbb*)	$\frac{1}{4} \times \frac{1}{2} = \frac{1}{8}$
Deformed brown (*CcBb*)	$\frac{1}{2} \times \frac{1}{2} = \frac{1}{4}$
Deformed white (*Ccbb*)	$\frac{1}{2} \times \frac{1}{2} = \frac{1}{4}$
Ratio of viable offspring:	1:1:2:2

6. Father's genotype must be *Pp* since polydactyly is dominant and he has had one normal child. Mother's genotype is *pp*. The chance of the next child having normal digits is $\frac{1}{2}$ or 50% because the mother can only donate a *p* allele and there is a 50% chance that the father will donate a *p* allele.

7. The genotypes of the puppies were $\frac{3}{8}$ *B_S_*, $\frac{3}{8}$ *B_ss*, $\frac{1}{8}$ *bbS_*, and $\frac{1}{8}$ *bbss*. Because recessive traits show up in the offspring, both parents had to have had at least one recessive allele for both genes. Black:chestnut occurs in a 6:2 or 3:1 ratio, indicating a heterozygous cross. Solid:spotted occurs in a 4:4 or 1:1 ratio, indicating a cross between a heterozygote and a homozygous recessive. Parental genotypes were *BbSs* × *Bbss*.

8. The 1:1 ratio of the first cross indicates a cross between a heterozygote and a homozygote. The second cross indicates that the hairless hamsters cannot have a homozygous genotype, because then only hairless hamsters would be produced. Let us say that hairless hamsters are *Hh* and normal-haired are *HH*. The second cross of heterozygotes should yield a 1:2:1 genotype ratio. The 1:2 ratio indicates that the homozygous recessive genotype may be lethal, with embryos that are *hh* never developing.

9. a. Tail length appears to be a quantitative character. A general rule for the number of alleles in a cross involving a quantitative character is one less than the number of phenotypic classes. Here there are five phenotypic classes, thus four alleles or two gene pairs. In this cross, the phenotypic ratio is approximately 1:4:6:4:1. The ratio base of 16 also indicates a dihybrid cross. Each dominant gene appears to add 5 cm in tail length. The 15-cm pigs are double recessives; the 35-cm pigs are double dominants.

b. A cross of a 15-cm and a 30-cm pig would be equivalent to *aabb* × *AABb*. Offspring would have either one or two dominant alleles and would have tail lengths of 20 or 25 cm.

10. a. The black parent would be *CC^ch*. The Himalayan parent could be either *C^hC^h* or *C^hc*. First list the alleles in order of dominance: *C*, *C^ch*, *C^h*, *c*. To approach this problem you should write down as much of the genotype as you can be sure about for each phenotype involved. The parents were black (*C_*) and Himalayan (*C^h_*). The offspring genotypes would be black (*C_*) and Chinchilla (*C^ch_*). Right away you can tell that the black parent must have been *CC^ch* because half of the offspring are Chinchilla; *C^ch* is dominant over *C^h* and *c* and could not have been present in the Himalayan parent. The genotype of this parent could be either *C^hC^h* or *C^hc*. A testcross with a white rabbit would be necessary to determine this genotype.

b. You cannot definitely determine the genotype of the parents in the second cross. You can, however, eliminate some genotypes. The black parent could not be *CC* or *CC^ch*, because both the *C* and *C^ch* alleles are dominant to *C^h*, and Himalayan offspring were produced. The Chinchilla parent could not be *C^chC^ch* for the same reason. One or both parents had to have *C^h* as their second allele; one, but not both, might have *c* as their second allele.

11. a. First figure out possible genotypes: _ _E_ = golden (any *B* combination with at least 1 *E*), *B_ee* = black, *bbee* = brown. All you know at the start is that both parents are _ _E_. Since you see black and brown offspring, you know that both parents had to be heterozygous for *Ee*, and at least one was heterozygous for *Bb* (to get black and brown offspring). Since black and brown are in a 1:1 ratio (like a testcross ratio), then one dog was *Bb* and the other was *bb*. You need to consider the results from the second cross to know which dog was *Bb*.

b. For the second cross, you now know that Dog 2 is *?bEe* and Dog 3 is *B?ee*. A ratio of approximately 3:1 black to brown looks like a cross of heterozygotes, so both parents must be *Bb*. So Dog 3 is *Bbee*, and Dog 2 must be *BbEe*. That means that Dog 1 must be *bbEe*.

12.

(At least one of these grandparents must have been *Tt*)

13. Since the parents were true-breeding for two characters, the F$_1$ would be dihybrids. Since all F$_1$ had red, terminal flowers, those two traits must be dominant and their genotypes could be represented as *RrTt*. One would predict an F$_2$ phenotypic ratio of 9 red, terminal:3 red, axial:3 white, terminal:1 white, axial. If 100 offspring were counted, one would expect approximately 19($^3/_{16}$ × 100) plants with red, axial flowers.

14. Because the F$_2$ generation is a ratio based on 16, one can conclude that two genes are involved in the inheritance of this flower color character. The parent plants were probably double homozygotes (*AABB* for red and *aabb* for white). The all-red F$_1$ must be *AaBb*, and the red color must require at least one dominant allele of both genes. The 9:6:1 ratio of the F$_2$ indicates that 9 plants had at least one dominant allele of both genes to produce the red color; 1 plant was doubly homozygous recessive and thus white; and the 6 pale purple were caused by having at least one dominant allele of only one of the two genes involved in flower color (*A_bb or aaB_*).

ANSWERS TO TEST YOUR KNOWLEDGE

Matching:

1. I	**3.** F	**5.** L	**7.** H	**9.** C
2. A	**4.** E	**6.** J	**8.** D	**10.** K

Multiple Choice:

1. c	**5.** b	**9.** b	**13.** d
2. a	**6.** d	**10.** c	**14.** d
3. c	**7.** c	**11.** d	**15.** c
4. d*	**8.** d	**12.** a	**16.** b

* There are three different ways to get this outcome: HHT, HTH, THH. Each outcome has a probability of $^1/_8$.

CHAPTER 15: THE CHROMOSOMAL BASIS OF INHERITANCE

INTERACTIVE QUESTIONS

15.1

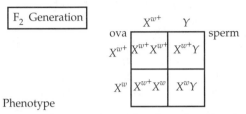

F_2 Generation		

ova / sperm: X^{w^+} and Y

	X^{w^+}	Y
X^{w^+}	$X^{w^+}X^{w^+}$	$X^{w^+}Y$
X^w	$X^{w^+}X^w$	X^wY

Phenotype

red-eyed female	red-eyed female	red-eyed male	white-eyed male

Genotype

$X^{w^+}X^{w^+}$	$X^{w^+}X^w$	$X^{w^+}Y$	X^wY

15.2 The gene is sex-linked, so a good notation is X^N, X^n, and Y so that you will remember that the Y does not carry the gene. Capital N indicates normal sight. Genotypes are:

1. X^NY
2. X^NX^n
3. X^NX^N or X^NX^n (probably X^NX^N since four sons are X^NY)
4. X^NX^n
5. X^NY
6. X^nY
7. X^NX^n

15.3 a. tall, purple-flowered and dwarf, white-flowered
b. tall, white-flowered and dwarf, purple-flowered

15.4 If linked genes have their loci close together on the same chromosome, they travel together during meiosis and *more* parental offspring are produced. Recombinants are the result of crossing over between nonsister chromatids of homologous chromosomes.

15.5 Solving a linkage problem is often a matter of trial and error. Sometimes it helps to lay out the loci with the greatest distance between them and fit the other genes between or on either side by adding or subtracting map unit distances.

```
                  ——— 15 ———
  m —— 9 —— j —— 6 —— 1 —— 6 —— k
                ——— 12 ———
```

15.6 a. An organism with a trisomy ($2n + 1$) has an extra copy of one chromosome, usually caused by a nondisjunction during meiosis. A triploid organism ($3n$) has an extra set of chromosomes,

possibly caused by a total nondisjunction in gamete formation.
b. The genetic balance caused by the trisomy would be more disrupted than that of an organism with a complete extra set of chromosomes.

15.7 translocation, deletion, and inversion

15.8 Aneuploidies of sex chromosomes appear to upset genetic balance less, perhaps because relatively few genes are located on the Y chromosome and extra X chromosomes are inactivated as Barr bodies.

SUGGESTED ANSWERS TO STRUCTURE YOUR KNOWLEDGE

1. Genes that are not linked assort independently, and the ratio of offspring from a testcross with a dihybrid heterozygote should be 1:1:1:1 (*AaBb* × *aabb* gives *AaBb*, *Aabb*, *aaBb*, *aabb* offspring and a recombination frequency of 50%—50% of offspring are recombinants). Genes that are linked and do not cross over should produce a 1:1 ratio in this testcross (*AaBb* and *aabb* because a heterozygote derived from a parental cross of *AABB* × *aabb* produces only *AB* and *ab* gametes). If crossovers between distant genes almost always occur, the heterozygote will produce equal quantities of *AB*, *Ab*, *aB*, and *ab* gametes, and the genotype ratio of offspring will be 1:1:1:1, the same as it is for unlinked genes. Because each 1% recombination frequency is equal to 1 map unit, this measurement ceases to be meaningful at relative distances of 50 or more map units. However, crosses with intermediate genes on the chromosome could establish both that the genes *A* and *B* are on the same chromosome and that they are a certain map unit distance apart.

2. A cross between a mutant female fly and a normal male should produce all normal females (who get a wild-type allele from their father) and all mutant male flies (who get the mutant allele on the X chromosome from their mother). $X^mX^m \times X^{m^+}Y$ produces $X^{m^+}X^m$ and X^mY offspring (normal females and mutant males).

3. The serious phenotypic effects that are associated with these chromosomal alterations indicate that normal development and functioning are dependent on genetic balance. Most genes

appear to be vital to an organism's existence, and extra copies of genes upset genetic balance. Inversions and translocations, which do not disrupt the balance of genes, can alter phenotype because of the effect of neighboring genes on gene functioning.

ANSWERS TO GENETICS PROBLEMS

1. **a.** The trait is recessive and probably sex-linked. Two sets of unaffected parents in the second generation have offspring with the trait, indicating that it must be recessive. More males than females display the trait. Females #3 and #5 must be carriers since their father has the trait, and they each pass it on to a son.
    b. Using the symbols X^T for the dominant allele and X^t for the recessive:

    1. X^tY
    2. X^TX^t
    3. X^TX^t
    4. X^TY
    5. X^TX^t
    6. X^TX^t or X^TX^T
    7. X^TX^t

    c. Individual #6 has a brother who has the trait and her mother's father had the trait, so her mother must be a carrier of the trait, meaning there is a $\frac{1}{2}$ probability that #6 is a carrier. Since she is mated to a phenotypically normal male, none of her daughters will show the trait (0 probability). Her sons have a $\frac{1}{2}$ chance of having the trait if she is a carrier. There is a $\frac{1}{2} \times \frac{1}{2} = \frac{1}{4}$ probability that her sons will have the trait. There is a $\frac{1}{2}$ chance that a child would be male, so the probability of an affected child is $\frac{1}{2} \times \frac{1}{4}$, or $\frac{1}{8}$.

2. c, a, b, d

3. The genes appear to be linked because the parental types appear most frequently in the offspring. Recombinant offspring represent 10 out of 40 total offspring for a recombination frequency of 25%, indicating that the genes are 25 map units apart.

4. One of the mother's X chromosomes carries the recessive lethal allele. One-half of male fetuses would be expected to inherit that chromosome and spontaneously abort. Assuming an equal sex ratio at conception, the ratio of girl to boy children would be 2:1, or six girls and three boys.

5. **a.** For female chicks to be black, they must have received a recessive allele from the male parent. If all female chicks are black, the male parent must have been Z^bZ^b. If the male parent was homozygous recessive, then all male offspring will receive a recessive allele, and the female parent would have to be Z^BW to produce all barred males.

    b. For female chicks to be both black and barred, the male parent must have been Z^BZ^b. If the female parent were Z^BW, only barred male chicks would be produced. To get an equal number of black and barred male chicks, the female parent must have been Z^bW.

ANSWERS TO TEST YOUR KNOWLEDGE

Multiple Choice:

1. e	**5.** a	**9.** a	**13.** c	**17.** e
2. a	**6.** e	**10.** e	**14.** a	**18.** d
3. b	**7.** b	**11.** d	**15.** e	**19.** c
4. d	**8.** c	**12.** b	**16.** d	**20.** d

CHAPTER 16: THE MOLECULAR BASIS OF INHERITANCE

INTERACTIVE QUESTIONS

16.1 a. They grew T2 with *E. coli* with radioactive sulfur to tag phage proteins.
b. T2 was grown with *E. coli* in the presence of radioactive phosphorus to tag phage DNA.
c. Radioactivity was found in the supernatant, indicating that the phage protein did not enter the bacterial cells.
d. In the samples with the labeled DNA, most of the radioactivity was found in the bacterial cell pellet.

e. They concluded that viral DNA is injected into the bacterial cells and serves as the hereditary material for viruses.

16.2 a. sugar-phosphate backbone
b. 3' end of chain
c. hydrogen bonds
d. cytosine
e. guanine
f. adenine (purine)
g. thymine (pyrimidine)
h. 5' end of chain

i. nucleotide
j. deoxyribose
k. phosphate group
l. 3.4 nm
m. 0.34 nm
n. 2 nm

16.3

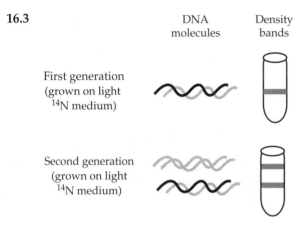

DNA molecules   Density bands

First generation (grown on light ^{14}N medium)

Second generation (grown on light ^{14}N medium)

16.4 The phosphate end of each strand is the 5′ end, and the hydroxyl group extending from the 3′ carbon of the sugar marks the 3′ end. See Figure 16.7 in the text.

16.5 a. single-strand binding protein
b. DNA pol III
c. leading strand
d. 5′ end of parental strand
e. 3′ end
f. helicase
g. RNA primer
h. primase
i. DNA pol III
j. Okazaki fragment
k. DNA pol I (replacing primer)
l. lagging strand
m. DNA ligase

16.6 No DNA nucleotides can be added after the RNA primer is removed because there is no 3′ end available for DNA polymerase. Thus, the daughter strand is shorter. Further rounds of replication will continue to produce shorter DNA molecules. See text Figure 16.19.

RNA Primer   Lagging Strand   Missing end

5′

3

3′ parental end

16.7 nucleosomes (10-nm fiber of nucleosomes and linker DNA); 30-nm fiber; looped domains (300-nm fiber); coiling and folding of looped domains into highly condensed metaphase chromosome

SUGGESTED ANSWERS TO STRUCTURE YOUR KNOWLEDGE

1. Watson and Crick used the X-ray diffraction photo of Franklin to deduce that DNA was a helix 2 nm wide, with nitrogenous bases stacked 0.34 nm apart, and making a full turn every 3.4 nm. Franklin had concluded that the sugar-phosphate backbones were on the outside of the helix with the bases extending inside. Using molecular models of wire, Watson and Crick experimented with various arrangements and finally paired a purine base with a pyrimidine base, which produced the proper diameter. Specificity of base pairing (A with T and C with G) is assured by hydrogen bonds.

2. Replication bubbles form where proteins recognize specific base sequences and open up the two strands. *Helicase*, an enzyme that works at the replication fork, unwinds the helix and separates the strands. *Single-strand binding proteins* support the separated strands while replication takes place. *Topoisomerase* eases the twisting ahead of the replication fork by breaking, untwisting, and re-joining DNA strands. *Primase* synthesizes a primer of about 10 RNA bases to start the new strand. After a proper base pairs up on the exposed template, *DNA polymerase III* joins the nucleotide to the 3′ end of the new strand. On the lagging strand, short Okazaki fragments are formed by primase and polymerase (again moving 5′ → 3′). *DNA polymerase I* replaces the primer with DNA, adding to the 3′ end of the leading strand or fragments. *Ligase* joins the 3′ end of one fragment to the 5′ end of its neighbor. Proofreading enzymes check for mispaired bases, and nucleases, DNA polymerase, and other enzymes repair damage or mismatches.

ANSWERS TO TEST YOUR KNOWLEDGE

Multiple Choice:

1. c	**6.** a	**11.** e	**16.** b	**21.** d
2. a	**7.** b	**12.** d	**17.** c	**22.** c
3. b	**8.** e	**13.** e	**18.** e	
4. d	**9.** c	**14.** a	**19.** c	
5. b	**10.** a	**15.** d	**20.** a	

CHAPTER 17: FROM GENE TO PROTEIN

INTERACTIVE QUESTIONS

17.1 DNA $\xrightarrow{\text{transcription}}$ RNA $\xrightarrow{\text{translation}}$ protein

17.2 Met Pro Asp Phe Lys stop

17.3 **a.** Initiation: Transcription factors bind to promoter and facilitate the binding of RNA polymerase II, forming a transcription initiation complex; RNA polymerase II separates DNA strands at initiation site.
b. Elongation: RNA polymerase II moves along DNA strand, connecting RNA nucleotides that have paired to the DNA template to the 3′ end of the growing RNA strand.
c. Termination: After polymerase transcribes past a polyadenylation signal sequence, the pre-mRNA is cut and released.

17.4 A 5′ cap consisting of a modified guanine nucleotide is added to the 5′ UTR. A poly-A tail consisting of up to 250 adenine nucleotides is attached to the 3′ UTR. Spliceosomes have cut out the introns and spliced the exons together.

17.5

DNA Triplet 3′ → 5′	mRNA Codon 5′ → 3′	Anticodon 3′ → 5′	Amino Acid
TAC	AUG	UAC	methionine
GGA	CCU	GGA	proline
TTC	AAG	UUC	lysine
ATC	UAG	AUC	stop

17.6 **1.** Codon recognition: An elongation factor (not shown) helps an aminoacyl tRNA into the A site where its anticodon base-pairs to the mRNA codon; two GTP increase accuracy and efficiency.
2. Peptide bond formation: Ribosome catalyzes peptide bond formation between new amino acid and polypeptide held in the P site.
3. Translocation: The empty tRNA in the P site is moved to the E site and released; the tRNA now holding the polypeptide is moved from the A to the P site, taking the mRNA with it; one GTP is required.
4. Termination: Release factor binds to stop codon in the A site. Free polypeptide is released from the P site. Ribosomal subunits and other assembly components separate.

a. amino end of growing polypeptide	**g.** peptide bond formation
b. aminoacyl tRNA	**h.** E site
c. large subunit	**i.** release factor
d. A site	**j.** stop codon
e. small subunit	**k.** P site of
f. 5′ end of mRNA	**l.** free polypeptide

17.7 A ribosome that is translating an mRNA that codes for a secretory or membrane protein will become bound to the ER when an SRP binds to the initial signal peptide and then to an ER receptor protein.

17.8 **a.** Silent: a base-pair substitution producing a codon that still codes for the same amino acid.
b. Missense: a base-pair substitution or frameshift mutation that results in a codon for a different amino acid.
c. Nonsense: a base-pair substitution or frameshift mutation that creates a stop codon and prematurely terminates translation.
d. Frameshift: an insertion or deletion of one, two, or more than three nucleotides that disrupts the reading frame and creates extensive missense and nonsense mutations.

SUGGESTED ANSWERS TO STRUCTURE YOUR KNOWLEDGE

1. Messenger RNA (mRNA) carries the code from DNA that specifies an amino acid sequence to ribosomes, where the RNA's sequence of nucleotides is translated into a polypeptide. Transfer RNA (tRNA) carries a specific amino acid to its position in a polypeptide based on matching its anticodon to an mRNA codon. Ribosomal RNA (rRNA) makes up about two-thirds of a ribosome and has specific binding and catalytic functions. Small nuclear RNA is part of spliceosomes and plays a catalytic role in splicing pre-mRNA.

The versatility of RNA molecules stems from their ability to base-pair with other RNA or DNA molecules, to form specific three-dimensional shapes by base-pairing within itself, and to act as a catalyst.

2.

	Transcription	Translation
Template	DNA	RNA
Location	nucleus (cytoplasm in prokaryotes)	cytoplasm; ribosomes can be free or attached to ER
Molecules involved	RNA nucleotides, DNA template strand, RNA polymerase, transcription factors	amino acids; tRNA; mRNA; ribosomes; ATP; GTP; enzymes; initiation, elongation, and release factors
Enzymes involved	RNA polymerases, spliceosomes (ribozymes)	aminoacyl-tRNA synthetase, ribosomal enzymes (ribozymes)
Control—start and stop	transcription factors locate promoter region with TATA box, polyadenylation signal sequence	initiation factors, initiation sequence (AUG), stop codons, release factor
Product	primary transcript (pre-mRNA)	polypeptide
Product processing	RNA processing: 5′ cap and poly-A tail, splicing of pre-mRNA—introns removed by snRNPs in spliceosomes	spontaneous folding, disulfide bridges, signal peptide removed, cleaving, quaternary structure, modification with sugars, etc.
Energy source	ribonucleoside triphosphate	ATP and GTP

3. The genetic code consists of the RNA triplets that code for amino acids. The order of these codons is specified by the sequence of nucleotides on DNA, which is transcribed into the codons found on mRNA and translated into their corresponding amino acids. There are 64 possible mRNA codons created from the four nucleotides used in the triplet code (4^3).

Redundancy of the code refers to the fact that several triplets may code for the same amino acid. Often these triplets differ only in the third nucleotide. The wobble phenomenon explains the fact that there are only about 45 different tRNA molecules that pair with the 61 possible codons (three codons are stop codons). The third nucleotide of many tRNAs can pair with more than one base. Because of the redundancy of the genetic code, these wobble tRNAs still place the correct amino acid in position.

The genetic code is nearly universal; each codon codes for exactly the same amino acid in almost all organisms. (Some exceptions have been found.) This universality points to an early evolution of the code in the history of life and the evolutionary relationship of all life on Earth.

4.

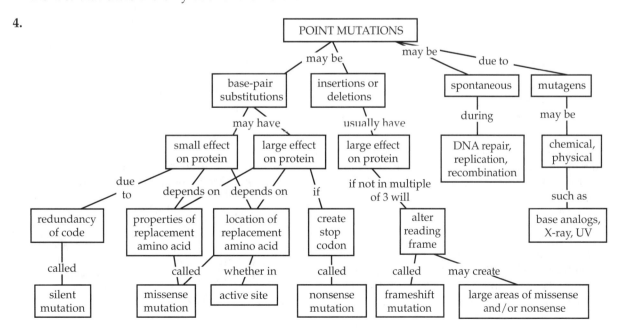

ANSWERS TO TEST YOUR KNOWLEDGE

Multiple Choice:

1. c	**4.** e	**7.** c	**10.** e	**13.** d	**16.** e	**19.** b	**22.** d
2. a	**5.** c	**8.** c	**11.** d	**14.** c	**17.** e	**20.** a	**23.** e
3. b	**6.** b	**9.** b	**12.** d	**15.** b	**18.** b	**21.** a	**24.** d

CHAPTER 18: REGULATION OF GENE EXPRESSION

INTERACTIVE QUESTIONS

18.1 a. regulatory gene
  b. promoter
  c. operator
  d. genes coding for enzymes
  e. operon
  f. RNA polymerase
  g. active repressor
  h. inducer (allolactose)
  i. mRNA for enzymes

18.2 a. anabolic; corepressor; on; inactive
  b. catabolic; inducers; off; active

18.3 a. Barr body—compacted X chromosome in cells of female
  b. Histone tail deacetylation would decrease transcription because it would make genes in the nucleosome less accessible.

18.4 a. distal control elements in enhancer
  b. activators
  c. DNA-bending protein
  d. promoter
  e. mediator proteins
  f. transcription factors
  g. TATA box
  h. RNA polymerase

18.5 a. Regulatory proteins may bind to sequences in the 5′ UTR and block attachment of ribosomes, thus decreasing gene expression. Sequences in the 3′ UTR may affect the length of time an mRNA remains intact, thus either increasing or decreasing gene expression.
  b. A cell marks a protein for destruction by attaching molecules of ubiquitin to the protein. Giant proteasomes chop up the marked proteins.

18.6 a. The primary miRNA transcript folds on itself by hydrogen bonding between complementary bases, forming loops called hairpins. Each hairpin is cut, trimmed by Dicer, and one strand is degraded. The single-stranded miRNA forms a complex with proteins. The complex binds to mRNA with complementary base sequences, and the mRNA is degraded or translation is blocked.
  b. MicroRNAs are coded for by RNA coding genes in a cell and processed as described in (**a**). Small interfering RNAs are longer, double-stranded RNA molecules that are introduced into a cell (by an experimenter or a virus), which are then processed by the cell and function like miRNAs.
  c. Evidence indicates that they do both. MicroRNAs affect translation by degrading mRNA or blocking translation. Several experiments have shown that small RNAs may silence transcription by changing chromatin structure.

18.7 a. A cell is said to be determined when its developmental fate is set. Its series of gene activations and inactivations has set it on the path to express the genes for tissue-specific proteins. When it produces these proteins and develops its characteristic structure, the cell has become differentiated.
  b. The action of MyoD must depend on a combination of regulatory proteins, some of which may be lacking in the cells that it is not able to transform into muscle cells.

18.8 Bicoid mRNA was shown to be localized at one end of the unfertilized egg; later in development, Bicoid protein was concentrated in the anterior cells of the embryo. Also, injection of bicoid mRNA into various regions of early embryos caused anterior structures to form at those sites.

18.9 a. Mutations may result in (1) more copies of the gene being present than normal (amplification); (2) translocation, which may bring the gene under the control of a more active promoter or control element; or (3) a change in a nucleotide sequence in either a promoter or enhancer that increases gene expression or in the gene that creates a more active or resilient protein.

b. Tumor-suppressor proteins may function in repair of damaged DNA, control of cell adhesion, or inhibition of the cell cycle.

SUGGESTED ANSWERS TO STRUCTURE YOUR KNOWLEDGE

1. **a.** operons
   b. promoter
   c. operator
   d. negative control
   e. repressor
   f. inactive
   g. active
   h. corepressor
   i. inducer
   j. anabolic
   k. activate
   l. inactivate
   m. catabolic
   n. cAMP
   o. lack of glucose

2. **a.** DNA packing into nucleosomes; histone tail acetylation increases, whereas deacetylation and methylation of tails decreases transcription; methylation of DNA may be involved in long-term inactivation of genes; small RNAs may promote heterochromatin
   b. Transcription factors (activators) bind with enhancers, then interact with mediator proteins and promoter region to form transcription initiation complex; repressor can inhibit transcription; steroid hormones or other chemical messages may bind with receptor proteins, producing transcription factors
   c. RNA processing (alternative splicing, 5′ cap and poly-A tail added); mRNA degradation by shortening of poly-A tail, removal of 5′ cap, and miRNA targeting
   d. Repressor proteins may prevent ribosome binding so mRNA can be stockpiled (awaiting fertilization in ovum); activation of initiation factors
   e. Protein processing by cleavage or modification; transport to target location; selective degradation by proteosomes of proteins marked with ubiquitin

3. Most cytoplasmic determinants are mRNA for transcription factors that are divided by the first few mitotic divisions. They are present in the cells and their translated product can enter the nucleus and regulate transcription. Inducers must communicate between cells. They are often proteins that bind to cell surface receptors and initiate a signal transduction pathway involving a cascade of enzyme activations, usually leading to the activation of transcription factors within the target cell.

ANSWERS TO TEST YOUR KNOWLEDGE

Multiple Choice:

1. b	**6.** b	**11.** e	**16.** e	**21.** b
2. c	**7.** e	**12.** b	**17.** e	**22.** e
3. e	**8.** c	**13.** a	**18.** a	
4. a	**9.** c	**14.** b	**19.** c	
5. b	**10.** e	**15.** d	**20.** d	

CHAPTER 19: VIRUSES

INTERACTIVE QUESTIONS

19.1 1. Phage attaches to host cell and injects DNA.
2. Phage DNA forms a circle. Certain factors determine which cycle is entered.
3. New phage DNA and proteins are synthesized and self-assemble into phages.
4. Bacterium lyses, releasing phages.
5. Phage DNA integrates into bacterial chromosome.
6. Bacterium reproduces, passing prophage to daughter cells.
7. Large population of infected bacteria forms.
8. Occasionally, prophage exits bacterial chromosome and begins lytic cycle.

a. phage DNA
b. bacterial chromosome
c. new phages
d. prophage
e. replicated bacterial chromosome with prophage

19.2 RNA → DNA → RNA; viral reverse transcriptase, host RNA polymerase

19.3 Viral particles spread easily through plasmodesmata, the cytoplasmic connections between plant cells. As there are no cures for plant viral diseases, reducing the spread of disease and breeding resistant varieties are the best approaches.

SUGGESTED ANSWERS TO STRUCTURE YOUR KNOWLEDGE

1.

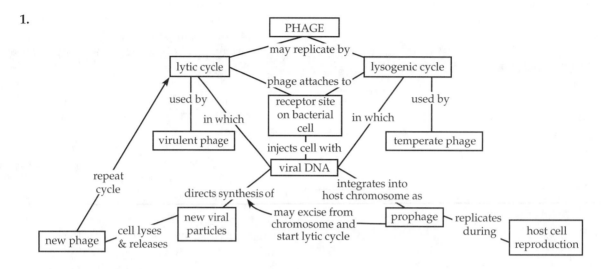

2. a. DNA
   b. RNA
   c. protein capsid
   d. host cell
   e. bacterium
   f. lytic or lysogenic cycle
   g. animal
   h. membraneous envelope
   i. reverse transcriptase
   j. plant
   k. viroids
   l. plasmids
   m. transposons

ANSWERS TO TEST YOUR KNOWLEDGE

Multiple Choice:

1. b	**4.** e	**7.** b	**10.** d
2. d	**5.** a	**8.** b	**11.** b
3. c	**6.** d	**9.** c	**12.** a

CHAPTER 20: BIOTECHNOLOGY

INTERACTIVE QUESTIONS

20.1 The third sequence, because it has the same sequence running in opposite directions. The enzyme would probably cut between G and A, producing AATT and TTAA sticky ends.

20.2 A genomic library contains copies of DNA segments from the entire genome. Thus, all genes should be represented, along with the regulatory sequences and introns. A cDNA library allows you to sequence only the exons of a gene, and also indicates which genes are expressed either in different cell types or at different stages of development in the same cell type.

20.3 a. bacterial plasmid
   b. *lacZ* gene
   c. restriction site
   d. *amp^R* (ampicillin resistance) gene
   e. gene of interest
   f. complementary sticky ends
   g. human DNA fragments
   h. recombinant plasmids
   i. recombinant bacteria
   j. plate with ampicillin and X-gal

1. Plasmid and human DNA are isolated.
2. Both DNAs are cut with the same restriction enzyme. Single cut in plasmid disrupts *lacZ* gene; multiple fragments of human DNA formed.
3. Fragments are mixed and some foreign fragments base-pair with plasmid. DNA ligase seals ends.
4. Recombinant plasmids transform bacteria with mutation in their *lacZ* gene.
5. Plate bacteria on agar containing ampicillin and X-gal.

Cells containing recombinant plasmid are identified by their ability to grow in presence of the antibiotic and by their white color. Blue colonies contain plasmids that resealed and thus have a functioning *lacZ* gene. Identify clones carrying the gene of interest with a nucleic acid probe.

20.4 a. By amplifying the gene prior to cloning, the later task of identifying clones carrying the desired gene is simplified.
b. There is a limit to the number of accurate copies that can be made due to the accumulation of relatively rare copying errors. Large quantities of a gene are better prepared by DNA cloning in cells.

20.5 Southern blotting procedure:
a. restriction enzyme treatment of samples
b. gel electrophoresis
c. DNA transfer by blotting onto membrane
d. hybridization with radioactive probe
e. autoradiography
   Crime samples contain blood from the victim and, presumably, the perpetrator. After identifying the bands from the victim, the remaining fragments match those of Suspect 2's genetic profile.

20.6 a. If similar sequences occur in known genes in other species, the function of a gene may be inferred.
b. RT-PCR: The mRNA from different tissue samples could be isolated. Reverse transcriptase would be used to make cDNA of all the mRNA, and PCR using specific primers could amplify only the gene of interest. Running the samples on a gel would show bands only in the tissues that were expressing the gene.
c. DNA microarray assay: As in RT-PCR, mRNA is isolated from different tissues, and cDNA is made and labeled with fluorescent dye. The cDNA is applied to a microarray (single-stranded DNA fragments of the genes of an organism arranged in a grid). The different cDNA will hybridize with the genes that were expressed in the tissue. The intensity with which the hybridized spots fluoresce indicates the relative amount of mRNA that was in the tissue.
d. Researchers analyzed the phenotypes of the worms that developed with each lacking gene and classified most of the genes into a few functional groups.

20.7 DNA methylation helps to regulate gene expression. An adult cell must have these epigenetic changes in its chromatin reprogrammed in order to support normal gene expression during development. The DNA of many cloned embryos has been found to be improperly methylated.

20.8 a. Before, members of the extended family had to be tested to determine the variant of the SNP or RFLP marker to which the disease-causing allele was linked. Then RFLP analysis was done with the patient's blood to determine which marker, and thus which allele, he or she had inherited. Now the cloned gene can be used as a probe, and hybridization analysis can detect the abnormal allele.
b. The major difficulty is to assure that proper control mechanisms are present so that the gene is expressed at the proper time, in the proper place, and to the proper degree. Insertion of the therapeutic gene must not harm other cell functions. Ethical considerations include whether genetic engineering should be done on germ cells, thus influencing the genetic makeup of future generations.

SUGGESTED ANSWERS TO STRUCTURE YOUR KNOWLEDGE

1. Agricultural applications: (1) improvement of the genomes of agricultural plants and animals to improve quality and yield, (2) "pharm" animals or plants to produce human proteins, (3) development of plant varieties that have genes for resistance to diseases, herbicides, and insects

   Medical applications: (1) development of vaccines, (2) diagnosis of genetic and infectious diseases, (3) treatment of genetic disorders, (4) production of insulin, human growth factor, TPA, and other useful products

2. **a.** Bacterial enzymes that cut DNA at restriction sites, creating "sticky ends" that can base-pair with other fragments. Use: make recombinant DNA, form restriction fragments used for many other techniques
   b. Mixture of molecules applied to gel in electric field; molecules separate, moving at different rates due to charge and size. Use: separate restriction fragments into pattern of distinct bands, fragments can be removed from gel and retain activity or can be identified with probes
   c. mRNA isolated from cell is treated with reverse transcriptase to produce a complementary DNA strand, and then a double-stranded DNA gene, minus introns and control regions. Use: creates genes that are easier to clone in bacteria, produces library of genes that are active in cell
   d. Radioactively or fluorescently labeled single-stranded DNA or mRNA used to base-pair with complementary sequence of DNA or RNA.

Use: locate gene in clone of bacteria, identify bands on gels, diagnose infectious diseases

e. DNA fragments separated by gel electrophoresis, transferred by blotting onto paper, labeled probe added, rinsed, autoradiography. Use: analyze DNA, genetic profile

f. Single-stranded DNA fragments are incubated with four nucleotides, DNA polymerase, and four labeled dideoxy nucleotides that interrupt synthesis, samples separated by size, sequence of nucleotides read from sequence of fluorescent tags. Use: determine nucleotide sequence

g. Polymerase chain reaction: DNA is mixed with heat-resistant DNA polymerase, nucleotides, and primers having complementary sequences for targeted DNA section and repeatedly heated to separate, cooled to pair with primers and replicate. Use: rapidly produce multiple copies of a gene or section of DNA *in vitro*

h. Restriction fragments analyzed by Southern blotting to compare different band patterns caused by DNA differences in restriction sites. Use: genetic profile for forensic use, diagnose genetic diseases

i. Fluorescently labeled cDNA is made from a cell's mRNA and applied to a DNA microarray of single-stranded DNA from many different genes attached to a glass grid. The intensity and location of fluorescence indicates gene expression in the cell. Use: test thousands of genes simultaneously to compare gene expression in different tissues or at different developmental stages or conditions.

j. Technique for determining expression of a particular gene: mRNA is isolated; cDNA is synthesized by reverse transcriptase and DNA polymerase; cDNA is amplified using primers specific for the gene; gel electrophoresis shows band in samples which had mRNA from that gene.

k. Technique for determining function of a gene: changes introduced in sequence of a cloned gene; mutated gene is inserted into a cell; phenotype of mutant is studied.

ANSWERS TO TEST YOUR KNOWLEDGE

Multiple Choice:

1. c	**6.** c	**11.** c	**16.** d	**21.** d
2. b	**7.** c	**12.** b	**17.** b	**22.** d
3. e	**8.** a	**13.** d	**18.** d	
4. e	**9.** a	**14.** a	**19.** a	
5. c	**10.** b	**15.** c	**20.** a	

CHAPTER 21: GENOMES AND THEIR EVOLUTION

INTERACTIVE QUESTIONS

21.1 The public consortium followed a hierarchy of three stages: (1) genetic (linkage) mapping that established about 200 markers/chromosome; (2) physical mapping that cloned and ordered smaller and smaller overlapping fragments (using YAC or BAC vectors for cloning fragments); and (3) DNA sequencing of each small fragment, followed by assembly of the overall sequence. The Celera whole-genome shotgun approach omitted the first two stages. Each chromosome was cut by several restriction enzymes into small fragments, which were sequenced, and powerful computers assembled the overlapping fragments to determine the overall sequence.

21.2 Comparing nucleotide sequences may reveal matches with other genes of known function and can identify similarities with closely or distantly related species that help trace a species' evolutionary history.

21.3 a. the archaean, *Archaeoglobus fulgidus*; humans
b. rice, *Oryza sativa*; the bacterium, *Haemophilus influenzae*

c. a plant, *Fritillaria assyriaca*; *H. influenzae*
d. 20,500; alternative splicing of exons can allow each gene to code for more than one polypeptide; post-translational processing can also alter polypeptides

21.4 They are first transcribed into RNA; thus, the original retrotransposon remains in place. Reverse transcriptase converts the RNA to DNA, which is inserted into another site in the genome.

21.5

1.	D	1.5
2.	H	24
3.	B	44
4.	A	3
5.	F	5
6.	C	15
7.	E	10
8.	G	17

21.6 The lysozyme gene, which codes for a bacterial infection-fighting enzyme, was present in the last common ancestor of birds and mammals. After their lineages split, the gene underwent a duplication event in the mammalian lineage, and a copy of the lysozyme gene evolved into a gene coding for a protein involved in milk production.

21.7 a. Exon shuffling can happen through an error in meiotic recombination, which may occur between two homologous transposable elements, or by the inclusion or tagging along of an exon with a transposable element, which moves a copy of the exon to a new location.
b. Exons often code for domains of a protein. Providing a new domain to a protein may enhance or change its function.

21.8 The homeobox codes for a DNA-binding homodomain, while other domains specific to each regulatory gene interact with transcription factors to recognize particular enhancers or promoters and thus control different batteries of developmental genes.

SUGGESTED ANSWERS TO STRUCTURE YOUR KNOWLEDGE

1. Many noncoding sequences are highly conserved across species, indicating that they perform some important, but as yet unidentified, function. As more data accumulates, the functions of these noncoding DNA sequences may become clear. One example is the newly recognized miRNAs that are involved in regulation of gene expression.
    a. Transposable elements include transposons, which are DNA segments, and retrotransposons, which are produced from RNA transcribed into DNA. These transposable elements are very common and move DNA to new locations in the genome.
    b. *Alu* elements are about 300 nucleotides long and make up about 10% of the human genome. Some are transcribed into RNA, but their function is unknown. They may provide alternate splice sites for RNA processing.
    c. L1 sequences are long retrotransposons that rarely move about. They are found in many introns and may help regulate gene expression.
    d. Simple sequence DNAs are highly repetitive tandem sequences found at centromeres and teleomeres of a chromosome. They appear to have structural functions in organizing chromatin.
    e. Pseudogenes are remnants of genes that are no longer functional because mutations have altered their regulatory sequences.

2. Chromosome duplication provides additional copies of genes that may undergo mutation and produce new proteins. Chromosomal rearrangements such as duplications, inversions, and translocations may create reproductive barriers between populations that lead to the formation of new species. Errors during meiosis can lead to duplications of genes or exons, or rearrangements of exons, providing genetic material that may take on related or novel functions. Transposable elements can facilitate recombination between different chromosomes, can disrupt genes or control elements, and can carry genes or exons to new locations.

ANSWERS TO TEST YOUR KNOWLEDGE

Multiple Choice:

1. c	**5.** d	**9.** a
2. b	**6.** d	**10.** b
3. c	**7.** e	**11.** e
4. a	**8.** d	**12.** d

CHAPTER 22: DESCENT WITH MODIFICATION: A DARWINIAN VIEW OF LIFE

INTERACTIVE QUESTIONS

22.1 a. 1. Ef
2. Ab
3. Be
4. C, d, g
5. Fa
6. Dc
7. Gg
b. 5, 1, 2, 4, 3, 7, 6

22.2 Observation 1: Members of a population vary greatly in their traits.
Observation 2: Traits are inherited from parent to offspring.
Observation 3: All species can produce more offspring than the environment can support.
Observation 4: Due to a lack of resources, many offspring do not survive.

22.3 Treatment with the drug 3TC prevents most HIV from reproducing when their enzyme reverse transcriptase inserts this cytosine mimic into its DNA, halting replication and production of new HIV. The 3TC-resistant HIV have a version of reverse transcriptase that discriminates between cytosine-bearing nucleotides and 3TC, and they are able to reproduce. With HIV's short generation time, it takes very little time for 3TC-resistant HIV to be strongly selected for by this drug and 100% of a patient's HIV population to be 3TC resistant.

22.4 a. biogeography
b. fossil record
c. homologies
d. island species and mainland species or neighboring species in different habitats
e. ancestral and transitional forms
f. homologous structures
g. functions and form
h. molecular comparisons
i. DNA and proteins
j. descent from a common ancestor

SUGGESTED ANSWERS TO STRUCTURE YOUR KNOWLEDGE

1. The two major components of Darwin's evolutionary theory are that all of life has descended

from common ancestral species and that this evolution has been the result of natural selection. The theory of natural selection is based on several key observations and inferences. Members of a population vary greatly in their traits, many of which are inherited from parent to offspring. The overproduction of offspring in conjunction with limited resources leads to the differential reproductive success of those organisms best suited to the local environment. The increased survival and reproduction of the best adapted individuals in a population leads to the gradual accumulation of favorable characteristics in a population.

ANSWERS TO TEST YOUR KNOWLEDGE

Multiple Choice:

1. c	**5.** b	**9.** d	**13.** c
2. e	**6.** a	**10.** d	**14.** c
3. a	**7.** e	**11.** c	**15.** b
4. e	**8.** d	**12.** d	

CHAPTER 23: THE EVOLUTION OF POPULATIONS

INTERACTIVE QUESTIONS

23.1 a. mutation, with some genetic recombination
b. sexual reproduction
c. Prokaryotes and viruses have very short generation times and a new beneficial mutation can increase in frequency rapidly in an asexually reproducing population. Although mutations are the source of new alleles, they are so infrequent that their contribution to genetic variation in a large, diploid population is minimal. However, the production and union of gametes produces zygotes with fresh combinations of alleles each generation.

23.2 a. The frequencies of genotypes are 0.49 *BB*, 0.42 *Bb*, and 0.09 *bb* ($^{98}/_{200}$, $^{84}/_{200}$, $^{18}/_{200}$).
b. The 98 *BB* mice contribute 196 *B* alleles, and the 84 *Bb* mice contribute 84 *B* alleles to the gene pool. These 84 *Bb* mice also contribute 84 *b* alleles, and the 18 *bb* mice contribute 36 *b* alleles. Of a total of 400 alleles, 280 are *B* and 120 are *b*. Allele

frequencies are 0.7 *B* and 0.3 *b*. Another way to determine allele frequencies is from genotype frequencies: Add the frequency of the homozygous dominant genotype to $^1/_2$ the frequency of the heterozygote for the frequency of *p*. For *q*, add the homozygous recessive frequency and $^1/_2$ the heterozygote frequency.

23.3 0.7; 0.3; 0.49; 0.42; 0.09

23.4 a. 0.36; 0.48; 0.16. Plug *p* (0.6) and *q* (0.4) into the expanded binomial: $p^2 + 2pq + q^2 = 1$.
b. 0.6; 0.4. Add the frequencies of *AA* + $^1/_2$ *Aa* to get *p*. Add the frequencies of *aa* + $^1/_2$ *Aa* to get *q*. Alternatively, to determine *q*, take the square root of the homozygous recessive frequency if you are sure the population is in Hardy-Weinberg equilibrium. The frequency of *p* is then 1 − *q*.

23.5 a. natural selection
b. genetic drift
c. gene flow

d. better reproductive success
e. small population
f. bottleneck effect
g. founder effect
h. genetic variation between populations

23.6 a. Diploidy—The sickle-cell allele is hidden from selection in heterozygotes.
b. Heterozygote advantage—Heterozygotes are resistant to malaria and have a selective advantage in areas where malaria is a major cause of death.

SUGGESTED ANSWERS TO STRUCTURE YOUR KNOWLEDGE

1. a. The Hardy-Weinberg principle states that allele frequencies within a population will remain constant from one generation to the next as long as only Mendelian segregation and sexual recombination of alleles are involved. An equilibrium in which neither allele nor genotype frequencies will change from generation to generation requires five conditions: The population is large, mutation and migration are negligible, mating is random, and no natural selection occurs. This Hardy-Weinberg equilibrium provides a null hypothesis to allow researchers to test for evolution.
b. $p^2 + 2pq + q^2 = 1$. In the Hardy-Weinberg equation, p and q refer to the frequencies of two alleles in the gene pool. The frequency of homozygous offspring is ($p \times p$) or p^2 and ($q \times q$) or q^2. Heterozygous individuals can be formed in two ways, depending on whether the ovum or sperm carries the p or q allele, and their frequency is equal to $2pq$.

2. Genetic variation is retained within a population by diploidy and balancing selection. Diploidy masks recessive alleles from selection when they occur in the heterozygote. Thus, less adaptive or even harmful alleles are maintained in the gene pool and are available should the environment change. Balancing selection maintains several alleles at a gene locus in a population. In situations in which there is heterozygote advantage, the two alleles will be retained in stable frequencies within the gene pool. Frequency-dependent selection, in which phenotypes present in higher numbers are selected against by predators or other factors, is another type of balancing selection.

ANSWERS TO TEST YOUR KNOWLEDGE

Multiple Choice:

1. e	**6.** d	**11.** e	**16.** b	**21.** d
2. a	**7.** e	**12.** c	**17.** c	
3. c	**8.** d	**13.** d	**18.** b	
4. c	**9.** b	**14.** a	**19.** b	
5. d	**10.** e	**15.** b	**20.** e	

CHAPTER 24: THE ORIGIN OF SPECIES

INTERACTIVE QUESTIONS

24.1 a. reduced hybrid fertility **b.** post-
c. gametic isolation **d.** pre-
e. mechanical isolation **f.** pre-
g. temporal isolation **h.** pre-
i. hybrid breakdown **j.** post-
k. behavioral isolation **l.** pre-
m. reduced hybrid viability **n.** post-
o. habitat isolation **p.** pre-

24.2 a. reproductive isolation
b. morphological
c. ecological
d. smallest group that shares a common ancestor, a single branch on the tree of life

24.3 a. 20 chromosomes
b. 24 chromosomes

24.4 a. In allopatric speciation, a new species forms while geographically isolated from the parent population. In sympatric speciation, some reproductive barrier isolates the gene pool of a subgroup of a population within the same geographic range as the parent population.
b. Reproductive barriers may evolve as a by-product of the genetic change associated with the isolated population's adaptation to a new environment, genetic drift, or sexual selection. Sympatric speciation by polyploidy is common in plants. A change in resource use or sexual selection may reproductively isolate a subset of an animal population.

24.5 a. Reinforcement: when hybrid offspring are less viable or reproduce less successfully, the reproductive barriers between the two species may strengthen. Traits that prevent or discourage cross-breeding will increase in frequency while traits that allow for hybridization will decrease in frequency as the hybrids carrying those traits produce fewer offspring.

b. Fusion: If hybrids are successful in mating with each other and with the parent species, reproductive barriers may weaken. Increased gene flow between the two species may eventually lead to their fusion into a single species.

c. Stability: Hybrids continue to be produced between the two species in the area of their overlap, but the gene pools of both parent species remain distinct. Several cases of such stability among hybridizing species have been documented.

24.6 The total time between speciation events has been estimated to range from 4,000 to 40 million years (with an average of 6.5 million years). Thus, even if the process of speciation is relatively rapid once it begins, it may be millions of years before that new species gives rise to another new species. The biodiversity of Earth may take a very long time to recover from the current increased extinction rate.

SUGGESTED ANSWERS TO STRUCTURE YOUR KNOWLEDGE

1. Speciation, which leads to an increase in biological diversity, is the process through which a new species evolves from a parent species. Microevolution involves changes in the gene pool of a population as a result of either chance events or natural selection. If the makeup of an isolated gene pool changes enough, microevolution may lead to speciation.

2. Reproductive barriers do not develop in order to create new species; they develop as a side consequence of the genetic changes that occur as a species adapts to a new environment or undergoes its own genetic drift. Gene flow between two species tends to break down reproductive barriers between them. Since gene flow is much less likely between allopatric species, one would predict that reproductive barriers are more likely to evolve. In the case of sympatric speciation by polyploidy, however, reproductive barriers arise in one generation.

3. Punctuated equilibria describes the common pattern seen in the fossil record in which species appear rather suddenly and then seem to change little for the rest of their existence. Thus, evolution seems often to occur in spurts of relatively rapid change interspersed with long periods of stasis.

ANSWERS TO TEST YOUR KNOWLEDGE

Multiple Choice:

1. c	**5.** c	**9.** c
2. d	**6.** c	**10.** a
3. d	**7.** b	**11.** d
4. e	**8.** e	**12.** e

CHAPTER 25: THE HISTORY OF LIFE ON EARTH

INTERACTIVE QUESTIONS

25.1 A cell needs metabolic machinery to provide the enzymes, energy, and building blocks to replicate its genetic information. This genetic information codes for that machinery. Thus both replication and metabolism are required for life.

25.2 17,190 years. Carbon-14 has a half-life of 5,730 years. In 5,730 years the ratio of C-14 to C-12 would be reduced by $\frac{1}{2}$; in 11,460 years it would be reduced by $\frac{1}{4}$; and in 17,190 years it would be reduced by $\frac{1}{8}$ (3 half-lives; $3 \times 5,730$).

25.3 The inner membranes of both mitochondria and chloroplasts have enzymes and transport systems homologous to those in the plasma membranes of modern prokaryotes. They replicate by a splitting process similar to that of some prokaryotes. They contain a single, circular DNA molecule that is not associated with histones. The ribosomes of mitochondria and plastids are more similar to prokaryotic ribosomes than to the cytoplasmic ribosomes of eukaryotic cells.

25.4 The change in nutrient procurement from herbivory, filter-feeding, and scavenging to predation: predators with claws and other prey-capturing adaptations as well as prey with heavy body armor appeared in the fossil record during the Cambrian explosion.

25.5 Marsupials probably migrated to Australia through South America and Antarctica while the continents were still joined. When Pangaea broke up, marsupials diversified on isolated Australia.

25.6 a. Early mammals were restricted in size and diversity because they were probably outcompeted or eaten by the larger and more diverse dinosaurs. After most dinosaurs became extinct, mammals diversified and filled the niches vacated by the dinosaurs.

b. The rise in predators in the Cambrian explosion, the diversification of terrestrial plants and vertebrates, the adaptive radiation of insects that ate or pollinated plants.

c. The Hawaiian Archipelago is a series of relatively young, isolated, and physically diverse islands whose thousands of endemic species are examples of adaptive radiation following multiple colonization and speciation events.

25.7 a. This study provided experimental evidence that a change in the sequence of a developmental gene could produce a major evolutionary change—the origin of the six-legged insect from a crustacean-like ancestor with many legs.

b. The coding sequence of the developmental gene was identical in the two populations; the gene was not expressed in the ventral spine region of developing lake sticklebacks. Thus, the loss of ventral spines results from a change in gene regulation, not from a change in the gene itself.

25.8 a. The successful species that last the longest before extinction and generate the most new species will determine the direction of an evolutionary trend, just as the individuals that produce the most offspring determine the direction of adaptation in a population.

b. Both microevolution and macroevolution are primarily driven by natural selection resulting from interactions between organisms and their environments. Evolutionary trends are ultimately dictated by environmental conditions; if conditions change, an evolutionary trend may end or change direction.

SUGGESTED ANSWERS TO STRUCTURE YOUR KNOWLEDGE

1. The abiotic synthesis of organic molecules; the abiotic synthesis of macromolecules; the spontaneous formation of protobionts; the origin of self-replicating molecules, probably RNA, that made inheritance and natural selection possible

2. **a.** Archaean **g.** Prokaryotes
    b. Proterozoic **h.** Atmospheric oxygen
    c. Phanerozoic **i.** Single-celled eukaryotes
    d. Paleozoic **j.** Multicellular eukaryotes
    e. Mesozoic **k.** Animals
    f. Cenozoic **l.** Colonization of land

3. *Continental drift:* When continents form one supercontinent, many habitats are destroyed or changed. As continents change latitude, their climates either warm or cool. The separation of continents creates major geographic separation events, making possible allopatric speciations. Indeed, continental drift helps explain much of biogeography.

   Mass extinctions empty many ecological roles, which may then be exploited by species that survived extinction.

   Adaptive radiations are multiple speciation events that fill newly formed or newly emptied niches. Thus, adaptive radiations may follow mass extinctions, the colonization of new regions, or the evolution of novel adaptations.

4. New morphological forms may evolve as a result of mutations in developmental genes or changes in the regulation of such genes. Such changes may affect the rate or timing of development or the spatial arrangement of body parts (e.g., *Hox* genes).

ANSWERS TO TEST YOUR KNOWLEDGE

Multiple Choice:

1. b	**5.** d	**9.** c	**13.** c
2. e	**6.** e	**10.** c	**14.** d
3. a	**7.** a	**11.** a	**15.** b
4. e	**8.** c	**12.** b	

CHAPTER 26: PHYLOGENY AND THE TREE OF LIFE

INTERACTIVE QUESTIONS

26.1 #1; #4; A and B

26.2 Adaptations to different environments may lead to large morphologic differences within related groups, and convergent evolution can produce similar structures in unrelated organisms in response to similar selective pressures. Examples are the varied Hawaiian silversword plants and the similar Australian and North American burrowing moles.

26.3 A C G T G C A C G
A   G T G   A G G

The second DNA segment shows two deletions and one base substitution.

26.4

26.5 The four-chambered heart of birds and mammals is analogous, not homologous. It evolved independently in the two groups. An abundance of evidence supports the hypothesis that birds and mammals evolved from different reptilian ancestors.

26.6 **a.** rRNA genes
b. mtDNA (mitochondrial DNA)
c. genes for olfactory receptors

26.7 Assuming that genes have a constant rate of mutation, in a protein that has a crucial function, mutations that are harmful would be quickly removed from the population. Therefore, fewer mutations would be neutral and remain in the genome. If the exact sequence of amino acids in a protein is less critical to survival, more mutations would be neutral and remain, and the sequence of the gene will change more than the sequence of the more crucial gene in the same amount of time.

SUGGESTED ANSWERS TO STRUCTURE YOUR KNOWLEDGE

1.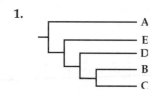

2. **a.** Bacteria;
b. Eukarya;
c. Archaea
#2 represents gene transfer between mitochondrial ancestor with ancestor of eukaryotes
#3 represents gene transfer between chloroplast ancestor with ancestor of green plants

ANSWERS TO TEST YOUR KNOWLEDGE

Multiple Choice:

1. c	**4.** d	**7.** a	**10.** d	**13.** d
2. e	**5.** c	**8.** e	**11.** a	**14.** b
3. b	**6.** d	**9.** b	**12.** e	

CHAPTER 27: BACTERIA AND ARCHAEA

INTERACTIVE QUESTIONS

27.1 **a.** spheres (cocci), rods (bacilli), or spirals
b. 0.5–5 μm in diameter
c. cell wall made of peptidoglycan; gram-negative bacteria also have outer lipopolysaccharide membrane; archaea lack peptidoglycan; sticky capsule for adherence; fimbriae and pili for attachment or genetic exchange
d. flagella; may show taxis to stimuli
e. infoldings of plasma membrane may be used in metabolic functions
f. circular DNA molecule with little associated protein found in nucleoid; may have plasmids with other genes
g. binary fission; rapid population growth; rapid adaptive evolution

27.2 **a.** circular chromosome
b. F plasmid, R plasmid
c. mutation

d. transformation
e. naked DNA
f. transduction
g. bacteriophage (phage)
h. conjugation
i. F^+ or HRF cell and F^- cell
j. adaptation to environment/evolution

27.3 **a.** endospore; thick-walled, resistant cell that is formed in response to a lack of nutrients and can persist for long periods of time, breaking dormancy when conditions are favorable; chemoheterotroph—it obtains both energy and carbon from organic compounds.
b. photoautotroph—uses light energy and CO_2 to synthesize organic compounds; the cell is a heterocyte, specialized for nitrogen fixation.

27.4 **a.** cyanobacteria
b. Genetic prospecting is a technique of sampling genetic material directly from the

environment. Before its development, researchers could study only those species that could be cultured in the lab. This technique has led to the discovery of new species and even new branches in prokaryotic phylogeny.

27.5 The photosynthesis of cyanobacteria brings carbon from CO_2 into the food chain and releases O_2. Cyanobacteria also fix atmospheric nitrogen, supplying other organisms with the nitrogen they need to make proteins.

SUGGESTED ANSWERS TO STRUCTURE YOUR KNOWLEDGE

1. Prokaryotes are adapted to live in the diverse habitats on (and in) Earth. They have a wide range of metabolic abilities, including the nutritional categories of chemoautotrophy and photoheterotrophy. Their genetic diversity, along with their short generation time, huge populations, and mechanisms of genetic exchange, enable rapid adaptation to changing environments. The formation of biofilms consisting of multiple species and their many symbiotic associations are further examples of the remarkable abilities of prokaryotes.

2. Decomposers—recycle nutrients; bioremediation: sewage treatment, clean up oil spills.
Nitrogen fixers—provide nitrogen to the soil by bacteria in nodules on legume roots or by cyanobacteria.
Mutualism—enteric bacteria, some digest food and produce vitamins.
Biotechnology—production of antibiotics, hormones, and many useful products.

ANSWERS TO TEST YOUR KNOWLEDGE

Multiple Choice:

1. b	4. e	7. e	10. e
2. c	5. a	8. b	
3. d	6. c	9. d	

Fill in the Blanks:

1. cocci	6. endospore
2. nucleoid	7. biofilms
3. Gram stain	8. exotoxins
4. fimbriae	9. bioremediation
5. taxis	10. anaerobic respiration

CHAPTER 28: PROTISTS

INTERACTIVE QUESTIONS

28.1 The four membranes indicate secondary endosymbiosis of a green alga (whose chloroplasts originated from the primary endosymbiont, a cyanobacterium). The inner two membranes belonged to the ancient cyanobacterium. The third membrane is from the engulfed green alga's plasma membrane and the outer membrane from the heterotrophic eukaryote's food vacuole. The vestigial nucleus indicates that this process occurred relatively recently. In more ancient secondary endosymbioses, many components of the engulfed alga have been lost.

28.2 Frequent global changes in the surface proteins of these parasites enable them to evade the host's immune system.

28.3 *Paramecium* is a ciliate in the clade Alveolata in the supergroup Chromalveolata.

   a. cilia
   b. oral groove
   c. food vacuoles, combine with lysosomes
   d. contractile vacuole

   e. anal pore
   f. macronucleus
   g. micronucleus

28.4 These seaweeds may have a holdfast that maintains a strong hold on intertidal rocks. Biochemical adaptations include cellulose cell walls containing gel-forming polysaccharides that protect the algae from the abrasive action of waves and reduce drying when exposed at low tides.

28.5 The calcareous tests of forams are long-lasting fossils in marine sediments and sedimentary rocks. Foram fossils are useful in correlating ages of sedimentary rocks in different areas.

28.6 a. diploid
   b. multinucleate mass, plasmodium
   c. amoeboid or flagellated cells that emerge from resistant spores
   d. haploid
   e. solitary amoeboid cells until aggregate to form fruiting body
   f. two amoebas fuse to form resistant zygote

SUGGESTED ANSWERS TO STRUCTURE YOUR KNOWLEDGE

1. **a.** cyanobacterium
    b. primary endosymbiosis
    c. red algae
    d. green algae
    e. secondary endosymbiosis
    f. dinoflagellates
    g. apicomplexans
    h. stramenopiles
    i. euglenids

ANSWERS TO TEST YOUR KNOWLEDGE

Matching:

1. C	**4.** G	**7.** A
2. F	**5.** H	**8.** D
3. B	**6.** E	

Multiple Choice:

1. d	**4.** d	**7.** b
2. a	**5.** c	**8.** e
3. c	**6.** d	**9.** b

CHAPTER 29: PLANT DIVERSITY I: HOW PLANTS COLONIZED LAND

INTERACTIVE QUESTIONS

29.1 Benefits: bright, unfiltered sunlight, plentiful CO_2 and soil minerals, few herbivores and pathogens. Challenges: lack of structural support, relative scarcity of water, and problem of dehydration.

29.2 a. gametes
b. mitosis
c. zygote
d. mitosis
e. sporophyte
f. spores
g. meiosis
h. gametophyte

29.3 Although lycophytes and pterophytes are both seedless plants, they do not form a monophyletic group. Pterophytes (ferns) share a more recent common ancestor with seed plants than with lycophytes.

29.4 a. gametophyte
b. archegonia
c. antheridia
d. swim
e. sporophyte
f. meiosis
g. capsule (sporangium)
h. gametophyte (germinate to form protonema)

29.5 a. meiosis
b. spore
c. gametophyte
d. antheridium
e. sperm
f. archegonium
g. egg
h. fertilization

i. zygote
j. gametophyte
k. young sporophyte
l. mature sporophyte
m. sporangia (in sori)
The gametophyte (upper portion of the diagram) is haploid; the sporophyte (lower portion) is diploid.

29.6 Lignified xylem provides the mechanical support for aerial growth. Vascular tissues convey water and minerals to photosynthesizing leaves and transport sugars throughout the plant. Roots both absorb water and minerals and anchor the plant.

SUGGESTED ANSWERS TO STRUCTURE YOUR KNOWLEDGE

1. alternation of generations with protected and nourished embryo, sporopollenin-walled spores produced in sporangia, apical meristems, multicellular gametangia, cuticle, secondary compounds (protection from UV radiation and herbivores), stomata, vascular tissue strengthened with lignin (vascular plants)

2. **a.** land plants
    b. nonvascular plants (bryophytes)
    c. vascular plants
    d. seedless vascular plants
    e. seed plants
    f. liverworts
    g. hornworts
    h. mosses
    i. pterophytes (ferns, horsetails, whisk ferns)
    j. gymnosperms
    k. angiosperms

ANSWERS TO TEST YOUR KNOWLEDGE

Multiple Choice:

1. c	**5.** d	**9.** d	**13.** b
2. e	**6.** b	**10.** e	**14.** e
3. d	**7.** d	**11.** c	
4. a	**8.** a	**12.** e	

CHAPTER 30: PLANT DIVERSITY II: THE EVOLUTION OF SEED PLANTS

INTERACTIVE QUESTIONS

30.1 Resistant, airborne, or animal-carried pollen grains can transport the male gametophyte, even over long distances, to the female gametophyte. A germinating pollen tube delivers sperm directly to an ovule, eliminating the need for a moist environment for sperm to swim to reach eggs.

30.2 a. integument (2n)
b. megasporangium (2n)
c. megaspore (n)
d. seed coat (from integument) (2n)
e. food supply (from female gametophyte tissue) (n)
f. embryo (new sporophyte) (2n)

30.3 a. The smaller pollen cones have many scalelike structures holding microsporangia. Microsporocytes (microspore mother cells) undergo meiosis to produce microspores that develop into pollen grains, the male gametophyte. Two sperm nuclei are eventually discharged from a pollen tube.
b. Ovulate cones have scales that hold two ovules, each containing a megasporangium. A megasporocyte (megaspore mother cell) undergoes meiosis to form four megaspores. The surviving megaspore develops into a multicellular female gametophyte containing a few archegonia, each with an egg.

30.4 a. sepals, petals, stamens, carpels
b. an embryo surrounded by food enclosed in a seed coat
c. mature ovary that may be modified to help disperse seeds

30.5 a. sepal
b. petal
c. anther (top of stamen)
d. stigma (top of carpel)
e. ovary
f. ovule with megasporocyte
g. microsporangium
h. microsporocytes
i. meiosis
j. microspore (n)
k. pollen grain (male gametophyte with generative cell and tube cell)
l. two sperm
m. surviving megaspore (n)
n. pollen tube
o. style
p. central cell with two nuclei
q. embryo sac (female gametophyte)
r. egg
s. fertilization
t. endosperm nucleus (3n)
u. zygote
v. seed coat
w. endosperm
x. embryo
y. germinating seed

30.6

Characteristic	Monocot	Eudicot
Number of cotyledons	one	two
Leaf venation	parallel	netlike
Vascular bundles in stems	scattered	in ring
Root system	fibrous	taproot
Flower parts	multiples of three	multiples of four or five
Openings in pollen grain	one	three

SUGGESTED ANSWERS TO STRUCTURE YOUR KNOWLEDGE

1. Phyla Ginkgophyta, Cycadophyta, Gnetophyta, and Coniferophyta

2. **a.** *Amborella*
   b. water lilies
   c. star anise and relatives

d. monocots
e. magnoliids
f. eudicots

3. reduced gametophytes develop from spores retained within the sporangia of the sporophyte; the female gametophyte protected and nourished in ovule retained on sporophyte plant; nonswimming sperm protected and dispersed in pollen; seed for protection, dispersal, and nourishment of embryo

4. flower as enclosed reproductive structure; animal pollinators; fruit for protection and dispersal of seeds

True or False:

1. False—add *or algae*.
2. True
3. False—change *gymnosperms* to *angiosperms*.
4. True
5. True
6. False—change *fruit* to *seed*, or change *an embryo . . .* to *a mature ovary*.
7. True
8. False—change *angiosperms* to *gymnosperms*, or change *haploid cells . . .* to *an embryo sac with a few haploid cells.*

ANSWERS TO TEST YOUR KNOWLEDGE

Multiple Choice:

1. e	**3.** a	**5.** c	**7.** b
2. b	**4.** d	**6.** c	**8.** e

CHAPTER 31: FUNGI

INTERACTIVE QUESTIONS

31.1 The fungus provides phosphate ions and other minerals to the plants; the plants supply the fungi with organic nutrients.

31.2 a. network of hyphae, typical body form
b. cross-walls between cells, with pores
c. multinucleated hyphae, aseptate
d. nuclei from two different parents exist together in hyphae
e. fusion of two parent hyphae
f. hyphae with two different nuclei paired up in cells
g. fusion of nuclei to form diploid state

31.3 a. Nucleariids
b. Fungi
c. Chytrids
d. Other fungi (zygomycetes, glomeromycetes, ascomycetes, basidiomycetes)

31.4 (1) ascomycete
(2) zygomycete
(3) basidiomycete

a. conidia (asexual spores)
b. ascocarp
c. ascospores
d. asci
e. sporangia
f. zygosporangium
g. basidiocarp
h. basidium
i. basidiospores

SUGGESTED ANSWERS TO STRUCTURE YOUR KNOWLEDGE

1. **a.** chytrids, aquatic and soil decomposers or parasites
b. flagellated zoospores
c. zygomycetes, molds and parasites; black bread mold; microsporidia
d. resistant zygosporangia in sexual stage; asexual spores in sporangium on aerial hyphae
e. glomeromycetes; all form mycorrhizae
f. 90% of all plants have arbuscular mycorrhizae
g. ascomycetes; sac fungi, yeasts, molds, morels, mycorrhizae
h. dikaryotic hyphae in ascocarp, karyogamy in asci, ascospores; asexual spores called conidia at tips of hyphae
i. basidiomycetes; club fungi, mushrooms, shelf fungi, rusts
j. fruiting body called basidiocarp, basidiospores produced in basidia

2. The role of fungi as decomposers of organic material is central to the recycling of chemicals between living organisms and their physical environment. Parasitic fungi have economic and health effects on humans. Plant pathogens cause

extensive loss of food crops and trees. Mycorrhizae are found on the roots of almost all vascular plants. These mutualistic associations greatly benefit the plant. Endophytes living inside leaves benefit plants by deterring herbivores and other ways. Lichens colonize barren habitats and prepare the way for plant succession.

ANSWERS TO TEST YOUR KNOWLEDGE

Fill in the Blanks:

1. septum
2. heterokaryon
3. chitin
4. enzymes
5. basidium
6. asci

7. mycorrhizae
8. zygosporangia
9. coenocytic
10. chytrids
11. arbuscular mycorrhizae
12. opisthokonts

Multiple Choice:

1. a	3. c	5. e	7. a	9. c
2. c	4. d	6. b	8. a	10. e

CHAPTER 32: INTRODUCTION TO ANIMAL DIVERSITY

INTERACTIVE QUESTIONS

32.1 a. A larva is a free-living, sexually immature stage in some animal life cycles that may differ from the adult in habitat, nutrition, and body form.
b. A juvenile resembles the adult but is not sexually mature; a larva differs from the adult form.
c. metamorphosis

32.2 Changes in predator–prey interactions; increased oxygen levels that could support more active animals; evolution of *Hox* complex of genes that regulate development

32.3 a. spiral and determinate
b. radial and indeterminate
c. split in solid mass of mesoderm
d. mesodermal buds from archenteron
e. mouth, second opening for anus
f. anus, second opening for mouth

c. Cnidaria: radial symmetry; diploblastic
d. Bilateria: bilateral symmetry; triploblastic
e. Deuterostomes: radial, indeterminate cleavage; coelom from outpockets of archenteron; blastopore forms anus
f. Protostomes: spiral, determinate cleavage; coelom from split in blocks of mesoderm; blastopore forms mouth
g. Lophotrochozoa: molecular similarities; named for lophophore feeding structure and trochophore larva of some phyla
h. Ecdysozoa: molecular similarities; named for molting of exoskeleton typical of some phyla

ANSWERS TO TEST YOUR KNOWLEDGE

Multiple Choice:

1. c	4. e	7. d
2. d	5. c	8. b
3. e	6. b	

SUGGESTED ANSWERS TO STRUCTURE YOUR KNOWLEDGE

1. **a.** Porifera (sponges); assymetrical; no true tissues
   b. Eumetazoa: "true" animals; have true tissues

CHAPTER 33: INVERTEBRATES

INTERACTIVE QUESTIONS

33.1 a. line body cavity (spongocoel); create water current, trap and engulf food particles, may produce gametes

b. wander through mesohyl; digest food and distribute nutrients, produce skeletal fibers, may produce gametes

33.2 a. polyp
b. medusa

c. mouth/anus
d. tentacle
e. mesoglea
f. gastrovascular cavity

33.3 a. Planarians have an extensively branched gastrovascular cavity; food is drawn in and undigested wastes are expelled through the mouth at the tip of a muscular pharynx.
b. Tapeworms are bathed in predigested food in their host's intestine.

33.4 a. 1. muscular foot for movement
2. visceral mass containing internal organs
3. mantle that may secrete shell and enclose visceral mass, forming a mantle cavity that houses gills

b. 1. Snails use their foot to creep slowly and their radula to rasp up algae.
2. Clams are sedentary suspension feeders that use siphons to draw water over the gills, where food is trapped in mucus and swept by cilia to the mouth.
3. Squid are active predators that capture and crush prey with jaws, aided by a poison in their saliva.

33.5 a. septum
b. intestine
c. metanephridium
d. chaetae
e. ventral nerve cord
f. dorsal and ventral blood vessels
g. longitudinal and circular muscles

33.6 Earthworms have longitudinal and circular muscle layers and a segmented body cavity that allows them to move by peristalsis, alternately elongating and contracting body sections and anchoring to the substrate with chaetae. Nematodes are not segmented and have only longitudinal muscles. These pull against the hydrostatic skeleton of the roundworm's tapered body, producing thrashing movements.

33.7 a. anterior feeding appendages modified as pincers or fangs
b. Many spiders trap their prey in webs they have spun from abdominal gland secretions. Spiders kill prey with poison-equipped, fang-like chelicerae. After secreting digestive juices, they suck up partially digested food.

33.8 a. mostly terrestrial
b. mostly aquatic
c. fly, walk, burrow
d. swim, float (planktonic)
e. tracheal system

f. gills or body surface
g. Malpighian tubules
h. diffusion across cuticle, glands regulate salt balance
i. one pair
j. two pairs
k. mandibles and modified mouthparts for chewing, piercing, or sucking; thorax has three pairs of walking legs and may have wings (extensions of cuticle)
l. mandibles and multiple mouthparts, walking legs on thorax, abdominal appendages

33.9 a. radial anatomy as adults, usually five spokes; sessile or slow moving
b. endoskeleton of calcareous plates
c. water vascular system with tube feet for locomotion, feeding, respiration

SUGGESTED ANSWERS TO STRUCTURE YOUR KNOWLEDGE

1. a. Eumetazoa
b. Bilateria
c. Deuterostomia
d. Lophotrochozoa
e. Ecdysozoa
f. Porifera (sponges); lack true tissues; choanocytes used in suspension feeding
g. Cnidaria (hydras, jellies, sea anemones, corals); diploblastic; radial symmetry; gastrovascular cavity; polyp and medusa form; cnidocytes
h. Platyhelminthes (flatworms: planaria, flukes, tapeworms); acoelomate; gastrovascular cavity
i. Rotifera (rotifers); pseudocoelomate; alimentary canal; crown of cilia around mouth; parthenogenesis
j. Lophophorates (brachiopods and bryozoans); coelomate; lophophore feeding structure; exoskeleton or bivalve shell
k. Mollusca (clams, snails, squids); coelomates; muscular foot; visceral mass; mantle; most with hard shell
l. Annelida (segmented worms: earthworms, polychaetes, leeches); coelomates; segmented body wall and organs
m. Nematoda (roundworms); pseudocoelomate; unsegmented, cylindrical worms; ecdysis
n. Arthropoda (crustaceans, insects, spiders, millipedes); coelomate; segmented body; jointed appendages; exoskeleton
o. Echinodermata (sea stars, sea urchins); coelomate; endoskeleton; water vascular system
p. Chordata (vertebrates, two invertebrate subphyla); coelomates; segmentation; notochord

ANSWERS TO TEST YOUR KNOWLEDGE

Matching:

1. C l	**4.** A h	**7.** F b	**10.** C i
2. B e	**5.** I d	**8.** B a	**11.** I m
3. F g	**6.** B j	**9.** D f	**12.** F c

Multiple Choice:

1. b	**4.** e	**7.** c	**10.** b	**13.** d
2. c	**5.** a	**8.** b	**11.** a	**14.** a
3. c	**6.** c	**9.** a	**12.** b	**15.** e

CHAPTER 34: VERTEBRATES

INTERACTIVE QUESTIONS

34.1 lancelet, Cephalochordata
  a. mouth
  b. *pharyngeal slits
  c. intestine
  d. segmental muscles
  e. anus
  f. *post-anal tail
  g. *dorsal, hollow nerve cord
  h. *notochord

   * Chordate characteristics

34.2 head with eyes and other sense organs, skull enclosing the brain, two clusters of *Hox* genes, other duplicated families of genes, neural crest, pharyngeal clefts that develop into gill slits, higher metabolism, more muscles, chambered heart, red blood cells with hemoglobin, and kidneys that remove wastes

34.3 a. Chordates
  b. Craniates
  c. Vertebrates
  d. Gnathostomes
  e. Osteichthyans
  f. Lobe-fins
  g. Tetrapods
  h. Amniotes
  i. lancelets (Cephalochordata)
  j. tunicates (Urochordata)
  k. hagfishes (Myxini)
  l. lampreys (Petromyzontida)
  m. sharks and rays (Chondrichthyes)
  n. ray-finned fishes (Actinopterygii)
  o. coelacanths (Actinistia)
  p. lungfishes (Dipnoi)
  q. frogs, salamanders (Amphibia)
  r. turtles, snakes, crocodiles, and birds (Reptilia)
  s. mammals (Mammalia)

34.4 more extensive skull, vertebrae (backbone) that encloses spinal cord, and, in aquatic vertebrates, fin rays

34.5 a. jaws, lateral line system, enlarged forebrain with better olfaction and vision, duplication of *Hox* genes
  b. increase the area for absorption and slow the passage of food through the short shark intestine
  c. common opening for the reproductive tract, excretory system, and digestive tract

34.6 See answer to Interactive Question 34.3.

34.7 a. Coelacanths, lungfishes, and terrestrial tetrapods
  b. It had the derived appendages of tetrapods with fully formed legs, ankles, and digits. But it retained primitive adaptations for living in water, such as gills and a finned tail. Its pectoral and pelvic girdles were too weak to support its body on land.

34.8 a. chorion—functions in gas exchange
  b. amnion—encases embryo in fluid, prevents dehydration, and cushions shocks
  c. allantois—stores metabolic wastes, functions with chorion in gas exchange
  d. yolk sac—expands over yolk; blood vessels transport stored nutrients into embryo

34.9 Archosaurs include the extinct dinosaurs and pterosaurs as well as crocodilians and birds. Lepidosaurs include extinct huge marine reptiles, tuatara, lizards, and snakes.

34.10 strong skeleton; no teeth and absence of some organs (reduces weight); feathers that shape wings into airfoil; strong keel on sternum to which flight muscles attach; high metabolic rate supported by efficient circulatory system, respiratory system, and endothermy; well-developed visual and motor areas of brain

34.11 a. hair; milk produced by mammary glands; endothermic with active metabolism; diaphragm to ventilate lungs; four-chambered heart; larger brains, increased learning capacity; differentiated teeth and remodeled jaw
  b. See answer to Interactive Question 34.3.

34.12 increased brain size; change in jaw shape and dentition; bipedal posture; reduced size difference between the sexes; capacity for language and symbolic thought; manufacture and use of complex tools

SUGGESTED ANSWERS TO STRUCTURE YOUR KNOWLEDGE

1. See Interactive Question 34.3 on page 493.

2. (1) Pharyngeal slits used for suspension feeding became adapted for gas exchange; (2) skeletal supports for gills became adapted for use as hinged jaws; (3) lungs of bony fish were transformed to air bladders, aiding in buoyancy; (4) fleshy fins supported by skeletal elements developed into limbs for terrestrial animals; (5) feathers, probably originally used for insulation, came to be used for flight; (6) dexterous hands important for an arboreal life evolved to manipulate tools.

3. Amphibians remain in damp habitats, burrowing in mud during droughts; some secrete foamy protection for eggs laid on land. Non-bird reptiles have scaly, waterproof skin and behavioral adaptations to modulate changing temperatures. The amniote egg provides an aquatic environment for the developing embryo. Gestation protects a developing embryo. All terrestrial groups are tetrapods, using limbs for locomotion on land.

ANSWERS TO TEST YOUR KNOWLEDGE

Fill in the Blanks:

1. Urochordata (tunicates)
2. Cephalochordata
3. somites
4. gnathostomes
5. operculum
6. amniotic egg
7. monotremes
8. diaphragm
9. anthropoids
10. *Australopithecus*

Multiple Choice:

1. a	5. a	9. b
2. c	6. d	10. e
3. e	7. a	11. d
4. e	8. c	12. c

CHAPTER 35: PLANT STRUCTURE, GROWTH, AND DEVELOPMENT

INTERACTIVE QUESTIONS

35.1
a. reproductive shoot (flower)
b. apical bud
c. node
d. internode
e. vegetative shoot
f. leaf
g. blade
h. petiole
i. axillary bud
j. stem
k. taproot
l. lateral roots
m. shoot system
n. root system

35.2
a. sclerenchyma (fibers and sclereids), tracheids, and vessel elements
b. sieve-tube elements (and the cells listed in a.)

35.3 Primary growth from apical meristems continues to add to the tips of roots and shoots, while secondary growth from the lateral meristems (vascular and cork cambiums) thickens and strengthens older regions of a woody plant

35.4
a. epidermis (dermal)—root hairs for absorption
b. cortex (ground)—uptake of minerals and water, food storage
c. vascular cylinder (vascular)—transport
d. endodermis—regulates water and mineral movement into stele
e. pericycle—origin of lateral roots
f. xylem—water and mineral transport
g. phloem—nutrient transport

35.5
a. cuticle
b. upper epidermis
c. palisade mesophyll
d. spongy mesophyll
e. guard cells
f. xylem
g. phloem
h. vein
i. stoma

35.6 H, F, A, B, G, D, C, E

35.7 Microtubules of the preprophase band leave behind ordered actin microfilaments that determine the orientation of cell division. Microtubules may orient the cellulose microfibrils in cell walls that will determine the direction in which cells can expand.

35.8 a.

Whorl	Genes Active	Organs in Normal	Genes Active Mutant A	Organs in Mutant A
1	A	sepals	C	carpels
2	AB	petals	BC	stamens
3	BC	stamens	BC	stamens
4	C	carpels	C	carpels

b. Gene *A* would be expressed alone in each whorl, producing a flower with four whorls of sepals.

SUGGESTED ANSWERS TO STRUCTURE YOUR KNOWLEDGE

1. Except for leaves and flowers, which stop growing when they reach maturity (determinate growth), plant roots and shoots continue to grow throughout the life of a plant. Although certain characteristic forms have been favored by natural selection in species adapted to particular environments, the specific environment in which a plant grows greatly influences its individual body form. For example, a plant may grow taller or produce different shapes of leaves depending on its light exposure. This developmental plasticity may help compensate for a plant's immobility.

2. This drawing is a eudicot stem, indicated by the ring of vascular bundles and the central pith.
   a. sclerenchyma
   b. phloem
   c. xylem
   d. cortex
   e. pith
   f. vascular bundle
   g. epidermis

ANSWERS TO TEST YOUR KNOWLEDGE

Matching:

1. H	5. I	9. J
2. F	6. A	10. G
3. B	7. C	
4. E	8. D	

Multiple Choice:

1. a	5. d	9. e	13. c
2. e	6. a	10. d	14. d
3. b	7. e	11. b	
4. c	8. c	12. c	

CHAPTER 36: RESOURCE ACQUISITION AND TRANSPORT IN VASCULAR PLANTS

INTERACTIVE QUESTIONS

36.1 a. At low light intensities, horizontal leaf orientation would maximize exposure to sunlight. In bright sun, a vertical orientation would shield leaves from too high a light intensity and allow light to penetrate to lower leaves.
b. Plant productivity would increase with an increasing leaf area index as more leaf area is available to capture sunlight. If this index gets too high, however, then lower branches and leaves are too shaded to photosynthesize and they may die through a process of self-pruning.

36.2 a. $\Psi_p = 0$, $\Psi_s = -0.6$, $\Psi = -0.6$
b. $\Psi_p = 0.6$, $\Psi_s = -0.6$, $\Psi = 0$. The turgor pressure of the cell that develops with movement of water into the cell finally offsets the solute potential of the cell. No net osmosis occurs, and the water potentials of both cell and solution are equal and = 0.
c. $\Psi_p = 0$, $\Psi_s = -0.8$, $\Psi = -0.8$. The cell would plasmolyze, losing water until the solute potential of the cell would equal that of the bathing solution (both –0.8). There would be no turgor pressure.

36.3 a. symplastic
b. apoplastic
c. Casparian strip
d. xylem vessels
e. stele (vascular cylinder)
f. endodermis
g. cortex
h. epidermis
i. root hair

36.4 a. The loss of water vapor from the air spaces through the stomata causes evaporation from the water film coating the mesophyll cells, which creates a tension resulting from the surface tension of the curving air–water interface.

b. Water molecules hold together due to hydrogen bonding, and a pull on one water molecule is transmitted throughout the column of water.

c. Water molecules adhere to the hydrophilic walls of narrow xylem cells, which helps to support the column of water against gravity.

d. The transpiration of water from the leaf creates a tension, or negative pressure, which produces a gradient of water potentials that extends from the leaf to the root.

36.5 In order to provide sufficient CO_2 for photosynthesis, stomata must be open and large amounts of water will be lost to transpiration. A sunny, windy, dry day will increase the rate of transpiration, perhaps to the point that insufficient water is available in the soil. Due to a lack of water, guard cells may lose turgor (and also be signaled by abscisic acid) and stomata will close, slowing transpiration. In reducing excessive water loss, however, the plant mesophyll cells now receive insufficient CO_2, and the photosynthetic rate will decline.

36.6 A storage organ, such as a root or tuber, stores carbohydrates during the summer and is a sugar sink, but when its starch is broken down to sugar in the spring to supply growing buds, it becomes a sugar source.

36.7 a. cotransport protein
   b. sucrose
   c. H^+
   d. proton pump
   e. ATP

SUGGESTED ANSWERS TO STRUCTURE YOUR KNOWLEDGE

1. Solutes may move across membranes by passive transport when they move down their electrochemical gradient. Transport proteins, which speed this passive transport, may be specific carrier proteins or selective channels. Active transport usually involves a proton pump that creates a membrane potential as positively charged hydrogen ions are moved out of the cell. Cations may now move down their electrochemical gradient. In cotransport, the inward diffusion of H^+ through a transport protein moves a solute against its concentration gradient.

2. The transpiration–cohesion–tension mechanism explains the ascent of xylem sap. The cohesion of water molecules transmits the pull resulting from transpiration throughout the column, and the adhesion of water to the hydrophilic walls of the xylem vessels aids the flow. Tension created by the evaporation of water and the resulting curvature of the air–water interface lowers the water potential in the leaf and sets up this bulk flow of water.

    A pressure flow mechanism also explains the movement of phloem sap, but the plant must expend energy to create the differences in water potential that drive translocation. The active accumulation of sugars in sieve-tube elements lowers water potential at the source end, resulting in an inflow of water. The removal of sugar from the sink end of a phloem tube is followed by the osmotic loss of water. The resulting difference in pressure between the source and sink end of the phloem tube causes the bulk flow of phloem sap.

ANSWERS TO TEST YOUR KNOWLEDGE

Multiple Choice:

1. b	**6.** d	**11.** a	**16.** e
2. e	**7.** d	**12.** c	**17.** d
3. d	**8.** c	**13.** e	**18.** b
4. a	**9.** a	**14.** b	
5. c	**10.** c	**15.** c	

CHAPTER 37: SOIL AND PLANT NUTRITION

INTERACTIVE QUESTIONS

37.1 A loam, which is a mixture of sand, silt, and clay, has enough large particles to provide air spaces and enough small particles to retain water and minerals. Cations bind to soil particles and are made available to roots by cation exchange.

Humus in a soil helps to retain water and provide a steady supply of mineral nutrients.

37.2 a. the variety of conservation-minded, environmentally safe, and profitable farming methods that permit the use of soil as a renewable resource

b. the percentage of nitrogen, phosphorus, and potassium that is in a fertilizer

37.3 a. component of chlorophyll, activates enzymes
b. component of nucleic acids, phospholipids, ATP, some coenzymes
c. component of cytochromes, cofactor for chlorophyll synthesis, activates enzymes

37.4 a. the yellowing of leaves due to a lack of chlorophyll, often indicative of a deficiency of magnesium
b. in the younger, growing parts of a plant

37.5 a. nitrogen-fixing bacteria
b. ammonifying bacteria
c. nitrifying bacteria

d. denitrifying bacteria
e. nitrate and nitrogenous organic compounds

37.6 a. Legume crops, which increase nitrogen in the soil, are often followed the next growing season with a nonlegume crop.
b. "Green manure" is a legume crop that is plowed under to improve soil fertility by adding both humus and nitrogen.

37.7 The fossil record shows that some of the earliest land plants had mycorrhizae, and these symbiotic associations may have been an important adaptation in helping plants to colonize nutrient-poor soils.

SUGGESTED ANSWERS TO STRUCTURE YOUR KNOWLEDGE

1.

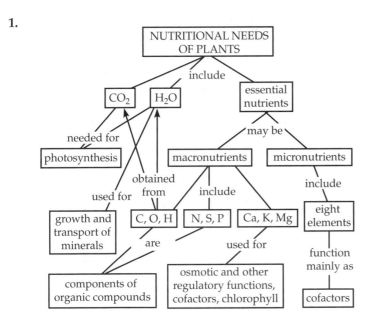

2. Root nodules are mutualistic associations between legumes and nitrogen-fixing bacteria of the genus *Rhizobium*. These bacteria receive nourishment from the plant and provide the plant with ammonium. Mycorrhizae are mutualistic associations between the roots of most plants and various species of fungi. The fungi stimulate root growth, increase surface area for absorption, and secrete antibiotics that may help protect the plant. In return, the fungi receive nourishment from the plant.

3. Both these types of symbiotic relationships involve chemical recognition between specific plant species and bacterial or fungal species. Both are mutualistic interactions.

ANSWERS TO TEST YOUR KNOWLEDGE

Multiple Choice:

1. b	**4.** b	**7.** e	**10.** b	**13.** c
2. a	**5.** a	**8.** d	**11.** e	**14.** e
3. e	**6.** e	**9.** c	**12.** d	

CHAPTER 38: ANGIOSPERM REPRODUCTION AND BIOTECHNOLOGY

INTERACTIVE QUESTIONS

38.1 **a.** stamen
b. anther
c. filament
d. petal
e. carpel
f. stigma
g. style
h. ovary
i. sepal
j. receptacle
k. ovule

38.2 **a.** a pollen grain with a patterned spore wall surrounding a tube cell and a generative cell (which moves into the tube cell and will divide to form two sperm)
b. the embryo sac, often containing seven cells and eight nuclei

38.3 **a.** Research has shown that a GABA gradient helps guide pollen tube growth toward the egg.
b. Double fertilization conserves resources by allowing for nutrient development only when fertilization has occurred.

38.4 **a.** seed coat
b. epicotyl
c. hypocotyl
d. cotyledons with endosperm
e. radicle
f. cotyledon (scutellum)

g. endosperm
h. epicotyl
i. hypocotyl
j. radicle
k. coleorhiza
l. coleoptile

38.5 The coleoptile pushes through the soil, and the shoot tip is protected as it grows up through the tubular sheath.

38.6 **a.** Buoyant fruits may be dispersed by water; winged fruits may "fly" seeds to new destinations; seeds in edible fruits may be dispersed in feces; and some animals may "plant" seeds in their underground caches.
b. Hormonal interactions induce softening of the pulp, a change in color, and an increase in sugar content.

38.7 **a.** Apomixis allows for the dispersal value of seeds, which in this case are clones of the parent plant.
b. Selfing is an advantage in certain crop plants because it ensures that a seed will develop.

38.8 Burning of fossil fuels releases CO_2 that was removed from the atmosphere millions of years ago. CO_2 that is released by the burning of biofuels is absorbed for photosynthesis as biofuel crops grow.

SUGGESTED ANSWERS TO STRUCTURE YOUR KNOWLEDGE

1. One version of the life cycle of an angiosperm. See also textbook Figures 38.2 and 38.3.

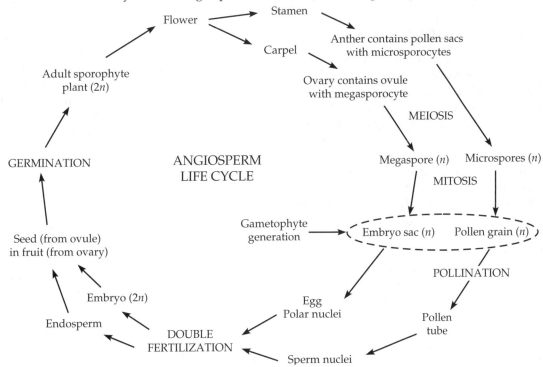

2. *Sexual:* Advantages include increased genetic variability, which provides the potential to adapt to changing conditions, and the dispersal and dormancy capabilities provided by seeds. Disadvantages are that the seedling stage is very vulnerable, sexual reproduction is very energy intensive (since most seeds and seedlings don't survive), and genetic recombination may separate adaptive traits.

    Asexual: Advantages include the hardiness of vegetative propagation and the maintenance of genetically well-adapted plants in a given environment. Disadvantages relate to the advantages of sexual reproduction: There is no genetic variability from which to "choose" should conditions change, and there are no seeds for dormancy or dispersal.

3. *Benefits:* Reduce use of chemical pesticides; weed crops with herbicides instead of erosion-producing tillage; increase nutritional value of crops; produce disease-resistant crops.

    Dangers: Introduce allergens into human foods; harm nontarget species with pesticides; create hybrid "superweeds" due to crop-to-weed transgene escape; produce other unanticipated harmful results that cannot be stopped once GMOs are released into the environment.

ANSWERS TO TEST YOUR KNOWLEDGE

Fill in the Blanks:

1.	ovary	6.	epicotyl
2.	sporophyte	7.	coleoptile
3.	biofuel	8.	scion
4.	embryo sac	9.	protoplast
5.	radicle	10.	callus

Multiple Choice:

1.	e	5.	e	9.	c	13.	e
2.	a	6.	a	10.	d	14.	a
3.	d	7.	a	11.	a	15.	e
4.	d	8.	e	12.	d		

CHAPTER 39: PLANT RESPONSES TO INTERNAL AND EXTERNAL SIGNALS

INTERACTIVE QUESTIONS

39.1 a. The signal is light; the receptor is a phytochrome located in the cytosol.

b. Light-activated phytochrome activates at least two pathways, one producing the second messenger cGMP and one leading to an increase in cytosolic Ca^{2+} levels. Both affect ion channels and activate specific kinases. Some kinase cascades may activate various transcription factors that change gene expression.

c. The plant response involves growth changes (slowing of stem elongation, expansion of leaves, elongation of roots). The de-etiolation process involves decreases in levels of growth-regulating hormones and the activation or new production of enzymes involved in producing chlorophyll or in photosynthesis.

39.2 a. ethylene
b. cytokinins
c. abscisic acid
d. auxin
e. brassinosteroids
f. gibberellins

39.3 The relative amount of red and far-red light is communicated to a plant by the ratio of the two forms of phytochrome. Canopy trees absorb red light, and a shaded tree will have a higher ratio of P_r to P_{fr}, inducing the tree to grow taller.

39.4 Free-running periods are circadian rhythms that vary from exactly 24 hours when an organism is kept in a constant environment without environmental cues.

39.5 a. not flower
b. flower
c. not flower
d. flower
e. not flower
f. flower
g. not flower
h. flower
i. not flower
j. flower

In the next to last light regimen, a flash of far-red light cancels the effect of the red flash, and the red flash in the last one cancels the effect of the far-red flash.

39.6 a. positive phototropism, negative gravitropism

b. thigmotropism; mechanical stimulation causing unequal growth rates of cells on opposite sides of the tendril

39.7 a. reduce transpirational water loss by stomatal closing and wilting; deep roots continue to grow in moister soil; heat-shock proteins produced to stabilize plant proteins

b. change membrane lipid composition by increasing unsaturated fatty acids; air tubes may develop in cortex of root

SUGGESTED ANSWERS TO STRUCTURE YOUR KNOWLEDGE

1. cytokinin spray used to keep flowers fresh; ethylene used to ripen stored fruit; synthetic auxins to induce seedless fruit set; gibberellins sprayed on Thompson seedless grapes; 2,4-D (auxin) used as herbicide

2.

3. Specific resistance begins with an *R-Avr* match. A pathogen molecule (elicitor) binds to a specific plant receptor (coded for by an *R* gene) and triggers a signal transduction pathway that initiates a hypersensitive response and production of an alarm signal. The HR response includes production of antimicrobial molecules, sealing off of infected areas, and destruction of infected cells. The distribution of the alarm signal, probably salicylic acid, stimulates distant cells to activate a systemic acquired resistance. Molecules are produced that protect the cell against a diversity of pathogens for several days.

ANSWERS TO TEST YOUR KNOWLEDGE

True or False:

1. False—change *abscisic acid* to *gibberellin*, or say *The removal of abscisic acid.*
2. True
3. False—change *thigmomorphogenesis* to *triple response;* or say *Thigmomorphogenesis is a change in growth form in response to mechanical stress.*
4. True
5. False—change *circadian rhythm* to *photoperiodism;* or change *day or night length* to *on a 24-hour cycle.*
6. True
7. False—change *A virulent* to *An avirulent*
8. False—change *cryptochrome* to *chromophore;* or say *Cryptochrome is a blue-light photoreceptor.*

Multiple Choice:

1. b	4. b	7. b	10. b	13. d
2. c	5. a	8. e	11. b	14. d
3. e	6. b	9. c	12. e	15. c

CHAPTER 40: BASIC PRINCIPLES OF ANIMAL FORM AND FUNCTION

INTERACTIVE QUESTIONS

40.1 Complexity allows for more specialization. Specialized organs can meet the animal's needs and control the composition of the interstitial fluid. A complex body is especially advantageous in the highly variable environments on land.

40.2 **a.** Stratified squamous; lining of esophagus; thick layer is protective, and new cells produced near basal lamina replace those abraded off.
b. Simple columnar; lining of intestines; cells with large cytoplasmic volumes are specialized for secretion or absorption.

40.3 **a.** osteon
b. central canal
c. bone
d. collagenous fiber
e. elastic fiber
f. loose connective tissue
g. red blood cells
h. plasma
i. blood

40.4 **a.** Cardiac; dark band is intercalated disk that relays electrical signal from cell to cell during heartbeat.
b. smooth muscle

40.5 Hormones are released and travel more slowly throughout the body in the bloodstream. Target tissues have receptors for specific hormones. Electrical signals travel rapidly through neurons directly to target cells. Hormone signals last longer and usually mediate responses of longer duration than the shorter responses to electrical signals. The endocrine system coordinates more gradual functions such as growth, reproduction, and metabolism. The nervous system regulates rapid responses such as locomotion and behavior.

40.6 Positive feedback is a mechanism in which a change in a variable serves to amplify rather than reverse the stimulus. An example is the increasing uterine contractions during childbirth. Homeostatic mechanisms tend to return a fluctuating variable to the normal range. Thus, positive feedback loops do not typically contribute to homeostasis.

40.7 Because the environmental temperature determines its body temperature, an ectotherm that lives where temperatures fluctuate very little will be homeothermic. If environmental temperatures are very high, an ectotherm may have a body temperature higher than that of an endotherm.

40.8 **a.** vasodilation of surface arterioles, activation of sweat glands, panting
b. vasoconstriction of surface vessels, erection of fur, shivering by contraction of skeletal muscles, nonshivering thermogenesis

40.9 **a.** In hibernation, the body temperature is maintained at a lower level, greatly conserving energy and allowing the animal to withstand long periods of cold temperature and decreased food supply.

b. In estivation, animals survive long stretches of elevated temperature and diminished water supply by entering a period of inactivity and lowered metabolism.

c. Many small mammals and birds with very high metabolic rates enter a daily torpor during periods when they are not feeding.

SUGGESTED ANSWERS TO STRUCTURE YOUR KNOWLEDGE

1.

Tissue	Structural Characteristics	General Functions	Specific Examples
Epithelial	Tightly packed cells; basal lamina; cuboidal, columnar, squamous shapes; simple, stratified, or pseudostratified	Protection, absorption, secretion; lines body surfaces	Mucous membrane, may be ciliated; glandular epithelium; lining of blood vessels
Connective	Few cells that secrete extracellular matrix of protein fibers in liquid, gel, or solid matrix	Connect and support other tissues	Loose connective; adipose; fibrous connective; bone; cartilage; blood
Muscle	Long cells (fibers) with filaments of actin and myosin	Contraction, movement	Skeletal (voluntary); smooth (involuntary); cardiac
Nervous	Neurons with cell bodies, axons, and dendrites; supportive glial cells	Sense stimuli, conduct impulses	Nerves, brain

2. All organs are covered with epithelial tissue, and internal lumens, ducts, and blood vessels are lined with epithelium. Most organs contain some types of connective tissue. Muscle tissue would be present if the organ must contract. Nervous tissue innervates most organs. The stomach is an example of an organ composed of many tissues. An epithelial mucosa lines the lumen; the submucosa contains blood vessels and loose connective tissue; the muscularis, a layer of circular and longitudinal smooth muscle, comes next; and the outer serosa is an epithelium.

3. Endotherms have thermoregulatory mechanisms that allow them to maintain a constant internal temperature across a wider range of environmental temperatures. While ectotherms may use behavioral mechanisms to reduce temperature fluctuations, they may still be exposed to greater internal fluctuations as environmental temperatures change.

4. Due to the higher energy cost of maintaining a warm body temperature, the BMR of an endotherm would always be higher than the SMR of a similar sized ectotherm. Energy for reproduction would vary by sex, the size and number of offspring produced, and the amount of parental care. The cost of temperature regulation for an endotherm would vary depending on the environment and the thermoregulation adaptations of the animal. The growth and activity expenditures would vary depending on whether the animal is growing and the amount of activity required for obtaining food, building shelter, or escaping predators. Even as the pieces of pie devoted to each activity change depending on the organism and the environment, the overall energy expenditure per unit mass is almost always higher for smaller animals. This inverse relationship for endotherms may be related to the greater surface area to volume ratio of small animals across which heat may be lost.

ANSWERS TO TEST YOUR KNOWLEDGE

Multiple Choice:

1. b	**6.** d	**11.** d	**16.** a
2. a	**7.** c	**12.** b	
3. d	**8.** a	**13.** d	
4. e	**9.** b	**14.** a	
5. b	**10.** e	**15.** e	

CHAPTER 41: ANIMAL NUTRITION

INTERACTIVE QUESTIONS

41.1 Vegetarians must eat a combination of plant foods that are complementary in amino acids, such as corn and beans.

41.2 a. Malnourishment is the absence from the diet of one or more essential nutrients; undernourishment is a deficiency in calories.
b. Calcium and phosphorus are required by vertebrates for bone construction.

41.3 The carbohydrates, proteins, nucleic acids, and fats that make up food are too large to pass through cell membranes, and the molecules in food are not identical to those an animal needs for its own tissues.

41.4 a. pharynx—sucks food through mouth
b. esophagus—passes food to crop
c. crop—stores and moistens food
d. gizzard—grinds food
e. intestine—digests and absorbs nutrients

The typhlosole increases the surface area of the intestine for absorption.

41.5 a. Polysaccharides (starch and glycogen) broken down in the mouth by salivary amylase to smaller polysaccharides and maltose; hydrolysis continues until salivary amylase is inactivated by the low pH in the stomach.
b. Proteins in the stomach are hydrolyzed by pepsin to smaller polypeptides.

41.6 a. The digestion of starch and glycogen into disaccharides is continued by pancreatic amylase. Disaccharidases, enzymes of the brush border, split disaccharides into monosaccharides.
b. Protein digestion is completed in the small intestine by pancreatic trypsin and chymotrypsin, enzymes specific for peptide bonds adjacent to certain amino acids; pancreatic carboxypeptidase; and enzymes of the brush border: carboxypeptidase and aminopeptidase, which split amino acids off from opposite ends of a polypeptide; and dipeptidases.
c. Pancreatic nucleases are a group of enzymes that hydrolyze DNA and RNA into their nucleotide monomers. Nucleotidases, nucleosidases, and phosphatases dismantle nucleotides.
d. The digestion of fats is aided by bile salts, which coat or emulsify tiny fat droplets so they do not coalesce, leaving a greater surface area for lipase to hydrolyze the fat molecules into glycerol, fatty acids, and glycerides.

41.7 Carnivore dentition is characterized by sharp incisors, fanglike canines, and jagged premolars and molars; herbivores have broad molars for grinding plant material. Carnivorous vertebrates often have large, expandable stomachs. The intestine of an herbivore is longer to facilitate digestion of plant material, and its cecum may contain cellulose-digesting microorganisms.

41.8 Human craving for fatty foods may have evolved from the feast-and-famine existence of our ancestors. Natural selection may have favored individuals who gorged on and stored high-energy molecules, as they were more likely to survive famines.

SUGGESTED ANSWERS TO STRUCTURE YOUR KNOWLEDGE

1. **a.** salivary glands—produce saliva
b. oral cavity—teeth and tongue mix food with saliva
c. pharynx—throat, opening to esophagus and trachea
d. esophagus—transport bolus to stomach
e. stomach—churns food with gastric juice
f. small intestine—digestion and absorption
g. large intestine (colon)—absorb water, compact feces
h. anus—exit of alimentary canal
i. rectum—stores feces until expelled
j. pancreas—produces enzymes, bicarbonate, and hormones
k. sphincter—controls opening to small intestine
l. gallbladder—stores bile from liver
m. liver—produces bile; processes nutrients; detoxifies

ANSWERS TO TEST YOUR KNOWLEDGE

Matching:

1. M	**4.** B	**7.** J	**10.** E
2. G	**5.** F	**8.** H	
3. D	**6.** C	**9.** N	

Multiple Choice:

1. d	**6.** d	**11.** b	**16.** e
2. c	**7.** d	**12.** a	**17.** d
3. e	**8.** c	**13.** d	**18.** c
4. b	**9.** b	**14.** c	
5. d	**10.** e	**15.** b	

CHAPTER 42: CIRCULATION AND GAS EXCHANGE

INTERACTIVE QUESTIONS

42.1 a. The heart pumps hemolymph through vessels into sinuses, and body movements squeeze the sinuses, forcing hemolymph through the body and back to the heart. When the heart relaxes, hemolymph is drawn back in through pores.

b. Open circulatory systems take less energy to operate. Nutrient and oxygen delivery is more efficient in closed circulatory systems, which also allow for greater regulation of blood distribution.

42.2 a. In the single circulation typical of fish, blood is pumped only once. It flows first through the gill capillaries, through a vessel to the systemic capillaries, and then back to the heart. In a double circulation, blood is pumped twice as it circulates through the body: once to the gas exchange organs, from which it returns to the heart, and again to the systemic circuit.

b. The amphibian heart has three chambers: two atria and a ventricle. A ridge in the ventricle helps to direct oxygen-rich blood from the left atrium into the systemic circuit and oxygen-poor blood from the right atrium into the pulmocutaneous circuit. In the four-chambered heart of mammals, the right side of the heart receives and then pumps oxygen-poor blood to the lungs, and the left side of the heart receives and then pumps oxygen-rich blood to the systemic capillaries.

42.3 a. capillaries of head and forelimbs
b. left pulmonary artery
c. aorta
d. capillaries of left lung
e. left pulmonary vein
f. left atrium
g. left ventricle
h. aorta
i. capillaries of abdominal organs and hind limbs

j. inferior vena cava
k. right ventricle
l. right atrium
m. right pulmonary vein
n. capillaries of right lung
o. right pulmonary artery
p. superior vena cava

The pulmonary veins, left side of the heart, and aorta and arteries leading to the systemic capillary beds carry oxygen-rich blood. See text Figure 42.6 for numbers. Note that oxygen-rich blood in the aorta splits and travels either to the head region or lower body, returning to the heart through the superior or inferior vena cava, respectively. Thus, numbers 7–10 do not represent the sequential flow of blood.

42.4 a. atrioventricular valve
b. semilunar valve
c. SA (sinoatrial) node
d. AV (atrioventricular) node

42.5 a. artery walls
b. contraction of ventricle
c. diastole
d. systole
e. diastolic pressure (lower number)
f. systolic pressure (higher number)
g. resistance
h. vasoconstriction
i. vasodilation
j. nerves, hormones

42.6 a. plasma
b. electrolytes
c. osmotic balance, buffering, muscle and nerve functioning
d. proteins
e. buffers, osmotic factors, lipid escorts, antibodies, clotting factors
f. nutrients, wastes, respiratory gases, hormones
g. erythrocytes (red blood cells)
h. transport O_2; help transport CO_2

    i. leukocytes (white blood cells)
    j. phagocytes
    k. defense and immunity
    l. platelets
    m. blood clotting

42.7 smoking, not exercising, and a diet rich in animal fats or *trans* fats

42.8 **a.** water
    b. gill
    c. gill cover and mouth pump water in mouth, across gills, and out of body
    d. air
    e. tracheoles of tracheal system
    f. body movements compress and expand air tubes
    g. air (and water)
    h. lungs and moist skin
    i. positive pressure breathing: fill mouth, close nostrils, raise floor of mouth, and blow up lungs
    j. air
    k. alveoli in lungs
    l. negative pressure breathing: lower diaphragm, raise ribs, thus increasing volume and decreasing pressure of lungs; air flows in

42.9 This graph shows the Bohr shift. The dissociation curve for hemoglobin is shifted to the right at a lower pH, meaning that at any given partial pressure of O_2, hemoglobin is less saturated with oxygen. A rapidly metabolizing tissue produces more CO_2, which lowers pH, and hemoglobin will unload more of its O_2 to that tissue.

SUGGESTED ANSWERS TO STRUCTURE YOUR KNOWLEDGE

1.   **a.** aorta
    b. pulmonary artery
    c. pulmonary veins
    d. left atrium
    e. left ventricle
    f. right ventricle
    g. inferior vena cava
    h. atrioventricular valve
    i. semilunar valve
    j. right atrium
    k. anterior vena cava

    Blood flow from venae cavae → right atrium → right ventricle → pulmonary artery → lung → pulmonary vein → left atrium → left ventricle → aorta

2.   Nasal cavity → pharynx → through glottis to larynx → trachea → bronchus → bronchiole → alveoli → interstitial fluid → alveolar capillary → hemoglobin molecule → venule → pulmonary vein → left atrium → left ventricle → aorta → renal artery → arteriole → capillary in kidney

ANSWERS TO TEST YOUR KNOWLEDGE

Multiple Choice:

1. c	**8.** b	**15.** a	**22.** e	**29.** e
2. c	**9.** a	**16.** d	**23.** d	**30.** c
3. d	**10.** e	**17.** b	**24.** e	**31.** b
4. e	**11.** d	**18.** e	**25.** b	**32.** d
5. a	**12.** d	**19.** c	**26.** d	
6. b	**13.** c	**20.** b	**27.** d	
7. a	**14.** a	**21.** b	**28.** d	

CHAPTER 43: THE IMMUNE SYSTEM

INTERACTIVE QUESTIONS

43.1 The mutant flies that were engineered to express defensin survived as well as the wild type flies whose immune system was intact. The mutant flies that made no antimicrobial peptides and those mutants that were engineered to produce drosomycin survived equally poorly following infection.

43.2 **a.** the most abundant white blood cells, which kill microbes by phagocytosis
    b. large phagocytes that engulf microbes
    c. leukocytes that attack parasitic worms with enzymes
    d. leukocytes located in skin and lymphatic tissue that stimulate acquired immunity
    e. leukocytes that attack the body's infected or abnormal cells
    f. release histamine to initiate inflammatory response
    g. causes vasodilation and increased permeability of blood vessels
    h. proteins released by virus-infected cells that stimulate neighboring cells to produce substances that inhibit viral replication
    i. set of serum proteins that cause lysis of microbes and are involved with innate and acquired defenses

j. enzyme in tears, saliva, and mucus that attacks bacterial cell walls

43.3 B cell receptors and antibodies recognize intact antigens that are either molecules on the surfaces of infectious agents or molecules free in the body. T cell receptors recognize pieces of antigens that have complexed with an MHC molecule inside a cell and are then presented on the cell surface. Cytotoxic T cells recognize foreign antigens that have been manufactured by a body cell and displayed with class I MHC molecules. Helper T cells recognize antigen fragments from microbes that have been engulfed by antigen-presenting cells (dendritic cells, macrophages, and B cells) and displayed with class II MHC molecules.

43.4 a. During development, the *V* and *J* gene segments of the genes coding for the polypeptide chains of the receptors are randomly combined, producing a large number of different antigen-binding specificities.
b. As lymphocytes mature in the thymus or bone marrow, they are tested for self-reactivity, and those that do react against self components are inactivated or destroyed.
c. Once a lymphocyte recognizes its antigen, the cell divides into a clone of effector and memory cells specific for that antigen.

43.5 a. antigen-presenting cell (dendritic cell)
b. bacterium
c. peptide antigen
d. class II MHC molecule
e. T cell receptor
f. CD4
g. helper T cell
h. cytokines stimulate helper T, cytotoxic T, and B cells
i. cytotoxic T cell
j. B cell
k. cell-mediated immunity (attack on infected cells)
l. humoral immunity (secretion of antibodies by plasma cells)

43.6 a. CD4
b. CD8
c. cytokines that stimulate cytotoxic T cells, B cells, and themselves
d. perforin molecules that form pores in the target cell and granzymes that initiate apoptosis of the target cell

43.7 a. neutralization—blocks viral binding sites, neutralizes toxins in body fluids
b. opsonization—binding of multiple antibodies to bacterium makes more recognizable to macrophages; cross-linking bacteria, virus particles, or antigens into aggregates facilitates phagocytosis
c. complement activation and formation of a membrane attack complex—pores in membrane of foreign cell allow water to rush in, causing the cell to lyse

43.8

Blood Type	Antigens on RBCs	Antibodies in Plasma	Can Receive Blood from	Can Donate Blood to
A	A	anti-B	A, O	A, AB
B	B	anti-A	B, O	B, AB
AB	AB	none	A, B, AB, O	AB
O	none	anti-A + B	O	A, B, AB, O

43.9 a. HIV destroys helper T cells, thus crippling the humoral and cell-mediated immune systems and leaving the body unable to fight the HIV virus and other opportunistic diseases such as Kaposi's sarcoma and *Pneumocystis* pneumonia.
b. HIV is readily passed through unprotected sex and needle sharing. The antigenic changes of HIV during replication complicate the development of a vaccine. Mutational changes also lead to drug resistant strains of HIV.

SUGGESTED ANSWERS TO STRUCTURE YOUR KNOWLEDGE

1.

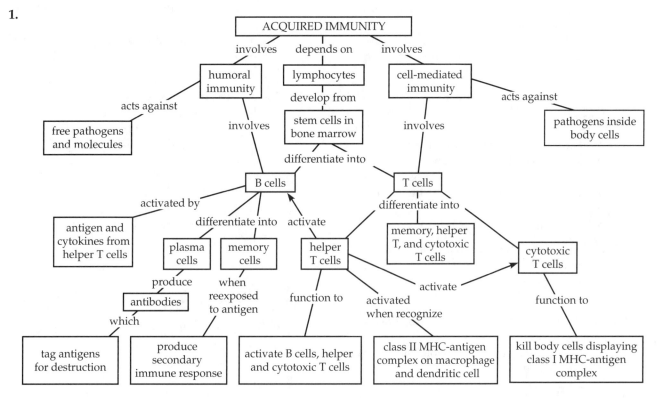

2. An antibody is a protein that consists of two identical light polypeptide chains and two identical heavy chains, held together by disulfide bonds in a Y-shaped molecule. The amino acid sequences in the variable sections of the light and heavy chains in the arms of the Y account for the specificity in binding between antibodies and epitopes of an antigen. The two arms of an antibody allow it to cross-link antigens, which enhances phagocytosis. The constant region of the antibody tail determines its effector function: five types of these constant regions correspond to five classes of antibodies.

3. Macrophages and dendritic cells of innate immunity phagocytose microbes in response to broad categories of molecules, but also present peptide antigens in their class II MHC molecules to helper T cells. This interaction, along with synthesis of cytokines by these phagocytes, helps to activate helper T cells. The complement system functions in both innate and acquired immunity. Some activated complement proteins promote inflammation or stimulate phagocytosis. The complement system may lyse microbes that are bearing antigen–antibody complexes.

ANSWERS TO TEST YOUR KNOWLEDGE

Multiple Choice:

1. d	7. a	13. d	19. e	25. e
2. c	8. c	14. c	20. c	26. a
3. e	9. a	15. d	21. e	27. c
4. c	10. b	16. d	22. b	28. d
5. b	11. a	17. a	23. b	
6. e	12. e	18. e	24. d	

CHAPTER 44: OSMOREGULATION AND EXCRETION

INTERACTIVE QUESTIONS

44.1 **a.** isoosmotic
b. conformer but may regulate specific solutes
c. slightly hyperosmotic
d. maintain urea and TMAO at high concentrations; rectal gland and kidneys remove salt
e. hypoosmotic
f. drink water to compensate for loss to hyperosmotic seawater; gills pump out salt; little urine excreted
g. hyperosmotic

h. copious dilute urine to compensate for osmotic gain; may pump salts in through gills
i. hyperosmotic
j. body coverings reduce evaporation; behavior adaptations; drinking and eating moist foods

44.2 a. An endotherm—it must eat more food to maintain its higher metabolic rate.

A carnivore—it eats more protein and thus produces more nitrogenous wastes.

b. *Uric acid* can be stored safely within the egg while the embryo develops. *Urea* is soluble and less toxic than ammonia and can be removed by the mother's blood. Urea is less expensive to produce than uric acid.

44.3 a. osmoregulation; remove excess water
b. both; remove excess water
c. both; conserve water
d. both; usually conserve water

44.4 a. water, salts, nitrogenous wastes, glucose, vitamins, and other small molecules
b. blood cells and large molecules such as plasma proteins

44.5 a. Bowman's capsule
b. glomerulus
c. proximal tubule
d. descending limb, loop of Henle
e. ascending limb, loop of Henle
f. distal tubule
g. collecting duct

(1) HCO_3^-, $NaCl^*$, H_2O, nutrients*, and K^+ are reabsorbed from the proximal tubule. H^{+*} and NH_3 are secreted into the proximal tubule.

(2) Water is reabsorbed from the descending limb of the loop of Henle.

(3) NaCl diffuses from the thin segment of the ascending limb and is pumped out* of the thick segment.

(4) $NaCl^*$, H_2O, and HCO_3^-* are reabsorbed from the distal tubule. K^{+*} and H^{+*} are secreted into the tubule.
5. $NaCl^*$ is reabsorbed; urea and H_2O diffuse out of the collecting tubule.

* indicates active transport.

44.6 a. Juxtamedullary nephrons; the longer the loop of Henle, the more hyperosmotic the urine can become.
b. Mammals and birds; loops of Henle are especially long in those taxa inhabiting dry habitats.

44.7 a. high blood osmolarity
b. ADH, antidiuretic hormone
c. distal tubules and collecting ducts
d. JGA, juxtaglomerular apparatus
e. renin
f. adrenal glands
g. aldosterone
h. Na^+ and water reabsorption
i. increase in blood pressure or volume
j. ANP, atrial natriuretic peptide

SUGGESTED ANSWERS TO STRUCTURE YOUR KNOWLEDGE

1. **a.** water, salts, glucose, amino acids, vitamins, urea, other small molecules
   b. ammonia (to offset acidity), drugs and poisons, H^+, K^+
   c. water, glucose, amino acids, vitamins, K^+, NaCl, bicarbonate

2. Two solutes, NaCl and urea, contribute to the high osmolarity of the medulla. As the filtrate moves up the ascending limb, NaCl first diffuses and is then pumped out, contributing to the gradient of NaCl in the medulla. As the filtrate flows through the collecting duct, some urea diffuses out and also contributes to the osmotic gradient. As the filtrate in the collecting duct moves through the high osmolarity of the medulla, water flows out of the collecting duct by osmosis, creating a more concentrated urine.

ANSWERS TO TEST YOUR KNOWLEDGE

Multiple Choice:

1. a	**5.** d	**9.** d	**13.** a	**17.** d
2. c	**6.** a	**10.** b	**14.** c	**18.** e
3. d	**7.** c	**11.** e	**15.** a	
4. b	**8.** e	**12.** c	**16.** c	

CHAPTER 45: HORMONES AND THE ENDOCRINE SYSTEM

INTERACTIVE QUESTIONS

45.1 a. water-soluble hormone
b. signal receptor
c. signal transduction produces cytoplasmic response
d. signal transduction leads to gene regulation
e. lipid-soluble hormone
f. signal receptor
g. hormone-receptor complex acts as transcription factor
h. gene regulation

45.2 a. insulin
b. glucagon
c. uptake of glucose
d. glycogen breakdown in liver
e. islets of Langerhans in pancreas
f. hydrolyze glycogen and release glucose
g. blood glucose level

45.3 In negative feedback, the response reduces the initial stimulus, thus turning off the response. In positive feedback, such as in the release of milk during nursing, the stimulus is reinforced, leading to an even greater response.

45.4 a. stimulates contraction of uterus and milk release
b. increases water reabsorption in kidney
c. stimulates thyroid gland
d. stimulates activity of gonads
e. stimulates adrenal cortex to produce and secrete hormones
f. various effects, depending on species, including stimulation of milk production
g. regulates skin pigment cells in amphibians
h. promotes growth (especially bones) through stimulating release of insulin-like growth factors and metabolic functions

45.5 Insufficient iodine results in reduced production of T3 and T4. Thus, there is no negative feedback to turn off production of TRH by the hypothalamus and TSH by the anterior pituitary. TSH continues to stimulate the thyroid, which enlarges.

45.6 a. Nervous stimulation from the hypothalamus to the adrenal medulla causes the secretion of epinephrine, which increases blood pressure, rate and volume of heart beat, breathing rate, and metabolic rate; stimulates glycogen hydrolysis and fatty acid release; and changes blood flow patterns. Norepinephrine release maintains blood pressure.

b. A releasing hormone from the hypothalamus stimulates ACTH release from the pituitary, which stimulates the adrenal cortex to release corticosteroids. Glucocorticoids increase blood glucose through conversion of proteins and fats. Mineralocorticoids increase blood volume and pressure by stimulating the kidney to reabsorb sodium ions and water.

SUGGESTED ANSWERS TO STRUCTURE YOUR KNOWLEDGE

1. Water-soluble chemical signals that bind to plasma membrane receptors initiate signal transduction pathways that may activate cellular enzymes or affect gene expression. Steroid hormones and other lipid-soluble signals bind with protein receptors inside a cell, and the hormone-receptor complex acts as a transcription factor to turn on (or off) specific genes.

2. In a simple hormone pathway, a stimulus triggers the release of a hormone, whose effect is to reduce the stimulus, which, by negative feedback, stops triggering the release of the hormone. Secretin, released into the bloodstream in response to an acid pH, triggers the release of bicarbonate from the pancreas, which raises the pH of the duodenum. The resulting rise in pH removes the stimulus for secretin release.

3. A hormone cascade pathway involves a sequence of signals in which the hypothalamus secretes releasing hormones that stimulate the anterior pituitary to secrete a tropic hormone, which then acts on target cells to secrete hormones that produce some physiological or developmental effect. For example, in response to a decrease in body temperature, the hypothalamus secretes TRH, which stimulates the anterior pituitary to secrete TSH, which acts on the thyroid gland to stimulate release of thyroid hormone, which increases metabolic rate. Thyroxine then acts by negative feedback to block the release of TSH and TRH.

ANSWERS TO TEST YOUR KNOWLEDGE

Matching:

1. I f
2. G a
3. F d
4. B h
5. A g
6. E b
7. L j
8. D j
9. C c, released from g
10. J c, released from g

Multiple Choice:

1. e	**5.** d	**9.** b	**13.** a
2. d	**6.** a	**10.** e	**14.** d
3. d	**7.** a	**11.** d	**15.** c
4. c	**8.** b	**12.** c	**16.** d

CHAPTER 46: ANIMAL REPRODUCTION

INTERACTIVE QUESTIONS

46.1 a. Perpetuation of successful genotypes in stable habitats; eliminates need to locate mate.
b. Varying genotypes and phenotypes of offspring may enhance reproductive success of parents in fluctuating environments.
c. Sexual reproduction because it involves union of egg and sperm and offspring will be genetically varied due to genetic recombination during meiosis and fertilization.

46.2 a. courtship behaviors that trigger the release of gametes and may provide a means for mate selection
b. chemical signals (pheromones) from individual releasing gametes
c. environmental signals (temperature, day length)

46.3 a. vas deferens
b. erectile tissue of penis
c. urethra
d. glans
e. prepuce (foreskin)
f. scrotum
g. testis
h. epididymis
i. bulbourethral gland
j. prostate gland
k. seminal vesicle
l. oviduct
m. follicles
n. corpus luteum
o. uterine wall (muscular layer)
p. endometrium (lining of uterus)
q. cervix
r. vagina
s. ovary

46.4 a. Spermatogonia continue to divide by mitosis and spermatogenesis occurs continuously, whereas a female's supply of primary oocytes may be present at birth.
b. Each meiotic division produces four sperm, but only one ovum (and polar bodies that disintegrate).

c. Spermatogenesis is an uninterrupted process. Oogenesis occurs in stages: prophase I before birth; after puberty, meiosis I and II up to metaphase II in a maturing follicle just before ovulation; and completion of meiosis II (in humans) when a sperm cell penetrates the oocyte.

46.5 a. GnRH (gonadotropin-releasing hormone)
b. FSH (follicle-stimulating hormone)
c. LH (luteinizing hormone)
d. androgens (primarily testosterone)
e. negative feedback loops involving testosterone and inhibin

46.6 a. LH
b. FSH
c. follicular phase
d. ovulation
e. luteal phase
f. estradiol
g. progesterone
h. menstrual flow phase
i. proliferative phase
j. secretory phase

46.7 a. secreted by embryo, maintains corpus luteum (and thus progesterone) in first trimester
b. secreted by corpus luteum and later by placenta, regulates growth and maintenance of placenta, formation of mucus plug in cervix, cessation of ovulation, growth of uterus
c. secreted by fetus and mother's posterior pituitary; stimulates uterine contractions and uterine secretion of prostaglandins during birth; controls release of milk during nursing

46.8 a. prevent release of gametes—sterilization, combination birth control pills (or injection, patch, or vaginal ring)
b. prevent fertilization—abstinence, rhythm method, condom, diaphragm, progestin minipill, coitus interruptus, cervical cap, spermicides
c. prevent implantation of embryo—morning after pills (MAP), IUD
The most effective methods are abstinence, sterilization, and chemical contraception. Least effective are the rhythm method and coitus interruptus.

SUGGESTED ANSWERS TO STRUCTURE YOUR KNOWLEDGE

1. Path of sperm from formation to fertilization

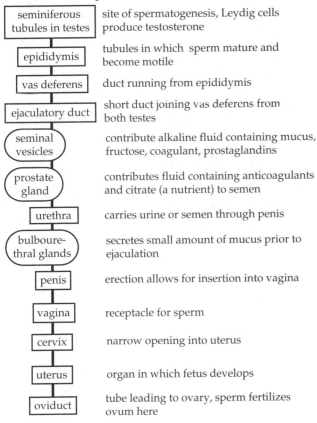

seminiferous tubules in testes	site of spermatogenesis, Leydig cells produce testosterone
epididymis	tubules in which sperm mature and become motile
vas deferens	duct running from epididymis
ejaculatory duct	short duct joining vas deferens from both testes
seminal vesicles	contribute alkaline fluid containing mucus, fructose, coagulant, prostaglandins
prostate gland	contributes fluid containing anticoagulants and citrate (a nutrient) to semen
urethra	carries urine or semen through penis
bulboure-thral glands	secretes small amount of mucus prior to ejaculation
penis	erection allows for insertion into vagina
vagina	receptacle for sperm
cervix	narrow opening into uterus
uterus	organ in which fetus develops
oviduct	tube leading to ovary, sperm fertilizes ovum here

2. **a.** GnRH is a releasing hormone of the hypothalamus that stimulates the anterior pituitary to secrete FSH and LH.
b. FSH stimulates growth of follicles.
c. The LH surge is caused by an increase in GnRH production that resulted from increasing levels of estradiol produced by the developing follicle.
d. The LH surge induces the maturation of the follicle and ovulation, and it transforms the ruptured follicle to the corpus luteum.
e. LH maintains the corpus luteum, which secretes estradiol and progesterone.
f. High levels of estradiol and progesterone act on the hypothalamus and pituitary to inhibit the secretion of LH and FSH.
g. The lack of LH causes the corpus luteum to disintegrate. Thus, the production of estradiol and progesterone ceases, which allows LH and FSH secretion to begin again.

3. Birth control pills, which contain a combination of synthetic estrogens and progestin, act by negative feedback to prevent the release of GnRH by the hypothalamus and FSH and LH by the pituitary, thus preventing the development of follicles and ovulation. RU-486 is an analog of progesterone that blocks progesterone receptors in the uterus, thus preventing progesterone from maintaining the endometrium.

ANSWERS TO TEST YOUR KNOWLEDGE

Fill in the Blanks:

1. budding	6. estrous cycle
2. pheromomes	7. *in vitro* fertilization
3. parthenogenesis	8. urethra
4. hermaphrodite	9. vasocongestion
5. cloaca	10. menopause

Multiple Choice:

1. d	5. e	9. b	13. e
2. d	6. e	10. b	14. c
3. a	7. b	11. c	15. b
4. c	8. a	12. a	16. b

CHAPTER 47: ANIMAL DEVELOPMENT

INTERACTIVE QUESTIONS

47.1 a. specific binding between receptors on the plasma membrane (extending through the vitelline layer) and proteins on acrosomal process
b. fast block to polyspermy caused by membrane depolarization during acrosomal reaction and slow block to polyspermy caused by formation of the fertilization envelope during cortical reaction

47.2 a. 8-cell stage
b. blastula
c. animal pole
d. gray crescent
e. vegetal pole
f. blastocoel

47.3 a. ectoderm
b. mesenchyme cells
c. endoderm
d. archenteron
e. blastopore
f. ectoderm

    g. mesoderm
    h. endoderm
    i. yolk plug in blastopore
    j. archenteron

47.4 a. neural tube
    b. neural crest cells
    c. somite
    d. archenteron
    e. coelom
    f. notochord

47.5 a. The **yolk sac** grows to enclose the yolk and develops blood vessels to carry nutrients to the embryo.
    b. The **amnion** encloses a fluid-filled sac, which provides an aqueous environment for development and acts as a shock absorber.
    c. The **allantois** serves as a receptacle for certain metabolic wastes.
    d. The **chorion** (pressed against the inside of the egg shell), in conjunction with the allantois, provides gas exchange for the developing embryo.

47.6 a. expanding region of trophoblast—chorion, fetal portion of placenta
    b. epiblast—three primary germ layers, amnion, some mesoderm becomes part of placenta, allantois
    c. hypoblast—yolk sac
    d. trophoblast

47.7 Cytoplasmic determinants in the sea urchin egg must have an unequal polar distribution. The developmental fates of cells in the animal and vegetal halves of the embryo are determined by the first horizontal division.

47.8 The developing limb would be a mirror image with posterior digits facing both forward and backward. This is the same effect as transplanting a donor ZPA to that location.

SUGGESTED ANSWERS TO STRUCTURE YOUR KNOWLEDGE

1.

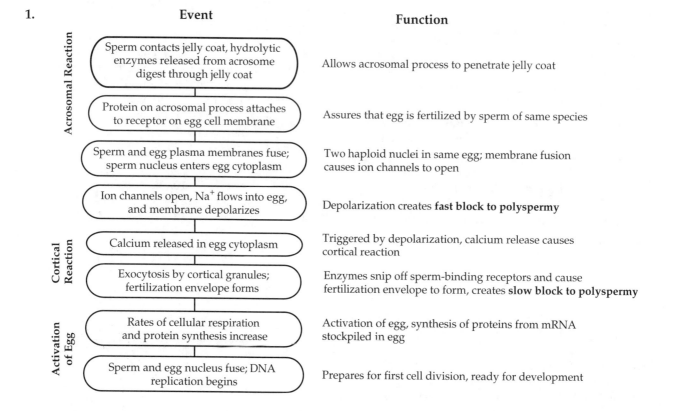

Event	Function
Acrosomal Reaction — Sperm contacts jelly coat, hydrolytic enzymes released from acrosome digest through jelly coat	Allows acrosomal process to penetrate jelly coat
Protein on acrosomal process attaches to receptor on egg cell membrane	Assures that egg is fertilized by sperm of same species
Sperm and egg plasma membranes fuse; sperm nucleus enters egg cytoplasm	Two haploid nuclei in same egg; membrane fusion causes ion channels to open
Ion channels open, Na$^+$ flows into egg, and membrane depolarizes	Depolarization creates **fast block to polyspermy**
Cortical Reaction — Calcium released in egg cytoplasm	Triggered by depolarization, calcium release causes cortical reaction
Exocytosis by cortical granules; fertilization envelope forms	Enzymes snip off sperm-binding receptors and cause fertilization envelope to form, creates **slow block to polyspermy**
Activation of Egg — Rates of cellular respiration and protein synthesis increase	Activation of egg, synthesis of proteins from mRNA stockpiled in egg
Sperm and egg nucleus fuse; DNA replication begins	Prepares for first cell division, ready for development

2.

	Cleavage	Blastula	Gastrula
Sea urchin	Holoblastic; equal-sized blastomeres	Hollow ball of cells, one cell thick, surrounding blastocoel	Invagination through blastopore to form endoderm of archenteron, migrating mesenchyme forms mesoderm
Frog	Holoblastic, but yolk impedes divisions; unequal-sized blastomeres	Blastocoel in animal hemisphere; walls several cells thick	Involution of cells at dorsal lip of blastopore, migrating cells form mesoderm and endoderm around archenteron
Bird	Meroblastic, only in cytoplasmic disk on top of yolk	Blastoderm; blastocoel cavity between epiblast and hypoblast	Migration of epiblast cells through primitive streak forms mesoderm and endoderm, lateral folds join to form archenteron
Mammal	Holoblastic, no apparent polarity to egg, blastomeres equal	Embryonic disc forms from innercell mass with epiblast and hypoblast; trophoblast surrounds embryo	Migration of epiblast cells through primitive streak forms mesoderm and endoderm

ANSWERS TO TEST YOUR KNOWLEDGE

Multiple Choice:

1. c	4. b	7. a	10. b	13. b	16 a
2. c	5. e	8. e	11. e	14. e	17 b
3. b	6. d	9. c	12. a	15. e	18 d

CHAPTER 48: NEURONS, SYNAPSES, AND SIGNALING

INTERACTIVE QUESTIONS

48.1 **a.** dendrites
b. cell body
c. axon hillock
d. axon
e. synaptic terminal

The impulse moves from the cell body along the axon to the synaptic terminals. Neurotransmitter is released into the synapse with another cell.

48.2 **a.** K^+ inside the cell; Na^+ outside the cell
b. The inside of the cell is more negative compared to outside the cell.
c. The membrane potential moved closer to E_K; more potassium channels must have opened, and/or sodium channels must have closed.

48.3 **a.** sodium channel
b. potassium channel
c. K^+
d. Na^+
e. sodium inactivation gate
f. membrane potential (mV)
g. time
h. resting potential
i. threshold
j. action potential

1. Resting state: Voltage-gated Na^+ and K^+ channels are closed.
2. Depolarization: Some Na^+ channels open; depolarization opens more Na^+ channels. If threshold reached, action potential is triggered.
3. Rising phase: Most Na^+ channels open and Na^+ influx makes inside of cell positive.
4. Falling phase: Na^+ channels become inactivated, K^+ channels open, K^+ leaves cell and inside of cell becomes negative.
5. Undershoot: Na^+ channels closed (many are still inactivated), K^+ channels still open, and membrane becomes hyperpolarized. As these channels close, membrane returns to resting potential.

48.4 **a.** Ca^{2+} flowing into presynaptic cell
b. synaptic terminal

   c. synaptic cleft
   d. synaptic vesicles containing neurotransmitter
   e. presynaptic membrane
   f. postsynaptic membrane
   g. receptor with bound neurotransmitter
   h. closed ligand-gated ion channel

48.5 a. Temporal
    b. Spatial

48.6 The type of postsynaptic receptor and its mode of action determine neurotransmitter function. Binding of acetylcholine to receptors in heart muscle activates a signal transduction pathway that makes it harder to generate an action potential, thus reducing the strength and rate of contraction. The acetylcholine receptor in skeletal muscle cells opens ion channels and depolarizes the muscle cell membrane.

SUGGESTED ANSWERS TO STRUCTURE YOUR KNOWLEDGE

1.

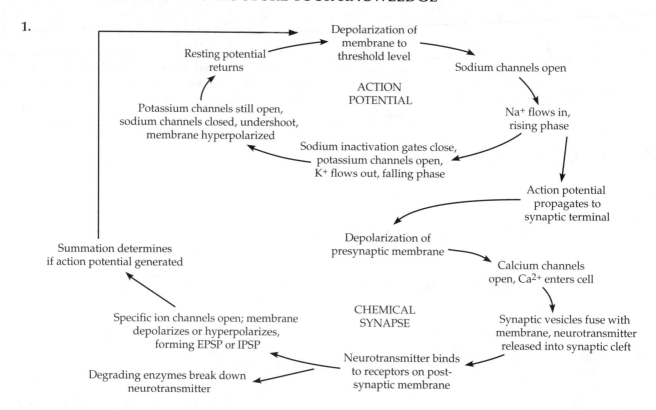

ANSWERS TO TEST YOUR KNOWLEDGE

Multiple Choice:

1. d	**4.** e	**7.** b	**10.** b	**13.** a
2. e	**5.** c	**8.** c	**11.** c	**14.** c
3. a	**6.** d	**9.** d	**12.** b	

CHAPTER 49: NERVOUS SYSTEMS

INTERACTIVE QUESTIONS

49.1 Sessile molluscs such as clams have little cephalization and only simple sense organs. Cephalopod molluscs such as squid have large brains, image-forming eyes, and giant axons, all of which contribute to their active predatory life and their ability to learn.

49.2 a. efferent neurons
    b. afferent (sensory) neurons
    c. autonomic nervous system
    d. motor system
    e. sensory organs and cells
    f. enteric
    g. parasympathetic

h. sympathetic

i. slows heart, conserves energy

j. acetylcholine

k. norepinephrine at target organ (acetylcholine by preganglionic neuron)

l. skeletal muscles

49.3 a. cerebral cortex 5

b. thalamus 3

c. cerebellum 1

d. medulla 4

e. pons 2

f. midbrain 7

g. hypothalamus 6

49.4 The primary somatosensory cortex receives information from somatosensory receptors in the body about touch, pain, pressure, temperature, and position of muscles and limbs. The motor cortex generates motor commands that travel to skeletal muscles. The neurons are distributed in an orderly fashion, from those associated with the legs and feet toward the midline of the parietal or frontal lobe to those associated with the head located along the sides. The proportion of the primary somatosensory cortex and primary motor cortex devoted to each part of the body is correlated with the importance of that area.

49.5 Glutamate released by the presynaptic neuron opens NMDA receptors, but they remain blocked by magnesium ions until the postsynaptic membrane is depolarized by activity at nearby synapses. The flow of Ca^{2+} through the now-open NMDA receptor channels trigger insertion of AMPA glutamate receptors in the membrane. With both types of receptors in the membrane, glutamate release by the presynaptic neuron is now sufficient to initiate action potentials without input from other synapses.

SUGGESTED ANSWERS TO STRUCTURE YOUR KNOWLEDGE

1.   a. system of clusters of neurons extending through the brainstem; regulates state of arousal; filters all sensory information going to the cortex

b. in the hypothalamus; functions as biological clock for circadian rhythms

c. clusters of neurons deep in white matter of cerebrum; centers for motor coordination and planning and learning movements

d. functional center in parts of thalamus, hypothalamus, and inner cerebral cortex (olfactory bulb, hippocampus, and amygdala); centers of human emotions and memories

e. two structures in inner cerebral cortex; involved in limbic system and memory; amygdala is involved in forming emotional memories; hippocampus functions in short-term memory and transfer to long-term memory

ANSWERS TO TEST YOUR KNOWLEDGE

Multiple Choice:

1. b	**5.** b	**9.** d	**13.** d
2. a	**6.** c	**10.** a	**14.** c
3. e	**7.** a	**11.** d	**15.** d
4. e	**8.** b	**12.** b	

CHAPTER 50: SENSORY AND MOTOR MECHANISMS

INTERACTIVE QUESTIONS

50.1 a. sensory reception

b. sensory transduction

c. transmission

d. perception

50.2 a. Most are concentrated in the skin; some are associated with other organs.

b. They inhibit prostaglandin synthesis.

50.3 a. auditory canal

b. malleus (hammer)

c. incus (anvil)

d. stapes (stirrup)

e. semicircular canals

f. auditory nerve

g. cochlea

h. round window

i. Eustachian tube

j. oval window

k. tympanic membrane (ear drum)

50.4 a. utricle and saccule for position and linear movement, semicircular canals for angular movement

b. lateral line system

c. statocysts containing statoliths

d. The balance organs of mammals have otoliths and those of invertebrates contain statoliths. These granules settle by gravity and stimulate hair cells to signal position and movement. Sensory hairs on the hair cells in the lateral line system of fishes are embedded in a gelatinous cap, which is bent by moving water. Hair cells at the base of the semicircular canals also project into a gelatinous cap, which is bent by moving fluid. All of these mechanoreceptors are hair cells.

50.5 find mates, recognize territory, navigate, communicate, and locate and taste food

50.6 a. sclera
 b. choroid
 c. retina
 d. fovea (center of visual field)
 e. optic nerve
 f. optic disk (blind spot)
 g. vitreous humor
 h. lens
 i. aqueous humor
 j. pupil
 k. iris
 l. cornea
 m. suspensory ligament
 n. ciliary body

50.7 a. small
 b. fovea

50.8 a. sarcomere
 b. Z line
 c. thin filaments (actin)
 d. thick filament (myosin)

The Z lines will be closer together as the overlap between thick and thin filaments increases. Binding of ATP releases myosin heads from the actin filament; hydrolysis of ATP converts the head to a high energy form that binds to actin, forming a cross-bridge, and then pulls the thin filament toward the center of the sarcomere.

50.9 a. fast-twitch glycolytic, fast-twitch oxidative, slow-twitch oxidative
 b. fast-twitch glycolytic
 c. slow-twitch oxidative

50.10 a. Hydrostatic skeleton: no extra materials needed, good shock absorber; not much protection or support for lifting animal off the ground; cnidarians, flatworms, nematodes, and annelids
 b. Exoskeleton: quite protective; in arthropods, must be molted in order to grow, restricts size; most molluscs and arthropods
 c. Endoskeleton: various types of joints allow for flexible movement in vertebrates, good

structural support; not very protective; sponges, echinoderms, chordates

50.11 a. swimming
 b. running
 c. flying
 d. smaller

SUGGESTED ANSWERS TO STRUCTURE YOUR KNOWLEDGE

1. Light enters the eye through the pupil and is focused by the lens onto the retina. When rhodopsin (visual pigment in rods) absorbs light energy, it sets off a signal transduction pathway that decreases the receptor cell's permeability to sodium and hyperpolarizes the membrane. The reduction in the release of neurotransmitter by a rod cell releases some connected bipolar cells from inhibition, which in turn generates action potentials in ganglion cells—the sensory neurons that form the optic nerve. Horizontal and amacrine cells provide lateral integration of visual information.

2. Sound waves are collected by the *pinna* and travel through the *auditory canal* to the *tympanic membrane*, where they are transmitted by the *malleus, incus*, and *stapes*. Vibration of the stapes against the *oval window* sets up pressure waves in the fluid in the *vestibular canal* within the *cochlea*, from which they are transmitted to the *tympanic canal* and then dissipated when they strike the *round window*. The pressure waves vibrate the *basilar membrane*, on which the *organ of Corti* is located within the *cochlear duct*. Tips of the hair cells arising from the organ of Corti are embedded in the *tectorial membrane*. When vibrated, they bend, triggering a depolarization, release of neurotransmitter, and initiation of action potentials in the sensory neurons of the *auditory nerve*, which leads to the *cerebral cortex*.

3. A motor neuron releases acetylcholine into the synapse with a muscle, initiating an action potential within the muscle fiber, which spreads into the center of the cell along the transverse tubules. The action potential changes the permeability of the sarcoplasmic reticulum membrane, which releases Ca^{2+} into the cytosol. The calcium binds with troponin, moving tropomyosin and exposing the myosin-binding sites of the actin molecules. Heads of myosin molecules in their high-energy form bind to these sites, forming cross-bridges. The myosin head bends, pulling the thin filament

toward the center of the sarcomere. When myosin binds to an ATP, the cross-bridge breaks. Hydrolysis of the ATP returns the myosin to its high-energy form, and it binds further along the actin molecule. This sequence continues as long as there is ATP and until calcium is pumped back into the sarcoplasmic reticulum, when the tropomyosin–troponin complex again blocks the actin sites.

ANSWERS TO TEST YOUR KNOWLEDGE

Multiple Choice:

1. c	6. c	11. a	16. a	21. a
2. b	7. c	12. d	17. c	22. a
3. c	8. a	13. b	18. c	
4. b	9. e	14. d	19. e	
5. d	10. b	15. b	20. c	

CHAPTER 51: ANIMAL BEHAVIOR

INTERACTIVE QUESTIONS

51.1 a. The stimulus of an increase in day length may result in reproductive behavior.
b. Production of viable offspring is most successful at this time due to warm temperature and abundant food supply; thus, this behavior has been selected for by natural selection.

51.2 The sow bugs were showing negative phototaxis (moving away from light) and a kinesis for relative humidity.

51.3 Mammals are mostly nocturnal; birds are diurnal.

51.4 a. imprinting
b. operant conditioning (a type of associative learning)
c. habituation
d. spatial learning
e. classical conditioning (associative learning)

51.5 Raising animals in the lab allows researchers to search for evidence of genetic bases of behavior by controlling for environmental influences. In some cases, crossbreeding experiments provide evidence of genetic control of behavior.

51.6 Researchers measured the average number of drops required to break whelk shells at various heights and calculated the average total effort required by computing the *total flight height* (average number of drops × height per drop). They measured the average flight height for crows in their whelk-breaking behavior (5.2 m) and found it was very close to the 5-meter height predicted based on an optimal compromise between energy gained versus energy expended.

51.7 If a male's parental care resulted in his greater reproductive success, then the genes for this behavior would increase in frequency in a population. But if the male were parenting offspring that were not his own, then he would not pass these "parental care" genes on to the next generation in a higher proportion because he was also helping to perpetuate the "non-parental care" genes of other males.

51.8 Orange throats cannot successfully defend all their females from sneaky yellow throats, so yellow throats mate more often and their frequency increases in the population. But blue throats restrict the access of yellow throats to their small number of females, and blue throat numbers will increase. The aggressive orange throats can take over blue throat territories, and once again, orange throat numbers will increase.

51.9 a. a parent or a sibling
b. The individual shares more genes in common with a parent or a sibling. The coefficient of relatedness is 0.5, whereas it is only 0.125 for a cousin. Thus, with Hamilton's rule $rB > C$, the benefit to the recipient is discounted less by a larger value of r.

SUGGESTED ANSWERS TO STRUCTURE YOUR KNOWLEDGE

1. The scientific study of animal behavior seeks to identify those behaviors that are innate and genetically programmed as well as those that are a product of experience and learning. Fixed-action patterns are clearly developmentally fixed, although experience may improve the performance of such behaviors. In other cases, genetics may set the parameters for an organism's behavior; however, experience can modify behavior, and learning is clearly evident.

2. According to the concept of fitness, an animal's behavior should help to increase its chance of survival and production of viable offspring. Survival behaviors that are genetically

programmed would be most likely to be passed on, and natural selection would refine innate behaviors for foraging, migrating, mating, and care of offspring. The evolution of cognitive ability would help individuals deal with novel situations. Social behaviors and communication increase an individual's fitness. Parental investment and certainty of paternity may result in differences in mating systems and parental care that influence an individual's reproductive success. Altruistic behavior may be explained on the basis of kin selection; the inclusive fitness of an animal increases if its altruistic behavior benefits related animals who share many genes.

ANSWERS TO TEST YOUR KNOWLEDGE

Multiple Choice:

1. d	**6.** c	**11.** a	**16.** d
2. b	**7.** b	**12.** c	**17.** e
3. a	**8.** a	**13.** e	**18.** a
4. b	**9.** d	**14.** d	
5. c	**10.** e	**15.** c	

CHAPTER 52: AN INTRODUCTION TO ECOLOGY AND THE BIOSPHERE

INTERACTIVE QUESTIONS

52.1 a. organismal: physiological, behavioral, and morphological adaptations of organisms to the environment
b. population: growth and regulation of population size
c. community: interactions that affect community structure
d. ecosystem: energy flow and chemical cycling between biotic and abiotic components
e. landscape: exchanges among ecosystems of a landscape or seascape
f. global: relationship of the distribution of organisms and regional energy and material exchanges

52.2 a. dispersal: an area may be beyond the dispersal ability of a species
b. behavior and habitat selection: insect larvae may be able to feed on more plants, but females oviposit on a single type of plant
c. biotic factors: the presence of predators, parasites, mutualists, or competitors may restrict a species' range
d. abiotic factors: sunlight, water, temperature, salinity, and soil characteristics may determine whether a species can inhabit an area

52.3 a. South-facing slopes in the northern hemisphere receive more sunlight and are warmer and drier than north-facing slopes.
b. Air temperature drops with an increase in elevation, and high-altitude communities may be similar to communities in higher latitudes.
c. The windward side of a mountain range receives much more rainfall than the leeward side. The warm, moist air rising over the mountain releases moisture, and the drier, cooler air absorbs moisture as it descends the other side.

52.4 a. + − −
b. − + +
c. + − −
d. − + −
e. − + +
f. + + +

52.5 _7_ abyssal   _2_ neritic
      _6_ aphotic   _3_ oceanic
      _8_ benthic   _5_ pelagic
      _1_ intertidal   _4_ photic

52.6 a. desert
b. temperate grassland
c. tropical forest
d. temperate broadleaf forest
e. northern coniferous forest
f. arctic and alpine tundra

SUGGESTED ANSWERS TO STRUCTURE YOUR KNOWLEDGE

1. **a.** Ecology is the study of the distribution and abundance of organisms and their relationship to the abiotic and biotic environments.
    b. The interactions of organisms with their environment can result in changes in the gene pool of a population, or evolution. Interactions occurring within an ecological time frame translate into adaptations that are evident on the scale of evolutionary time.

2. **a.** Biomes are characteristic communities, usually identified by the predominant vegetation and climate, that range over broad geographic areas.
    b. Convergent evolution, common adaptations of different organisms to similar environments, accounts for similarities in life forms within geographically separated biomes.

ANSWERS TO TEST YOUR KNOWLEDGE

Multiple Choice:

1. b	**5.** a	**9.** c	**13.** e
2. d	**6.** b	**10.** b	**14.** d
3. d	**7.** e	**11.** d	
4. e	**8.** c	**12.** e	

Matching:

1. C	**3.** D	**5.** A	**7.** B
2. H	**4.** E	**6.** G	**8.** F

CHAPTER 53: POPULATION ECOLOGY

INTERACTIVE QUESTIONS

53.1 If 10 of the 30 voles in the second trapping were marked, she would estimate that 1/3 of the vole population is marked. Since 25 voles were initially marked, the total population size is estimated at 75 (25/0.333). Another method is to use the formula: N = (number marked × total second catch)/number marked recaptures.

53.2 a. Type I is typical of populations that produce relatively few offspring and provide parental care, such as humans and many large mammals.
b. Type II has a constant death rate over the organism's life span, such as Belding's ground squirrels and some other rodents, some annual plants, various invertebrates, and some lizards.
c. Type III is typical of populations that produce many offspring, most of which die off rapidly, such as many fishes and marine invertebrates, and long-lived plants.

53.3 a. Selection would most likely favor iteroparity, with fewer, larger, better-provisioned or cared-for offspring.
b. An organism has limited resources to divide between growth, survival, and reproduction.

53.4 a. exponential growth; $dN/dt = r_{max}N$
b. logistic growth; $dN/dt = r_{max}N(K - N)/K$; K is 1,500

53.5 a. r-selected;
b. K-selected

53.6 a. nutrients, space for nests, accumulation of toxic wastes, predation, intrinsic limiting factors
b. extremes in weather, natural disasters, fires, sunspot activity (which then influences plants)

SUGGESTED ANSWERS TO STRUCTURE YOUR KNOWLEDGE

1.

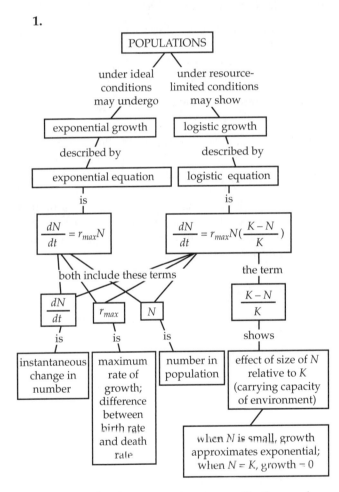

2. Reproductive success is measured in the number of offspring that survive and live to reproduce. Many "choices" are available in life history traits: big-bang versus repeated reproduction, age at first reproduction, number of reproductive episodes, number of offspring, parental investment in size of offspring or care. There are always trade-offs between reproduction and survival due to limited energy budgets. Considering that diverse environments (with both biotic and abiotic factors) and different population densities (*K*- or *r*-selection) create different selection pressures on a population, there is no one "best" reproductive strategy.

ANSWERS TO TEST YOUR KNOWLEDGE

Multiple Choice:

1. a	6. a	11. e	16. e
2. c	7. c	12. e	17. d
3. d	8. d	13. b	18. c
4. b	9. c	14. a	
5. e	10. d	15. a	

CHAPTER 54: COMMUNITY ECOLOGY

INTERACTIVE QUESTIONS

54.1 The realized niche of species I is smaller than its fundamental niche when it is in competition with species II. In these trees, species II's fundamental and realized niches are the same.

54.2 a. Batesian mimicry
   b. Müllerian mimicry

54.3 a. mutualism: mycorrhizae, flowering plants and pollinators, ants on acacia trees, cellulose-digesting microorganisms in termites and ruminants
   b. commensalism: cattle egrets and cattle that flush insects (although cattle may benefit when ectoparasites are eaten or when the birds warn of predators)
   c. competition: weeds and garden plants compete for nutrients and water
   d. predation: animal predators killing prey
   e. herbivory: herbivores eating parts of plants
   f. parasitism: endo- and ectoparasites feed in or on host

54.4 Pool 2 has more species and thus a greater species richness. Pool 1 has the higher diversity index.

Pool 1: $H = -[(0.4)(\ln 0.4) + (0.3)(\ln 0.3) + (0.3)(\ln 0.3)] = -(-0.37 + -0.36 + -0.36) = 1.09$

Pool 2: $H = -(-0.25 + -0.23 + -0.23 + 0.23) = 0.94$

54.5 energetic hypothesis

54.6 The $+/-$ cascade that would be needed to end with a decrease in algae would require an increase in zooplankton, which would require a decrease in primary predators caused by an increase in top predators. More top predators could be added to the lake or primary predators could be removed.

54.7 a. Soil nitrogen levels begin quite low, but rise due to the symbiotic nitrogen-fixing bacteria of alder.
   b. The soil pH changes from about 7.0 to 4.0 as the acid spruce leaves decompose.

54.8 More interspecific interactions would have had time to develop, and this longer span of evolutionary time would allow for more speciation events to have occurred.

54.9 a. immigration—large island
   b. immigration—small island
   c. extinction—small island
   d. extinction—large island

   The small island equilibrium number of species is projected down from the intersection of lines b and c. The large island equilibrium number is higher and is projected down from the intersection of a and d.

54.10 They are testing for the presence of the H5N1 strain of avian flu virus in migrating waterfowl to monitor when the disease enters North America.

SUGGESTED ANSWERS TO STRUCTURE YOUR KNOWLEDGE

   1. a. competition $(-/-)$
      b. resource partitioning and character displacement
      c. slightly different niches
      d. keystone predator
      e. herbivory
      f. mutualism $(+/+)$
      g. food chain or food web

   2 a. No two species with the same niche can coexist in a habitat; the more competitive species will cause the local elimination of the other.

b. Food chains are limited to a few links because of the inefficiency of energy transfer (only about 10%) from one trophic level to the next.

c. Food chains are more stable with fewer links because the effect of environmental disruptions becomes magnified in higher trophic levels.

d. Each trophic level controls the next higher level; adding nutrients will increase the biomass of all other trophic levels.

e. Community organization is controlled from the top by predation, with a cascade of +/− effects down the trophic levels with changes in predator numbers.

f. Most communities are constantly changing in composition due to the effects of disturbances.

g. The rates of immigration and extinction on "islands" of habitat are affected by the size of the island, the closeness to the "mainland," and the number of species currently on the island. When these rates are equal, an equilibrium of species is reached.

ANSWERS TO TEST YOUR KNOWLEDGE

Multiple Choice:

1. d	5. b	9. d	13. a	17. e
2. d	6. b	10. c	14. a	18. c
3. b	7. e	11. d	15. d	
4. c	8. b	12. d	16. b	

CHAPTER 55: ECOSYSTEMS

INTERACTIVE QUESTIONS

55.1 The element may limit production in that ecosystem.

55.2 Whereas energy makes a one-way trip through ecosystems, chemical elements move through the trophic levels and are then recycled back to producers.

55.3 a. tropical rain forest, coral reef, swamp and marsh, estuary
b. desert, tundra, lake and stream, open ocean
c. The open ocean covers 65% of Earth's surface area.
d. Upwellings in these cold seas bring nitrogen and phosphorus to the surface. The lack of these nutrients limits production in many tropical waters.

55.4 a. Much of a bird or a mammal's assimilated energy is used to maintain a warm body temperature and is not available for net secondary production (growth and reproduction).
b. $1/1000$ (10% of 10% of 10%) or 0.1%

55.5 a. nitrogen-fixing bacteria in root nodules
b. nitrogen-fixing soil bacteria
c. decomposers
d. ammonium (NH_4^+)
e. nitrifying bacteria
f. nitrates (NO_3^-)
g. denitrifying bacteria

55.6 Water is essential to all organisms and its availability influences primary production and decomposition. Carbon forms the backbone for all organic molecules. Nitrogen is a component of amino acids and nucleic acids. Phosphorus is found in nucleic acids, phospholipids, and ATP.

55.7 a. tropical rain forests, because decomposition proceeds rapidly in the warm, wet climate, and nutrients are rapidly assimilated by new growth
b. lakes and oceans; lack of oxygen slows decomposition and, unless there are upwellings, nutrients are not available to producers
c. Nutrients are not recycled and may leave the ecosystem through runoff.

55.8 melting of ice caps and flooding of coastlines; change in precipitation patterns such that many currently important agricultural areas may become arid; changes in species composition or increased extinction rate as temperatures rise faster than communities can adjust and species (particularly plants) can migrate

SUGGESTED ANSWERS TO STRUCTURE YOUR KNOWLEDGE

1.

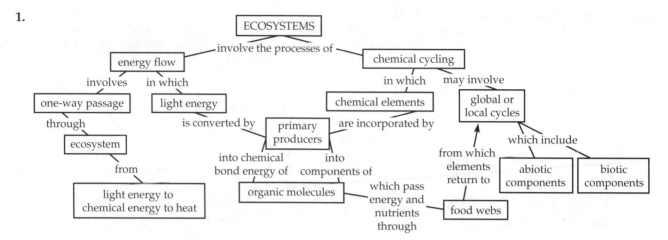

2. **a.** Deforestation can cause an increase in water runoff with an accompanying loss of soil and minerals. Clear-cutting of tropical forests results in loss of species diversity and a reduction in productivity of the area.

    b. Dumping of wastes and runoff from agricultural lands has resulted in eutrophication of many lakes, killing fish and other organisms.

    c. Toxic chemicals introduced into the environment have been incorporated into the food chain. As a result of biological magnification, these substances pose threats to top-level consumers.

    d. An increase in atmospheric CO_2 from fossil fuel and wood combustion adds to the greenhouse effect. The resulting increase in temperature may have far-reaching effects on climate, sea level, and species composition.

    e. Agriculture interrupts nutrient cycling and requires the addition of synthetic fertilizers, which upset the nitrogen cycle, can cause eutrophication, and add to the greenhouse effect, ozone depletion, and acid precipitation.

    f. Acid precipitation, caused by fossil fuel combustion, is harming forests and lakes.

    g. Chlorine-containing pollutants are thinning the ozone layer, allowing more harmful UV radiation to reach Earth.

ANSWERS TO TEST YOUR KNOWLEDGE

Multiple Choice:

1. e	**5.** d	**9.** d	**13.** e
2. c	**6.** e	**10.** a	**14.** d
3. c	**7.** b	**11.** e	**15.** b
4. e	**8.** c	**12.** c	**16.** d

CHAPTER 56: CONSERVATION BIOLOGY AND RESTORATION ECOLOGY

INTERACTIVE QUESTIONS

56.1 Many examples are provided in the text.

   a. Coral reefs have been damaged by human activities; 40–50% of the reefs could be lost in the next few decades. About a third of all marine fish species utilize these reefs.

   b. The introduction of the brown tree snake to Guam resulted in the extinction of multiple species of birds and lizards.

   c. Commercial harvest or illegal hunting have reduced populations of whales, the African elephant, and numerous fishes.

56.2 Smaller, because usually not all individuals in a population successfully breed. Thus, the MVP should refer to the number of reproductively active individuals in a population.

56.3 The loss of genetic variation within a small population due to inbreeding and genetic drift can force it into an extinction vortex in which the population grows smaller and smaller. Promoting migration between small populations or introducing individuals from other populations to increase genetic variation is proposed as an urgent conservation need.

56.4 This proactive approach relies on early detection of population decline, identification of the species' habitat needs, testing to determine which factor is contributing to the decline, recommending corrective measures, and monitoring results.

56.5 Corridors promote dispersal between populations and may be essential to species that migrate between different habitats. However, they may also contribute to the spread of diseases.

56.6 Large preserves are required for large, far-ranging animals that require extensive habitats. They also have proportionally less border area and thus have fewer edge effects. An advantage of smaller preserves that collectively have the same area as a large one is the slower spread of disease within a population.

56.7 Prokaryotes that can metabolize uranium are being used to treat contaminated groundwater. Plants that can fix nitrogen are often used in biological augmentation to enrich nutrient-poor soils and facilitate the recolonization of native species.

SUGGESTED ANSWERS TO STRUCTURE YOUR KNOWLEDGE

1. habitat destruction, introduced species, and overexploitation of wild species

2. Biodiversity is a natural resource from which we obtain medicines, crops, fibers, and other products; many potentially valuable species will become extinct before they are known to scientists. A loss of biodiversity may disrupt ecosystem processes in harmful ways. Humans have evolved within the context of living communities and may be affected in unknown ways by changes in our ecosystem.

3. Fragmentation of habitats produces more interfaces or *edges* between different ecosystems (between forests and cleared areas, between deserts and housing developments). The species that inhabit edges are able to use both types of ecosystems. As edges proliferate, edge-adapted species may become more dominant than the species in the adjoining habitats. Strips of quality habitat that connect fragmented habitat patches may serve as *movement corridors* that promote dispersal between isolated populations and help maintain genetic variation.

ANSWERS TO TEST YOUR KNOWLEDGE

Multiple Choice:

1. e	**4.** b	**7.** e	**10.** d
2. b	**5.** d	**8.** a	**11.** b
3. c	**6.** a	**9.** c	**12.** c